Advances in
ORGANOMETALLIC CHEMISTRY

VOLUME 46

Advances in Organometallic Chemistry

EDITED BY

F. GORDON A. STONE

DEPARTMENT OF CHEMISTRY
BAYLOR UNIVERSITY
WACO, TEXAS

ROBERT WEST

DEPARTMENT OF CHEMISTRY
UNIVERSITY OF WISCONSIN
MADISON, WISCONSIN

VOLUME 46

ACADEMIC PRESS

A Harcourt Science and Technology Company

San Diego London Boston New York
Sydney Tokyo Toronto

This book is printed on acid-free paper. 

Academic Press
A Harcourt Science and Technology Company
525 B Street, Suite 1900, San Diego, California 92101-4495, USA
http://www.academicpress.com

Academic Press
Harcourt Place, 32 Jamestown Road, London NW1 7BY, UK
http://www.academicpress.com

International Standard Book Number: 0-12-000771-1

PRINTED IN THE UNITED STATES OF AMERICA
00 01 02 03 04 05 QW 9 8 7 6 5 4 3 2 1

Contents

Transition-Metal Systems Bearing a Nucleophilic Carbene Ancillary Ligand: from Thermochemistry to Catalysis

LALEH JAFARPOUR and STEVEN P. NOLAN

Organometallic Chemistry of Transition Metal Porphyrin Complexes

PENELOPE J. BROTHERS

Contributors

Numbers in parentheses indicate the pages on which the authors' contributions begin.

PENELOPE J. BROTHERS (223), Department of Chemistry, The University of Auckland, Auckland, New Zealand

BARRETT EICHLER (1), Department of Chemistry, University of California-Davis, Davis, California 95615

MARK G. HUMPHREY (47), Department of Chemistry, Australian National University, Canberra, ACT 0200 Australia

LALEH JAFARPOUR (181), Department of Chemistry, University of New Orleans, New Orleans, Louisiana 70148-2920

IL NAM JUNG (145), Organosilicon Chemistry Laboratory, Korea Institute of Science and Technology, Seoul 130-650, Korea

NIGEL T. LUCAS (47), Department of Chemistry, Australian National University, Canberra, ACT 0200 Australia

STEVEN P. NOLAN (181), Department of Chemistry, University of New Orleans, New Orleans, Louisiana 70148-2920

SUSAN M. WATERMAN (47), Department of Chemistry, Australian National University, Canberra, ACT 0200 Australia

ROBERT WEST (1), Department of Chemistry, University of Wisconsin, Madison, Wisconsin 53706-1396

BOK RYUL YOO (145), Organosilicon Chemistry Laboratory, Korea Institute of Science and Technology, Seoul 130-650, Korea

ADVANCES IN ORGANOMETALLIC CHEMISTRY, VOL. 46

Chemistry of Group 14 Heteroallenes

BARRETT EICHLER

Department of Chemistry
University of California-Davis
Davis, California 95616

ROBERT WEST

Department of Chemistry
University of Wisconsin
Madison, Wisconsin 53706

I

INTRODUCTION

An allene (or 1,2-propadiene) is a moiety with two cumulated double bonds between three atoms ($R_2C{=}C{=}CR_2$). The central atom of the allene is therefore sp-hybridized. The π-bond of one double bond is orthogonal to the other and this unusual π-bonding arrangement can lead to unique electronic effects (Fig. 1). This also results in steric properties at the ends of the allene by forcing the substituents to also be orthogonal to each other.

This review will focus on allenes which have at least one carbon atom replaced by a heavier group 14 atom, commonly referred to as a heteroallene. Group 14 heteroallenes have appeared in the literature over the last 20 years, and stable examples of this moiety have been synthesized since 1992. Heteroallenes that do not have a group 14 heteroatom will not be discussed, although it is useful to consider phosphaallenes, which have been reviewed by Regitz in 1990.[1] To date, heteroallenes with the heteroatom at the end of the allene, the one position, have been easier to synthesize because of their thermodynamic stability compared to those with the heteroatom as the middle atom, the two position.

0065-3055/01 $35.00

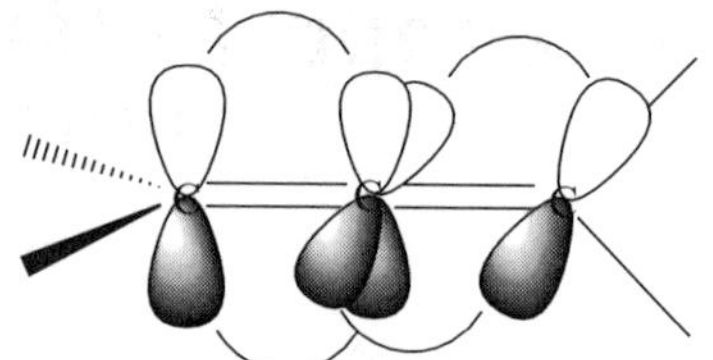

FIG. 1. π-Orbital arrangement of an allene.

II

BONDING THEORY

Heteroallene structures can be regarded as depending on the extent of contribution from each of two bonding arrangements (Fig. 2). Bonding Model A depicts both the heteroatom and the carbon atom in the triplet state, forming one σ-bond and one π-bond to make a formal double bond. Lappert *et al.*[2a] first postulated bonding Model B, and illustrated bonding between the heteroatom and the carbon atom in singlet states that "may be described as a 'double' π-donor-acceptor interaction," according to Grützmacher *et al.*[2b] Differences between these two models

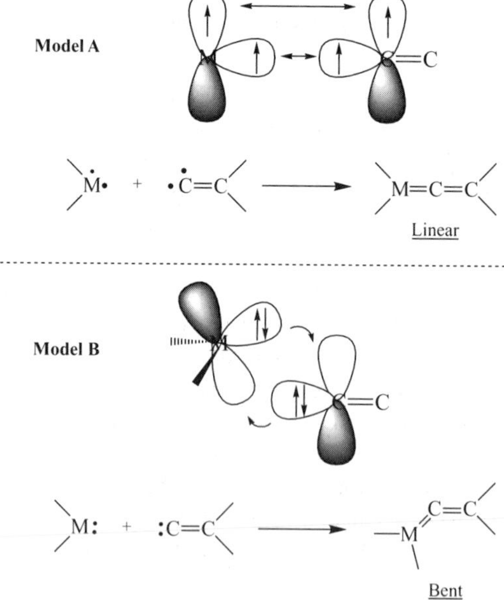

FIG. 2. Two bonding models of 1-heteroallenes.

TABLE I

RELATIVE ENERGIES OF C_2H_4Si ISOMERS (kcal mol^{-1})

$H_2C{=}Si{=}CH_2$	$H_2Si{=}C{=}CH_2$	$H_3Si{-}C{\equiv}CH$	Ref.
45.7	–	0	4
49.0	29.8	0	5
50.1	28.9	0	6
–	25.7	0	7
–	ca. 55	0	12f

leads to obvious differences in geometry in order to maximize orbital overlap. Model A should lead to a linear allene backbone and Model B should produce a structure bent at the central carbon, as well as pyramidalization at the heteroatom. The key issue in determining which model applies to a particular system is the singlet–triplet energy difference[3] for both atoms involved in bonding. Typically, the triplet state is favored only for carbon, whereas silicon, germanium, tin, and lead favor the singlet state. The filled orbitals of the heavier atoms are progressively richer in s-character and therefore the singlet state becomes more energetically favorable as one moves down the periodic table. Thus, the pure carbon allene bonding is dominated by the triplet contribution, and so the $C_1{=}C_2{=}C_3$ bond angle should be 180°. In the heteroallenes, the bond angle may deviate from 180°, and the bending is predicted to increase as the size of the heteroatom increases. The empirical results, which agree very well with this view, will be discussed later.

Several groups have reported ab initio calculations of C_2H_4Si isomers[4–7,9f]; some of the results are listed in Table I. The most stable structure is ethynylsilane. Relative to this molecule, 1-silapropadiene is less stable by about 25–30 kcal mol^{-1} (Ref. 12(f) places the energy of the parent silaallene ca. 55 kcal mol^{-1} above ethynylsilane) and 2-silapropadiene is even more unstable, lying ca. 50 kcal mol^{-1} above ethynylsilane. This is consistent with the fact that 1-heteroallenes have been isolated, but 2-heteroallenes are still unknown.

A recent calculation[10] of the relative energies of CH_2SiO isomers shows that a bent silylene-carbon monoxide adduct (lone pair donation from carbon to empty p-orbital on silicon) is the most stable, and the planar/linear $H_2Si{=}C{=}O$ isomer (strong double bond between silicon and carbon) lies 16.6 kcal mol^{-1} higher in energy (Fig. 3). Earlier energy calculations of $(CH_3)_2SiCO$ provide contrasting results depending on the methods used—MNDO/AM1 calculations predict the minimum energy structure is the planar/linear silaketene, but ab initio calculations favor the pyramidal silylene-CO adduct. Although no stable heteroketenes have been synthesized to date, structural, spectroscopic, and reactivity data from related heteroketenimines (Sects. IIIC3 and IV) also suggest that the silylene-CO adduct is more stable.

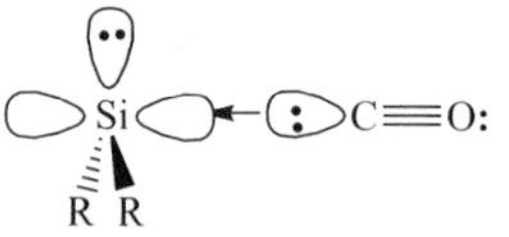

silylene-carbon monoxide adduct 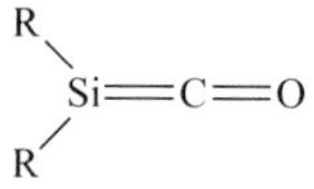   linear/planar silaketene

FIG. 3.

It is instructive to examine the atomic charges predicted for various propadienes. The normal polarization of the Si=C double bond in silenes is calculated to be strongly polarized, $Si^{+}-C^{-}$, with large net charges of +0.46 on silicon and −0.67 on carbon (Table II). In allene itself, the C=C bonds are also substantially polarized, with net negative charge on the outer two carbon atoms.[6] In 2-silaallenes, the net charges on silicon and carbon are predicted to be quite similar to those in silenes, but in 1-silaallenes, the net charges on all three atoms are greatly reduced. The normal allene polarization evidently cancels out much of the Si=C polarization. Thus, 1-silaallenes are far less polar than silenes and 2-silallenes, and so may be less reactive toward polar reagents. Calculations have been reported only for silaketenes (see below), but similar trends are likely for the other group 14 heteroallenes.

Apeloig and co-workers have pointed out that decreased polarity of the Si=C double bond is also calculated for silenes with oxygen substitution, i.e., $H_2Si{=}CH(OSiH_3)$.[9] In this case, the reduced net charges are due to resonance electron donation by oxygen. The calculations are in accord with the greater Si=C bond length, increased ^{29}Si shielding, and decreased ^{13}C shielding found in C-oxygen-substituted silenes, compared with silenes lacking oxygen substituents.

Atomic charges were calculated[11] for parent silaketene, H_2SiCO, using generalized atomic polar tensor (GAPT) population analysis. Two geometries were investigated—the "doubly bonded" planar and the "silylene-CO adduct" bent. The most important point relative to the examination of heteroallenes is that the planar structure has a significantly more negative charge on silicon (−0.05, C = +1.27) than does the bent structure (+0.32, C = +0.81). Based on the charge calculations for silaallenes, a slight negative charge on silicon seems unlikely, making the bent structure a better model than the planar structure.

TABLE II

CALCULATED Si=C BOND LENGTHS AND NET ATOMIC CHARGES

	$r_{Si=C}$,pm	Si, Chg.	C*, Chg.	Ref.
$H_2Si{=}CH_2$	171.8	+0.46	−0.67	9
$H_2Si{=}C^*{=}CH_2$	170.2	+0.17	−0.10	10
$H_2C^*{=}Si{=}CH_2$	170.1	+0.40	−0.63	10

III

SYNTHESIS AND REACTIONS

A. *Transient 1-Silaallenes*

1. *Photolysis, Thermolysis, and Pyrolysis*

Extensive studies of the photolysis and thermolysis of alkynylsilanes and silacyclopropenes were carried out by Kumada and Ishikawa, beginning in 1977. The first report of a group 14 heteroallene in the literature was the proposal by this group of a transient 1-silaallene as a product of the photolysis of a 1-alkynyldisilane (**1a**)[12a] (Scheme 1). When this precursor was irradiated in the presence of methanol, a 1,3-silyl migration occurred. The major products (40%) were methoxysilaethenes, whose existence can be explained as being methanol adducts of 1-silacyclopropene (**2**). Two additional methoxysilanes were isolated in a combined yield of 21% and were rationalized as methanol trapping products of the intermediate silaallene **3.** A similar photolysis of **1a** in the presence of acetone produced the transient acetone adduct of the 1-silaallene, a 2-silaoxetane,

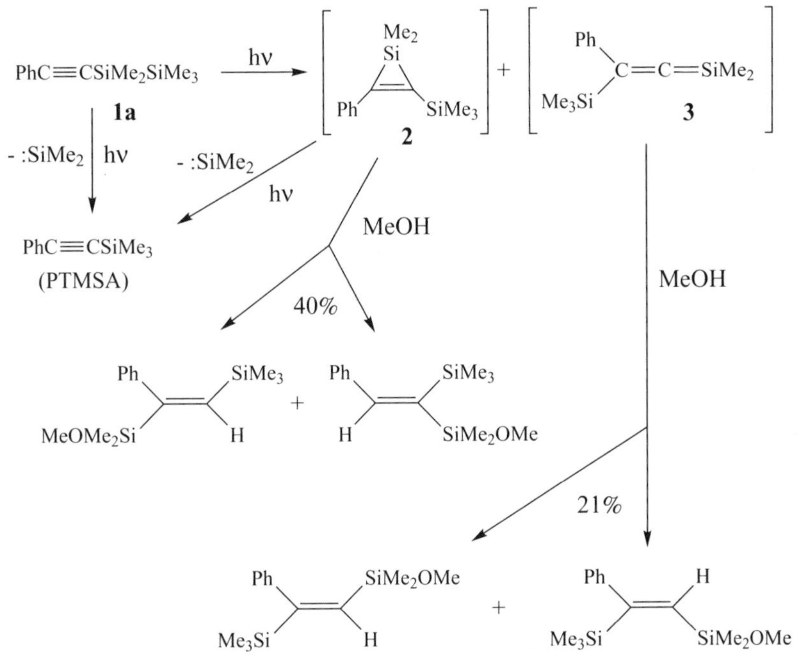

PhC≡CSiMe₂SiMe₃ (**1a**) —hν→ [1-silacyclopropene **2**: Me_2Si ring with Ph and $SiMe_3$] + [Ph(Me₃Si)C=C=SiMe₂ **3**]

1a —(- :SiMe₂, hν)→ PhC≡CSiMe₃ (PTMSA); **2** —(- :SiMe₂, hν)→ PTMSA

2 —MeOH (40%)→ Ph(MeOMe₂Si)C=CH(SiMe₃) + Ph(H)C=C(SiMe₃)(SiMe₂OMe)

3 —MeOH (21%)→ Ph(Me₃Si)C=CH(SiMe₂OMe) + Ph(Me₃Si)C=C(H)(SiMe₂OMe)

SCHEME 1.

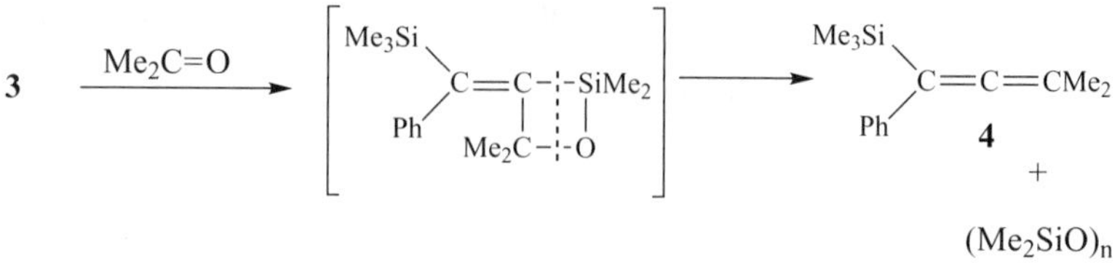

$$\mathbf{3} \xrightarrow{Me_2C{=}O} \left[\begin{array}{c} Me_3Si \\ \quad C{=}C{-}SiMe_2 \\ Ph \quad | \qquad | \\ \quad Me_2C{-}O \end{array} \right] \longrightarrow \begin{array}{c} Me_3Si \\ \quad C{=}C{=}CMe_2 \\ Ph \\ \mathbf{4} \\ + \\ (Me_2SiO)_n \end{array}$$

SCHEME 2.

which decomposed to an allene (**4**—5% isolated yield) and a siloxane (Scheme 2). This type of decomposition has been seen with other 2-silaoxetanes made from the ketone cycloadducts of silenes.[13] It was proposed that photolysis of both **1a** and **2** can eliminate :$SiMe_2$ and phenyltrimethylsilylacetylene (PTMSA). Although this silylene elimination was not observed for the reaction in Scheme 1, PTMSA was isolated in 5% yield in another reaction using methanol, and in 10% yield from the acetone trapping reaction in Scheme 2.

Further studies[12b–g] spanning nearly 15 years suggested that photolysis of other alkynylpolysilanes can, but do not necessarily, form 1-silaallenes. In the paper[12b] following the original communication, six 1-alkynylpolysilanes were irradiated in the presence of methanol, but only four of the six (**5a–c, e**) gave methanol adducts of 1-silaallenes [**6**—Eq. (1)]. There is no clear-cut substituent pattern leading to 1-silaallene production, but it seems that those precursors having more phenyl groups lead to higher yields of 1-silaallene trapping products (cis + trans yields: **5a** = 16%, **5b** = 34%, **5c** = 44%, **5e** = 28%). The alkynylpolysilanes **5a–c** and **e** were also irradiated from 3 to 9 h in the absence of methanol followed by immediate methanolysis of the photolysis products. None of the methanol adducts of 1-silaallenes was found from these experiments, indicating that the 1-silaallenes, if formed, survived for only a very short time in solution.

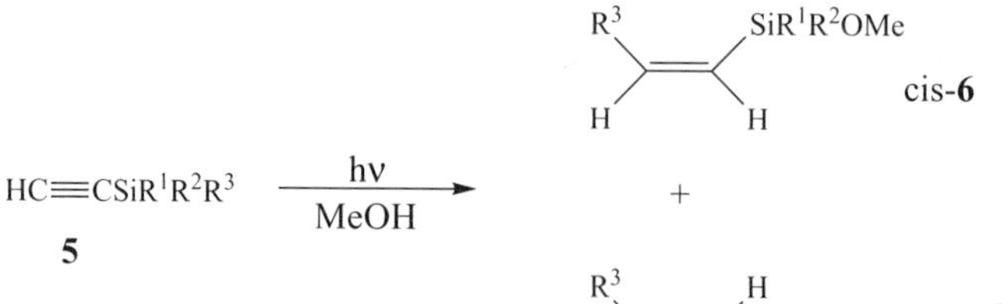

$$\underset{\mathbf{5}}{HC{\equiv}CSiR^1R^2R^3} \xrightarrow[MeOH]{h\nu} \underset{\text{cis-}\mathbf{6}}{(R^3)(H)C{=}C(H)(SiR^1R^2OMe)} + \underset{\text{trans-}\mathbf{6}}{(R^3)(H)C{=}C(SiR^1R^2OMe)(H)} \qquad (1)$$

(a) $R^1 = R^2 = Me$, $R^3 = SiMe_2Ph$
(b) $R^1 = Me$, $R^2 = Ph$, $R^3 = SiMe_3$
(c) $R^1 = R^2 = Ph$, $R^3 = SiMe_3$
(d) $R^1 = R^2 = Me$, $R^3 = SiMe_2SiMe_3$
(e) $R^1 = Me$, $R^2 = R^3 = SiMe_3$
(f) $R^1 = R^2 = R^3 = SiMe_3$

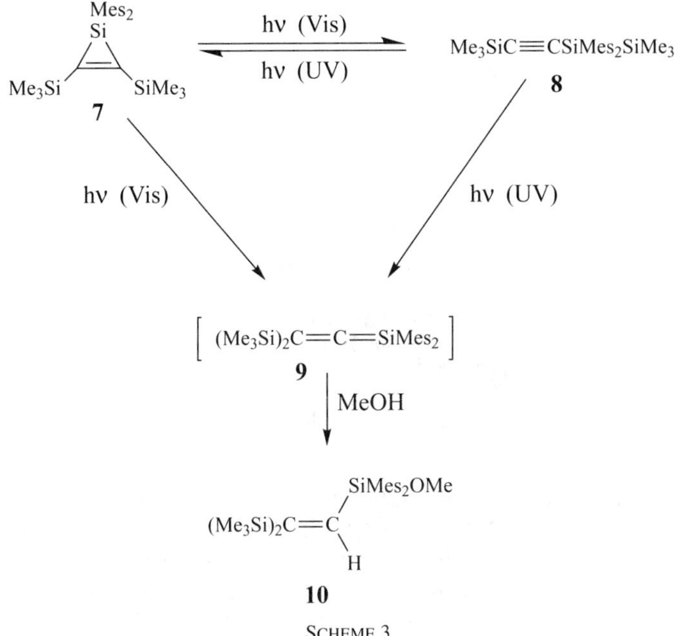

SCHEME 3.

In another study,[12c] Kumada and Ishikawa proposed that the stable 1-silacyclopropene **7,** when irradiated in the presence of methanol, equilibrated with 1-alkynylpolysilane **8** and produced the methanol adduct (**10**) of the 1-silaallene **9** in 36% yield (Scheme 3). From previous studies using ultraviolet(UV) radiation as the photon source, it was shown that **8** was the most likely precursor to 1-silaallene **9.** Compound **7** was never indicated as a direct source of **9.** From the observation that **8** is stable to visible light, whereas **7** is not, the authors inferred that the silaallene can be created from both **7** and **8.** When **7** was irradiated with UV light for 10 h, **7** and **8** were found in a constant 2/3 ratio irrespective of the time of irradiation, showing that **7** and **8** were equilibrating with each other. Irradiation of **7** with UV light in the presence of methanol gave methanol adduct (**10**) of 1-silaallene **9** in 32% yield. But when a Pyrex filter (allowing only visible light to reach **7**) was included in the photolysis with methanol present, compounds **8** and **10** were produced in 60 and 16% yields, respectively, clearly indicating that **7** was converting to **9** when photolyzed.

Experiments reported in 1982[12d] by the same group provided the first example of heteroallene dimerization. In this work, head-to-head dimerization of 1-silaallenes **11a–c** (Scheme 4) was observed, forming 1,2-disilacyclobutanes **12a–c** with two exocyclic double bonds.

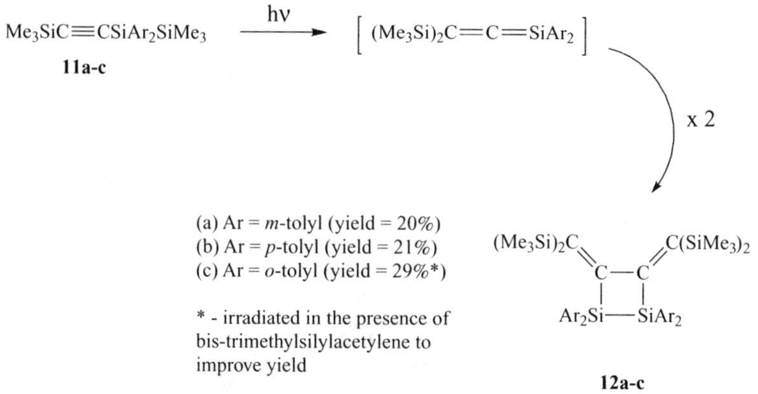

$Me_3SiC{\equiv}CSiAr_2SiMe_3$ **11a-c** $\xrightarrow{h\nu}$ $[(Me_3Si)_2C{=}C{=}SiAr_2]$ $\xrightarrow{\times 2}$ **12a-c**

(a) Ar = *m*-tolyl (yield = 20%)
(b) Ar = *p*-tolyl (yield = 21%)
(c) Ar = *o*-tolyl (yield = 29%*)

* - irradiated in the presence of bis-trimethylsilylacetylene to improve yield

$(Me_3Si)_2C{=}C{-}C{=}C(SiMe_3)_2$; $Ar_2Si{-}SiAr_2$

SCHEME 4.

When alkynyldisilanes **13a** and **b** were photolyzed[12g] in the presence of freshly generated dimesitylsilylene (Mes_2Si:), the silylene added to the Si=C double bond of 1-silaallenes **14a** and **b** to form disilacyclopropanes **15a** and **b** (Scheme 5). Even without the independently generated silylene, photolysis of **13b** produced **15b** in 8% yield, but compound **13a** gave only traces of **15a.** In the case of **15b,** the dimesitylsilylene most likely originated from silacyclopropene **16.**

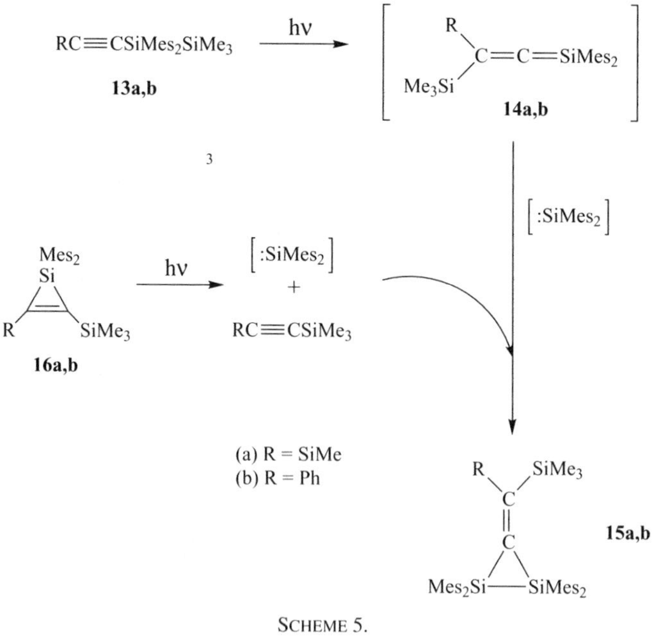

$RC{\equiv}CSiMes_2SiMe_3$ **13a,b** $\xrightarrow{h\nu}$ $[R(Me_3Si)C{=}C{=}SiMes_2]$ **14a,b** $\xrightarrow{[:SiMes_2]}$ **15a,b**

3

16a,b (Mes_2Si, R, $SiMe_3$) $\xrightarrow{h\nu}$ $[:SiMes_2]$ + $RC{\equiv}CSiMe_3$

(a) R = SiMe
(b) R = Ph

R, $SiMe_3$, C=C, $Mes_2Si{-}SiMes_2$ **15a,b**

SCHEME 5.

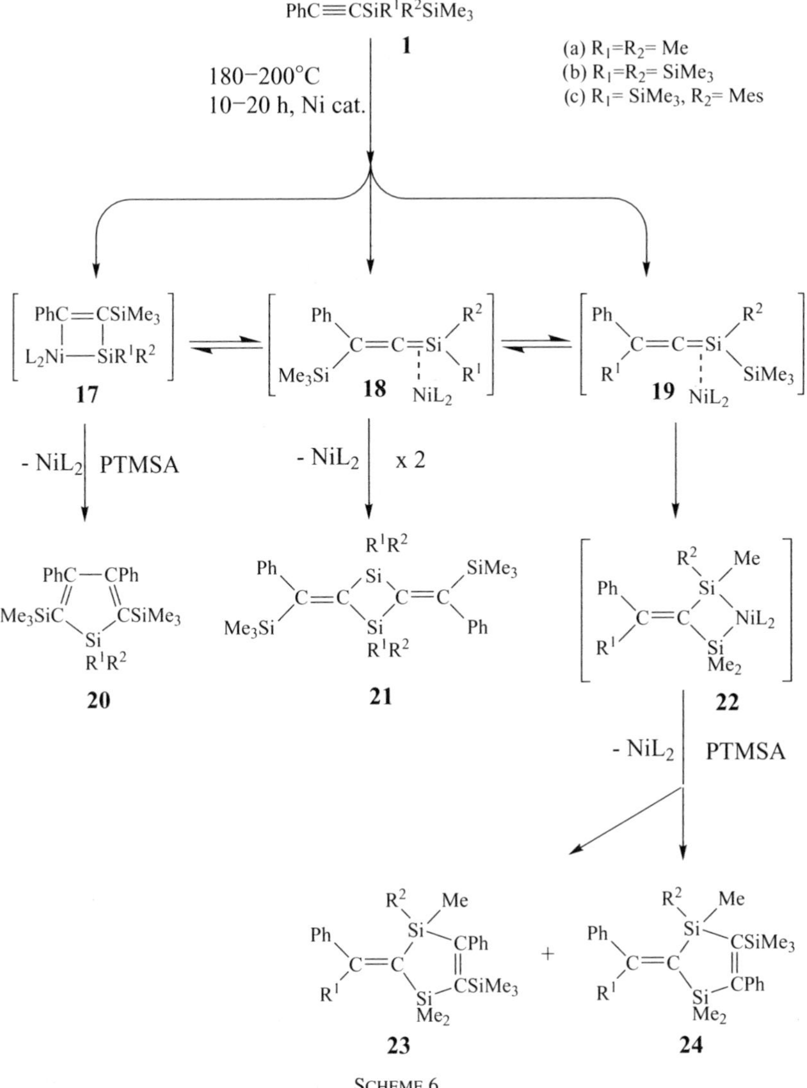

SCHEME 6.

The Kumada/Ishikawa group also investigated thermolytic reactions of alkynylpolysilanes and silacyclopropenes in the presence of nickel catalysts and implicated a 1-silaallene–nickel complex as an intermediate in the reaction pathway to the observed products.[12h–k] When alkynylpolysilanes **1a–c** (Schemes 6 and 7) were heated to 180–200°C for 20 h in the presence of a catalytic amount of $NiCl_2(PEt_3)_2$ (**1a,b**) or $Ni(PEt_3)_4$ (**1c**) and two equivalents of PTMSA, products

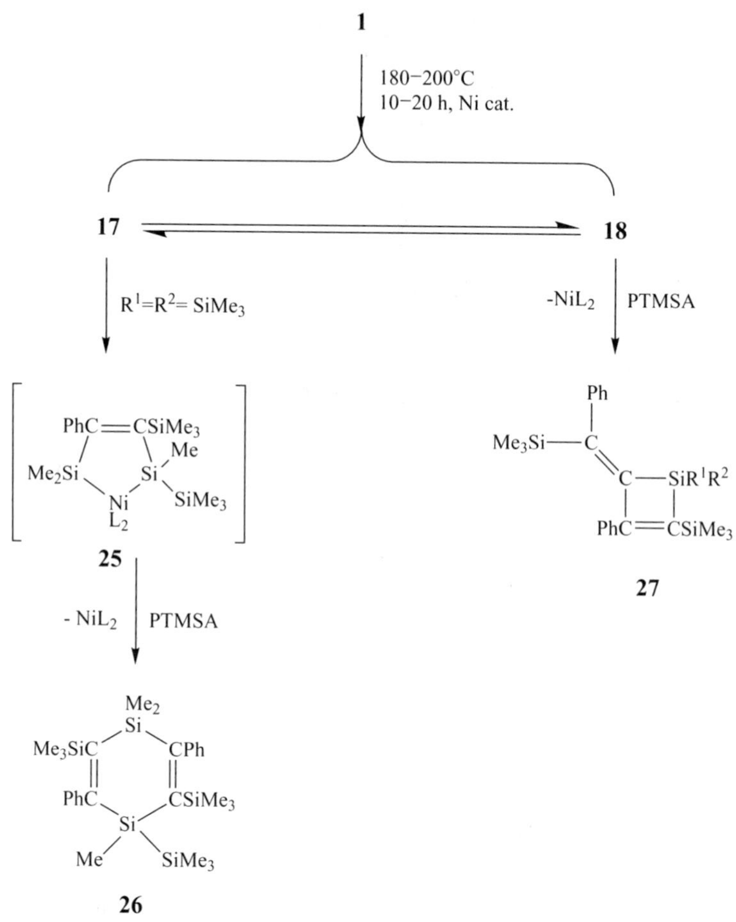

SCHEME 7.

21–30 were observed.[12h] Intermediates **17** (a nickelsilacyclobutene) and **18** and **19** (two isomeric 1-silaallene–nickel complexes) were proposed to be in equilibrium with each other and were formed initially by a 1,3-trimethylsilyl shift. Intermediate **17a** added PTMSA (Scheme 6) with loss of NiL_2 to form silole **20a** (32%), whereas intermediate **18a** simply underwent head-to-tail dimerization ($-NiL_2$, 15%). Intermediate **19a** performed an unusual rearrangement where the nickel moves a $-SiMe_2$ fragment from a $-SiMe_3$ group to the α-carbon, replacing the $-SiMe_2$ fragment with the remaining methyl group, forming intermediate **22a.** This then added PTMSA to give a perfectly statistical distribution of isomers **23a** (23%) and **24a** (23%). The greater steric bulk of **1b** (two $SiMe_3$ groups) over that

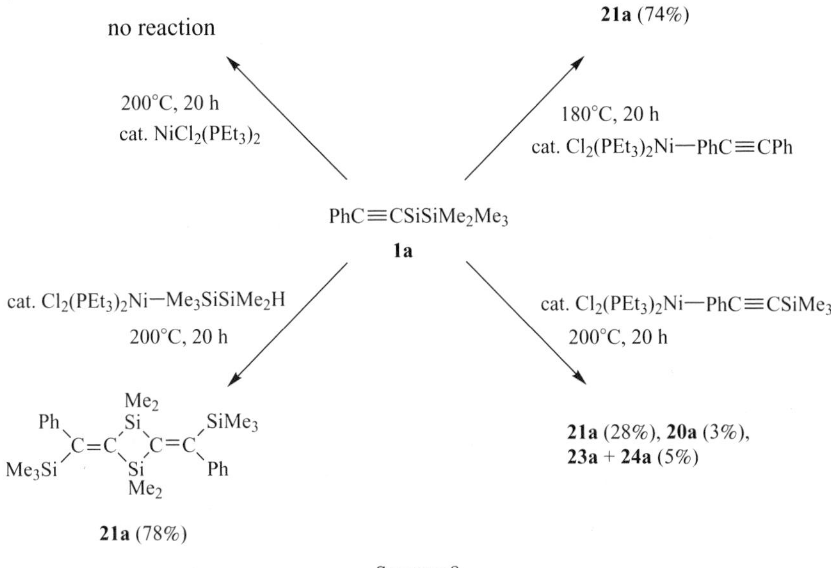

SCHEME 8.

of **1a** (two Me groups) caused the alkynylpolysilane to follow a completely different mechanistic pathway (Scheme 7). Compound **1b** did not produce any of the same compounds as **1a,** instead forming **26** (58%) and **27a** (19%). Compound **26** presumably formed via intermediate **17b,** which rearranged in a manner similar to the formation of intermediate **22,** and then added PTMSA. Compound **27b** is the adduct of PTMSA with 1-silaallene intermediate **18b.**

Thermolysis of **1a** with various nickel catalysts gave some surprising results (Scheme 8).[12h] Possibly the most unexpected finding was that the thermolysis of **1a** with $NiCl_2(PEt_3)_2$ in the absence of PTMSA left **1a** unchanged, but when it was heated with a catalytic amount of $NiCl_2(PEt_3)_2$-$PhCCSiMe_3$, dimerization product **21a** (28%) and PTMSA adducts **20a** (3%) and **23** + **24a** (5%) were obtained. Again in the absence of PTMSA, heating **1a** with $NiCl_2(PEt_3)_2$-PhCCPh gave **21a** (74%) and with $NiCl_2$ $(PEt_3)_2$-$Me_3SiSiMe_2H$ also gave **21a** (78%).

To see if silacyclobutenes (**28**) would react in the same manner as the alkynylpolysilanes (**1**) and respond similarly to steric differences, compounds **28a** and **b** were heated in the presence of $NiCl_2(PEt_3)_2$ and PTMSA (Scheme 9).[12i] Compound **28a** gave silole **20a** in 94% yield,[12j] but the only isolable products from the thermolysis of **28b** were **26** (51%) and **27b** (36%). The products from the thermolysis of silacyclopropene **28b** were very similar to that for alkynylsilane **1b,** but for some reason there were many fewer products for the thermolysis of **28a** than for alkynylsilane **1a.** These results suggest that the more sterically hindered **1b**

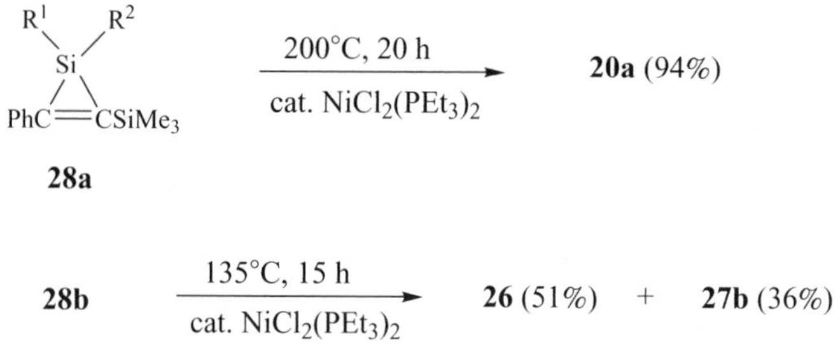

$$\underset{\mathbf{28a}}{\text{PhC}=\text{CSiMe}_3 \text{ (silirene, Si}R^1R^2)} \xrightarrow[\text{cat. NiCl}_2(\text{PEt}_3)_2]{200°\text{C, 20 h}} \mathbf{20a}\ (94\%)$$

$$\mathbf{28b} \xrightarrow[\text{cat. NiCl}_2(\text{PEt}_3)_2]{135°\text{C, 15 h}} \mathbf{26}\ (51\%) + \mathbf{27b}\ (36\%)$$

SCHEME 9.

and **28b** prefer the 1-silaallene intermediate **18b,** whereas the smaller precursors **1a** and **28a** prefer the nickelsilacyclobutene intermediate **17a,** although electronic differences cannot be ruled out as important factors.

To further illustrate the effect of steric hindrance, a bulky mesityl group was added to the alkynylpolysilane (**1c**) (Eq. (2)).[12k] When **1c** was heated in the presence of $Ni(PEt_3)_4$ and PTMSA, compounds **27c** (77%) and **29** (11%—a compound similar to **26**) were obtained, which is a product distribution similar to that of bulky **1b.**

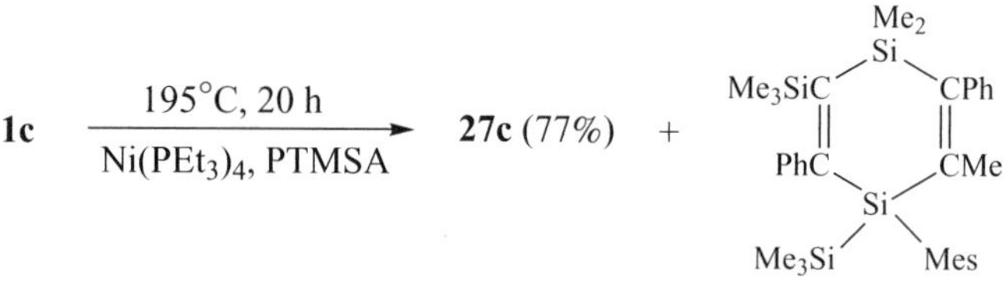

$$\mathbf{1c} \xrightarrow[\text{Ni(PEt}_3)_4\text{, PTMSA}]{195°\text{C, 20 h}} \mathbf{27c}\ (77\%) + \underset{\mathbf{29}}{\text{[Me}_2\text{Si; Me}_3\text{SiC}=\text{CPh; PhC}=\text{CMe; Si(Me}_3\text{Si)(Mes)]}} \qquad (2)$$

Numerous other studies[14] by Ishikawa and co-workers, with or without nickel catalysts, have reinforced the importance of 1-silaallenes and nickel-complexed 1-silaallenes as intermediates in the pathways of the photolyses and thermolyses of alkynylsilanes.

Barton and co-workers[15] performed flash vacuum pyrolysis (FVP) on trimethylsilylvinylmethylchlorosilane (**30**), resulting in the production of trimethylchlorosilane (30%), trimethylvinylsilane (11.5%), and most interestingly, ethynylmethylsilane (**34,** 11.9%). A proposed mechanism for the synthesis of **34** (Scheme 10) begins with the loss of trimethylchlorosilane to form silylene **31,** which can rearrange either to silaallene **32** or to silirene **33,** both of which can lead to the isolated ethynylsilane.

Maier *et al.*[7] studied the FVP of another vinylsilane, **35** (Scheme 11). They proposed that, instead of leading directly to the observed products, silaallene **37**

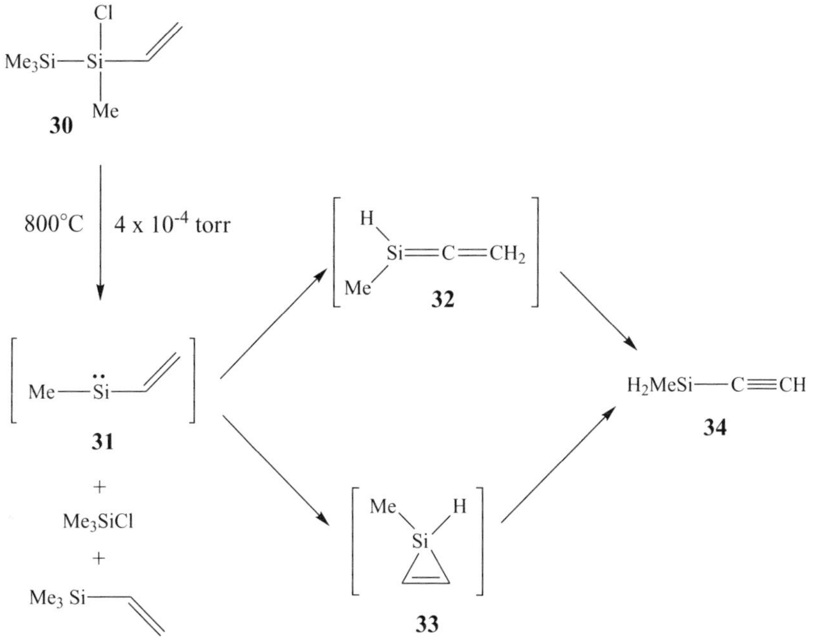

SCHEME 10.

was in an equilibrium process with silylene **36** and that the silylene directly made silacyclopropene **38,** which photolyzed to ethynylsilane **39.**

2. *Flash Photolysis Studies*

The early photolysis studies of Kumada and Ishikawa have been greatly augmented by recent investigations of the laser flash photolyses by Leigh and his students, in which the silapropadienes have been characterized spectroscopically.[16,17] Thus, the flash photolysis of **40** using a KrF excimer laser produced transient compounds **41** and **42,** along with a non-decaying species, **43**[11] (Scheme 12). These products were identified on the basis of their UV absorption spectra and reactivity. Silaallene **41** is a minor (~15%) photoproduct, but could be identified because it is relatively long lived compared with silylene **42,** and has rather strong electronic absorption bands at 275 and 325 nm. Compound **41** reacted with MeOH, *t*-BuOH, HOAc, acetone, and O_2; absolute rate constants were obtained for these reactions by quenching studies. Also investigated by the group was the flash photolysis of **1a,** leading to **2** and **3,** the transient products postulated earlier by Kumada and Ishikawa, as well as dimethylsilylene and the stable product PTMSA.[12] Quenching reactions and UV spectroscopy again identified the products. Silaallene **3** reacts

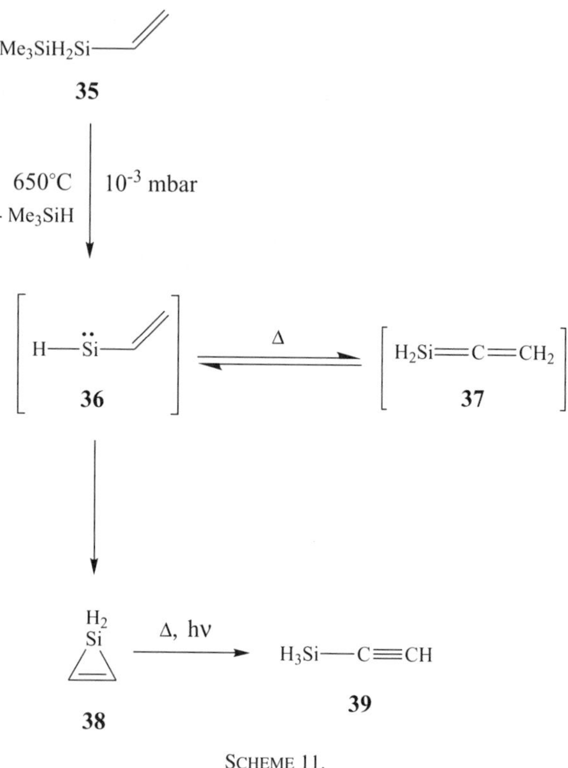

SCHEME 11.

more rapidly with quenching agents than does **41,** by factors of 20 to 1000. The difference is suggested to result from a combination of steric effects and hyperconjugative stabilization of the Si=C bond in **41** by the trimethylsilyl substituent.

B. *Transient 1-Silaketenes*

Using CO-saturated hydrocarbon matrices, Pearsall and West[18] photolyzed silylene precursors at 77 K and monitored CO coordination to the silylenes by UV-vis spectroscopy (Scheme 13). Bis(trimethylsilyl)silanes **44a–c** or Si_6Me_{12} were irradiated at 254 nm to create silylenes **45a–d,** which reacted with CO, causing new peaks to ca. 290 and 350 nm, which were attributed to complex **46a–d,** a resonance structure of silaketene **47a–d.** Silylene adducts form fairly weak bonds, as seen by warming of the matrices. In the case of silylene adducts where one R = Mes, the CO dissociates and the corresponding disilene **48a–c** peaks in the UV-vis spectra observed upon warming (R_2 = Me most likely produced silane rings Si_3Me_6, etc.).

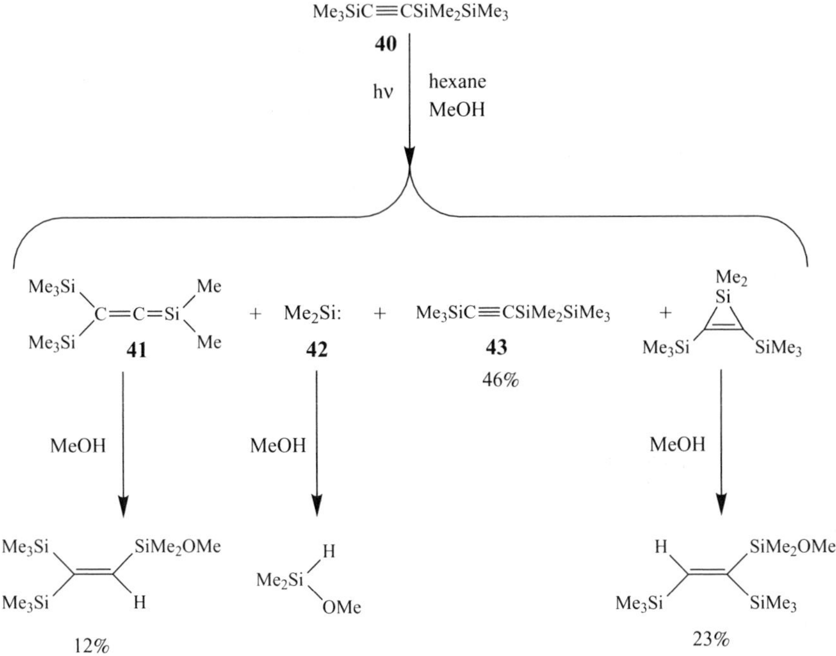

SCHEME 12.

A separate study of the interaction of $:SiMe_2$ with CO in an argon matrix (Scheme 14) was carried out by Arrington *et al.*[19] Dodecamethylcyclohexasilane or dimethyldiazidosilane were irradiated in the presence of CO at 15 K and produced a silylene-CO adduct. This species was detected by infrared (CO stretch = 1962 cm^{-1}) and UV-vis (peak at 342 nm) spectroscopies and the intensities of the peaks increased upon warming, indicating that more of **46d** was being formed at higher temperatures owing to the increased mobility of the reactants in the matrix.

Maier and co-workers[8] condensed formaldehyde and elemental silicon at 12 K in an argon matrix and photolyzed the mixture to form silaketene H_2SiCO, which is similar in structure to the silylene-CO adduct mentioned above. The reactants first form siloxiranylidene **49** (which equilibrates with an unknown species postulated as the planar/linear silaketene **50** when exposed to 313-nm-wavelength light) and then forms complex **51** when photolyzed at 366 nm (Scheme 15). This species could also be formed by photolyzing diazidosilane **52** in the presence of CO, and complex **51** equilibrates with SiCO (**53**) and H_2. The CO infrared shift for this bent structure was calculated at 2129 cm^{-1}, which is shifted $-80\,cm^{-1}$ from the calculated value of free CO, at 2210 cm^{-1}. The experimentally observed value was reported at 2038–2047 cm^{-1} at 12 K.

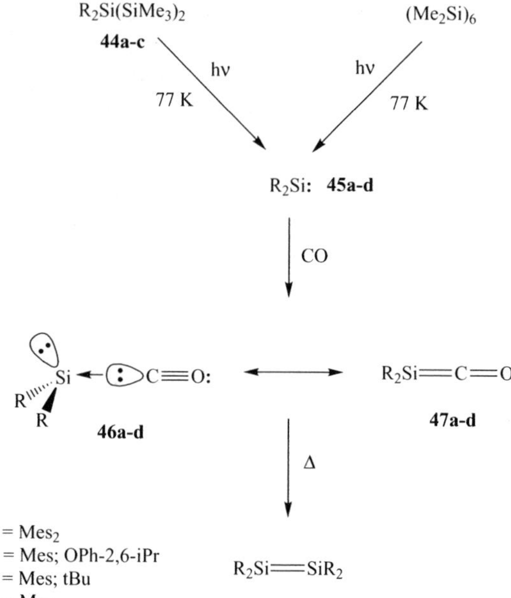

SCHEME 13.

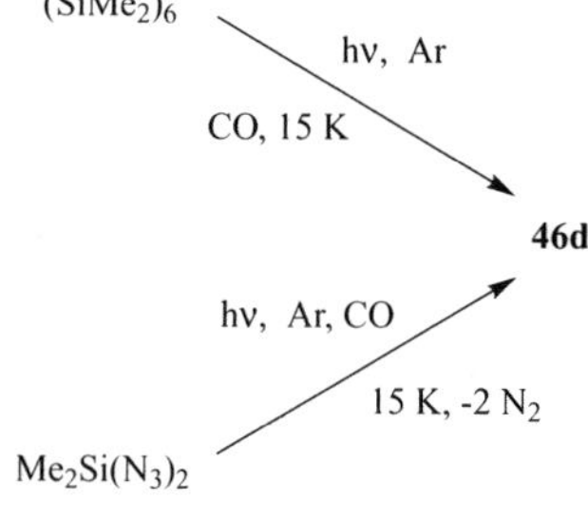

SCHEME 14.

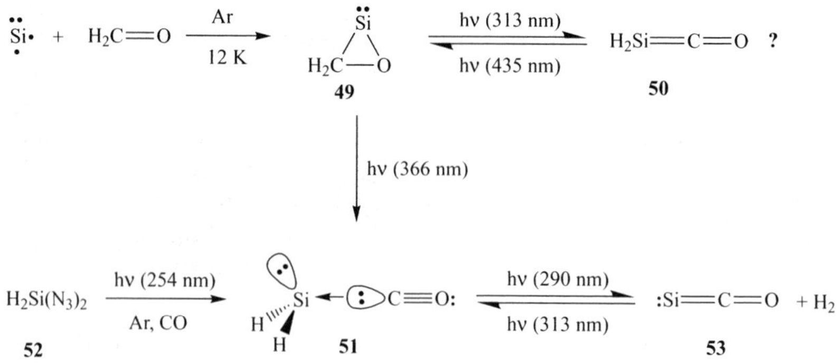

$\ddot{Si}\cdot$ + $H_2C{=}O$ $\xrightarrow[12\ K]{Ar}$ **49** $\underset{h\nu\ (435\ nm)}{\overset{h\nu\ (313\ nm)}{\rightleftharpoons}}$ $H_2Si{=}C{=}O$? **50**

49 $\xrightarrow{h\nu\ (366\ nm)}$ **51**

$H_2Si(N_3)_2$ **52** $\xrightarrow[Ar,\ CO]{h\nu\ (254\ nm)}$ $H_2Si \leftarrow C{\equiv}O$: **51** $\underset{h\nu\ (313\ nm)}{\overset{h\nu\ (290\ nm)}{\rightleftharpoons}}$ $:Si{=}C{=}O$ + H_2 **53**

SCHEME 15.

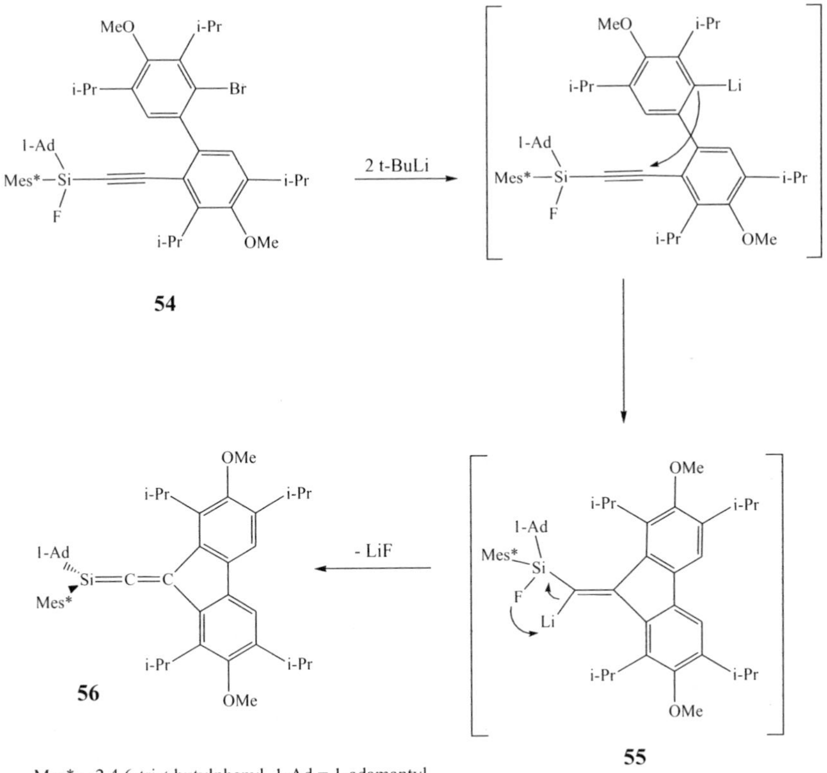

Mes* = 2,4,6-tri-*t*-butylphenyl, 1-Ad = 1-adamantyl

SCHEME 16.

C. *Stable Heteroallenes*

1. *1-Silaallenes*

The first stable silaallene, **56,** was synthesized in 1993[20–22] by the intramolecular attack of an organolithium reagent at the β-carbon of a fluoroalkynylsilane (Scheme 16). Addition of two equivalents of *t*-butyllithium in toluene at 0°C to compound **54** gave intermediate **55.** The α-lithiofluorosilane then eliminated lithium fluoride at room temperature to form the 1-silaallene **56,** which was so sterically hindered that it did not react with ethanol even at reflux temperatures. 1-Silaallene **56** was the first, and so far the only, multiply bonded silicon species to be unreactive toward air and water. The X-ray crystal structure and NMR spectra of **56** is discussed in Sect. IVA.

In 1997,[23] the intermolecular addition of organolithium reagents to fluoroalkynylsilanes was used to synthesize three novel, stable 1-silaallenes. In this

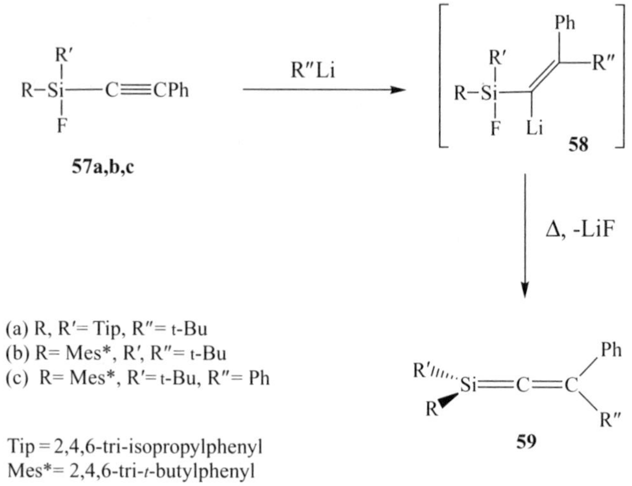

$$R{-}Si(R')(F){-}C{\equiv}CPh \xrightarrow{R''Li} [R{-}Si(R')(F){-}C(Li){=}C(Ph)R''] \; \mathbf{58} \xrightarrow{\Delta,\ -LiF} R'RSi{=}C{=}C(Ph)R'' \; \mathbf{59}$$

57a,b,c

(a) R, R′= Tip, R″= t-Bu
(b) R= Mes*, R′, R″= t-Bu
(c) R= Mes*, R′= t-Bu, R″= Ph

Tip = 2,4,6-tri-isopropylphenyl
Mes*= 2,4,6-tri-*t*-butylphenyl

SCHEME 17.

salt-elimination approach, one equivalent of the organolithium reagent is added to a solution of a sterically hindered (to prevent attack at silicon) fluoroalkynylsilane **57.** The organic part of the organolithium compound adds to the β-carbon of the alkyne and the lithium is transferred to the α-carbon, forming α-lithiosilane **58** (Scheme 17). Compound **58a** was stable at 0°C in solution and was trapped by methanol, forming the α-hydrovinylsilane **60a** (Eq. (3)). The TMEDA complex of compound **58a** was studied by X-ray crystallography, clearly showing that this intermediate is present in solution prior to the formation of 1-silaallene **59.** These intermediates were not detected for 1-silaallenes **59b** or **59c.** Further warming (25°C for 2 h for **59a;** 0°C for 1 h for **59b;** 25°C for 3 days for **59c**) led to elimination of lithium fluoride, providing 1-silaallenes **59a–c.**

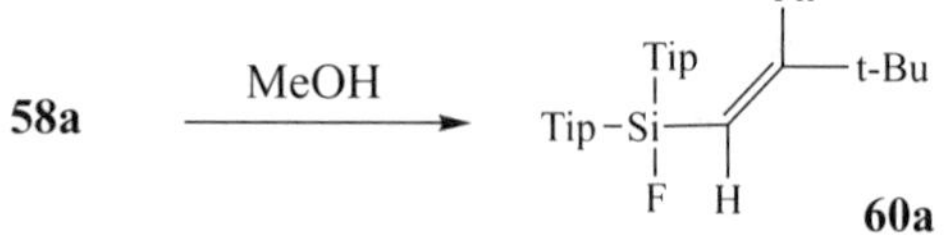

$$\mathbf{58a} \xrightarrow{MeOH} Tip{-}Si(Tip)(F){-}C(H){=}C(Ph)(t\text{-}Bu) \; \mathbf{60a} \tag{3}$$

1-Silaallene **59a** decomposes to silacyclobutane **61** (Eq. (4)) upon heating to 135°C in solution. This sole decomposition product was created from the insertion of the Si=C double bond of the silaallene into one of the *ortho* tertiary carbon isopropyl C–H bonds, which are preferred (or more acidic) over the primary carbon methyl C–H bonds on the isopropyl groups, which would form a less strained five-membered ring.

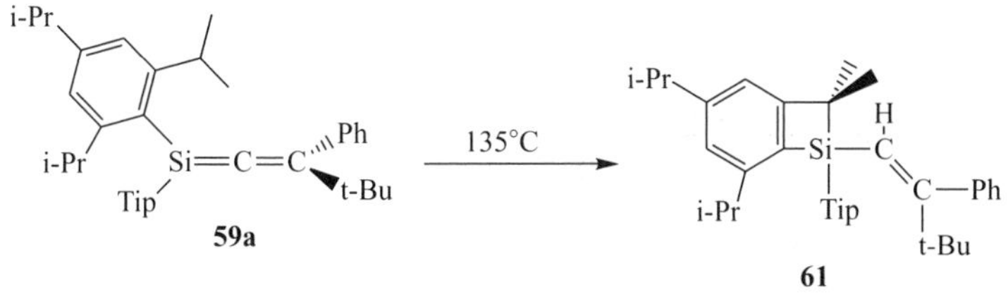

(4)

Only one type of reaction occurred for 1-silaallenes **59b** and **59c** (Eq. (5)). The greater steric bulk of the substituents on silicon led to a similar insertion of the Si=C double bond into a C—H bond on one of the *ortho t*-butyl groups on the Mes* group, forming the five-membered ring **62.** In this reaction, there is no longer a tertiary C—H bond to insert into, giving only one option for insertion into a C—H bond, that of rearrangement to the silacyclopentane. The insertion occurred in three separate pathways for **59b.** The first was observed when **59b** was heated to 90°C overnight. The second method for synthesis of **62b** was to stir **59b** with excess ethanol at room temperature for several hours. The same rearrangement occurred upon mixing at −78°C when a catalytic amount of acid in excess ethanol was added to **59b.** Excess deuterated ethanol and 5 mol% D_2SO_4 added to **59b** incorporated a deuterium atom in the α-vinylic position (>95%, **63**), suggesting initial protonation (deuteration) at the 1-silaallene central carbon (Scheme 18), thereby creating a reactive silicenium ion, which inserts into an *ortho t*-butyl C—H bond, liberating H^+ and making **64.** 1-Silaallene **59c** is less sterically hindered and forms **62c** under mild conditions of excess ethanol at room temperature in only a few seconds.

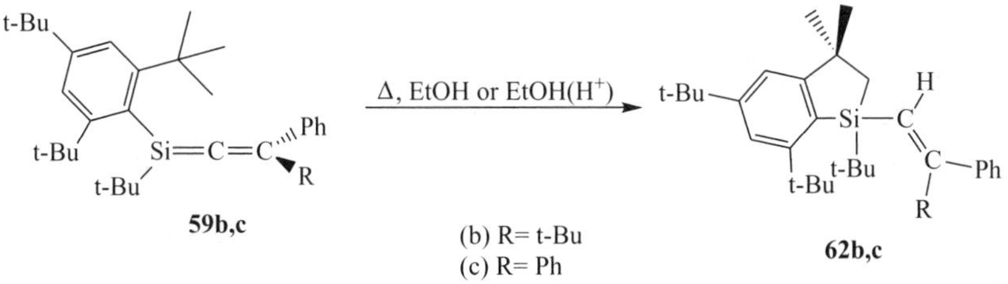

(5)

Compound **59a** underwent intermolecular reactions characteristic of silenes (Scheme 19). Water added instantly across the Si=C double bond of the 1-silaallene is expected to give vinylhydroxysilane **65** in 71% yield, and methanol was added

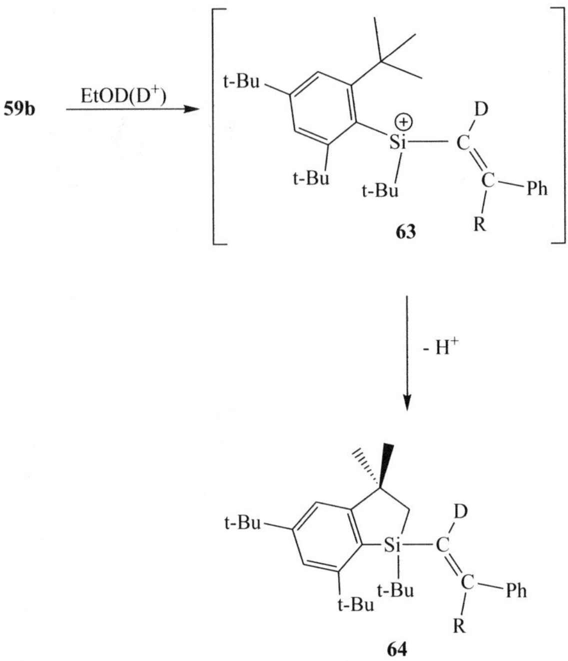

SCHEME 18.

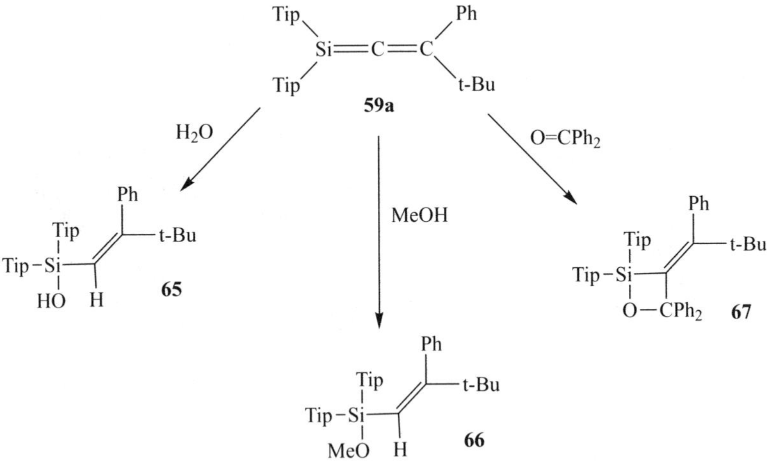

SCHEME 19.

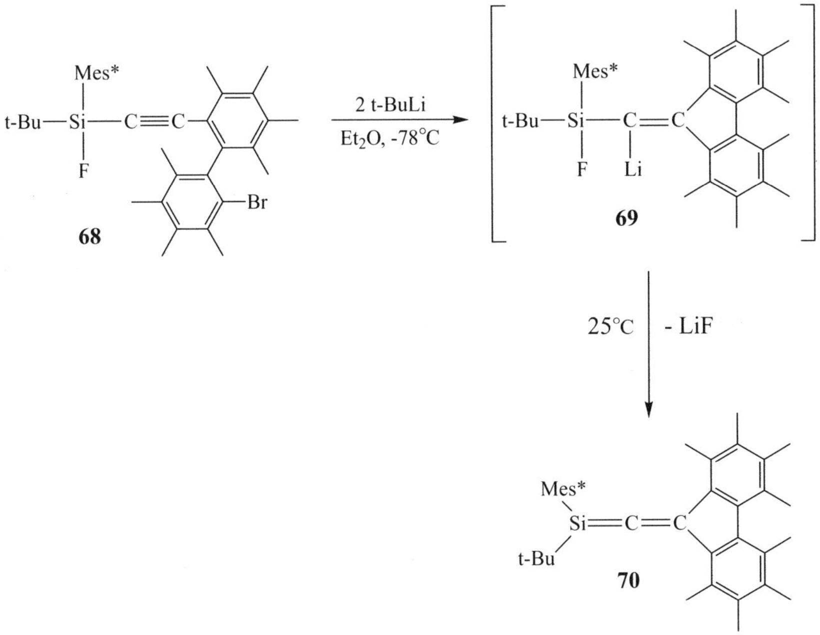

SCHEME 20.

to provide vinylmethoxysilane **66** in 78% yield. Benzophenone also reacted with **59a** to give a mixture of E- and Z-isomers of 1,2-oxasiletane **67** in 22% isolated yield.

Two more novel silaallenes were reported in 1995[21] and 1999[22]. 1-Silaallene **70**[21,22] was synthesized (Scheme 20) from alkynylfluorosilane **68** and two equivalents of *t*-butyllithium to give intermediate **69,** which eliminates lithium fluoride upon warming to room temperature for 2 to 6 h yielding bright yellow silaallene **70.** Compound **70** is stable in refluxing neutral or slightly basic ethanol for 3 days, but undergoes a rearrangement (Scheme 21) similar to that of silaallenes **59b** and **59c** when submitted to slightly acidic conditions in refluxing ethanol, giving **71.** Silaallene **70** also photolyzes to insert the Si=C double bond into the C—H bond of the nearest methyl group on the fluorenyl moiety to form the strained silane **72.**

Silaallene **73**[22] was synthesized in an manner analogous to that of **70.** Compound **73** was stable at room temperature over 1 month, but in the presence of any protic source (i.e., water, methanol), it underwent a rearrangement different than that observed for **70,** inserting into a methyl C—H bond (**74**) on the octamethylfluorenyl moiety rather than into one of the groups on silicon (Eq. (6)). It is believed that the favored mode of rearrangement for these groups is that of silaallene **73,** but **70**

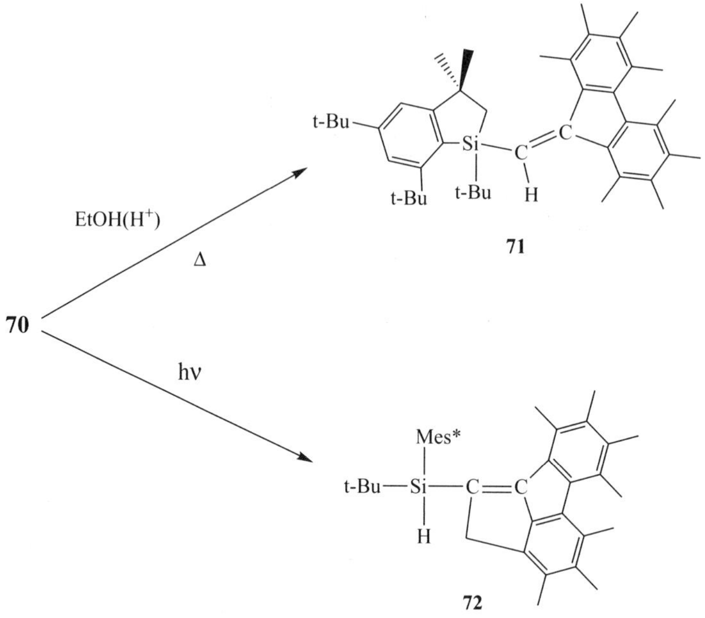

SCHEME 21.

is too sterically hindered to do so, and therefore the Si=C double bond is near enough only to the group on silicon to react to form **71.**

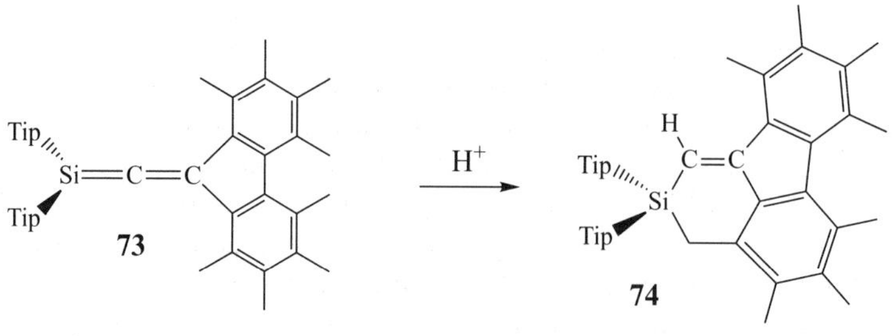

(6)

2. *1-Germaallenes*

West *et al.* have recently described the synthesis[24] and reactions[25] of a 1-germaallene. Germaallene **76** (Eq. (7)) is analogous to silaallene **59a** and is synthesized by intermolecular addition of *t*-butyllithium to precursor **75,** followed by salt elimination at −78°C. This germaallene is not stable above 0°C in solution, but remains intact until heated above 90°C in the solid state. In either case, the germaallene performs an insertion of the Ge═C double bond into a C—H bond on an *ortho* isopropyl group, forming four-(**77**) and five-membered (**78**) rings (Eq. (8)) in a ratio of approximately 3:2, depending on which C—H bond is inserted into. Compound **76** also added external reagents such as water and alcohols across the Ge═C double bond to make compounds **79a–c** (Eq. (9)). Benzophenone was too large to add to the germaallene at 0°C, but benzaldehyde added readily to provide oxagermetane **80** (Scheme 22). Germaallene **76** also reacts with acetone to give the unexpected trapping product **79a.**

$Tip_2Ge(F)—C≡C—Ph$ (**75**) —(t-BuLi, -78°C)→ $Tip_2Ge═C═C(Ph)(t\text{-}Bu)$ (**76**) (7)

76 —(0°C (solution), > 90°c (solid))→ **77** (60%) + **78** (40%) (8)

76 —(ROH)→ $Tip_2Ge(OR)—C(H)═C(t\text{-}Bu)—Ph$ **79** (9)

(a) R= H
(b) R= Me
(c) R= Et

Another germaallene was also reported in 1998 by Okazaki *et al.*[26b] Initially, the report of a germaallene trap with chalcogens, alkylidenetelluragermirane **86a,** appeared in 1997.[26a] The germylene precursor **82** is made *in situ* from dichlorogermane **81** and two equivalents of lithium naphthalenide (Scheme 23). The addition

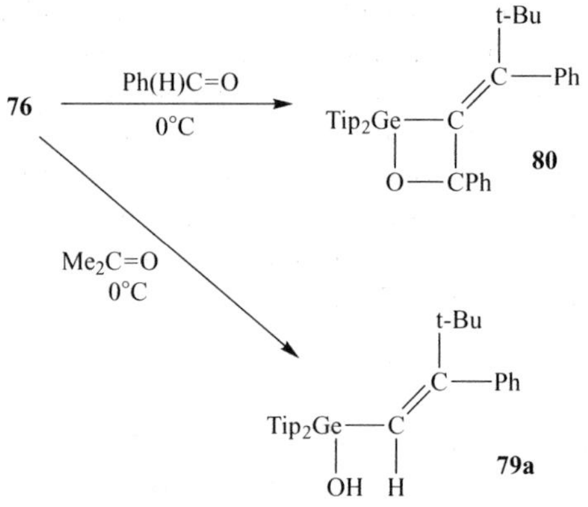

SCHEME 22.

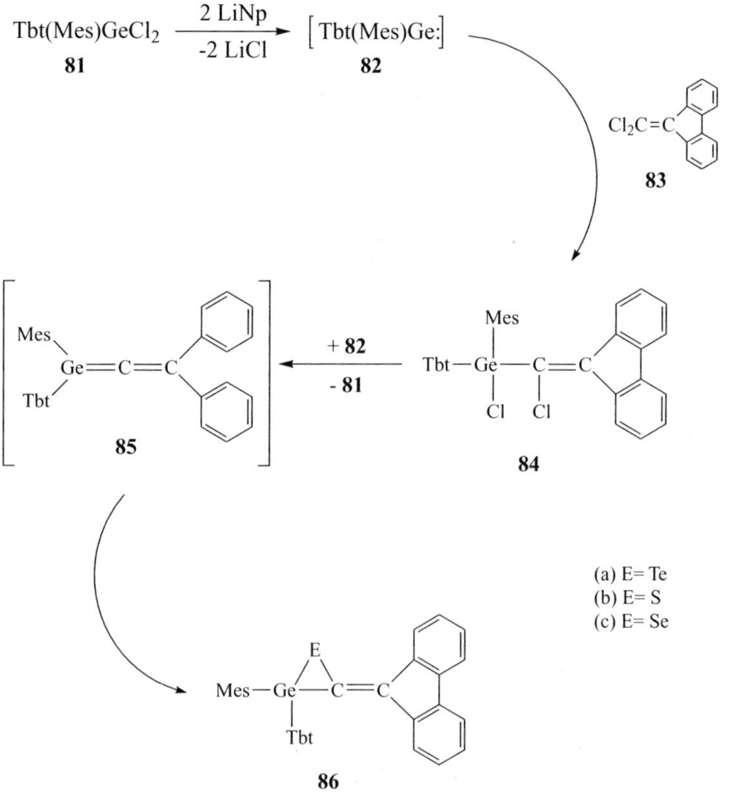

Tbt= 2,4,6-tris[bis(trimethylsilyl)methyl]phenyl

SCHEME 23.

of dichloromethylenefluorene **83** and tributylphosphine telluride to the solution, leads first to compound **84,** which was isolated in 5% yield and can be dechlorinated by the addition of germylene **82** to give 1-germaallene **85,** regenerating dichlorogermane **81** in the process. The germaallene then abstracts a tellurium atom from tributylphosphine telluride to make tributylphosphine and the final isolated alkylidenetelluragermirane **86a** in 10% yield as orange crystals. This reaction also produced the alkylidenethiagermirane (**86b**) and the alkylideneselenagermirane (**86c**) from elemental sulfur and selenium, respectively.

Germaallene **85** can be synthesized in two other ways.[26b] In the first method (Eq. (10)), compound **86a** was reacted with excess hexamethyl phosphorus triamide in order to reversibly abstract the tellurium atom away from the germaallene. The second method (Eq. (11)) utilized dehalogenation by addition of two equivalents of *t*-butyllithium to **84** at −72°C. Germaallene **85** reacted with methanol (Scheme 24) to give methoxyvinylgermane **87** and with mesitonitrile oxide to give compound **88.** Germaallene **85** also underwent cyclization in a manner similar to that of germaallene **76** to make four-membered ring **89,** upon storage at room temperature for 4 days (50% complete) or at 80°C for 13.5 h.

Te; Mes—Ge—C=C; Tbt; **86a** $\xrightarrow[(Me_2N)_3P{=}Te]{(Me_2N)_3P,\ C_6D_6,\ 25°C}$ Mes, Tbt Ge=C=C; **85** (10)

Mes; Tbt—Ge—C=C; Cl Cl; **84** $\xrightarrow[-72°C]{2\ t\text{-BuLi, THF}}$ Mes, Tbt Ge=C=C; **85** (11)

3. *Heteroketenimines*

The addition of isocyanates (isoelectronic to CO) to group 14 carbene analogs was investigated in order to see if the resulting molecules would have an allene-like framework. In the first study, Weidenbruch *et al.*[27a,b] attempted to synthesize stable 1-silaketenimines starting from silylenes and isocyanates, but only produced dimerization products of silaketenimines. In this investigation, di-*t*-butylsilylene **91,** generated by photolysis of hexa-*t*-butyltrisilacyclopropane (**90**), was allowed to react with four isocyanates (**92a–d**) (Scheme 25). The reaction presumably forms

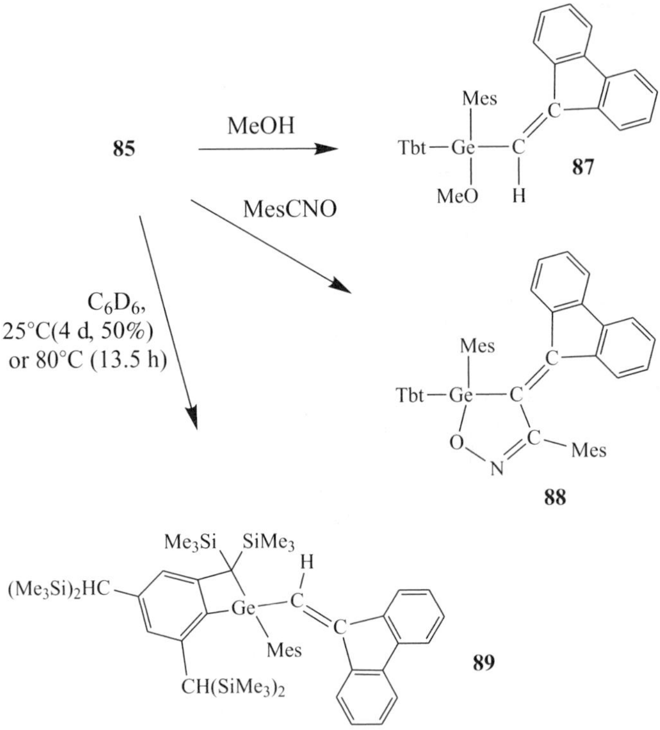

SCHEME 24.

1-silaketenimines **93a–d,** which (depending on substituent R) either rearrange to a 1,3-alkyl shifted cyanotrialkylsilane or dimerize. Only the head-to-tail dimer was isolated when R = phenyl (**94a**—65%) and 2,4,6-tri-methylphenyl (**94b**—45%), but when the steric bulk was increased to R = 2,4,6-tri-isopropylphenyl to hinder dimerization, a mixture of the head-to-tail dimer (**94c**—76%) and the head-to-head dimer (**95c**—minor product) was produced. When even more steric hindrance was employed (also to prove that the isocyanate was not simply adding twice to the disilene formed in the photolysis to make the same dimer product) by changing R to 2,4,6-tri-*t*-butylphenyl (Mes*), the silaketenimine did not dimerize, but rather rearranged to give compound **96.** The mechanism is not well understood, but a similar isomerization with a 2,4,6-tri-*t*-butylphenyl group was also reported for chlorostannanes by the same group.[27c]

The first stable group 14 heteroallene, a 1-stannaketenimine (**99**), was reported by Grützmacher *et al.* in 1992.[2b] Compound **99** was synthesized in 91% yield by adding diarylstannylene **97** and mesityl isocyanide **98** in hexane (Eq. (12)). The bonding in **99** can be described as a stannylene-isocyanide adduct rather than a

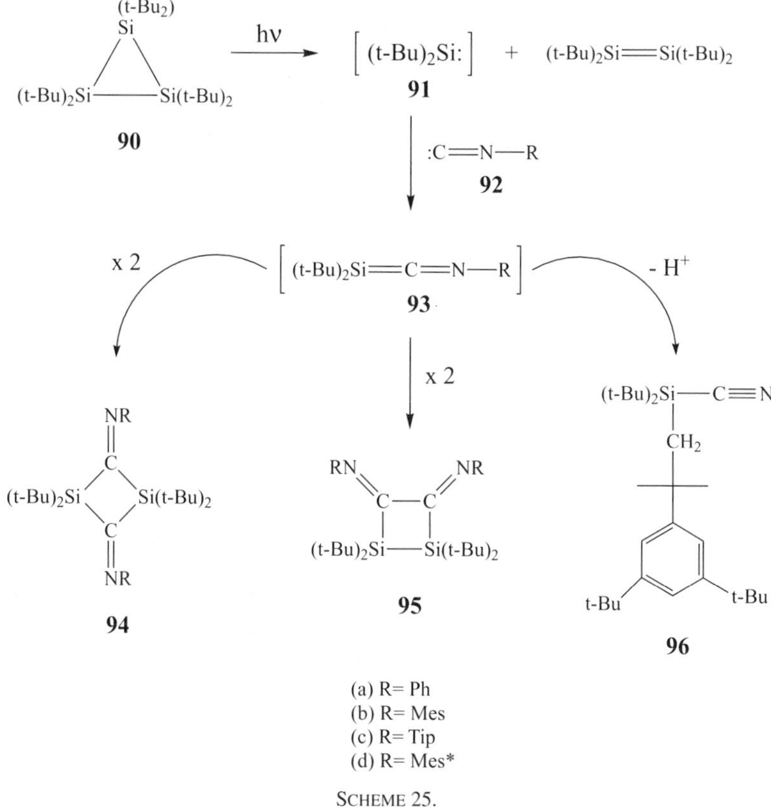

SCHEME 25.

stannaethene and this is illustrated through the reactivity of **99** (Scheme 26). When 2,3-dimethylbutadiene was added to **99,** mesityl isocyanide **98** was displaced and compound **100** was isolated, which is the addition product of the butadiene to stannylene **97.** A similar situation occurred when *t*-butanol is added to **99;** *t*-butanol not only expels **98,** but also replaces both of the aryl groups on tin with *t*-butoxy groups to give stannylene **101.**

$$\underset{\mathbf{97}}{R_2Sn:} \quad + \quad \underset{\mathbf{98}}{:C{=}N{-}Mes} \longrightarrow \underset{\mathbf{99}}{R_2Sn{=}C{=}N{-}Mes} \qquad (12)$$

$R = 2,4,6\text{-}(CF_3)_3C_6H_2$

The reactivity of a silylene **103** with isocyanides was probed by Okazaki *et al.* in 1997.[28] When disilene **102** is heated to 60°C in THF or C_6D_6, it dissociates

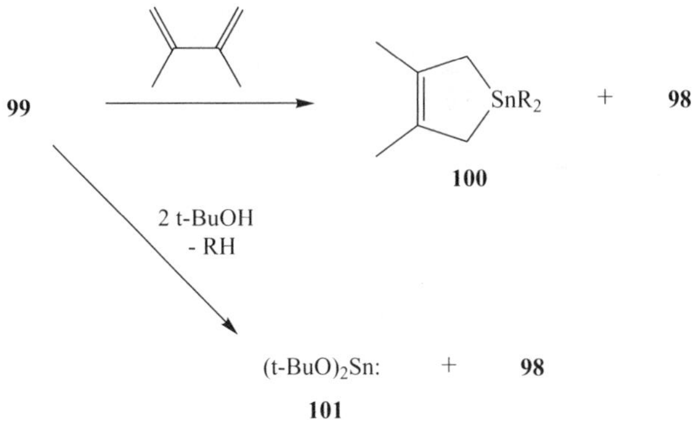

99 → SnR_2 + **98**

100

2 t-BuOH
- RH

$(t\text{-}BuO)_2Sn$: + **98**

101

SCHEME 26.

into two equivalents of silylene **103** (Scheme 27) and in the presence of various isocyanides **104a–c,** 1-silaketenimines **105a–c** are formed, on the basis of NMR results and trapping reactions. Similar to the 1-stannaketenimine just mentioned, data and calculations suggest that **105a–c** are best thought of as silylene-isocyanide adducts and a better depiction of **105a–c** is shown in Scheme 28. Although stable in the absence of external reagents, compounds **105a–c** reacted with triethylsilane to give good yields of the corresponding isocyanides and silane **106,** indicating a facile displacement of the isocyanide and subsequent addition of silylene **103** across the triethylsilane Si—H bond. The lability of the isocyanides is also demonstrated by the reaction of **105a–c** with MeOH, which provides methoxysilane **107** and, in

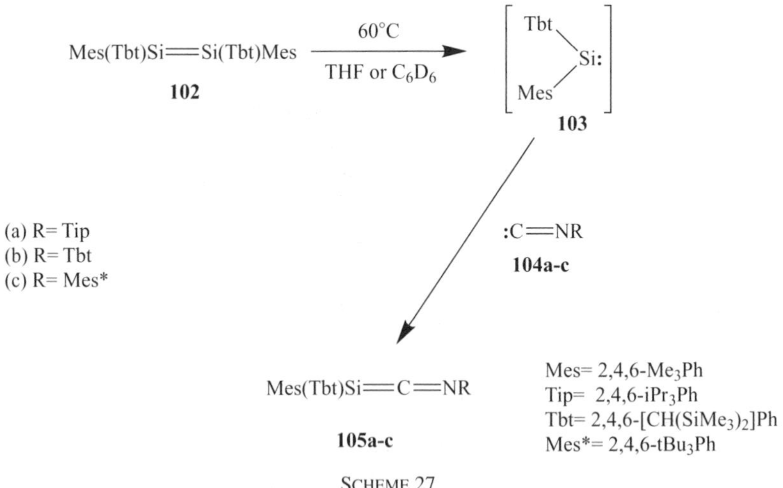

SCHEME 27.

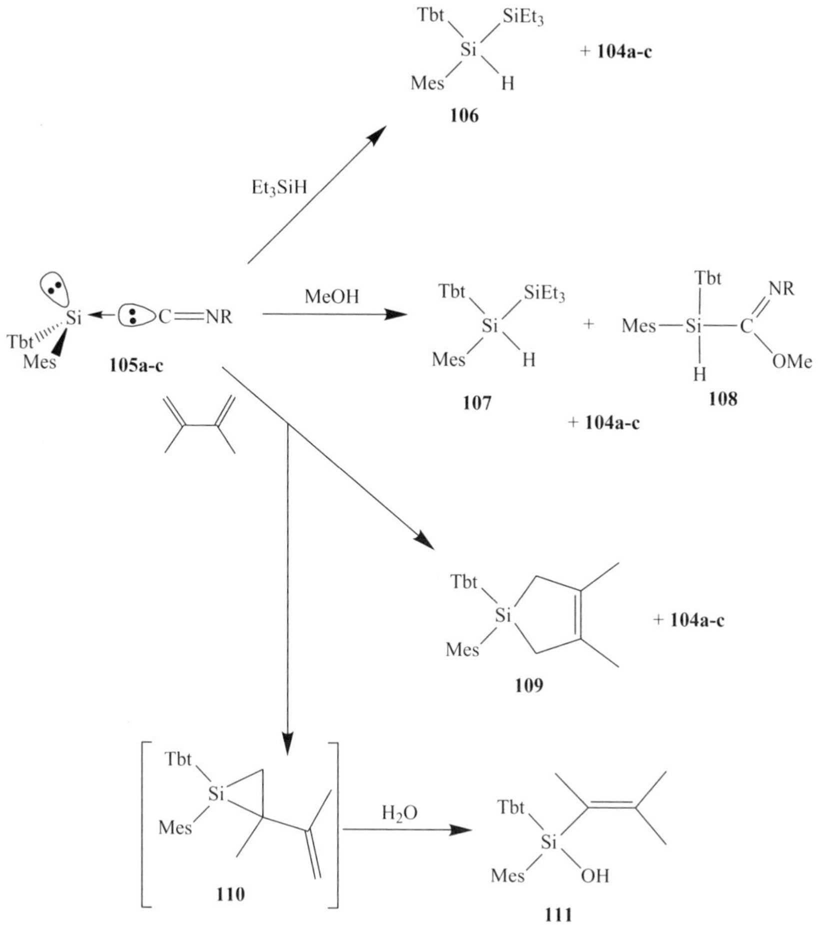

SCHEME 28.

the case of only **105a,** a small amount of compound **108.** The isolation of this compound is unusual for 1-heteroallenes in that the oxygen of the methoxy group normally attaches to the heteroatom (i.e., silicon) and the hydrogen adds to the central carbon atom. The authors suggest that for both **107** and **108,** the initial step of the mechanism is protonation of the silicon atom followed by attack of the methoxy group at the silicon for **107** (eliminating isocyanide **105a**) and at the carbon for **108.** Regardless of the mechanism, compound **108** is important in proving the existence of **105a,** since it is the only trapping product which retains the isocyanide portion of the silaketenimine. In a manner identical to that for stannaketenimine, silaketenimines **105a–c** replaced the isocyanides with

2,3-dimethylbutadiene to produce compound **109.** Another compound, silanol **111,** was isolated from the reaction mixture, the existence of which is proposed to be the hydrolysis of strained compound **110** during work-up, although direct attack of compound **109** by water cannot be ruled out.

4. *Heterophosphaallenes*

Escudié and co-workers[29] synthesized a metastable 1-germa-3-phosphaallene (**114**) in 1996 by the salt-elimination method, a process they called "debromofluorination" (Scheme 29). In a reaction that was followed by ^{31}P NMR, one equivalent of *n*-butyllithium was added to **112** at $-90°$C to form intermediate **113.** Upon warming to approximately $-60°$C, lithium fluoride was eliminated, forming germaphosphaallene **114** in 65–70% yield. Compound **114** was stable up to $-50°$C, whereupon it dimerized in the absence of trapping agents. This process not only gave the expected head-to-tail dimer **115** (between the two Ge=C bonds), but also an unexpected dimerization product (**116**) resulting from the reaction

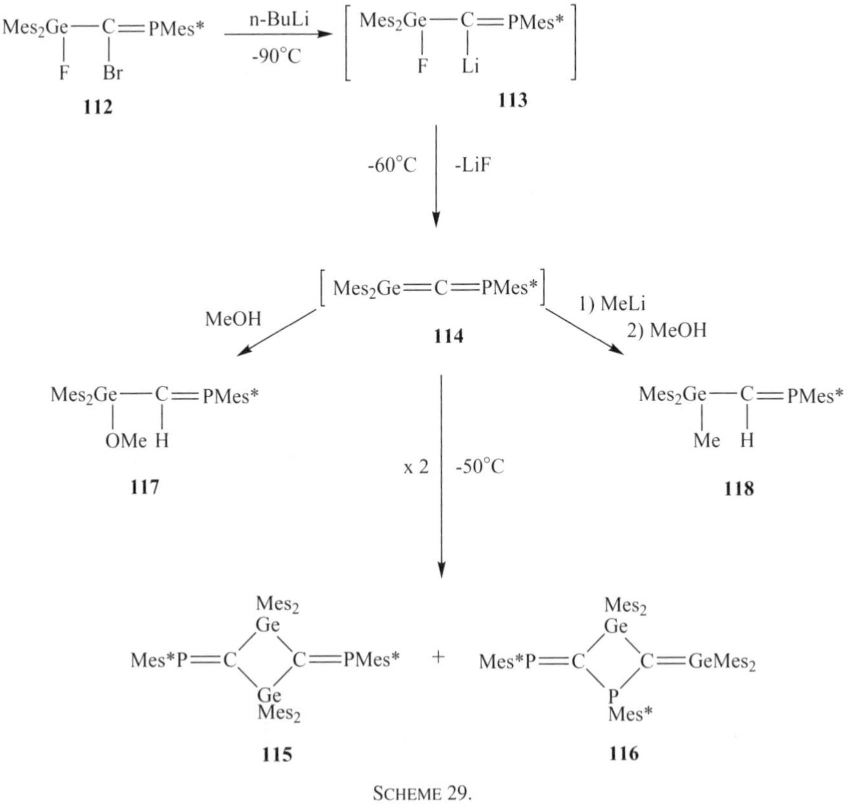

$Mes_2Ge(F)$—$C(Br)$=PMes* (**112**) → n-BuLi, -90°C → [$Mes_2Ge(F)$—$C(Li)$=PMes*] (**113**)

113 → -60°C, -LiF → [Mes_2Ge=C=PMes*] (**114**)

114 → MeOH → $Mes_2Ge(OMe)$—$C(H)$=PMes* (**117**)

114 → 1) MeLi, 2) MeOH → $Mes_2Ge(Me)$—$C(H)$=PMes* (**118**)

114 → x 2, -50°C → **115** + **116**

115: Mes*P=C<($GeMes_2$)$_2$>C=PMes*

116: Mes*P=C<($GeMes_2$)(PMes*)>C=$GeMes_2$

SCHEME 29.

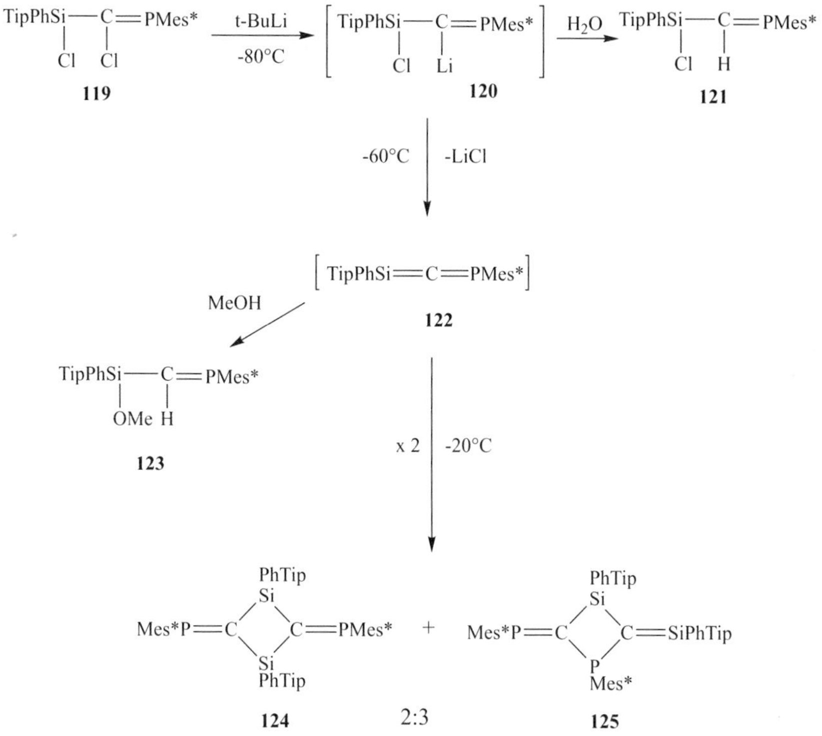

SCHEME 30.

of a Ge=C bond with a P=C bond of another molecule. Two trapping experiments were performed on germaphosphaallene **114:** one with methanol, giving the expected methoxygermane **117,** the other with methyllithium followed by quenching with methanol, providing methylgermane **118.**

Escudié *et al.*[30] followed up the 1-germa-3-phosphallene work by making a similar silicon analog, a 1-sila-3-phosphaallene (**112**). Compound **112** was synthesized by a salt-elimination method similar to that for the germaphosphaallene, as shown in Scheme 30. Dehalogenation of **119** at carbon by *t*-butyllithium at −80°C gave α-lithiochlorosilane **120,** which was observed by ^{31}P NMR (as evidenced by the downfield shift compared to precursor **119**) and was isolated as the hydrolysis product **121.** Upon warming **120** to −60°C, the elimination of lithium chloride was complete, giving silaphosphaallene **122.** Methanol addition to **122** at −60°C afforded methoxysilane **123** in 50% yield. When the silaphosphaallene was warmed above −20°C in the absence of trapping agents, it dimerized in a manner similar to the dimerization of germaphosphaallene **124.** This gave a head-to-tail dimer (**124**) between two Si=C bonds and the product of dimerization at one Si—C bond and one P=C bond (**125**) in a 2:3 ratio.

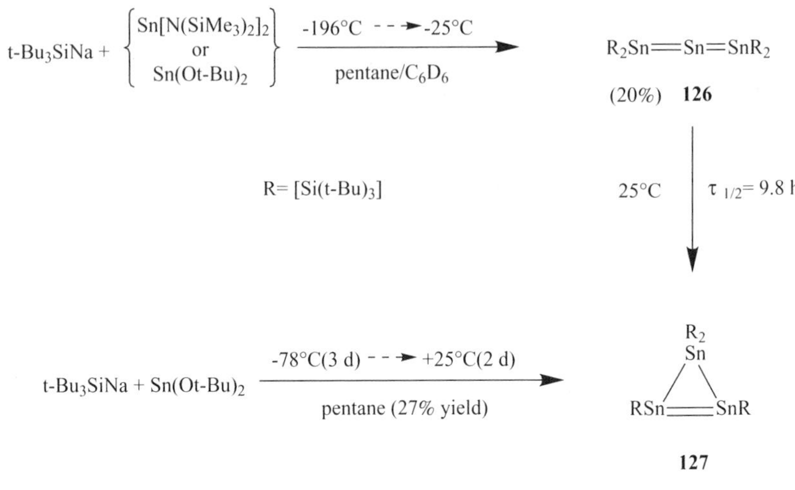

SCHEME 31.

5. *A 1,2,3-Tristannaallene*

Wiberg *et al.*[31] have just recently published the synthesis of the only heteroallene (**126**) to date in which all of the atoms in the allene backbone are heavy group 14 elements, and more specifically, are all tin atoms. Compound **126** was synthesized by adding tri-*t*-butylsilyl sodium to $Sn(OtBu)_2$ or $Sn[N(SiMe_3)_2]_2$ at $-196°C$ and warming to $-25°C$ to provide a mixture of compounds, from which the tristannaallene is separated by fractional crystallization as blue crystals (Scheme 31). Care is taken not to warm **126,** because it rearranges to form cyclotristannene **127** ($\tau_{1/2} = 9.8$ h), which can be independently synthesized by stirring tri-*t*-butylsilyl sodium and $Sn(OtBu)_2$ in pentane at $-78°C$ for 3 days and then at 25°C for another 2 days. Like the stannaketenimine mentioned earlier, the tendency of tin to form stannylene complexes manifests itself in the bent structure of **126** (Fig. 4), which can be considered as a mixture of the resonance structures **126a–d** and, according to the authors, is best described as structures **126b** and **126c.** The synthesis and isolation of this 1,2,3-tristannaallene illustrates that it is feasible to synthesize other heteroallenes consisting of more than one heavy group 14 atom.

6. *A Transition Metal Complex of a 1-Silaallene*

One other study of group 14 heteroallenes involving transition metals was reported in 1995. Jones *et al.*[32] described the isolation of a ruthenium complex of a 1-silaallene (**132**—Scheme 32). The 1-silaallene also interacts with a hydrogen atom as well as the ruthenium metal center. Jones *et al.* describe this view

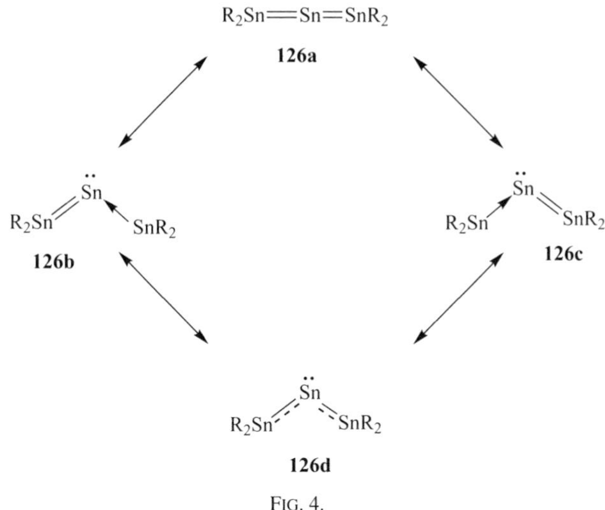

FIG. 4.

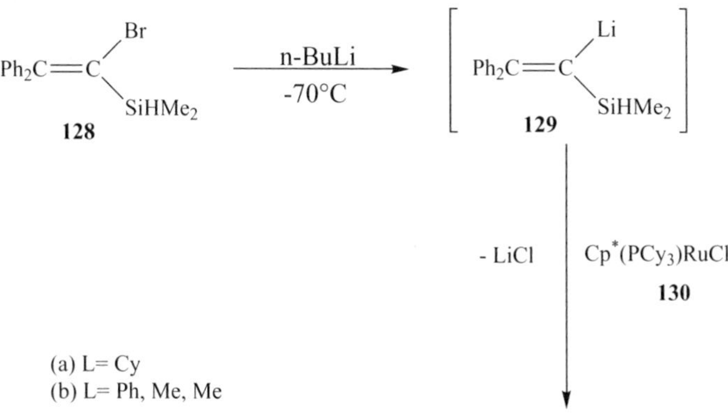

SCHEME 32.

as "a 1-silaallene that is stabilized by both metal ligation and interaction with a metal-hydrogen bond." Compound **128** was lithiated at low temperature to give intermediate **129,** to which Cp*(PCy_3)RuCl (**130**) was added, yielding intermediate **131** which produced **132a** (42%) after the hydrogen atom from the silicon atom was arrested by the ruthenium atom. An X-ray crystal structure of **132a** was obtained and will be discussed later. Compound **132a** was stable (more so than similar silene complexes) upon warming to 45°C in C_6D_6 for 1 day, and did not react with CO. Ligand (PL_3) exchange could be performed by replacing PCy_3 with PMe_2Ph (with warming, 81%) to form **132b** without harming the silaallene moiety. Heating above 45°C led to multiple decomposition products unreported by the authors.

IV

PHYSICAL PROPERTIES

A. *X-ray Structure Determination*

The crystal structures[33] of a variety of heteroallenes have been solved and provide valuable information toward the understanding of the bonding in this series of molecules.

If one were to ask a first-year organic chemistry student what the bond angle for an allene (C=C=C) should be, they would most likely answer "180 degrees." All things considered, they would most likely be correct, and an example of an all-carbon allene[34] (**133**) with a bond angle of 179.0° is listed in Table III. It must be noted that due to crystal packing forces and steric restraints, the bond angles and bond lengths of these molecules may deviate slightly from the ideal. As mentioned before, on descending the group (C $<$ Si $<$ Ge $<$ Sn $<$ Pb), the singlet state is more favored over the triplet state for the atom in the one position of the allene. This means that the M=C=E (M=Si, Ge, Sn; E=C, N) heteroallene angle will be bent more upon substitution with a heavier atom. Experiment agrees with theory in this case as bending angles systematically shrink from carbon (179°) to tin (154°) (Table III).

In three studies from the early 1980s,[5,6,12f] it was suggested that 1-silaallenes would have a linear Si=C=C framework. Later calculations by Trinquier and Malrieu[3] predicted qualitatively that the 1-silaallene framework would be nonlinear, but how much the moiety would deviate from 180° was unclear. The first quantitative values were seen with the isolation and structural determination of the two 1-silaallenes **56** and **59a,** which have very similar Si=C=C angles of 173.5° and 172.0°, respectively (Table III). This is an average deviation of 7.3° from linearity—significant, but relatively small compared to the deviations shown by the germanium and tin substituted allenes.

TABLE III

SUMMARY OF CRYSTAL STRUCTURE DATA FOR GROUP 14 1-HETEROALLENES

	M=C=E Bond Angle (deg.)	Sum of Angles Around M (deg.)	M=C Bond Length (Å)	C=E Bond Length (Å)	Ref.
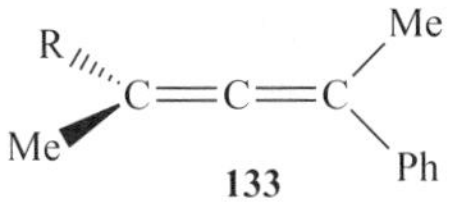 R(Me)C=C=C(Me)Ph **133**; R= (Me)C=C=C(Ph)Me	179.0	360.0	1.313(3)	1.318(3)	34
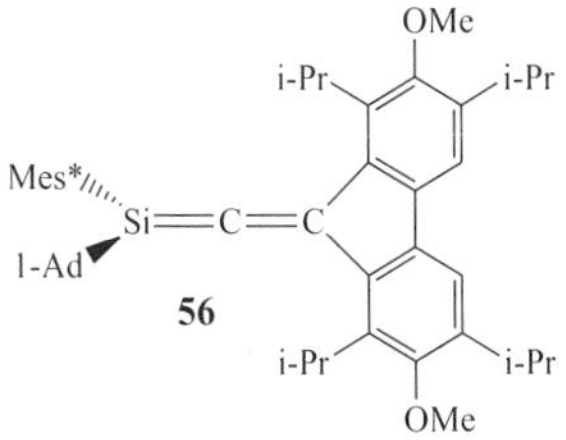 (Mes*)(1-Ad)Si=C=C(fluorenylidene: i-Pr, i-Pr, OMe; i-Pr, i-Pr, OMe) **56**	173.5	360.0	1.704(4)	1.324(5)	20
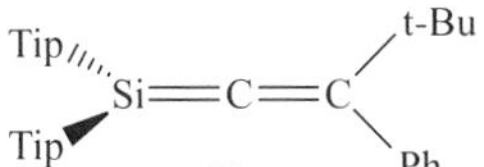 $Tip_2Si=C=C(t\text{-}Bu)Ph$ **59a**	172.0	357.2	1.693(3)	1.325(4)	23
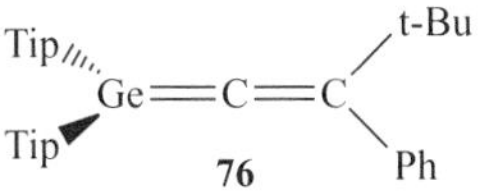 $Tip_2Ge=C=C(t\text{-}Bu)Ph$ **76**	159.2	348.4	1.783(2)	1.314(2)	24
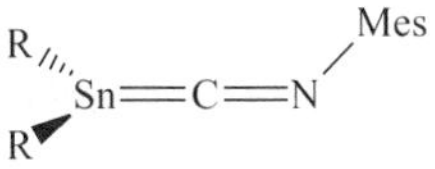 $R_2Sn=C=N\text{-}Mes$ **99**; R= 2,4,6-$(CF_3)_3C_6H_2$	153.9	290.9	2.397(3)	1.158(3)	2b
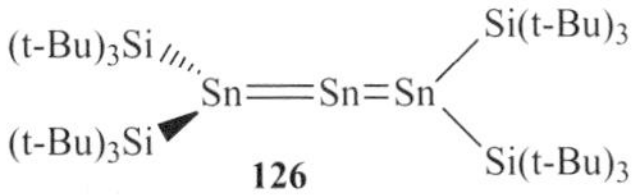 $((t\text{-}Bu)_3Si)_2Sn=Sn=Sn(Si(t\text{-}Bu)_3)_2$ **126**	155.9	344.0	2.68(avg. Sn=Sn)		31

The 1-silaallene calculations preceded the isolation of the first stable silaallene by 12 years, but despite the isolation of a stannaketenimine in 1992[2b] and a germaallene in 1998,[24] no calculations of bond angles or lengths have been performed yet for analogs larger than silicon. Given some guidelines and values provided by Trinquier and Malrieu,[3] a simple calculation can be performed (Scheme 33) to

$$0.5(E_{\sigma+\pi}) > \Sigma\Delta E_{S\text{-}T} \dashrightarrow \text{linear (condition 1)}$$

$$0.5(E_{\sigma+\pi}) < \Sigma\Delta E_{S\text{-}T} \dashrightarrow \text{bent (condition 2)}$$

$$H_2Ge\!: \quad + \quad :C{=}CH_2 \longrightarrow H_2Ge{=}C{=}CH_2$$

$$\Sigma\Delta E_{S\text{-}T} = \quad 22.5\ \text{kcal/mol} \qquad 46\ \text{kcal/mol} \qquad\qquad 68.5\ \text{kcal/mol}$$

$$E_{\sigma+\pi}(Ge{=}C) = 90 - 105\ \text{kcal/mol}$$

$$0.5(E_{\sigma+\pi}) = 45 - 53\ \text{kcal/mol}$$

$$45 - 53\ \text{kcal/mol} < 68.5\ \text{kcal/mol} \longrightarrow \text{bent!}$$

SCHEME 33.

predict whether or not a parent group 14 1-heteroallene will be linear or bent. Consider the example of a 1-germaallene. According to the above authors, if $0.5(E_{\sigma+\pi}) > \Sigma\Delta E_{S-T}$ (condition 1), the structure will be linear, and if $0.5(E_{\sigma+\pi})$ $\Sigma\Delta E_{S-T}$ (condition 2), the structure will be bent. $E_{\sigma+\pi}$ is the strength of a typical $\sigma+\pi$ planar double bond that is to be formed and $\Sigma\Delta E_{S-T}$ is the sum of the singlet-triplet energy separations for the two fragments being brought together. Using values from the Malrieu/Trinquier paper, $\Sigma\Delta E_{S-T}$ (68.5 kcal/mol) is larger than $0.5(E_{\sigma+\pi})$ (45–53 kcal/mol) for the parent 1-germallene, satisfying condition 2. Therefore, germallenes should be bent. This is the case for germallene **76,** which has an M=C=C angle of 159.2°. This deviates from linearity by 20.8°, more than twice that for the silaallenes. Stannaketenimine **99** deviates by 26.1°, although the nitrogen atom in the 3-position may affect this angle. Tristannaallene **126** is bent to a similar degree, 155.8°, although the emormous supersilyl [Si(t-Bu)$_3$] groups may prevent more severe deviation from a linear geometry, and Wiberg[31] has proposed that there may be a contribution to bending from the lone pair on the central tin atom.

In order to achieve maximum orbital overlap between atoms M and C in bonding model B (Fig. 2), the plane made by R_2M bends away from the vector made by the M=C bond (Fig. 5), resulting in pyramidalization of atom M. A simple test of the extent of pyramidalization at a particular atom can be performed by summing all of the bond angles around M and comparing the result to the ideal of 360° (for sp^2 atoms). The sum of the bond angles in allene **133** around the 1-position carbon total up to be exactly 360.0°, as expected. An unexpected result is that silaallene **56** also totals 360.0° around silicon, whereas silaallene **59a** bond angles add up to 357.2°. Germaallene **76** deviates somewhat more from planarity with a total of 348.4°. The sum of the bond angles around tin in stannaketenimine **99** comes to a meager

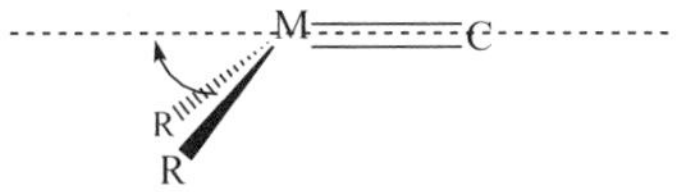

FIG. 5. Pyramidalization of M.

290.9°, indicating a severe distortion from planarity, even surpassing the pyramidalization of an idealized sp^3-hybridized atom ($3 \times 109.5° = 328.5°$). Tristannaallene **126** averages 344.0° for the angles around the 1- and 3-tins. These angles are much smaller than for the stannaketenimine, but pyramidalization around tin may be prevented due to the severe steric constraints caused by the large tri-*t*-butylsilyl ligands.

Trans bending of the substituents and bending at the central carbon for silaketene (H_2SiCO) become dominant for its calculated structure.[8] The Si—C—O angle is calculated to be 167.4° and the angle created between the Si—C bond and the vector bisecting the two Si—H bonds becomes a very sharp 89.0°. The calculated bond lengths for this yet unseen moiety are 1.891 Å (Si—C) and 1.148 Å (C—O).

The reason group 14 heteroallenes were not isolated until 1992 is undoubtedly because of the reactivity of the M=C (or M=M in the case of the tristannallene) double bond. Four Si=C double bond lengths were calculated in the early 1980s for 1-silaallenes; 1.696 Å by Lien and Hopkinson,[5] 1.703 Å by Gordon and Koob,[6] 1.702 Å by Krogh-Jespersen,[10] and 1.62 Å by Ishikawa *et al.*[12f] In excellent agreement with the majority of the theoretical results, silaallene **59a** has a Si=C bond length of 1.693(3) Å and silaallene **56** has a bond length of 1.704(4) Å. Although no calculations have been performed for germa- or stannaallenes, comparisons can be made to germenes and stannenes. Germaallene **76** was found to have a Ge=C double bond length of 1.783(2) Å, which falls right in the middle of values observed for germenes **134** [1.827(4)],[35a,c] **135** [1.803(4) Å],[36] and **136** [1.771(16) Å][37] (Table IV). Amazingly, the Sn=C double bond length [2.397(3) Å] in stannaketenimine **99** is longer than both of the Sn—C single bonds [2.306(2), 2.314(3) Å] to the *ipso* carbons of the R groups and is considerably longer than most other reported stannene Sn=C bond lengths [**137–140;** 2.03–2.38 Å, Table V]. The long Sn—C bond length in **99** reaffirms that the bonding interaction between tin and carbon is weak and is best described as a stannylene-carbene adduct. Tristannaallene **126** has an average Sn=Sn distance (two Sn=Sn bonds per molecule and two diastereomers) of 2.68 Å, which is the shortest Sn—Sn bond length next to its rearrangement product cyclotristannene **127** (2.59 Å) (Table VI).

The synthesis of three silaketenimines **105a–c** prompted Tokitoh and Okazaki[28] to calculate the optimized geometry of a model compound, $Ph_2SiCNPh$. This model reinforced that **105a–c** are truly Lewis acid–base pairs, with the isocyanide donating its carbon lone pair to an empty *p*-orbital perpendicular to the lone pair

TABLE IV

GERMENE BOND LENGTHS

	M=C Lengths (Å)	Reference
134: $((Me_3Si)_2N)_2Ge{=}C$ (ring B(t-Bu), $C(SiMe_3)_2$, B(t-Bu))	1.827(4)	35a
135: $Mes_2Ge{=}C$ (fluorenylidene)	1.803(4)	236
136: Tbt[a](Tip)Ge=C (S, S ring) Ge(Tbt)(Tip)	1.771(16)	37

[a] Tbt = 2,4,6-tris[bis(trimethylsilyl)methyl]phenyl.

and two phenyl substituents on silicon. A long [longer than even an average Si—C_{sp3} distance (1.863 Å)][38] bond length 1.882 Å is predicted for the Si—C(isocyanide) bond, and the Si—C=N bond angle is bent (163.4°). The average C(Ph_{ipso})—Si—C(isocyanide) angle is 99.4°. These bonds and angles, along with a weak Si—C(isocyanide) binding energy (25.1 kcal/mol), reaffirm that this is a silylene-isocyanide adduct.

The C=C double bond of the heteroallene moiety is often ignored in discussions because of its lack of reactivity, but should not be overlooked in a thorough investigation of these compounds. In calculating Si=C double bonds for 1-silaallenes, the three aforementioned groups also calculated the C=C bond lengths; 1.312 Å by Lien and Hopkinson,[5] 1.296 Å by Gordon and Koob,[6] 1.295 Å by Krogh-Jespersen,[10] and 1.29 Å by Ishikawa *et al.*[12f] Although isolated 1-silaallene C=C double bond lengths are slightly longer than the calculated lengths, they are at least consistent with lengths of 1.324(5) Å for **56** and 1.325(4) Å for **59a.** Germaallene **76** has a C=C double bond length [1.314(2) Å] only 0.01 Å shorter than its silicon analog (**59a**). As noted by Grützmacher and co-workers, the C=N bond length [1.158(3) Å] for stannaketenimine **99** is "typical for isocyanides."[2b]

The structures of two group 14 heteroallenes that are complexed to other atoms have been determined—a ruthenium complex of a 1-silaallene (**132a**) and a

TABLE V

STANNENE BOND LENGTHS

	M=C Length (Å)	Reference
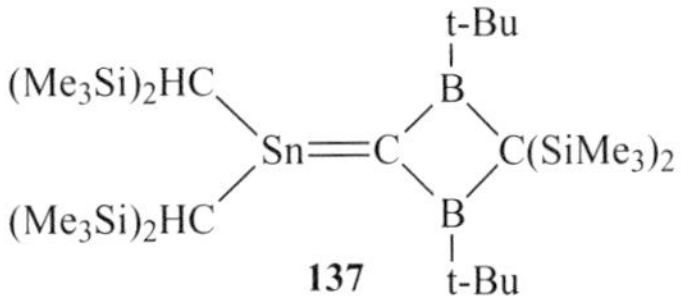 $(Me_3Si)_2HC$, $(Me_3Si)_2HC$ Sn=C; B(t-Bu), B(t-Bu); $C(SiMe_3)_2$ **137**	2.025(4)	35(b)
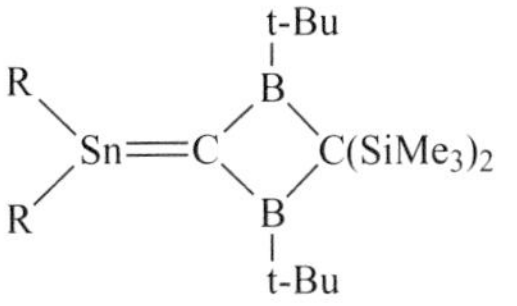 R, R Sn=C; B(t-Bu), B(t-Bu); $C(SiMe_3)_2$ **138** (R = 2-tBu-4,5,6-Me_3C_6H)	2.032(5)	47b
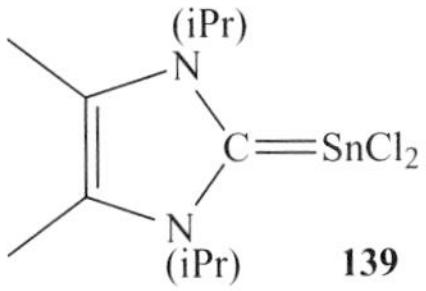 N(iPr), N(iPr); C=$SnCl_2$ **139**	2.290(5)	47c
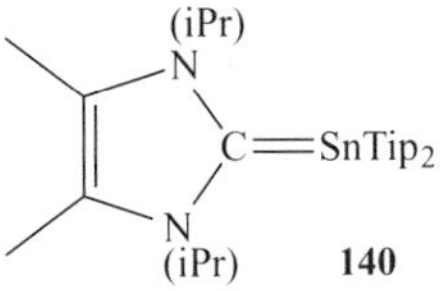 N(iPr), N(iPr); C=$SnTip_2$ **140**	2.379(5)	47a

tellurium complex of a 1-germaallene (**86a**). Compound **132a** has a silicon–carbon bond length of 1.805(6) Å, which is considerably longer than the bond lengths for silaallenes **56** (1.704 Å) and **59a** (1.693 Å), but is still shorter than normal Si–C(sp^2) single bonds (1.85–1.90 Å).[32] The Si=C=C bond angle 128.6(4)° suggests an sp^2-hybridized central carbon, not an sp-hybridized allenic carbon. The authors propose a 3-center 2-electron bond between ruthenium, silicon, and hydrogen atoms, indicating that this molecule is not only stabilized by a ruthenium atom but also by a hydrogen atom.

Compound **86a** has a germanium–carbon bond length of 1.88(2) Å, which is shorter than Ge–C single bonds of germirane derivatives (1.92–2.07 Å),[39] but is longer than the Ge=C double bond lengths listed in Table IV (1.77–1.83 Å). The relatively short Ge=C bond and the fact that the three C–Ge–C bond angles sum to 354.4° (which suggests considerable sp^2 character at germanium) led

TABLE VI

DISTANNENE BOND LENGTHS

	M=M Lengths (Å)	Reference
	2.59(avg.)	31
127 $[(Me_3Si)_2HC]_2Sn{=}Sn[CH(SiMe_3)_2]_2$	2.768(1)	48a
141 $[(Me_3Si)_3Si]_2Sn{=}Sn[Si(SiMe_3)_3]_2$	2.8247(6)	48b
142 $(Me_3Si)_3Si(R)Sn{=}Sn(R)Si(SiMe_3)_3$ **143** [R = 2,4,6-$(CF_3)_3C_6H_2$]	2.833(1)	48c
$R_2Sn{=}SnR_2$ **144** (R = 2-tBu-4,5,6-Me_3C_6H)	2.910(1)	48d
$R_2Sn{=}SnCl_2$ **145** {R = $[C(H)SiMe_3(C_9H_5N)]_2$}	2.961(1)	48e

the authors to propose that this molecule is in fact a π-complex of a tellurium atom with the Ge=C double bond of a 1-germaallene. The Ge=C=C bond angle is 149(1)°, almost directly in between the bonding angles for sp- (180°) and sp^2-hybridized (120°) carbons, and is only 10° larger than the Ge=C=C bond angle of 1-germaallene **76.**

B. *NMR Spectroscopy*

Possibly the most characteristic piece of information one can obtain to prove the existence of a 1-heteroallene is the central carbon ^{13}C NMR chemical shift. This carbon chemical shift is very deshielded, typically being greater than 200 ppm,[40] which stands out from most other carbon resonances in a normal organic molecule. Most of the group 14 1-heteroallenes listed in Table VII have shifts greater than 200 ppm. Also, as the heteroatom becomes larger, the resonance moves farther downfield.

TABLE VII

^{13}C CHEMICAL SHIFTS OF CENTRAL CARBON OR GROUP 14 1-HETEROALLENES

1-Heteroallene	^{13}C shift of Central Carbon (ppm)	Reference
56	225.7	20
59a	225.7	23
59b	216.3	23
59c	227.9	23
70	228.2	22
73	237.1	22
76	235.1	24
85	243.6	26b
105a	209.2	28
105b	196.6	28
105c	178.5	28
114	280.9	29
122	269.1	30

The 1-silaallenes in Table VII have chemical shifts in the range of 216 to 237 ppm. It is interesting to note how changing a tert-butyl group to a phenyl group on the carbon end of silaallenes **59b** and **59c** can change the chemical shift by greater than 10 ppm, but when the two phenyls of **59c** are changed to a fluorenyl group (**70**), the chemical shift moves by only 0.3 ppm. Germaallene **76** is the exact germanium analog of silaallene **59a,** and this substitution of germanium for silicon relocates the central carbon shift almost 10 ppm downfield to 235.1 ppm. Adding a phosphorous to the 3-position of a germaallene (giving germaphosphaallene **114**) moves the central carbon chemical shift a substantial 45 ppm to 280.9 ppm.* Changing the germanium atom to a silicon (**112**) moves the silicon back upfield to 269.1 ppm. The central ^{13}C shift for model silaketenimine Ph_2SiCNR was calculated to be 178 ppm, which agreed especially well with the shift for **105c** (178.5, **105a** = 209.2 ppm, **105b** = 196.6 ppm).

Reiterating the idea that the π-complexes **132a** and **86a** are not truly heteroallenes, one must consider the central carbon ^{13}C NMR chemical shifts. Silaallene complex **132a** starts to approach allene status with a chemical shift of 175.5 ppm, but the most deshielded carbon for alkylidenetelluragermirane **86a** is only 153.04 ppm.

Another useful tool for analyzing the structure of heteroallenes containing silicon is ^{29}Si NMR. Typically, sp^2-hybridized silicon atoms have a chemical shift

*Although it is not a group 14 heteroallene, arsaphosphaallene[42] Mes*As=C=PMes* (+299.5 ppm) over-emphasizes the point that the central carbons of heteroallenes are greatly deshielded. For reference, 1-phosphaallene[43] Mes*P=C=CPh_2 has a central carbon shift of +237.6 ppm and 1,3-diphosphaallene[44] Mes*P=C=PMes* has a shift of +276.2 ppm.

TABLE VIII

^{29}Si Chemical Shifts of Silenes, Disilenes, and Silaallenes

		sp^2-Hybridized ^{29}Si Shift (ppm)	Reference
$(Me_3Si)_2Si{=}C(OSiMe_3)(t\text{-Bu})$	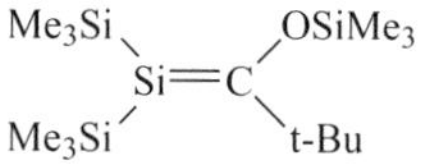 **146**	41.2	44a
$Me_2Si{=}C(SiMe_3)(SiMe(t\text{-Bu})_2)$	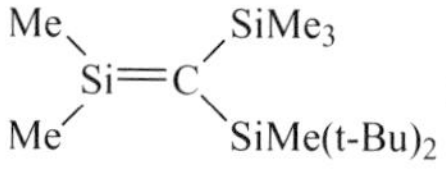 **147**	144.2	44b
$(t\text{-Bu})(Mes)Si{=}Si(Mes)(t\text{-Bu})$	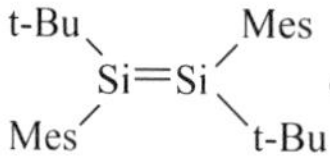 **cis,trans-148**	94.7 (cis), 90.3 (trans)	45a
$Mes_2Si{=}SiMes_2$	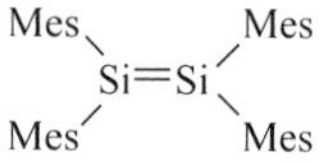 **149**	63.7	45b
	59a	13.1	23
	73	16.2	22
	56	48.4	20
	59b	55.1	23
	59c	58.7	23
	70	48.0	22
	122	57.7	30

downfield of +10 ppm (with respect to tetramethylsilane) and are deshielded with respect to sp^3-hybridized silicons, which usually have resonances upfield of +10 ppm. This is evidenced by two examples each of silenes[44] and disilenes[45] (Table VIII), and silaallenes are no exception. 1-Silaallenes have been compared to Brook's silene (**146**)[44a] through the resonance structure in Fig. 6, which puts more electron density at the silicon atom than does Wiberg's silene (**147**),[44b] thereby shifting the resonance farther upfield than the analogous silene. The silaallene chemical shifts are definitely closer to Brook's silene than Wiberg's.

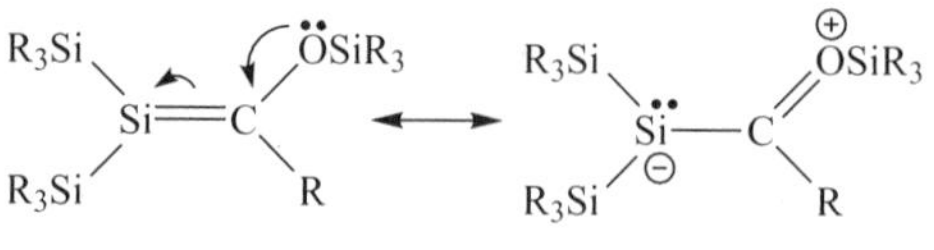

$(R_3Si)_2Si{=}C(\ddot{O}SiR_3)R \longleftrightarrow (R_3Si)_2\ddot{Si}^{\ominus}{-}C({=}\overset{\oplus}{O}SiR_3)R$

FIG. 6. Resonance structure of Brook's silene.

Replacing a carbon with a nitrogen in the 3-position of a 1-silaallene causes a dramatic change upfield of the ^{29}Si chemical shift. The calculated ^{29}Si shift in model silaketenimine Ph_2SiCNR ($\delta = -38.9$ ppm) was close to the observed values measured in different solvents and at various temperatures for silaketenimines **105a–c** (−47.9 to −57.9 ppm).

The type of substituent—alkyl vs aryl—seems to have a significant effect on the chemical shifts of sp^2-hybridized silicon atoms such as disilenes, and silaallenes also follow this trend. The disilene[45] (**148**) with one tert-butyl (alkyl) group and one mesityl (aryl) group has a ^{29}Si chemical shift near 90 ppm, but when there are two aryl groups (**149**),[45a] in this case mesityl, the chemical shift moves upfield by approximately 30 to 63.7 ppm. This large shift is also observed in silaallenes by changing alkyl/aryl substituents to diaryl substituents. Silaallene **59b,** with one alkyl and one aryl group on silicon, has a ^{29}Si chemical shift of 55.1, whereas silaallene **59a,** with two aryl groups, shifts 40 ppm upfield (13.1 ppm), similar to that of the disilenes. ^{29}Si resonances for all of the alkyl/aryl substituted silaallenes (**56, 59b, 59c, 70**) lie in the narrow range of 48 to 59 ppm; those for the two diaryl substituted silaallenes are considerably upfield at 13.1 (**59a**) and 16.2 (**73**) ppm.

Like the ^{13}C NMR chemical shifts for the central carbon of heteroallenes, changing the composition of the allene fragment can have as much of an effect (if not more) on the ^{29}Si chemical shift as does changing a substituent. Switching the carbon at position 3 to a phosphorus (silaphosphaallene **122**) moves the ^{29}Si chemical shift downfield to 75.7 ppm, which is the most deshielded ^{29}Si shift for a silaallene to date, even though it has two aryl groups on silicon, which should keep the shift upfield.

The central tin ^{119}Sn shift in tristannaallene **126** supports that this tin has considerable stannylene character, as its value is +2233 ppm, which is in the range of monomeric stannylenes SnR_2 [R_2 = $(Me_3Si)_2CCH_2CH_2C(SiMe_3)_2$ (δ = +2323),[46a] R = $[CH(SiMe_3)_2]$ (δ = +2328),[46b] R/R = Trip/Tbt (δ = +2208)].[46c] The terminal tin atoms have a chemical shift (+503 ppm) near other sp^2-hybridized, three-coordinate tins [**140** (δ = +710 ppm),[47a] **137** (δ = +835 ppm),[35b] 138 (δ = +374 ppm),[47b] $R_2Sn{=}SnR_2$ (R = Tip, δ = +427 ppm)[47d]].

V

FUTURE PROSPECTS

How might this field develop in the future? At this time, the chemistry of 1-silaallenes is fairly well understood, and that of 1-germaallenes is at least partially explored. Perhaps the series can be extended to 1-stannaallenes, but these are predicted to have quite limited stability. The isolation of the 1,2,3-tristannaallene shows that any combination of heteroatoms may be combined to form allenes and the series may be extended to silicon or germanium. 2-Heteroallenes are predicted to be less stable than their 1-isomers, but nevertheless may be isolable. 1,3-Diheteroallenes, on the other hand, should have stability comparable to the known 1-heteroallenes, and are therefore attractive targets for synthesis. Examples already known are the 1,3-phosphasilaallene and phosphagermaallene described in Sect. IIIC4. Since stable silaketenimines and a stannaketenimine have been synthesized, it seems that germaketenimines should also be available. The synthesis of tin heteroallenes indicates that lead analogues may soon be included in this class of interesting molecules.

As yet there is no firm evidence for heteroallenes of the group 13 elements; these are likely to be investigated in the future. Finally, the flash photolytic studies of silaallenes which have provided much insight into their formation (see Sect. IIIA2) will probably be continued and expanded to include other members of the heteroallene family.

References

(1) Appel, R. In *Multiple Bonds and Low Coordination in Phosphorus Chemistry;* Regitz, M.; Scherer, H. J.; Eds.; Georg Thieme Verlag: Stuttgart, New York, 1990; p. 157.

(2) (a) Davidson, P. J.; Harris, D. H.; Lappert, M. F. *J. Chem. Soc., Dalton Trans.,* **1976,** 2268; (b) Grützmacher, H.; Freitag, S.; Herbst-Irmer, R.; Sheldrick, G. S. *Angew. Chem. Int. Ed. Engl.* **1992,** *31,* 437.

(3) Trinquier, G.; Malrieu, J.-P. *J. Am. Chem. Soc.* **1987,** *109,* 5303.

(4) Barthelat, J. C.; Trinquier, G.; Bertrand, G. *J. Am. Chem. Soc.* **1979,** *101,* 3785.

(5) Lien, M. H.; Hopkinson, A. C. *Chem. Phys. Lett.* **1981,** *80,* 114.

(6) Gordon, M. S.; Koob, R. D. *J. Am. Chem. Soc.* **1981,** *103,* 2939.

(7) Maier, G.; Pacl, H.; Reisenauer, H. P. *Angew. Chem. Int. Ed. Engl.* **1995,** *34,* 1439.

(8) Maier, G.; Reisenauer, H. P.; Egenolf, H. *Organometallics* **1999,** *18,* 2155.

(9) Apeloig, Y.; Karni, M. *J. Am. Chem. Soc.* **1984,** *106,* 6676.

(10) Krogh-Jespersen, K. *J. Comp. Chem.* **1982,** *3,* 571.

(11) Cioslowski, J.; Hamilton, T.; Scuseria, G.; Hess., B. A. Jr.; Hu, J.; Schaad, L. J.; Dupuis, M. *J. Am. Chem. Soc.* **1990,** *112,* 4183.

(12) (a) Ishikawa, M.; Fuchikami, T.; Kumada, M. *J. Am. Chem. Soc.* **1977,** *99,* 245; (b) Ishikawa, M.; Sugisawa, H.; Yamamoto, K.; Kumada, M. *J. Organomet. Chem.* **1979,** *179,* 377; (c) Ishikawa, M.; Nishimura, K.; Sugisawa, H.; Kumada, M. *J. Organomet. Chem.* **1980,** *194,* 147; (d) Ishikawa, M.; Nishimura, K.; Ochiai, H.; Kumada, M. *J. Organomet. Chem.* **1982,** *236,* 7; (e) Ishikawa, M.; Kovar, D.; Fuchikami, T.; Nishimura, K.; Kumada, M.; Higuchi, T.; Miyamoto, S. *J. Am. Chem. Soc.* **1981,** *103,* 2324; (f) Ishikawa, M.; Sugisawa, H.; Fuchikami, T.; Kumada, M.; Yamabe, T.; Kawakami, H.; Fukui, K.; Ueki, Y.; Shizuka, H. *J. Am. Chem. Soc.* **1982,** *104,* 2872; (g) Ishikawa, M.; Sugisawa, H.; Kumada, M.; Higuchi, T.; Matsui, K.; Hirotsu, K.; Iyoda, J. *Organometallics* **1983,** *2,* 174; (h) Ishikawa, M.; Matsuzawa, S.; Hirotsu, K.; Kamitori, S.; Higuchi, T. *Organometallics* **1984,** *3,* 1930; (i) Ishikawa, M.; Matsuzawa, S.; Higuchi, T.; Kamitori, S.; Hirotsu, K. *Organometallics* **1985,** *4,* 2040; (j) Ishikawa, M.; Sugisawa, H.; Harata, O.; Kumada, M. *J. Organomet. Chem.* **1981,** *217,* 43; (k) Ishikawa, M.; Ohshita, J.; Ito, Y. *Organometallics* **1986,** *5,* 1518.

(13) Brook, A. G.; Brook, M. A. *Adv. Organomet. Chem.* **1996,** *39,* 71.

(14) (a) Ishikawa, M.; Fuchikami, T.; Kumada, M. *J. Am. Chem. Soc.* **1979,** *101,* 1348; (b) Ishikawa, M.; Sugisawa, H.; Fuchikami, T.; Kumada, M.; Yamabe, T.; Kawakami, H.; Fukui, K.; Ueki, Y.; Shizuka, H. *J. Am. Chem. Soc.* **1982,** *104,* 2872; (c) Ishikawa, M.; Matsuzawa, S.; Sugisawa, H.; Yano, F.; Kamitori, S.; Higuchi, T. *J. Am. Chem. Soc.* **1985,** *107,* 7706; (d) Ishikawa, M.; Ohshita, J.; Ito, Y.; Iyoda, J. *J. Am. Chem. Soc.* **1986,** *108,* 7417; (e) Ohshita, J.; Isomura, Y.; Ishikawa, M.; *Organometallics* **1989,** *8,* 2050; (f) Ishikawa, M.; Nomura, Y.; Tozaki, E.; Kunai, A.; Ohshita, J. *J. Organomet. Chem.* **1990,** *399,* 205; (g) Ishikawa, M.; Yuzuriha, Y.; Horio, T.; Kunai, A. *J. Organomet. Chem.* **1991,** *402,* C20; (h) Ishikawa, M.; Horio, T.; Yuzuriha, Y.; Kunai, A.; Tsukihara, T.; Naitou, H. *Organometallics* **1992,** *11,* 597; (i) Ohshita, J.; Naka, A.; Ishikawa, M. *Organometallics* **1992,** *11,* 602; (j) Kunai, A.; Yuzuriha, Y.; Naka, A.; Ishikawa, M. *J. Organomet. Chem.* **1993,** *455,* 77; (k) Kunai, A.; Matsuo, Y.; Ohshita, J.; Ishikawa, M.; Aso, Y.; Otsubo, T.; Ogura, F. *Organometallics* **1995,** *14,* 1204.

(15) Barton, T. J.; Burns, G. T.; Goure, W. F.; Wulff, W. D. *J. Am. Chem. Soc.* **1982,** *104,* 1149.

(16) Kerst, C.; Rogers, C. W.; Ruffolo, R.; Leigh, W. I. *J. Am. Chem. Soc.* **1997,** *119,* 466.

(17) Kerst, C.; Ruffolo, R.; Leigh, W. I. *Organometallics* **1997,** *16,* 5804.

(18) Pearsall, M.-A.; West, R. *J. Am. Chem. Soc.* **1988,** *110,* 7228.

(19) Arrington, C. A.; Petty, J. T.; Payne, S. E.; Haskins, W. C. K. *J. Am. Chem. Soc.* **1988,** *110,* 6240.

(20) Miracle, G.; Ball, J. L.; Powell, D. R.; West, R. *J. Am. Chem. Soc.* **1993,** *115,* 11598.

(21) Miracle, G. E.; Ball, J. L.; Bielmeier, S. R.; Powell, D. R. West, R. In *Progress in Organosilicon Chemistry;* Marciniec, B.; Chojnowski, J. Eds.; Gordon and Breach Science Publishers: Basel, 1995; p. 83.

(22) Eichler, B. E.; Miracle, G. E.; Powell, D. R.; West, R. *Main Group Chem.* **1999,** *22,* 147.

(23) Trommer, M.; Miracle, G.; Eichler, B.; Powell, D. R.; West, R. *Organometallics* **1997,** *16,* 5737.

(24) Eichler, B. E.; Powell, D. R.; West, R. *Organometallics* **1998,** *17,* 2147.

(25) Eichler, B. E.; Powell, D. R.; West, R. *Organometallics* **1999,** *18,* 540.

(26) (a) Kishikawa, K.; Tokitoh, N.; Okazaki, R. *Organometallics* **1997,** *16,* 5127; (b) Tokitoh, N.; Kishikawa, K.; Okazaki, R. *Chem. Lett.* **1998,** 811.

(27) (a) Weidenbruch, M.; Brand-Roth, B.; Pohl, S.; Saak, W. *Angew. Chem. Int. Ed. Engl.* **1990,** *29,* 90; (b) Weidenbruch, M.; Brand-Roth, B.; Pohl, S.; Saak, W. *Polyhedron* **1991,** *10,* 1147; (c) Weidenbruch, M.; Schäfers, S.; Pohl, S.; Saak, W.; Peters, K.; von Schnering, H. G. *J. Organomet. Chem.* **1988,** *346,* 171; (d) *Z. Anorg. Allg. Chem.,* **1989,** *570,* 75.

(28) Takeda, N.; Suzuki, H.; Tokitoh, N.; Okazaki, R. *J. Am. Chem. Soc.* **1997,** *119,* 1456.

(29) Ramdane, H.; Ranaivonjatovo, H.; Escudié, J.; Mathieu, S.; Knouzi, N. *Organometallics* **1996,** *15,* 3070.

(30) Rigon, L.; Ranaivonjatovo, H.; Escudié, J.; Dubourg, A.; Declercq, J.-P. *Chem. Eur. J.* **1999,** *5,* 774.

(31) Wiberg, N.; Lerner, H.-W.; Vasisht, S.-K.; Wagner, S.; Karaghiostoff, K.; Nöth, H.; Ponikwar, W. *Eur. J. Inorg. Chem.* **1999,** 1211.

(32) Yin, J.; Klosin, J.; Abboud, K. A.; Jones, W. M. *J. Am. Chem. Soc.* **1995,** *117,* 3298.

(33) For a recent review of structures of main group multiple bonds, see: Power, P. P. *Chem. Rev.* **1999,** *99,* 3463.

(34) Groth, P. *Acta Chem. Scand.* **1973,** *27,* 3302.

(35) (a) Meyer, H.; Baum, G.; Massa, W.; Berndt, A. *Angew. Chem. Int. Ed. Engl.* **1987,** *26,* 798; (b) Meyer, H.; Baum, G.; Massa, W.; Berger, S.; Berndt, A. *Angew. Chem. Int. Ed. Engl.* **1987,** *26,* 546; (c) Berndt, A.; Meyer, H.; Baum, G.; Massa, W.; Berger, S.; *Pure Appl. Chem.* **1987,** *59,* 1011.

(36) Lazraq, M.; Escudié, J.; Couret, C.; Satgé, J.; Dräger, M.; Dammel, R. *Angew. Chem. Int. Ed. Engl.* **1988,** *27,* 828.

(37) Tokitoh, N.; Kishikawa, K.; Okazaki, R. *J. Chem. Soc., Chem. Commun.,* **1995,** 1425.

(38) Allen, F. H.; Kennard, O.; Watson, D. G.; Brammer, L.; Orpen, A. G.; Taylor, R. *J. Chem. Soc., Perkin Trans. 2,* **1987,** S1.
(39) Ando, W.; Ohgaki, H.; Kabe, Y. *Angew. Chem. Int. Ed. Engl.* **1994,** *33,* 659.
(40) Runge W. In *The Chemistry of Allenes;* Landor, S. R., Eds.; Academic Press Inc.: London, **1982;** p. 833.
(41) Ranaivonjatovo, H.; Ramdane, H.; Gornitzka, H.; Escudié, J.; Satgé, J. *Organometallics* **1998,** *17,* 1631.
(42) Yoshifuji, M.; Toyota, K.; Shibayama, K.; Inamoto, N. *Tetrahedron Lett.* **1984,** *25,* 1809.
(43) Gouygou, M.; Koenig, M.; Escudié, J.; Couret, C. *Heteroatom Chem.* **1991,** *2,* 221.
(44) (a) Brook, A. G.; Harris, J. W.; Lennon, J.; El Sheikh, M. *J. Am. Chem. Soc.* **1979,** *101,* 83; (b) Wiberg, N.; Wagner, G.; Müller, G. *Angew. Chem. Int. Ed. Engl.* **1985,** *24,* 229.
(45) (a) Fink, M. J.; Michalczyk, M. J.; Haller, K. J.; Michl, J.; West, R. *Organometallics* **1984,** *3,* 793; (b) Murakami, S.; Collins, S.; Masamune, S. *Tetrahedron Lett.* **1984,** *25,* 2131; (c) Michalczyk, M. J.; West, R.; Michl, J. *Organometallics* **1985,** *4,* 826.
(46) (a) Kira, R.; Yanchibara, R.; Hirano, R.; Kabato, C.; Sakurai, H. *J. Am. Chem. Soc.* **1991,** *113,* 7785; (b) Zilm, W.; Lawless, G. A.; Merill, R. M.; Millar, J. M.; Webb, G. G. *J. Am. Chem. Soc.* **1987,** *109,* 7236; (c) Tokitoh, N.; Saito, M.; Okazaki, R. *J. Am. Chem. Soc.* **1993,** *115,* 2065.
(47) (a) Schäfer, A.; Weidenbruch, M.; Saak, W.; Pohl, S. *J. Chem. Soc., Chem. Commun.* **1995,** 1157; (b) Weidenbruch, M.; Kilian, H.; Stürmann, M.; Pohl, S.; Saak, W.; Marsmann, H.; Steiner, D.; Berndt, A. *J. Organomet. Chem.* **1997,** *530,* 255; (c) Kuhn, N.; Kratz, T.; Bläser, D.; Boese, R. *Chem. Ber.* **1995,** *128,* 245; (d) Masamune, S.; Sita, L. R. *J. Am. Chem. Soc.* **1985,** *107,* 6390.
(48) (a) Goldberg, D. E.; Hitchcock, P. B.; Lappert, M. F.; Thomas, K. M.; Fjelberg, T.; Haaland, A.; Schilling, B. E. R. *J. Chem. Soc., Dalton Trans.* **1986,** 2387; (b) Klinkhammer, K. W.; Schwarz, W. *Angew. Chem. Int. Ed. Engl.* **1995,** *34,* 1334; (c) Klinkhammer, K. W.; Fässler, T. F.; Grützmacher, H. *Angew. Chem. Int. Ed. Engl.* **1998,** *37,* 124.

ADVANCES IN ORGANOMETALLIC CHEMISTRY, VOL. 46

"Very Mixed"-Metal Carbonyl Clusters

SUSAN M. WATERMAN, NIGEL T. LUCAS, and MARK G. HUMPHREY

Department of Chemistry
Australian National University
Canberra, ACT 0200, Australia

I

INTRODUCTION

Transition metal carbonyl clusters have attracted interest for a number of reasons.[1] The oil crisis of the mid-1970s encouraged interest in clusters as pre-catalysts or models for catalysis, and this has been an enduring theme in cluster research since that time. The multimetallic coordination of organic molecules at clusters facilitates substrate transformations not readily achievable at mononuclear complexes. The aggregation of metal atoms within a metal cluster core can afford molecules with a large number of accessible oxidation states which may have the potential to function as "electron reservoirs." Fluxionality at metal clusters may provide an effective model of substrate mobility at surfaces active as heterogeneous catalysts. As clusters become progressively larger in size, they may be expected to adopt metallic character, and the intermediacy of a "metametallic" state with potentially interesting physical properties has been proposed.[2]

The pre-eminent factor controlling the development of areas of cluster chemistry has been the existence or otherwise of efficient routes into the clusters themselves.

0065-3055/01 $35.00

Most early syntheses were of the "heat it and hope" type which, not surprisingly, has been effective for a range of homometallic clusters but is less useful for mixed-metal clusters. In the last 20 years, general procedures to synthesize mixed-metal clusters have become available, due largely to the pioneering theoretical analyses by Hoffmann and Mingos and others, and the experimental advances of Stone and Vahrenkamp, and their research groups.[3–7] Access to a range of mixed-metal clusters affords the possibility to assess the significance of the heterometallic environment for a number of properties (substrate activation, site selectivity, ligand mobility, electrochemical responses, etc.), but thus far most reports have focused on synthetic and structural studies; systematic reactivity studies and investigations of physical properties are comparatively rare.

A significant number of mixed-metal clusters contain metals from the same group or adjacent groups, but far fewer mixed-metal clusters incorporating disparate metals have been reported. This is surprising, as synthetic procedures to afford such clusters are now well established, and a number of the properties listed above would be expected to show sharp differences from homometallic clusters if the heterometal is very different. For example, coupling an electropositive metal and electronegative metal to give a polar metal–metal bond may enhance substrate activation, and metalloselectivity for a range of reagents, and should affect activation energies for fluxional processes (which may facilitate their discrimination when more than one process is possible). Many heterogeneously catalyzed transformations couple early to mid-transition metals with late transition metals, and clusters comprised of these metals may be effective precatalysts.

A number of reviews of heterometallic clusters have appeared.[8–12] However, the focus has largely been on mixed-metal clusters containing similar metals, or on synthetic and structural aspects, although a review summarizing synthetic and reactivity aspects of group 6–group 9 heterometallic clusters has recently appeared.[13] This review, containing published examples to 1998 inclusive, focuses on the reactivity and physical properties of "very mixed"-metal carbonyl clusters, defined as those containing transition metals separated by three or more d-block groups, for groups 4–10. While this definition is somewhat arbitrary, it ensures that platinum group metals (PGM) are coupled to non-PGM.

II

REACTIVITY STUDIES

A. *General Comments*

Ligand activation and transformation at heterometallic clusters have been reviewed, but few examples of "very mixed"-metal clusters effecting these transformations were identified at the time;[14] almost all reports of ligand transformations at "very mixed"-metal clusters summarized in this Chapter were published subsequent to the earlier review. As mentioned above, major interests in the reactivity of "very mixed"-metal clusters are the possibility of directing reagents to particular sites (metallo-, bond-, or face-specificity), enhanced substrate activation, and substrate transformation utilizing the polar metal–metal bonds or unique heterometallic environment. While the reactivity classifications in the sections below are self-explanatory, some overlap is unavoidable; substitution reactions frequently proceed by associative mechanisms, with intermediates corresponding to ligand addition, and certain types of metal exchange reactions can occur by way of core-expanded intermediates. A number of studies modeling catalysis involve ligand transformations, and so have been mentioned briefly in both relevant sections.

B. *Ligand Substitution*

The presence of differing metals introduces the possibility of metalloselectivity into ligand substitution; this selectivity should be enhanced upon accentuating the disparity between the metals. The introduction of differing metals into a cluster core also reduces the effective symmetry over that of related homometallic clusters, rendering coordination sites for incoming ligands inequivalent, and affording the prospect of site- as well as metallo-selectivity. Attempts by Vahrenkamp and others to define a "hierarchy of site reactivities" in mixed-metal clusters have met with some success, but although a sequence of substitution labilities by core metal can be defined, isolated products often differ from those predicted due to donor ligand mobility.[15] The majority of ligand substitution studies of "very mixed"-metal clusters have involved phosphines or phosphites, and these are summarized in Section II.B.1. All other ligands are considered in Section II.B.2.

1. *P-Donor Ligands*

Phosphines and phosphites are among the most fundamental of organometallic reagents. Their availability with a broad range of steric and electronic properties provides the possibility of carrying out systematic investigations of site- and metallo-selectivity, but there are few studies which have defined the substitution sites of a range of phosphines at a "very mixed"-metal cluster; instead, phosphines have frequently been employed solely to enhance the prospects of obtaining samples suitable for single crystal X-ray studies. The results of phosphine and phosphite substitution at "very mixed"-metal clusters are summarized in Table I.

Studies of phosphine substitution have thus far focused almost exclusively on tri- or tetranuclear clusters, with almost all trimetallic examples pseudotetrahedral

TABLE I

PHOSPHINE/PHOSPHITE SUBSTITUTION AT VERY MIXED-METAL CLUSTERS

Product Cluster	Metal Coordinating *P*-Donor Ligand	Ref.
$[MoCo_2(\mu_3\text{-}CCO_2menthyl)(CO)_7(L)(\eta^5\text{-}C_5H_5)]$ [L = $P(OMe)_3$, PCy_3]	Co	16
$[MoCo_2(\mu_3\text{-}CCO_2menthyl)(CO)_6(arphos)(\eta^5\text{-}L)]$ (L = $C_5H_4Pr^i$, C_9H_7)	Co	16
$[MoCo_2(\mu_3\text{-}CCO_2Pr^i)(\mu\text{-}L)(CO)_6(\eta^5\text{-}C_5Me_5)]$ (L = dppe, arphos)	Co	17
$[WCo_2(\mu_3\text{-}CMe)(CO)_7(PHPh_2)(\eta^5\text{-}C_5H_5)]$	Co	18
$[WCo_2(\mu_3\text{-}CC_6H_4Me\text{-}4)(CO)_7(PMe_2Ph)(\eta^5\text{-}L)]$ (L = C_5H_5, $C_5H_4SiMe_3$)	Co	19
$[WCo_2(\mu_3\text{-}CC_6H_4Me\text{-}4)(CO)_6(L)_2(\eta^5\text{-}L')]$ (L = PMe_2Ph, L′ = C_5H_5, $C_5H_4SiMe_3$; L = $P(OMe)_3$, L′ = C_5H_5)	Co	19
$[WCo_2(\mu_3\text{-}CC_6H_4Me\text{-}4)(\mu\text{-}dppm)(CO)_6(\eta^5\text{-}L)]$ (L = C_5H_5, $C_5H_4SiMe_3$)	Co	19
$[WCo_2(\mu\text{-}H)(\mu_3\text{-}CC_6H_4Me\text{-}4)(\mu\text{-}PPh_2)(CO)_5(PMe_2Ph)(\eta^5\text{-}C_5H_5)]$	Co	20
$[WCo_2(\mu\text{-}\eta^4\text{-}CEtCEtCEtCEt)(\mu\text{-}\eta^2\text{-}CEtCEt)(CO)_7\{P(OMe)_3\}]$	Co	21
$[W_2Ir(\mu_3\text{-}CC_6H_4Me\text{-}4)(\mu\text{-}CC_6H_4Me\text{-}4)(Cl)(CO)_3(PPh_3)(\eta^5\text{-}C_5H_5)_2]$	Ir	22
$[MoMCo(\mu_3\text{-}S)(CO)_7(L)(\eta^5\text{-}C_5H_5)]$ [M = Fe, Ru, L = (*S*)-PMePhPr; M = Ru, L = P{O-(-)-menthyl}Ph_2]	Co	23, 24
$[MoFeCo(\mu_3\text{-}S)(\mu\text{-}dppe)(CO)_6(\eta^5\text{-}C_5H_4Me)]$	Fe, Co	25
$[MFeCo(\mu_3\text{-}PMe)(CO)_7(L)(\eta^5\text{-}L')]$ (M = Mo, W, L = PPh_3, PMe_2Ph, L′ = C_5H_5; M = W, L = PMe_3, PMe_2Ph, L′ = C_5H_5, C_5Me_5)	Co	26, 27
$[MoCoNi(\mu_3\text{-}CR)(CO)_4(L)(\eta^5\text{-}C_5H_5)_2]$ [L = (*S*)-PMePhPr, PMe_2Ph; R = Me, C(O)Ph]	Co	28, 29
$[MRuCo(H)(\mu_3\text{-}CR)(CO)_7(L)(\eta^5\text{-}C_5H_5)]$ (M = Mo, W, R = Me, L = PMe_3, PMe_2Ph, $PMePh_2$, PPh_3; M = Mo, W, R = Me, Ph, L = PPh_3)	Ru	30, 31

$[Mo_2Co_2(\mu_4\text{-}S)(\mu_3\text{-}S)_2(CO)_3(L)(\eta^5\text{-}C_5H_4Me)_2]$ [L = $P(OMe)_3$, PH_2Ph, $PHPh_2$, PPh_3, P^nBu_3]	Co	32
$[Mo_2Co_2(\mu_4\text{-}S)(\mu_3\text{-}S)_2(CO)_2(L)_2(\eta^5\text{-}C_5H_4Me)_2]$ [L = $P(OMe)_3$, $PHPh_2$]	Co	32
$[Mo_2Co_2(\mu_4\text{-}S)(\mu_3\text{-}S)_2(CO)_2(\eta^2\text{-}L)(\eta^5\text{-}C_5H_4Me)_2]$ (L = dppm, dppe, dmpe)	Co (chelated)	33
$[Mo_2Co_2(\mu_4\text{-}S)(\mu_3\text{-}S)_2(CO)(\eta^1\text{-}dmpe)(\eta^2\text{-}dmpe)(\eta^5\text{-}C_5H_4Me)_2]$	Co (chelated)	33
$[MIr_3(\mu\text{-}CO)_3(CO)_{8\text{-}n}(L)_n(\eta^5\text{-}C_5H_5)]$ [M = Mo, L = PPh_3, PMe_3, n = 1, 2; M = W, L = PPh_3, PMe_3, PMe_2Ph, $PMePh_2$, $P(OPh)_3$, $P(OMe)_3$, n = 1–3]	Ir	34–37
$[WIr_3(\mu\text{-}CO)_3(\mu\text{-}L)(CO)_6(\eta^5\text{-}C_5H_5)]$ (L = dppm, dppe, dppa)	Ir	38
$[Mo_2Ir_2(\mu\text{-}CO)_3(CO)_{7\text{-}n}(L)_n(\eta^5\text{-}C_5H_5)_2]$ (L = PPh_3, PMe_3, n = 1, 2)	Ir	39
$[MFeCo_2(\mu_3\text{-}PMe)(\mu\text{-}AsMe_2)(CO)_7(PPh_3)(\eta^5\text{-}C_5H_5)]$ (M = Mo, W)	Co	26
$[W_2RuPt(\mu_3\text{-}CC_6H_4Me\text{-}4)(\mu\text{-}CC_6H_4Me\text{-}4)(CO)_6(PMe_2Ph)(\eta^5\text{-}C_5H_5)_2]$	Ru	40
$[W_2Pt_2(\mu_3\text{-}CC_6H_4Me\text{-}4)(\mu\text{-}CC_6H_4Me\text{-}4)(CO)_4(PMePh_2)_2(\eta^5\text{-}C_5H_5)_2]$	Pt	41
$[RePt_3(\mu\text{-}dppm)_3(CO)_3\{P(OR)_3\}]^+$ (R = Me, Ph)	Re	42
$[RePt_3(\mu_3\text{-}O)_2(\mu\text{-}dppm)_3(CO)_2\{P(OR)_3\}]^+$ (R = Me, Ph)	Re	43,44
$[Mo_2W_3Pt_6(\mu_3\text{-}CC_6H_4Me\text{-}4)_2(\mu_3\text{-}CMe)_3(CO)_{10}(PMe_2Ph)_4(\eta^5\text{-}C_5H_5)_5]$	Pt	45,46

by virtue of a μ_3-capping ligand. The group 6–group 9 clusters $[MCo_2(\mu_3\text{-}CR)(CO)_8(\eta^5\text{-}L)]$ (M = Mo, W; R = Me, C_6H_4Me-4, CO_2 menthyl; L = C_5H_5, $C_5H_4Pr^i$, $C_5H_4SiMe_3$, C_9H_7) have been reacted with phosphites and monodentate and bidentate phosphines to afford cobalt-ligated products,[16,18,19] although for most of the derivatives the specific substitution sites were not defined. NMR studies have clarified the structures of the (+)-menthylester-containing derivatives (Fig. 1), with the bidentate arphos and monodentate $P(OMe)_3$ ligating at equatorial sites (with respect to the $MoCo_2$ plane) and the bulky PCy_3 proposed to ligate at an axial site; significantly, the cobalt atoms are diastereotopic, and metalloselectivity between the cobalt atoms increases on increasing the ligand cone angle from the phosphite (no discrimination) to the phosphine (3:1 mixture of diastereomers obtained). Similarly, $[MCo_2(\mu_3\text{-}CCO_2Pr^i)(CO)_8(\eta^5\text{-}C_5Me_5)]$ reacted with the bidentate ligands dppe and arphos to afford di-equatorially cobalt-ligated $[MCo_2(\mu_3\text{-}CCO_2Pr^i)(\mu\text{-}L)(CO)_6(\eta^5\text{-}C_5Me_5)]$.[17] The related $[WCo_2(\mu_3\text{-}CMe)(CO)_7(PHPh_2)(\eta^5\text{-}C_5H_5)]$ was not isolated from the reaction between $[WCo_2(\mu_3\text{-}CMe)(CO)_8(\eta^5\text{-}C_5H_5)]$

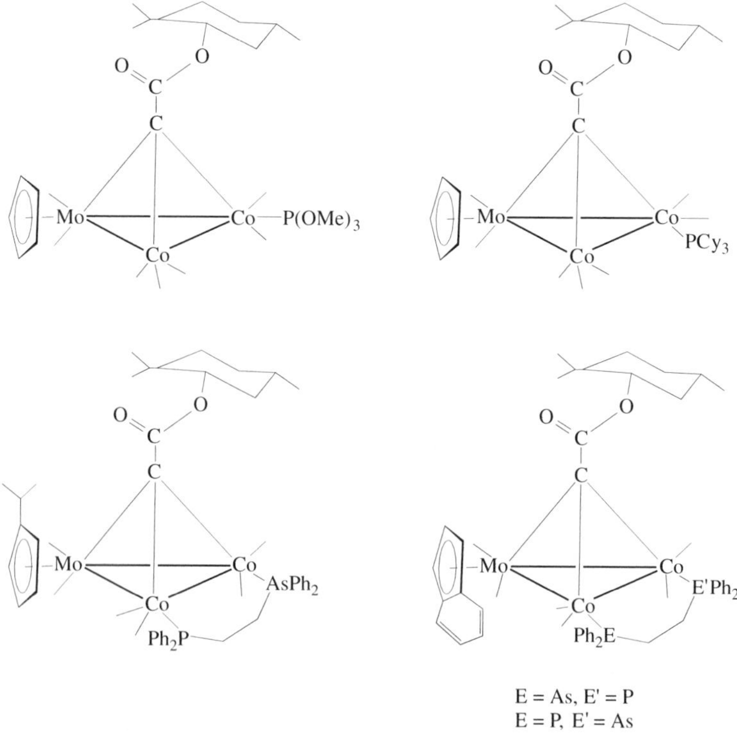

FIG. 1. *P*-donor ligand derivatives of (+)-menthylester methylidyne-capped clusters.

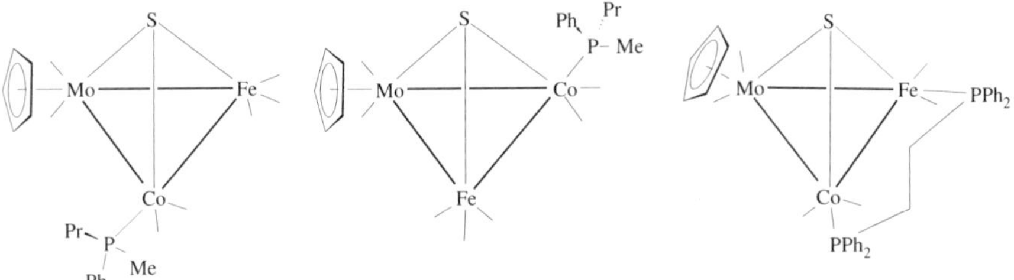

FIG. 2. *P*-donor ligand derivatives of [MoFeCo(μ_3-S)(CO)$_8$(η^5-C$_5$H$_5$)].

and PHPh$_2$, but is a presumed reaction intermediate en route to P—H activation (Section II.D.2.). The product of P—H cleavage in this case, namely [WCo$_2$(μ-H)(μ_3-CC$_6$H$_4$Me-4)(μ-PPh$_2$)(CO)$_6$(η^5-C$_5$H$_5$)], reacted with PMe$_2$Ph to afford a Co-ligated product.[20]

[WCo$_2$(μ-η^4-CEtCEtCEtCEt)(μ-η^2-CEtCEt)(CO)$_8$] reacted with P(OMe)$_3$ by displacement of carbonyl at the later transition metal.[21] Similarly, [W$_2$Ir(μ_3-CC$_6$H$_4$Me-4)(μ-CC$_6$H$_4$Me-4)(Cl)(CO)$_4$(η^5-C$_5$H$_5$)$_2$] reacted with PPh$_3$ to afford an iridium-ligated product, but the site of substitution was not determined.[22] The other trimetallic clusters to have been investigated are all chiral by virtue of four differing core constituents. The group 6–group 8–group 9 clusters [MFeCo(μ_3-S)(CO)$_8$(η^5-C$_5$H$_4$R)] (M = Mo, W; R = H, Me) have been reacted with monodentate and bidentate phosphines (Fig. 2).[23,25] The initial studies with these phosphines were undertaken with the goal of resolving the cluster enantiomers. A racemic mixture of clusters and the optically active phosphine (*R*)-PMePhPr afforded diastereomeric products which could be separated by crystallization. The phosphine was then removed by carbonylation to give the cluster enantiomers with an optical purity of 98–100%.[23] Reaction of the ruthenium-containing analogue [MoRuCo(μ_3-S)(CO)$_8$(η^5-C$_5$H$_5$)] with optically active PMePhPr and P(Omenthyl)Ph$_2$ gave mixtures of diastereoisomers [MoRuCo(μ_3-S)(CO)$_7$(L)(η^5-C$_5$H$_5$)] [L = (*S*)-PMePhPr, P{O-(-)-menthyl}Ph$_2$]; in this case, the diastereoisomers could be separated chromatographically, but the pure enantiomers could not be recovered. More recently, reaction of the (methylcyclopentadienyl)tungsten-iron-cobalt example with dppe has been studied. In contrast to the monodentate phosphine derivative above, in which the phosphine ligates at the cobalt atom *trans* to the Co—Fe vector, the bidentate ligand bridges the Co—Fe linkage and coordinates *trans* to the Fe—Mo and Co—S bonds. The related phosphinidene clusters [MFeCo(μ_3-PMe)(CO)$_7$(L)(η^5-C$_5$H$_5$)] (M = Mo, W; L = PMe$_2$Ph, PPh$_3$) resulted from degradation of [MFeCo$_2$(μ_3-PMe)(μ-AsMe$_2$)(CO)$_8$(η^5-C$_5$H$_5$)] with the appropriate ligand; cobalt ligation was proposed on the basis of spectral data, but the substitution site was not ascertained.[26] The analogous [WFeCo(μ_3-PMe)(CO)$_7$(L)(η^5-L′)] (L = PMe$_3$, PMe$_2$Ph; L′ = C$_5$H$_5$, C$_5$Me$_5$) was formed by

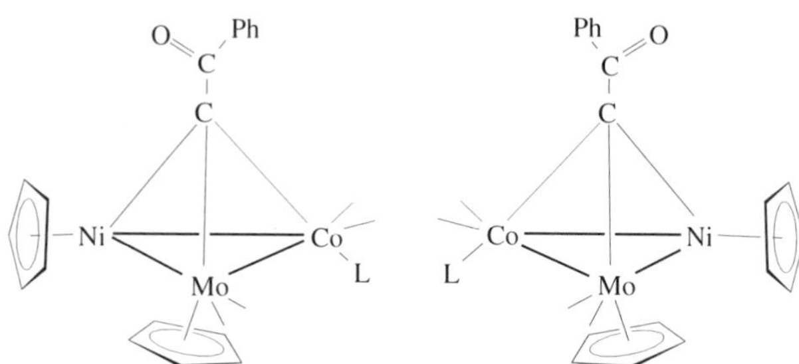

FIG. 3. *P*-donor ligand derivatives of $[MoCoNi\{\mu_3\text{-}CC(O)Ph\}(CO)_5(\eta^5\text{-}C_5H_5)_2]$: L = (*S*)-PMePhPr, PMe_2Ph.

reaction of $[WFeCo(\mu_3\text{-}PMe)(CO)_8(\eta^5\text{-}L')]$ with L. The reaction proceeded by way of an isolable adduct in which two equivalents of ligand have been added and two metal–metal bonds have been cleaved; heating the adduct in vacuum formed a *closo* cluster product by elimination of CO and L.[27] Substitution at chiral alkylidyne clusters has also been demonstrated. The clusters $[MoCoNi(\mu_3\text{-}CR)(CO)_5(\eta^5\text{-}C_5H_5)_2]$ [R = Me, C(O)Ph] reacted with both PMe_2Ph[28] and (*R*)-PMePhPr[29] to afford cobalt-substituted derivatives, not unexpected as the other metal atoms are ligated by cyclopentadienyl groups; the incoming ligand was shown to be *trans* to the Co—C vector for the benzoyl-containing cluster (Fig. 3). Phosphine substitution at $[MRuCo(H)(\mu_3\text{-}CMe)(CO)_8(\eta^5\text{-}C_5H_5)]$ (M = Mo, W) surprisingly occurred at the group 8 metal rather than the group 9 metal to afford $[MRuCo(H)(\mu_3\text{-}CMe)(CO)_7(L)(\eta^5\text{-}C_5H_5)]$ (L = PMe_3, PMe_2Ph, $PMePh_2$, PPh_3); the iron- and osmium-containing analogues were resistant to thermal CO substitution.[30] The alkyl-coordinated clusters $[MRuCo(\mu_3\text{-}CR)(CO)_7\{\eta^2\text{-}CMe(CO_2Me)NHC(O)Me\}(\eta^5\text{-}C_5H_5)]$ (M = Mo, W; R = Me, Ph) reacted with both PPh_3 and CO by way of deinsertion of acetamido acrylic acid methyl ester to form, in the case of phosphine, a ruthenium-ligated product.[31]

The dimolybdenum-dicobalt cluster $[Mo_2Co_2(\mu_4\text{-}S)(\mu_3\text{-}S)_2(CO)_4(\eta^5\text{-}C_5H_4Me)_2]$, which desulfurized organic thiols (Section II.D.2.), reacted with phosphines by ligand substitution at cobalt.[32,47] Kinetic studies showed that reaction proceeded by two elementary steps: initial formation of an adduct followed by loss of CO (Fig. 4). The extent of substitution is dominated by electronic considerations; while phosphine substitution proceeded to give bis-substituted derivatives, reaction with isocyanides afforded tris-substituted products (see Section II.B.2.). These substitution reactions proceeded at room temperature: heating the PH_2Ph-substituted product led to double P—H bond cleavage and coordination of μ_3-phosphinidene (Section II.D.2.). The same cluster reacted with bidentate phosphines to afford cobalt-chelated products (Fig. 5), with excess dmpe affording a bis-substituted derivative.[33] In contrast, the related cluster $[Mo_2Co_2(\mu_3\text{-}S)_4(CO)_2(\eta^5\text{-}C_5H_4Me)_2]$

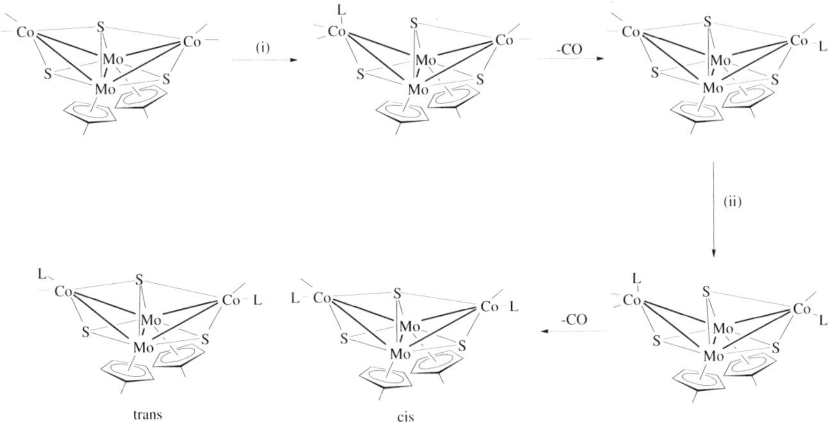

FIG. 4. Monodentate *P*-donor ligand substitution at $[Mo_2Co_2(\mu_4\text{-}S)(\mu_3\text{-}S)_2(CO)_4(\eta^5\text{-}C_5H_4Me)_2]$: (i) L = $P(OMe)_3$, PH_2Ph, $PHPh_2$, PBu^n_3, PPh_3, PMe_3. (ii) L = $P(OMe)_3$, $PHPh_2$.

reacted with dppe to form a product in which the bidentate ligand is believed to span the Co—Co vector, and in the presence of excess dppe afforded a polymeric product, but the products were incompletely characterized.[33]

The tungsten-triiridium cluster $[WIr_3(CO)_{11}(\eta^5\text{-}C_5H_5)]$ and its molybdenum analogue have also been extensively investigated (Fig. 6). Reaction of

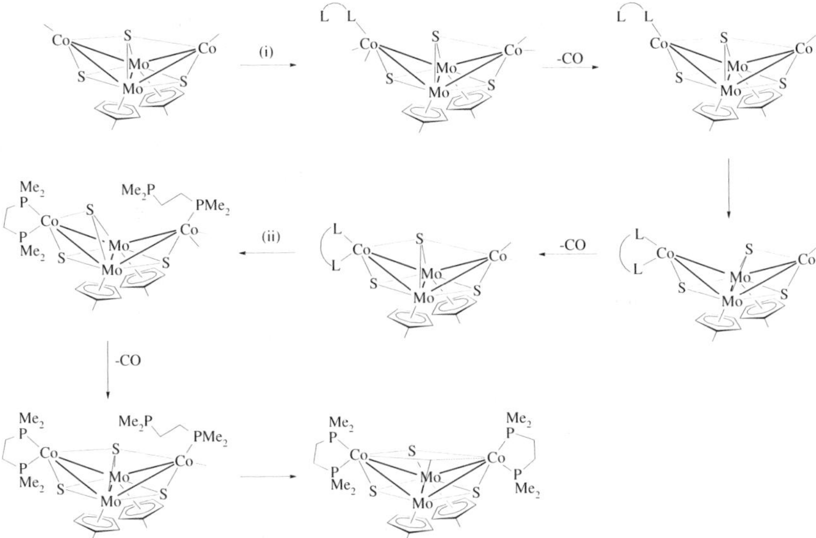

FIG. 5. Bidentate *P*-donor ligand substitution at $[Mo_2Co_2(\mu_4\text{-}S)(\mu_3\text{-}S)_2(CO)_4(\eta^5\text{-}C_5H_4Me)_2]$: (i) L ∩ L = dppm, dppe, dmpe. (ii) L ∩ L = dmpe.

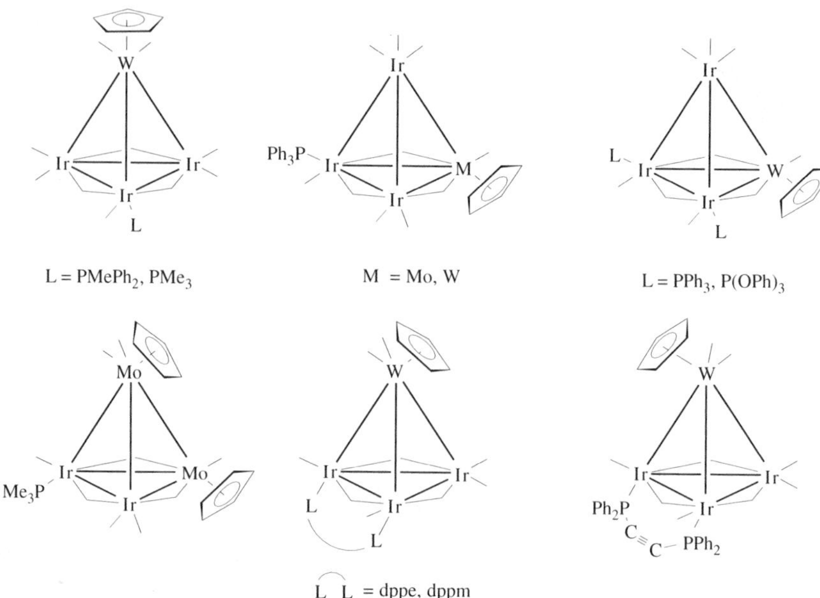

FIG. 6. *P*-donor ligand derivatives of mixed molybdenum/tungsten–iridium clusters.

$[WIr_3(CO)_{11}(\eta^5\text{-}C_5H_5)]$ with monodentate phosphines proceeded at room temperature in a stepwise manner to afford the mono-, bis-, and tris-substituted iridium-ligated derivatives as mixtures of interconverting isomers in solution, the identities of which were ascertained by crystallographic and spectroscopic studies.[34,38] Phosphites reacted similarly to afford mixtures of analogous products.[36] The 4d metal-containing homologue $[MoIr_3(\mu\text{-}CO)_3(CO)_8(\eta^5\text{-}C_5H_5)]$ reacted with less control, affording mixtures of clusters with varying extent of substitution, and reacted with greater steric constraints; only mono- and bis-substitution were observed.[35] The isostructural $[Mo_2Ir_2(\mu_3\text{-}CO)(\mu\text{-}CO)_5(CO)_4(\eta^5\text{-}C_5H_5)_2]$ reacted with monodentate phosphines at iridium to afford mono- and bis-substituted products.[39] Bidentate phosphines reacted with $[WIr_3(CO)_{11}(\eta^5\text{-}C_5H_5)]$ at iridium to afford diaxially ligated edge-bridged products, an unexpected result with the linear diphosphine dppa.

$[W_2Pt_2(\mu_3\text{-}CC_6H_4Me\text{-}4)(\mu\text{-}CC_6H_4Me\text{-}4)(CO)_4(\eta^4\text{-cod})(\eta^5\text{-}C_5H_5)_2]$ reacted with $PMePh_2$ at platinum by displacement of the "lightly stabilizing" cod ligand to afford $[W_2Pt_2(\mu_3\text{-}CC_6H_4Me\text{-}4)(\mu\text{-}CC_6H_4Me\text{-}4)(CO)_4(PMePh_2)_2(\eta^5\text{-}C_5H_5)_2]$,[41] and the ruthenium-ligated product (Fig. 7) was obtained on reaction of $[W_2RuPt(\mu_3\text{-}CC_6H_4Me\text{-}4)(\mu\text{-}CC_6H_4Me\text{-}4)(CO)_7(\eta^5\text{-}C_5H_5)_2]$ with PMe_2Ph.[40] The clusters $[MFeCo_2(\mu_3\text{-}E)(\mu\text{-}AsMe_2)(CO)_8(\eta^5\text{-}C_5H_5)]$ (M = Mo, W; E = S, PMe)

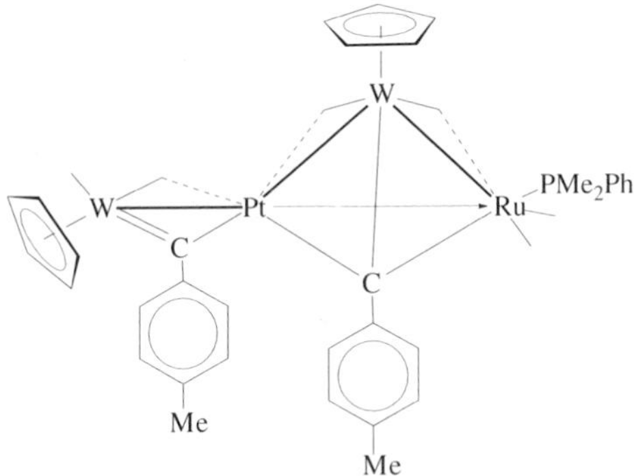

FIG. 7. $[W_2RuPt(\mu_3\text{-}CC_6H_4Me\text{-}4)(\mu\text{-}CC_6H_4Me\text{-}4)(CO)_6(PMe_2Ph)(\eta^5\text{-}C_5H_5)_2]$.

reacted with donor ligands RNC, $P(OR)_3$ and PR_3 to afford degradation products; the only products to retain the cluster core nuclearity were the cobalt-ligated $[MFeCo_2(\mu_3\text{-}PMe)(\mu\text{-}AsMe_2)(CO)_7(PPh_3)(\eta^5\text{-}C_5H_5)]$ (Fig. 8), although even these clusters were accompanied by other lower nuclearity products,[26] and the incoming phosphine could be readily replaced by CO. The chain complexes $[M_2Pt(CO)_6(NCPh)_2(\eta^5\text{-}C_5H_5)_2]$ (M = Cr, Mo, W) contain "lightly stabilizing" benzonitrile ligands and might be expected to undergo facile ligand substitution, but reaction with phosphines was somewhat complex, with core-expanded products being obtained (Section II.E.2.).[48–51]

The vast majority of phosphine/phosphite-substituted products involve *P*-ligand ligation at late transition metals. In contrast, phosphite ligands displaced rhenium-coordinated CO or acetylene in $[RePt_3(\mu\text{-dppm})_3(CO)_3(L)]^+$,[42] which are the

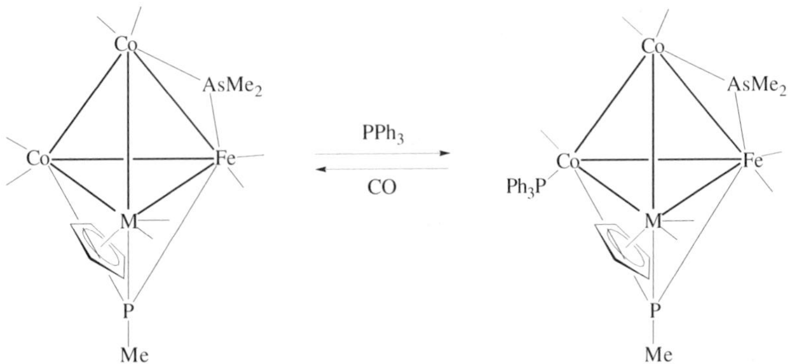

FIG. 8. Interconversion of $[MFeCo_2(\mu_3\text{-}PMe)(\mu\text{-}AsMe_2)(CO)_8(\eta^5\text{-}C_5H_5)]$ and $[MFeCo_2(\mu_3\text{-}PMe)(\mu\text{-}AsMe_2)(CO)_7(PPh_3)(\eta^5\text{-}C_5H_5)]$: M = Mo, W.

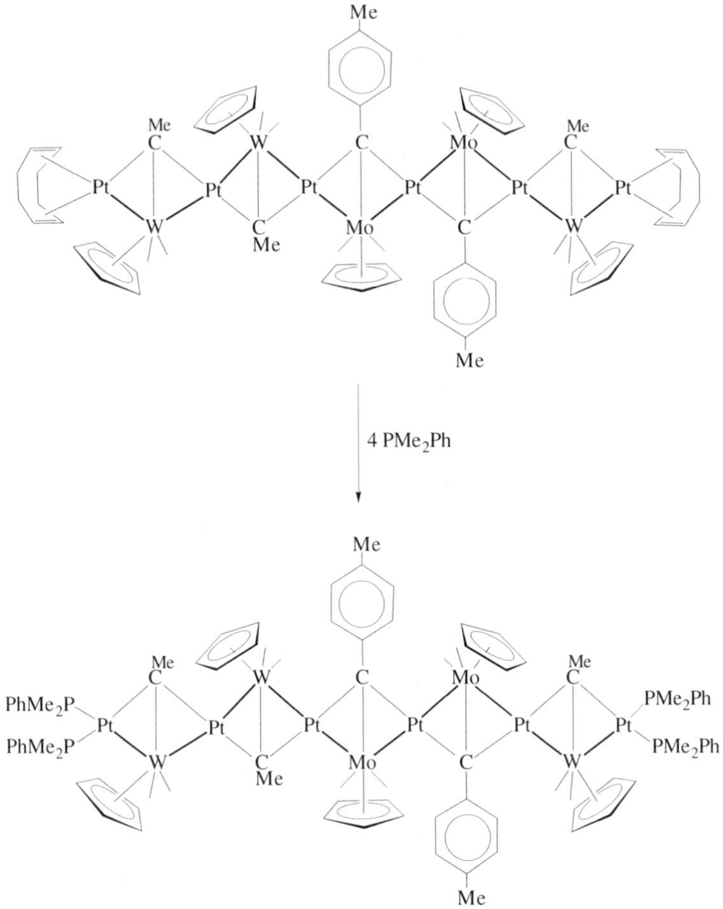

FIG. 9. *P*-donor ligand substitution at $[Mo_2W_3Pt_6(\mu_3\text{-}CC_6H_4Me\text{-}4)_2(\mu_3\text{-}CMe)_3(CO)_{10}(\eta^4\text{-}cod)_2(\eta^5\text{-}C_5H_5)_5]$.

addition products from reaction of L with $[RePt_3(\mu\text{-}dppm)_3(CO)_3]^+$. The CO/acetylene adducts were formed as the starting cluster is comparatively electron poor; the more electron-rich $[RePt_3(\mu_3\text{-}O)_2(\mu\text{-}dppm)_3(CO)_3]^+$ underwent substitution at rhenium, rather than addition, on reaction with phosphite.[43,44] Ligand substitution studies on larger "very mixed"-metal clusters are very rare, one example being the formation of $[Mo_2W_3Pt_6(\mu_3\text{-}CC_6H_4Me\text{-}4)_2(\mu_3\text{-}CMe)_3(CO)_{10}(PMe_2Ph)_4(\eta^5\text{-}C_5H_5)_5]$ from displacement of the weakly-bound cod ligands in $[Mo_2W_3Pt_6(\mu_3\text{-}CC_6H_4Me\text{-}4)_2(\mu_3\text{-}CMe)_3(CO)_{10}(\eta^4\text{-}cod)_2(\eta^5\text{-}C_5H_5)_5]$ by PMe_2Ph (Fig. 9).[46]

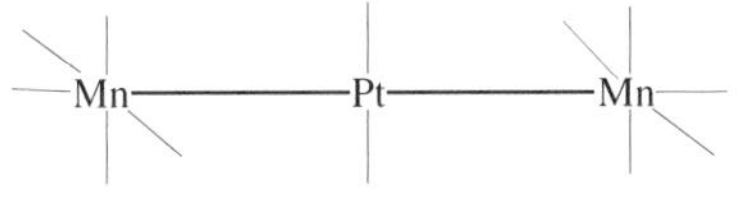

FIG. 10. $[Mn_2Pt(CO)_{12}]$.

2. *Other Ligands*

Almost all other examples of ligand replacement have occurred at the late transition metal and involved the introduction or replacement of CO (Table II). Carbonylation of clusters occurs easily with readily displaced ligands, but CO can also replace phosphines under some circumstances: for example, carbonylation of $[Mo_2Co_2(\mu_3\text{-PPh})(\mu_3\text{-S})_3(CO)(PH_2Ph)(\eta^5\text{-}C_5H_4Me)_2]$ afforded $[Mo_2Co_2(\mu_3\text{-PPh})(\mu_3\text{-S})_3(CO)_2(\eta^5\text{-}C_5H_4Me)_2]$,[52] and carbonylation of $[WFeCo_2(\mu_3\text{-PMe})(\mu\text{-AsMe}_2)(CO)_7(PPh_3)(\eta^5\text{-}C_5H_5)]$ proceeded to afford the substitution product $[WFeCo_2(\mu_3\text{-PMe})(\mu\text{-AsMe}_2)(CO)_8(\eta^5\text{-}C_5H_5)]$ under mild conditions (a slow stream of CO passed through a solution of the cluster for a few minutes); however, after 2 h the latter cluster had degraded affording trinuclear cluster products.[26] Nitriles and dienes are "lightly stabilizing" ligands at transition metal carbonyl clusters; not surprisingly, reaction of $[Mn_2Pt(CO)_{10}(NCPh)_2]$ with CO proceeded in a facile fashion to afford $[Mn_2Pt(CO)_{12}]$ (Fig. 10),[53] and the η^4-cod ligand in $[Cr_2Rh(\mu_3\text{-S})_2(\mu\text{-SBu}^t)(\eta^4\text{-cod})(\eta^5\text{-}C_5H_5)_2]$ was displaced by carbonyls to afford $[Cr_2Rh(\mu_3\text{-S})_2(\mu\text{-SBu}^t)(CO)_2(\eta^5\text{-}C_5H_5)_2]$.[54] In contrast, replacement of the chloro ligand at $[MPd_2(\mu_3\text{-CO})_2(\mu\text{-dppm})_2Cl(\eta^5\text{-}C_5H_5)]$ (M = Mo, W) required the assistance of a halide acceptor (Tl[PF$_6$]), proceeding to give $[MPd_2(\mu_3\text{-CO})_2(\mu\text{-dppm})_2(CO)(\eta^5\text{-}C_5H_5)]^+$ (Fig. 11).[55] In the tungsten-containing examples, both the chloro ligand in the precursor and the terminal carbonyl ligand in

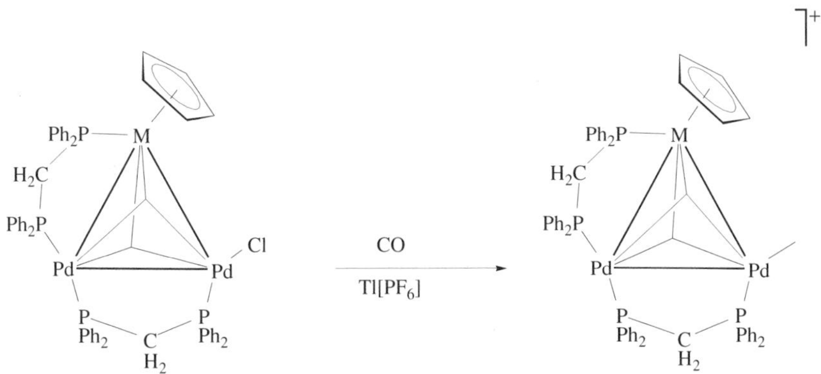

FIG. 11. Synthesis of $[MPd_2(\mu_3\text{-CO})_2(\mu\text{-dppm})_2(CO)(\eta^5\text{-}C_5H_5)]^+$: M = Mo, W.

TABLE II

OTHER SUBSTITUTION AT VERY MIXED-METAL CLUSTERS

Product Cluster	Formal Replacement	Face/Bond/Metal Coordinating Incoming Ligand	Ref.
$[WFeCo_2(\mu_3\text{-PMe})(\mu\text{-AsMe}_2)(CO)_8(\eta^5\text{-C}_5H_5)]$	$PPh_3 \rightarrow CO$	Co	26
$[Mo_2Co_2(\mu_3\text{-S})_3(\mu_3\text{-PPh})(CO)_2(\eta^5\text{-C}_5H_4Me)_2]$	$PH_2Ph \rightarrow CO$	Co	52
$[Mn_2Pt(CO)_{12}]$	$PhCN \rightarrow CO$	Pt	53
$[Cr_2Rh(\mu_3\text{-S})_2(\mu\text{-SBu}^t)(CO)_2(\eta^5\text{-C}_5H_5)_2]$	η^4-cod $\rightarrow$ 2CO	Rh	54
$[MPd_2(\mu_3\text{-CO})_2(\mu\text{-dppm})_2(CO)(\eta^5\text{-C}_5H_5)]^+$ (M = Mo, W)	$Cl^- \rightarrow CO$	Pd	55
$[WPd_2(\mu_3\text{-CO})_2(\mu\text{-dppm})_2(Br)(\eta^5\text{-C}_5H_5)]$	$CO \rightarrow Br^-$	Pd	55
	$Cl^- \rightarrow Br^-$	Pd	
$[MoCo_2(\mu_3\text{-CCO}_2Pr^i)(\mu\text{-CO})(CO)_2(\eta^5\text{-C}_5H_5)_3]$	$5CO \rightarrow 2Cp$	Co	56
$[Mo_2Co_2(\mu_4\text{-S})(\mu_3\text{-S})_2(CO)_{4\text{-n}}(CNR)_n(\eta^5\text{-C}_5H_4Me)_2]$ (R = Bu^t, Me; n = 1 − 3)	$CO \rightarrow RNC$	Co	47, 57
$[Mo_2Co_2(\mu_3\text{-S})_4(CO)_{2\text{-n}}(CNR)_n(\eta^5\text{-C}_5H_4Me)_2]$ (n = 1, R = Bu^t, Me; n = 2, R = Bu^t)	$CO \rightarrow RNC$	Co	57, 58
$[MoPd_2(\mu_3\text{-CO})(\mu\text{-CO})_2(\mu\text{-L})(PR_3)_2(\eta^5\text{-C}_5H_5)]$	$Cp \rightarrow L$ ($L = CH_3CO_2$, Cl)	Pd–Pd bond	59
$[Mo_2Co_2(\mu_3\text{-S})_4(CO)_2(\eta^5\text{-C}_5H_4Me)_2]$	$2CO \rightarrow S$	$MoCo_2$ face	58
$[Mo_2Co_2(\mu_3\text{-}\eta^2\text{-PhC}_2H)(\mu_3\text{-S})_3(CO)_2(\eta^5\text{-C}_5H_4Me)_2]$	$2CO \rightarrow PhC_2H$	$MoCo_2$ face	60
$[WCo_2(\mu_3\text{-CC}_6H_4Me\text{-4})(CO)_6(diars)(\eta^5\text{-L})]$ ($L = C_5H_5$, $C_5H_4SiMe_3$)	$2CO \rightarrow$ diars	Co	19
$[WCo_2(\mu_3\text{-CC}_6H_4Me\text{-4})(\mu\text{-}\eta^2\text{-Me}_3SiC_2SiMe_3)(CO)_6(\eta^5\text{-C}_5H_5)]$	$2CO \rightarrow Me_3SiC_2SiMe_3$	W–Co bond	19
$[WFeRh(\mu_3\text{-CR})(\mu\text{-MeC}_2Me)(CO)_4(\eta\text{-C}_9H_7)\{HB(pz)_3\}]$ (R = Me, C_6H_4Me-4)]	$2CO \rightarrow MeC_2Me$	W–Rh bond	61
$[WIr_3(\mu_3\text{-}\eta^2\text{-PhC}_2Ph)_2(CO)_7(\eta^5\text{-C}_5H_5)]$	$4CO \rightarrow 2PhC_2Ph$	WIr_2, Ir_3 faces	62
$[Mo_2Co_2(\mu_3\text{-S})_4(X)_2(\eta^5\text{-C}_5H_4Et)_2]$ (X = SPh, I, Cl, Br)	$2CO \rightarrow 2X$	Co	63, 64
$[MoRh_3\{\mu_3\text{-AsRhCl}_2(\eta^5\text{-C}_5H_5)\}(\mu_3\text{-CO})_2(\eta^5\text{-C}_5H_5)_4]$	$CO \rightarrow 2Cl$	Non-cluster Rh	65
$[RePt_3(\mu\text{-dppm})_3(O)_3]^+$	$3CO \rightarrow 3O$	Re	66, 67

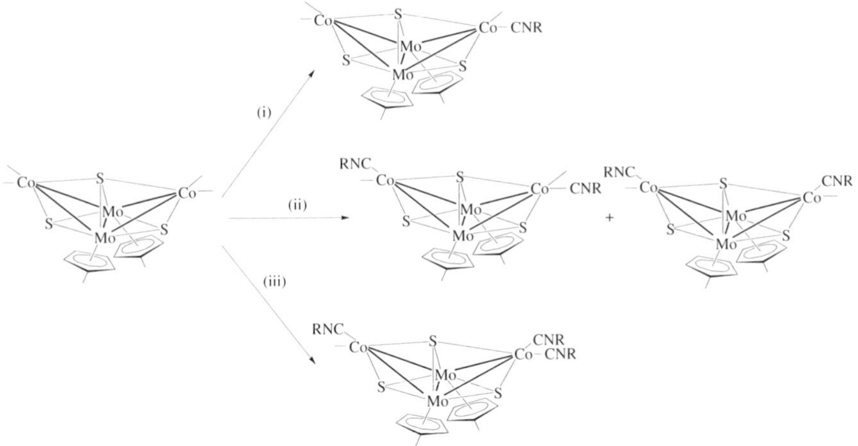

FIG. 12. Isonitrile substitution at $[Mo_2Co_2(\mu_4\text{-}S)(\mu_3\text{-}S)_2(CO)_4(\eta^5\text{-}C_5H_4Me)_2]$: (i) RNC, R = Me, Bu^t. (ii) 2RNC, R = Bu^t. (iii) 3RNC, R = Me, Bu^t.

the product can be displaced by bromide. Metal exchange procedures to form "very mixed"-metal clusters (Section II.E.1.) have utilized $[Ni(\eta^5\text{-}C_5H_5)_2]$ and $[Ni(CO)(\eta^5\text{-}C_5H_5)]_2$ as sources of "$Ni(\eta^5\text{-}C_5H_5)$," but the former also delivers the cyclopentadienyl group which can displace carbonyl ligands; reaction of $[MoCo_2(\mu_3\text{-}CCO_2Pr^i)(CO)_8(\eta^5\text{-}C_5H_5)]$ with either $[Ni(\eta^5\text{-}C_5H_5)_2]$ or the more logical cyclopentadiene afforded $[MoCo_2(\mu_3\text{-}CCO_2Pr^i)(\mu\text{-}CO)(CO)_2(\eta^5\text{-}C_5H_5)_3]$.[56]

The carbonyl ligands in $[Mo_2Co_2(\mu_4\text{-}S)(\mu_3\text{-}S)_2(CO)_4(\eta^5\text{-}C_5H_4Me)_2]$ and $[Mo_2Co_2(\mu_3\text{-}S)_4(CO)_2(\eta^5\text{-}C_5H_4Me)_2]$ can be displaced by isocyanides,[47,57,58] though reaction of the former only proceeded to form the tris-substituted product (Figs. 12, 13). Both clusters desulfurized isothiocyanates, with the resultant isocyanides forming substitution products.[68]

Carbonyl sulfide can be used as a source of sulfido ligands. Thus, reaction of $[Mo_2Co_2(\mu_4\text{-}S)(\mu_3\text{-}S)_2(CO)_4(\eta^5\text{-}C_5H_4Me)_2]$ with COS afforded $[Mo_2Co_2(\mu_3\text{-}S)_4(CO)_2(\eta^5\text{-}C_5H_4Me)_2]$, a process which can be reversed upon carbonylation (Fig. 14).[58] The same cluster added one equivalent of phenylacetylene in a $\mu_3\text{-}\eta^2$-fashion across a $MoCo_2$ face with loss of two CO ligands and rearrangement of the μ_4-sulfido ligand into a μ_3-coordination mode.[60] The tetrahedral cluster $[WIr_3(CO)_{11}(\eta^5\text{-}C_5H_5)]$, in contrast, added two equivalents of diphenylacetylene, one at a heterometallic WIr_2 face, and the other at the unique homometallic Ir_3 face.[62] Studies of phosphine and phosphite substitution at $[WCo_2(\mu_3\text{-}CC_6H_4Me\text{-}4)(CO)_8(\eta^5\text{-}L)]$ ($L = C_5H_5$, $C_5H_4SiMe_3$) summarized in Section II.B.1. have been extended to embrace diars and (for $L = C_5H_5$) $Me_3SiC{\equiv}CSiMe_3$.[19] For diars, a cobalt-ligated product was obtained, but the specific substitution sites were not ascertained, whereas the alkyne was shown to coordinate in a $\mu\text{-}\eta^2$-fashion across

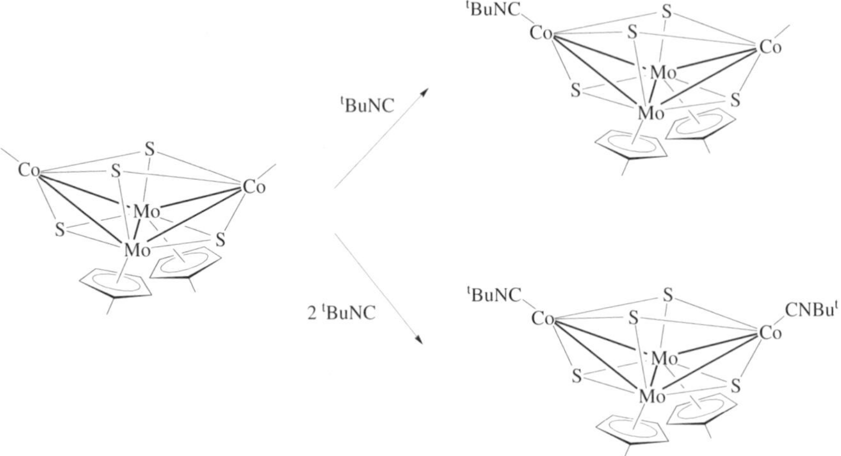

FIG. 13. Isonitrile substitution at $[Mo_2Co_2(\mu_3\text{-}S)_4(CO)_2(\eta^5\text{-}C_5H_4Me)_2]$.

a W—Co linkage. But-2-yne replacement of two CO ligands at $[WFeRh(\mu_3\text{-}CR)(\mu\text{-}CO)(CO)_5\{HB(pz)_3\}(\eta^5\text{-}C_9H_7)]$ (R = Me, C_6H_4Me-4) occurred to afford a μ-η^2-$\perp$-ligating internal acetylene bridging the W—Rh bond.[61] The η^3-coordinated edge-bridging cyclopentadienyl ligand in $[MoPd_2(\mu_3\text{-}CO)(\mu\text{-}\eta^3\text{-}C_5H_5)(\mu\text{-}CO)_2(PR_3)_2(\eta^5\text{-}C_5H_5)]$ (R = Pr^i, Et) was replaced on reaction with CH_3CO_2H or $SiClMe_3$, affording $[MoPd_2(\mu_3\text{-}CO)(\mu\text{-}CO)_2(\mu\text{-}L)(PR_3)_2(\eta^5\text{-}C_5H_5)]$ (L = CH_3CO_2, Cl).[59]

Other reports of ligand replacement at "very mixed"-metal clusters involve a formal oxidation state change. Reaction of $[Mo_2Co_2(\mu_3\text{-}S)_4(CO)_2(\eta^5\text{-}C_5H_4Et)_2]$ with halogens or diphenyl disulfide afforded $[Mo_2Co_2(\mu_3\text{-}S)_4(X)_2(\eta^5\text{-}C_5H_4Et)_2]$ (X = Cl, Br, I, SPh),[63,64] with a formal oxidation at the cobalt atoms, and cleavage of the Co—Co linkage (Fig. 15). All product clusters are paramagnetic in the solid state (but less so in solution), with higher spin states disfavored as the

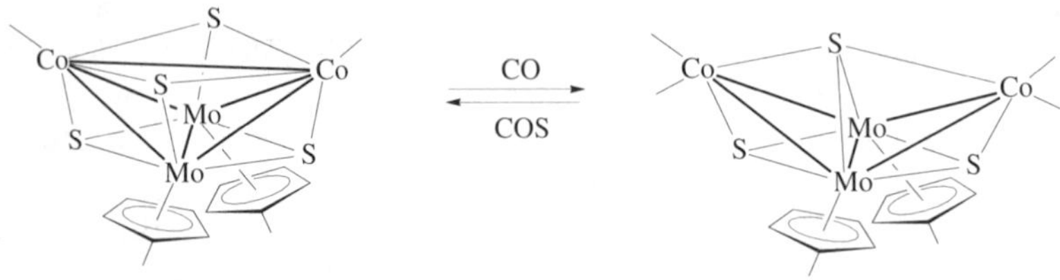

FIG. 14. Interconversion of $[Mo_2Co_2(\mu_3\text{-}S)_4(CO)_2(\eta^5\text{-}C_5H_4Me)_2]$ and $[Mo_2Co_2(\mu_4\text{-}S)(\mu_3\text{-}S)_2(CO)_4(\eta^5\text{-}C_5H_4Me)_2]$.

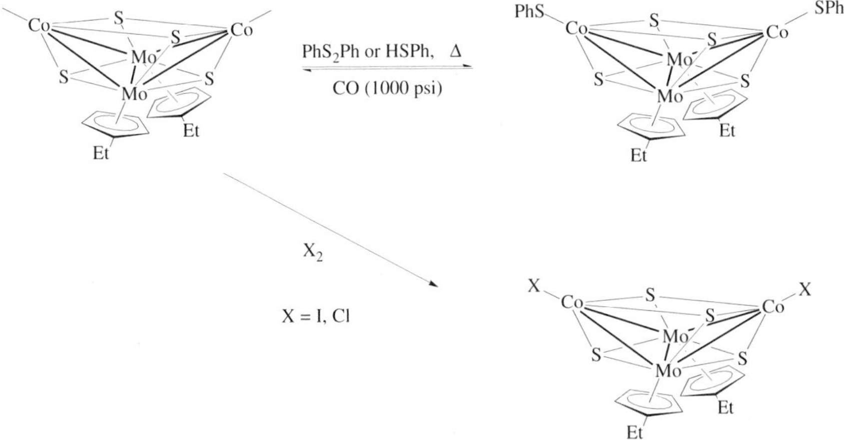

FIG. 15. Oxidative addition reactions at $[Mo_2Co_2(\mu_3\text{-}S)_4(CO)_2(\eta^5\text{-}C_5H_4Et)_2]$.

π-donor ability of X increases. Oxidation at the late transition metal was also seen upon heating $[MoRh_3\{\mu_3\text{-}AsRh(CO)(\eta^5\text{-}C_5H_5)\}(\mu_3\text{-}CO)_2(\eta^5\text{-}C_5H_5)_3]$ in $CHCl_3$, with the "action" occurring at the non-cluster rhodium atom (Fig. 16).[65] In contrast, the earlier transition metal was oxidized in a stepwise fashion upon reacting $[RePt_3(\mu\text{-dppm})_3(CO)_3]^+$ with molecular oxygen, with (overall) a formal increase of +6 in oxidation state (Fig. 17).[66] The oxidation product isolated was sensitive to the reaction conditions, with hydrogen peroxide or molecular oxygen/photolysis affording products lacking metal–metal bonds.[67]

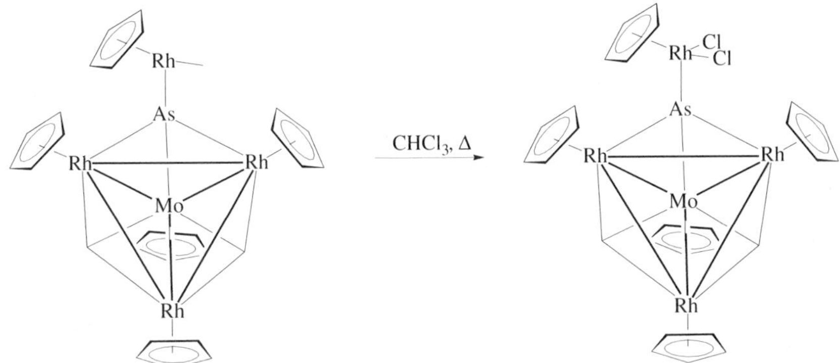

FIG. 16. Chlorination of $[MoRh_3\{\mu_3\text{-}AsRh(CO)(\eta^5\text{-}C_5H_5)\}(\mu_3\text{-}CO)_2(\eta^5\text{-}C_5H_5)_3]$.

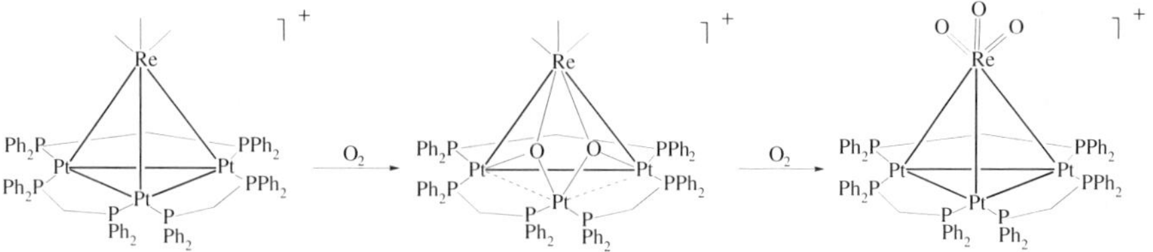

FIG. 17. Oxidation of $[RePt_3(\mu\text{-dppm})_3(CO)_3]^+$.

C. *Ligand Addition*

A range of electrophiles have been added to "very mixed"-metal clusters without ligand displacement, and these are summarized in this section. The classification of the reactions of clusters with nucleophiles as ligand substitution or ligand addition is frequently arbitrary, though, with the former often proceeding by way of the latter. The examples of the latter collected in this section are restricted to those in which the formal electron count on the cluster increases, sometimes accompanied by compensating cluster bond cleavage(s). Table III collects examples of ligand addition reactions at "very mixed"-metal clusters.

The reactions of $[W_2Pt(\mu\text{-PPh}_2)_2(CO)_5(\eta^5\text{-C}_5H_5)_2]$ with two sources of H^+, and the isolobal $[Au(PPh_3)]^+$, have been contrasted; HBF_4 and $[Au(PPh_3)]PF_6$ afforded adducts with the electrophile bridging a W—Pt linkage, but with differing stereochemistry with respect to the bridging groups across the other W—Pt bond, whereas HCl afforded a product with a terminal Pt-bound hydrido ligand (Fig. 18).[69] The cluster anion $[MoCo_2(\mu_3\text{-CC}_6H_4Me\text{-}4)(\mu\text{-PPh}_2)(CO)_6(\eta^5\text{-C}_5H_5)]^-$ reacted with HBF_4 by addition of H^+ across the heterometallic Mo—Co linkage.[70] Protonation at the related $[WCo_2(\mu\text{-H})(\mu_3\text{-CMe})(\mu\text{-PPh}_2)(CO)_6(\eta^5\text{-}$

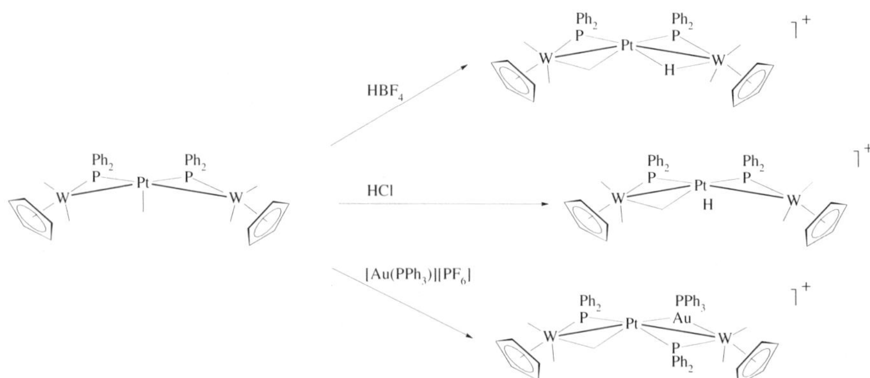

FIG. 18. Reactions of $[W_2Pt(\mu\text{-PPh}_2)_2(CO)_5(\eta^5\text{-C}_5H_5)_2]$ with electrophiles.

TABLE III

ADDITION AT VERY MIXED-METAL CLUSTERS

Product Cluster	Reagent	Face/Bond/Metal Coordinating Incoming Ligand	Ref.
$[W_2Pt(\mu\text{-}H)(\mu\text{-}CO)(\mu\text{-}PPh_2)_2(CO)_4(\eta^5\text{-}C_5H_5)_2]^+$	HBF_4	W–Pt bond	69
$[W_2Pt(H)(\mu\text{-}CO)(\mu\text{-}PPh_2)_2(CO)_4(\eta^5\text{-}C_5H_5)_2]^+$	HCl	Pt	69
$[W_2Pt(\mu\text{-}AuPPh_3)(\mu\text{-}CO)(\mu\text{-}PPh_2)_2(CO)_4(\eta^5\text{-}C_5H_5)_2]^+$	$[AuPPh_3]PF_6$	W–Pt bond	69
$[MoCo_2(\mu\text{-}H)(\mu_3\text{-}CC_6H_4Me\text{-}4)(\mu\text{-}PPh_2)(CO)_6(\eta^5\text{-}C_5H_5)]$	HBF_4	Mo–Co bond	70
$[WCo_2(\mu\text{-}H)_2(\mu_3\text{-}CMe)(\mu\text{-}PPh_2)(CO)_6(\eta^5\text{-}C_5H_5)]^+$	HBF_4	W–Co bond	18
$[Re_4Pt(\mu\text{-}H)_6(CO)_{16}]$	MeOH	Re–Pt bond	71
$[Re_3Pt(\mu\text{-}H)_3(CO)_{14}]$	CF_3SO_3H	Re–Pt bond	72
$[MnRe_2Pt(\mu\text{-}H)_3(CO)_{14}]$	CF_3SO_3H	Re–Pt, Mn–Pt bonds	73
$[Mo_2Co_2(\mu_4\text{-}\eta^2\text{-}NO)(\mu_3\text{-}\eta^2\text{-}PBu^tC_6H_4\text{-}2\text{-}PBu^t)(CO)_6(\eta^5\text{-}C_5H_5)_2]$	$NOBF_4$	Mo_2Co_2 butterfly	74
$[Mo_2Co_2(\mu_4\text{-}S)(\mu_3\text{-}S)_2(CO)_4(PMe_3)(\eta^5\text{-}C_5H_4Me)_2]$	PMe_3	Co	32, 47
$[W_2Rh_2(\mu_3\text{-}CMe)\{\mu\text{-}CMeC(O)\}(\mu\text{-}PPh_2)_2(\mu\text{-}CO)(CO)_3(\eta^5\text{-}C_5H_5)_2]$	CO	Rh	75
$[Mo_2Ir_2(\mu_4\text{-}\eta^2\text{-}RC_2R')(\mu\text{-}CO)_4(CO)_4(\eta^5\text{-}C_5H_5)_2]$ ($R = R' = Ph, H$; $R = H, R' = Ph, 4\text{-}C_6H_4NO_2, 4,4'\text{-}C_6H_4C{\equiv}CC_6H_4NO_2, CH_2Br$)	RC_2R'	Mo_2Ir_2 butterfly	39
$[MoCo_3(\mu_4\text{-}\eta^2\text{-}HC_2Ph)(\mu\text{-}CO)_2(CO)_7(\eta^5\text{-}C_5H_4Me)]$	HC_2Ph	$MoCo_3$ butterfly	76
$[W_2Co_2(\mu_4\text{-}\eta^2\text{-}HC_2Ph)(\mu\text{-}CO)_4(CO)_4(\eta^5\text{-}C_5H_4Me)_2]$	HC_2Ph	W_2Co_2 butterfly	76
$[W_2Ir_2(\mu_4\text{-}\eta^2\text{-}RC_2R')_2(\mu\text{-}CO)_4(CO)_4(\eta^5\text{-}C_5H_5)_2]$ ($R = R' = Ph, C_6H_4Me\text{-}4, CF_3, CO_2Et$; $R = Ph, R' = CO_2Et, Me$)	RC_2R'	W_2Ir_2 butterfly	77, 78
$[RePt_3(\mu_3\text{-}L)(\mu\text{-}dppm)_3(CO)_3]^+$ ($L = Cl^-, Br^-, I^-$)	L	Pt_3 face	79
$[RePt_3(\mu\text{-}dppm)_3(CO)_3(L)]^+$ [$L = CO, P(OR)_3$ ($R = Me, Ph$), RNC ($R = Bu^t, Me, Cy, Xy$), RSH ($R = Et, Bu^t, 3\text{-}MeC_6H_4$), RC_2H ($R = H, Ph$)]	L	Re	42, 79, 80
$[RePt_3(\mu_3\text{-}L)(\mu\text{-}dppm)_3(O)_3]^+$ [$L = CO, Hg, Tl(acac), SnX_3^-$ ($X = F, Cl$), Cl^-, Br^-, I^-]	L	Pt_3 face	79–81
$[RePt_3(\mu\text{-}dppm)_3\{P(OMe)_3\}(O)_3]^+$	$P(OMe)_3$	Pt	80, 81

$C_5H_5)]$ was shown by ^1H NMR to also occur at a heterometallic (W—Co) linkage, but the product $[WCo_2(\mu\text{-}H)_2(\mu_3\text{-}CMe)(\mu\text{-}PPh_2)(CO)_6(\eta^5\text{-}C_5H_5)]^+$ is unstable in the absence of HBF_4.[18] The pentanuclear "bow-tie" cluster $[Re_4Pt(\mu\text{-}H)_5(CO)_{16}]^-$ can also be protonated reversibly, with the incoming electrophile adding at the only non-hydrido-bridged Re—Pt linkage.[71] The spiked triangular cluster anion $[Re_3Pt(\mu\text{-}H)_2(CO)_{14}]^-$ can be protonated with stoichiometric CF_3SO_3H to afford $[Re_3Pt(\mu\text{-}H)_3(CO)_{14}]$; the reverse reaction proceeded with a variety of bases, the reaction with methoxide proceeding via the intermediacy of a carbomethoxy derivative.[72] Most of the electrophile addition chemistry has therefore occurred at the heterometallic linkage. Consistent with this, protonation of $[MnRe_2Pt(\mu\text{-}H)_2(CO)_{14}]^-$ occurred to give two isomers, probably corresponding to the incoming hydrido ligand bridging either the Mn—Pt or the unbridged Re—Pt linkage.[73] Addition of $NOBF_4$ to the tetrahedral cluster $[Mo_2Co_2(\mu_3\text{-}\eta^2\text{-}PBu^tC_6H_4\text{-}2\text{-}PBu^t)(\mu\text{-}CO)(CO)_6(\eta^5\text{-}C_5H_5)_2]$ proceeded by cleavage of the Mo—Mo bond to afford the butterfly cluster $[Mo_2Co_2(\mu_4\text{-}\eta^2\text{-}NO)(\mu_3\text{-}\eta^2\text{-}PBu^tC_6H_4\text{-}2\text{-}PBu^t)(CO)_6(\eta^5\text{-}C_5H_5)_2]$; under the same conditions, the isostructural $[MoCo_3(\mu_3\text{-}\eta^2\text{-}PBu^tC_6H_4\text{-}2\text{-}PBu^t)(\mu\text{-}CO)(CO)_7(\eta^5\text{-}C_5H_5)]$ failed to react.[74]

As mentioned in Section II.B.1., ligand substitution at $[Mo_2Co_2(\mu_4\text{-}S)(\mu_3\text{-}S)_2(CO)_4(\eta^5\text{-}C_5H_4Me)_2]$ occurred by way of an adduct at cobalt, with a compensating Co—S bond cleavage (Fig. 4),[32,57] although for PMe_3 reaction did not proceed past the adduct.[47] The unsaturated cluster $[W_2Rh_2(\mu_3\text{-}CMe)\{\mu\text{-}CMeC(O)\}(\mu\text{-}PPh_2)_2(\mu\text{-}CO)(CO)_2(\eta^5\text{-}C_5H_5)_2]$, with formal W=Rh double bonds, added one molecule of CO at the terminal rhodium atom to afford $[W_2Rh_2(\mu_3\text{-}CMe)\{\mu\text{-}CMeC(O)\}(\mu\text{-}PPh_2)_2(\mu\text{-}CO)(CO)_3(\eta^5\text{-}C_5H_5)_2]$ (Fig. 19). The product subsequently isomerized by P—C bond formation, and then eliminated CO to regenerate the formal unsaturation.[75,82] The acetylene chemistry of tetrahedral mixed group 6–group 9 clusters has been the subject of several studies. For example, $[Mo_2Ir_2(\mu\text{-}CO)_3(CO)_7(\eta^5\text{-}C_5H_5)_2]$ reacted with acetylenes to afford $[Mo_2Ir_2(\mu_4\text{-}\eta^2\text{-}RC_2R')(\mu\text{-}CO)_4(CO)_4(\eta^5\text{-}C_5H_5)_2]$ ($R = R' = Ph$, H; $R = H$, $R' = Ph$, 4-$C_6H_4NO_2$, 4,4′-$C_6H_4C{\equiv}CC_6H_4NO_2$, CH_2Br), with a butterfly metal core geometry formed

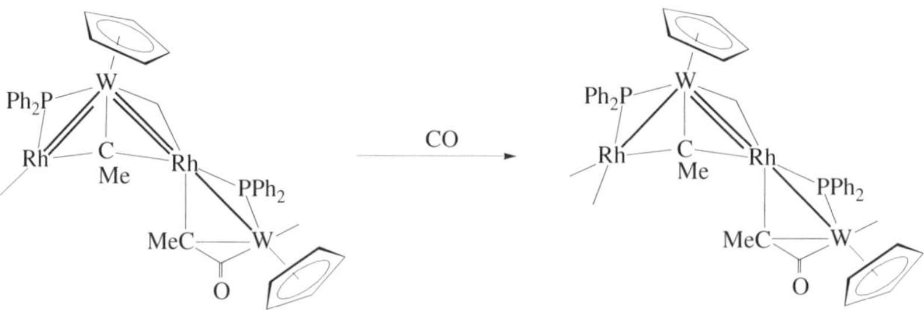

FIG. 19. Carbonylation of $[W_2Rh_2(\mu_3\text{-}CMe)\{\mu\text{-}CMeC(O)\}(\mu\text{-}PPh_2)_2(\mu\text{-}CO)(CO)_2(\eta^5\text{-}C_5H_5)_2]$.

by Mo—Mo cleavage, and the acetylene lying parallel to the Ir—Ir vector completing a $Mo_2Ir_2C_2$ octahedron; qualitative analysis of reaction rates revealed the trends acetylene > terminal alkyne > internal alkyne and 4-nitrophenylacetylene > phenylacetylene, ascribed to a combination of steric and electronic effects.[39] The related clusters $[W_2Ir_2(CO)_{10}(\eta^5\text{-}C_5H_5)_2]$[78] and $[W_2Co_2(\mu\text{-}CO)_3(CO)_7(\eta^5\text{-}C_5H_4Me)_2]$[76] reacted with acetylenes to afford analogous products, and a similarly ligated μ_4-η^2-alkyne occupying a butterfly cleft was obtained from the reaction between $[MoCo_3(\mu\text{-}CO)_3(CO)_8(\eta^5\text{-}C_5H_4Me)]$ and phenylacetylene.[76] In contrast, $[WIr_3(CO)_{11}(\eta^5\text{-}C_5H_5)]$ reacted with diphenylacetylene to afford *inter alia* $[WIr_3(\mu_3\text{-}\eta^2\text{-}PhC_2Ph)_2(CO)_7(\eta^5\text{-}C_5H_5)]$ with the acetylenes face-capping the Ir_3 and one of the WIr_2 faces (Section II.B.2.).[62]

A number of reactivity studies of the coordinatively unsaturated (54 e) clusters $[RePt_3(\mu\text{-dppm})_3(CO)_3]^+$ and $[RePt_3(\mu\text{-dppm})_3(O)_3]^+$, in which the apical rheniums differ by 6 in formal oxidation state, have been reported. The cluster $[RePt_3(\mu\text{-dppm})_3(O)_3]^+$ (itself formed by reaction of $[RePt_3(\mu\text{-dppm})_3(CO)_3]^+$ with O_2, via the addition product $[RePt_3(\mu_3\text{-}O)_2(\mu\text{-dppm})_3(CO)_3]^+$ [83]: Section II.B.2.) was more reactive to ligand addition than its carbonyl-ligated analogue $[RePt_3(\mu\text{-dppm})_3(CO)_3]^+$. $[RePt_3(\mu\text{-dppm})_3(CO)_3]^+$ reacted with neutral donor ligands at the rhenium atom, and with halide ions by capping the triplatinum face (Fig. 20).[42,79,80] Cluster $[RePt_3(\mu\text{-dppm})_3(O)_3]^+$ similarly reacted with halides, and with Hg, Tl(acac), and SnX_3^-, by capping the triplatinum face, but carbonylation also occurred at the triplatinum face, and reaction with phosphite occurred at platinum, the latter two both contrasting with the chemistry at the "low oxidation state" analogue (Fig. 20).[79–81] The related cluster cation $[RePt_3(\mu_3\text{-}O)_2(\mu\text{-dppm})_3(CO)_3]^+$, the intermediate in the transformation of tricarbonyl cluster to trioxo cluster (see above), reacted with phosphites to afford $[RePt_3(\mu_3\text{-}O)_2(\mu\text{-dppm})_3(CO)_2\{P(OR)_3\}]^+$ (R = Me, Ph);[43,44] unlike the tricarbonyl or trioxo analogues, the more electron rich dicarbonyl-dioxo cluster afforded the substitution rather than the addition products (Section II.B.1.).

D. *Ligand Transformations*

Metal clusters have been shown to transform organic substrates in a large number of ways, many of which are not possible at monometallic complexes;[1,14] two or more metal atoms in specific geometric relationships are frequently required to effect bond cleavage and formation and stabilize the resulting ligand fragments. The bond polarity in mixed-metal clusters which may enhance substrate activation should be maximized in progressing to "very mixed"-metal systems. For example, coupling oxophilic and carbophilic metals together in a "very mixed"-metal cluster should facilitate C-heteroatom cleavage by formation of strong M—C and M-heteroatom linkages, but this is one area that has been little exploited. Ligand

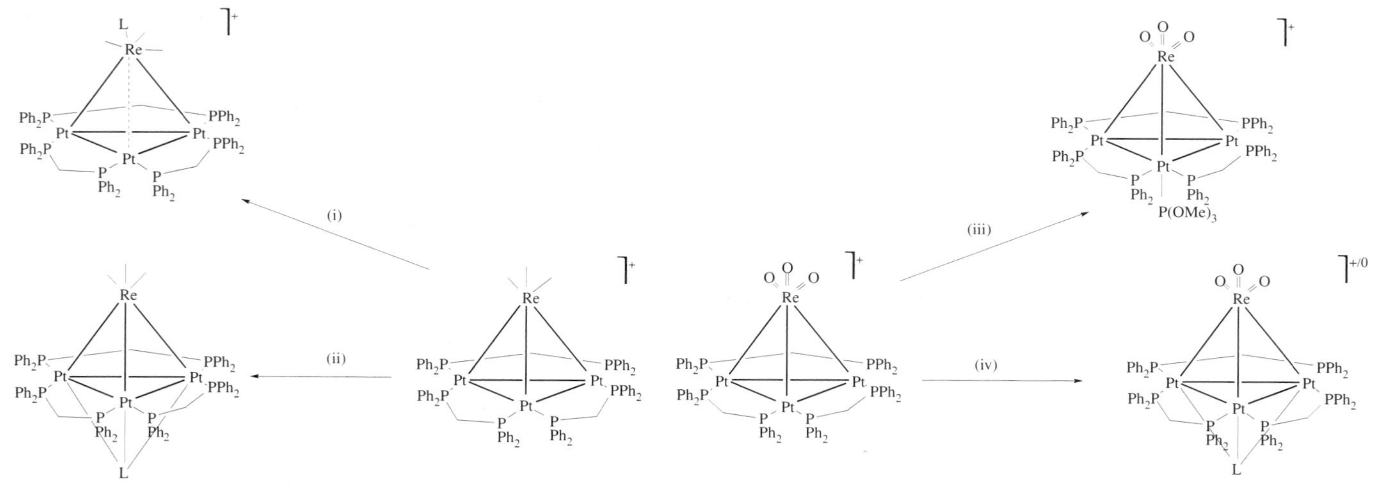

FIG. 20. Reactivity studies of $[RePt_3(\mu\text{-dppm})_3(CO)_3]^+$ and $[RePt_3(\mu\text{-dppm})_3(O)_3]^+$: (i) L = CO, $P(OMe)_3$, $P(OPh)_3$, MeNC, Bu^{t}NC, CyNC, XyNC, EtSH, Bu^{t}SH, $HSC_6H_4Me\text{-}3$, HC_2H, HC_2Ph. (ii) L = Cl^-, Br^-, I^-. (iii) L = $P(OMe)_3$. (iv) L = CO, Hg, Tl(acac), Cl^-, Br^-, I^-.

FIG. 21. Protonation of $[WRePt(\mu\text{-}CC_6H_4Me\text{-}4)(CO)_9(PMe_3)_2]$.

transformations at *C*-donor ligands are summarized in Section II.D.1., while modifications of other ligands are reviewed in Section II.D.2.

1. *C-Donor Ligands*

Transformations at *C*-donor ligands are collected in Table IV. Several examples of C—H bond activation or C—H bond formation at "very mixed"-metal clusters have appeared. Protonation at most clusters occurs at a M—M bond, but in $[WRePt(\mu\text{-}CC_6H_4Me\text{-}4)(CO)_9(PMe_3)_2]$ protonation ultimately occurred at the alkylidyne carbon, with the consequent electron deficiency at tungsten relieved by formation of an η^2-aryl linkage (Fig. 21).[84] Other examples of C—H formation have utilized H_2 as the hydrogen source.[85,86] Coordinated vinylidene can be converted into alkylidyne at a trimetallic face, but the reverse reaction appears the easier, occurring at a much wider range of heterotrimetallic faces (Fig. 22).[85] The coordinated vinylidene can be formed by thermolyzing μ_3-η^2-ligated alkyne, the reaction proceeding by a formal 1,2-H shift (Fig. 23).[15,86] The remaining example of C—H formation occurred by formal insertion of an alkene into an Ru—H bond to form a σ-alkyl linkage (Fig. 24); unlike other examples in this section, the "action" occurred at a single metal.[31]

Thus far, all examples of C—C activation and/or C—C formation have employed alkynes, either alone or in concert with cluster-bound alkylidyne or CO.

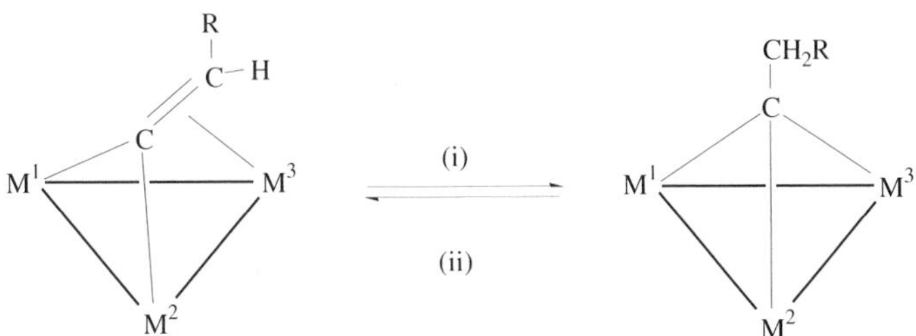

FIG. 22. Interconversion of vinylidene and alkylidyne at trinuclear clusters: (i) H_2, R = H; M_3 = FeCoMo, RuCoMo, RuCoW. (ii) Δ -H_2, R = H, Me; M_3 = FeCoMo, RuCoMo, RuCoW, OsCoMo, OsCoW; co-ligands omitted for clarity.

TABLE IV

C-DONOR LIGAND TRANSFORMATIONS AT VERY MIXED-METAL CLUSTERS

Transformation	Product Cluster	Face/Bond/Metal Coordinating Transformed Ligand	Ref.
C−H formation: $CC_6H_4Me\text{-}4 + H \rightarrow CHC_6H_4Me\text{-}4$	$[WRePt(\mu\text{-}\eta^3\text{-}CHC_6H_4Me\text{-}4)(CO)_9(PMe_3)_2]^+$	W−Pt bond	84
C−H formation: $C{=}CHR + H_2 \rightleftarrows CCH_2R + H$	$[MM'Co(\mu_3\text{-}\eta^2\text{-}C{=}CHR)(CO)_8(\eta^5\text{-}C_5H_5)]$ (MM′Co = MoFeCo, MoRuCo, WRuCo, MoOsCo, WOsCo; R = H, Me)	MM′Co face	85
C−H activation, C−H formation: $HC{\equiv}CR \rightleftarrows C{=}CHR + H_2 \rightleftarrows CCH_2R + H$	$[MoRuCo(\mu_3\text{-}\eta^2\text{-}CCHR)(CO)_8(\eta^5\text{-}C_5H_5)]$ $[MoRuCo(\mu\text{-}H)(\mu_3\text{-}CCH_2R)(CO)_8(\eta^5\text{-}C_5H_5)]$ (R = H, Bu^t)	MoRuCo face	86
Ru−H activation, C−H, Ru−C formation: $Ru{-}H + H_2C{=}CRR' \rightarrow Ru{-}CRR'Me$	$[MRuCo(CO)_7\{\eta^2\text{-}CMe(CO_2Me)NHC(O)Me\}(\eta^5\text{-}C_5H_5)]$ (M = Mo, W)	Ru	31
C−C formation: $2\ PhC{\equiv}CH \rightarrow CPhCHCHCPh$	$[Mo_2Co_2(\mu_3\text{-}S)_3(\mu_3\text{-}\eta^2\text{-}HC_2Ph)(CO)_2(\eta^5\text{-}C_5H_4Me)_2]$ $[Mo_2Co_2(\mu_3\text{-}S)_3(\mu_3\text{-}\eta^4\text{-}CPhCHCHCPh)(\eta^5\text{-}C_5H_4Me)_2]$	$MoCo_2$ face	60
C−C formation: $2\ C{=}CHPh \rightarrow PhC{=}CCH{=}CHPh$	$[Re_2Ni_2(\mu_4\text{-}\eta^2\text{-}PhC{=}CCH{=}CHPh)(CO)_6(\eta^5\text{-}C_5H_5)_2]$	Re_2Ni_2 butterfly	87
C−C activation, C−C formation: $2\ PhC{\equiv}CPh \rightarrow CPh + CPhCPhCPh$	$[W_2Ir_2(\mu_3\text{-}CPh)(\mu_3\text{-}\eta^2\text{-}CPhCPhCPh)(\mu\text{-}CO)_2(CO)_4(\eta^5\text{-}C_5H_5)_2]$	W_2Ir_2 butterfly	77, 78
C−C activation, C−C formation: $3\ PhC{\equiv}CPh \rightarrow 2\ CPh + CPhCPhCPhCPh$	$[WIr_3(\mu_3\text{-}CPh)(\mu\text{-}CPh)(\mu\text{-}\eta^4\text{-}CPhCPhCPhCPh)(CO)_5(\eta^5\text{-}C_5H_5)]$	WIr_3 butterfly	62
C−C formation: $CH + PhC{\equiv}CPh \rightarrow CHCPhCPh$	$[Mo_2Co(\mu_3\text{-}\eta^3\text{-}CHCPhCPh)(\mu\text{-}\eta^2\text{-}PhC_2Ph)(\mu_3\text{-}CO)(CO)_2(\eta^5\text{-}C_5H_5)_2]$	Mo_2Co face	88
C−C formation, C−C cleavage: $CH + 2\ PhC{\equiv}CPh \rightarrow CPhCHCPhCPh + CPh$	$[Mo_2Co(\mu_3\text{-}CPh)(\mu_3\text{-}\eta^5\text{-}CPhCHCPhCPh)(CO)_2(\eta^5\text{-}C_5H_5)_2]$	Mo_2Co face	88
C−C formation: $CH + EtC{\equiv}CEt + CO \rightarrow CHCEtCEtCO$	$[Mo_2Co(CHCEtCEtCO)(CO)_5(\eta^5\text{-}C_5H_5)_2]$	Mo_2Co face	88

C−C formation: $CC_6H_4Me\text{-}4 + RC\equiv CR + CO \rightarrow C(C_6H_4Me\text{-}4)CRCRC(O)$	$[WCo_2\{\mu_3\text{-}C(C_6H_4Me\text{-}4)CRCRCO\}(\mu\text{-}CO)(CO)_4\{PPh_2(cis\text{-}CR{=}CHR)\}(\eta^5\text{-}C_5H_5)]$ (R = Me, Et)	WCo_2 face	20
C−O activation: $CCO_2Et \rightarrow CCO + OEt$	$[MFeCo(\mu_3\text{-}\eta^2\text{-}CCO)(\mu\text{-}CO)(CO)_7(\eta^5\text{-}C_5H_4Me)]$ (M = Mo, W)	MFeCo face	89
N−C activation: $N_2CHCO_2R \rightarrow N_2 + CHCO_2R$	$[W_2Ir_2\{\mu_3\text{-}\eta^2\text{-}CHC(O)OR\}(\mu\text{-}CHCO_2R)(\mu\text{-}CO)(CO)_6(\eta^5\text{-}C_5H_5)_2]$ (R=Me, Et)	Ir−Ir bond, WIr_2 face	90, 91
C−H, C−Cl activation: $CH_2Cl_2 \rightarrow CH + H + 2Cl$	$[Mo_2Co_2(\mu_3\text{-}CH)(\mu_3\text{-}S)_3(\eta^2\text{-}dmpe)_2(\eta^5\text{-}C_5H_4Me)_2]^+$	$MoCo_2$ face	33
C=O reduction: $C_5H_4C(O)R \rightarrow C_5H_4CHR(OH)$	$[MoFeCo(\mu_3\text{-}S)(CO)_8\{\eta^5\text{-}C_5H_4CH(OH)R\}]$ (R = H, Me)	Mo	92
	$[\{MFeCo(\mu_3\text{-}S)(CO)_8(\eta^5\text{-}C_5H_4)\}_2(\mu\text{-}CHOH\text{-}4\text{-}C_6H_4CHOH)]$ (M = Mo, W)	M	93
	$[MoRuCo(\mu_3\text{-}Se)(CO)_8\{\eta^5\text{-}C_5H_4CH(OH)Me\}]$	Mo	94
	$[MFeNi(\mu_3\text{-}S)(CO)_5(\eta^5\text{-}C_5H_5)\{\eta^5\text{-}C_5H_4CH(OH)R\}]$ (M = Mo, W; R = H, Me)	M	95
O−H activation, O−C formation: $C_5H_4CH(OH)Me + Et^+ \rightarrow C_5H_4CH(OEt)Me + H^+$	$[MFeNi(\mu_3\text{-}S)(CO)_5(\eta^5\text{-}C_5H_5)\{\eta^5\text{-}C_5H_4CH(OEt)Me\}]$ (M = Mo, W)	M	95
C=O activation, C=N formation:	$[MFeNi(\mu_3\text{-}S)(CO)_5(\eta^5\text{-}C_5H_5)\{\eta^5\text{-}C_5H_4C(Me){=}NNH\text{-}2,4\text{-}C_6H_3(NO_2)_2\}]$ (M = Mo, W)	M	95

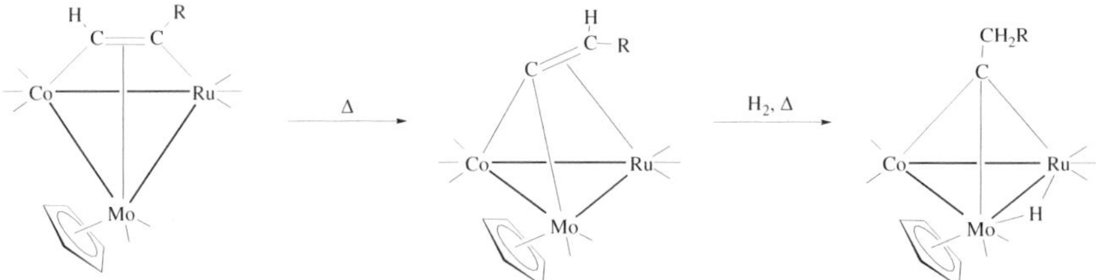

FIG. 23. Transformation of alkyne through vinylidene to alkylidyne: R = H, Bu^t.

Phenylvinylidene was dimerized in a head-to-head manner on thermolyzing $[Re_2Ni_2(\mu_4\text{-}\eta^2\text{-C=CHPh})(\mu\text{-CO})(CO)_6(\eta^5\text{-}C_5H_5)_2]$, a butterfly cluster with nickel atoms in the wing-tip positions;[87] the product $[Re_2Ni_2(\mu_4\text{-}\eta^2\text{-PhC=CCH=CHPh})(CO)_6(\eta^5\text{-}C_5H_5)_2]$ is a butterfly cluster with rhenium atoms in the wing-tip positions. Phenylacetylene was dimerized across a Co_2Mo face of a sulfur-rich dicobalt-dimolybdenum cluster, also in a head-to-head fashion, the reaction proceeding by stepwise addition of the acetylene molecules (Fig. 25).[60] In contrast, reaction of excess internal acetylene PhC≡CPh at a tetrahedral ditungsten-diiridium cluster proceeded by C—C cleavage as well as C—C formation, to afford a butterfly cluster (resulting from W—Ir cleavage) with cluster-bound allyl and alkylidyne units. The major product of the same reaction was a mono-acetylene adduct resulting from insertion into a W—W bond (Section II.C.), which could not be converted to the allyl(alkylidyne)-containing cluster (Fig. 26).[77,78] Modification in core composition has the potential to dramatically affect product selection. Thus, replacing one $CpW(CO)_2$ unit with an isolobal $Ir(CO)_3$ fragment afforded an isostructural cluster which reacted with excess of the same acetylene to afford a product with two cluster-bound acetylenes, and a butterfly cluster resulting from Ir—Ir cleavage with two alkylidyne ligands and a dimer of diphenylacetylene; again, these products could not be interconverted (Fig. 27).[62]

Reactions of alkynes with trinuclear group 6–group 9 clusters incorporating bridging alkylidyne ligands proceeded by coupling the *C*-donor ligands.

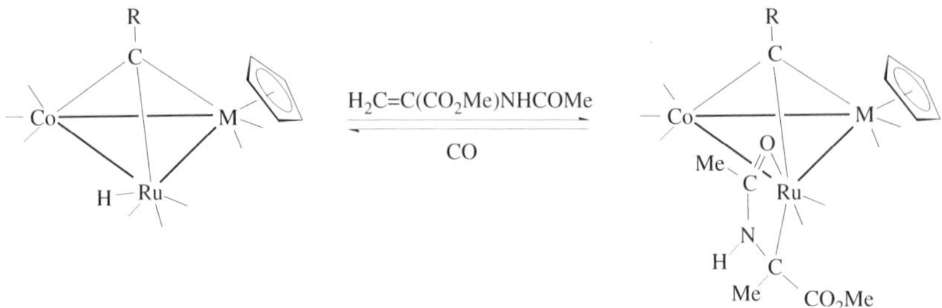

FIG. 24. Reversible insertion of alkyne into an Ru—H bond: M = Mo, W; R = Me, Ph.

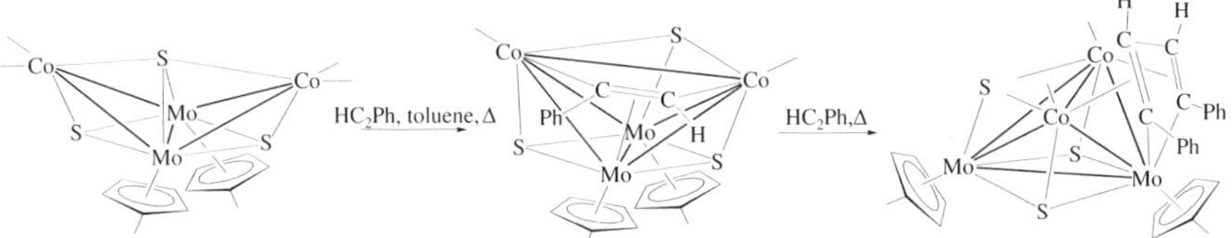

FIG. 25. Dimerization of phenylacetylene at $[Mo_2Co_2(\mu_4\text{-}S)(\mu_3\text{-}S)_2(CO)_4(\eta^5\text{-}C_5H_4Me)_2]$.

Diphenylacetylene and 3-hexyne reacted with $[Mo_2Co(\mu_3\text{-}CH)(CO)_7(\eta^5\text{-}C_5H_5)_2]$ to afford chain-lengthened organic ligands with allylic ligation, but whereas the former also gave a product resulting from PhC≡CPh cleavage and C—C bond formation, the latter gave a product from coupling the allyl unit with CO (Fig. 28).[88] 2-Butyne and 3-hexyne reacted in a similar fashion at tungsten-dicobalt alkylidyne clusters as at the dimolybdenum-cobalt cluster above; coupling of acetylene, $CC_6H_4Me\text{-}4$, and CO gave a WCo_2-supported C_4 fragment, although the presence of bridging phosphido led to a side reaction involving P—C formation (Fig. 29).[20] Alkylidyne groups have been cleaved, as well as coupled, when a sufficiently reactive substituent is present; reaction of $[MCo_2(\mu_3\text{-}CCO_2Et)(CO)_8(\eta^5\text{-}C_5H_4Me)]$ (M = Mo, W) with $[Fe(CO)_4]^{2-}/H^+$ afforded $[MCo_2(\mu_3\text{-}\eta^2\text{-}CCO)(\mu\text{-}CO)(CO)_7(\eta^5\text{-}C_5H_4Me)]$ by cleavage of the ethoxy group, in addition to

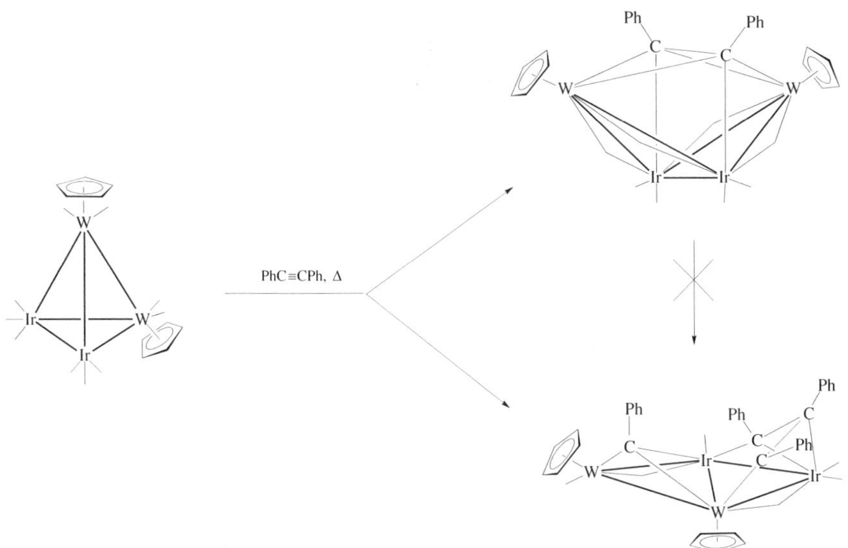

FIG. 26. Reaction of $[W_2Ir_2(CO)_{10}(\eta^5\text{-}C_5H_5)_2]$ with diphenylacetylene.

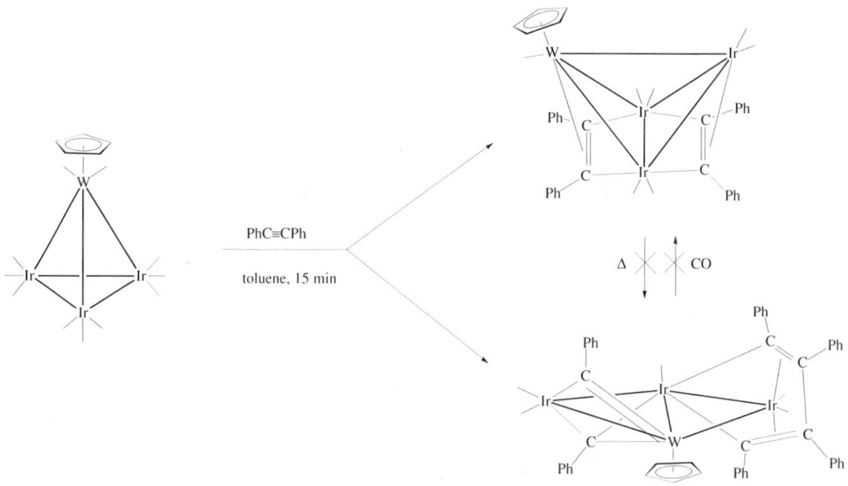

FIG. 27. Reaction of $[WIr_3(CO)_{11}(\eta^5\text{-}C_5H_5)]$ with diphenylacetylene.

the expected metal-exchange product.[89] The chloromethylidyne ligand in $[Co_3(\mu_3\text{-}CCl)(CO)_9]$ underwent C—Cl cleavage as well as metal exchange to afford $[MoCo_2(\mu_3\text{-}CH)(CO)_8(\eta^5\text{-}C_5H_5)]$.[96]

Alkyldiazocarboxylates reacted with $[W_2Ir_2(CO)_{10}(\eta^5\text{-}C_5H_5)_2]$ by C—N cleavage and coordination of the resultant carbene units in two distinct environments; one $CHCO_2R$ ligand bridges an Ir—Ir bond, while the other caps a W_2Ir face by bridging a W—Ir bond and the ester carbonyl coordinating to the oxophilic tungsten atom (Fig. 30).[90,91] This product could not be obtained by C—C cleavage; both

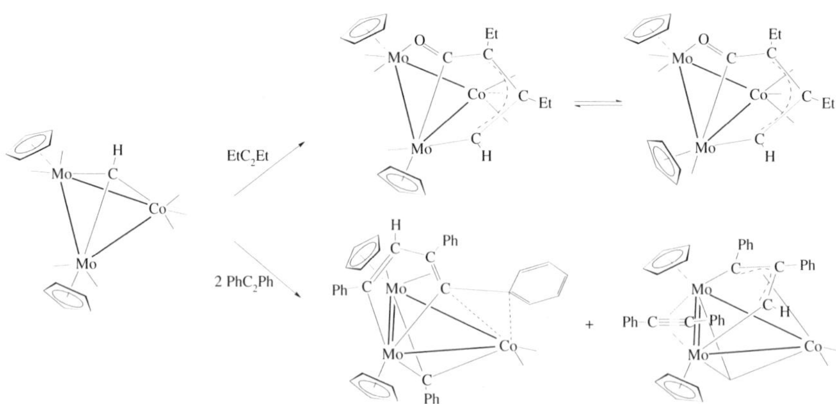

FIG. 28. Activation of internal acetylenes at $[Mo_2Co(\mu_3\text{-}CH)(CO)_7(\eta^5\text{-}C_5H_5)_2]$.

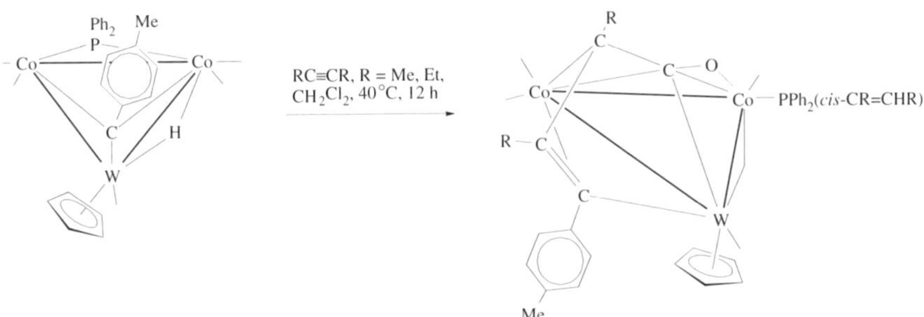

FIG. 29. C–C and P–C bond formation at [$WCo_2(\mu$-H)(μ_3-CC_6H_4Me-4)(μ-PPh_2)$(CO)_5(\eta^5$-$C_5H_5)$].

reaction of [$W_2Ir_2(CO)_{10}(\eta^5$-$C_5H_5)_2$] with $RO_2CCH{=}CHCO_2R$, and attempted hydrogenation of [$W_2Ir_2\{\mu_4$-η^2-$EtO_2CC_2CO_2Et\}(CO)_8$], were unsuccessful. Solvent dichloromethane has been activated; reaction with [$Mo_2Co_2(\mu_4$-S)(μ_3-S)$_2$(μ_3-CO)(η^2-dmpe)$_2$(η^5-$C_5H_4Me)_2$] proceeded by C–H and double C–Cl activation to afford the methylidyne-containing cluster [$Mo_2Co_2(\mu_3$-CH)(μ_3-S)$_3$(η^2-dmpe)$_2$(η^5-$C_5H_4Me)_2]^+$.[33]

The examples above involve the cluster mediating the transformation of organic substrates and stabilizing the resultant residues by coordination. Several cluster-coordinated functionalized cyclopentadienyl groups have also been modified while maintaining their η^5-ligation. For example, the cyclopentadienyl substituents in [MoFeCo(μ_3-S)$(CO)_8(\eta^5$-$C_5H_4R)$] [R = CHO, C(O)Me], [{MFeCo(μ_3-S)$(CO)_8$ (η^5-$C_5H_4)\}_2\{\mu$-C(O)-4-C_6H_4C(O)}] (M = Mo, W), [MoRuCo(μ_3-Se)$(CO)_8\{\eta^5$-C_5H_4C(O)Me}], and [MFeNi(μ_3-S)$(CO)_5(\eta^5$-$C_5H_5)\{\eta^5$-C_5H_4C(O)R}] (M = Mo, W; R = H, Me) were reduced by $NaBH_4$ to afford [MoFeCo(μ_3-S)$(CO)_8(\eta^5$-$C_5H_4R')$] [R′ = CH_2OH, CH(OH)Me], [{MFeCo(μ_3-S)$(CO)_8(\eta^5$-$C_5H_4)\}_2$

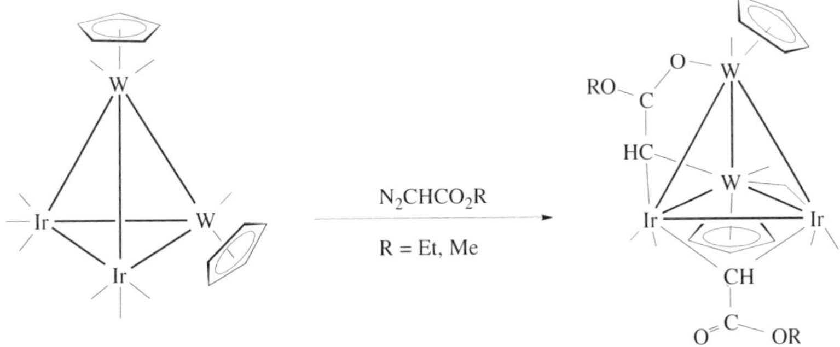

FIG. 30. Reaction of [$W_2Ir_2(CO)_{10}(\eta^5$-$C_5H_5)_2$] with alkyldiazocarboxylates.

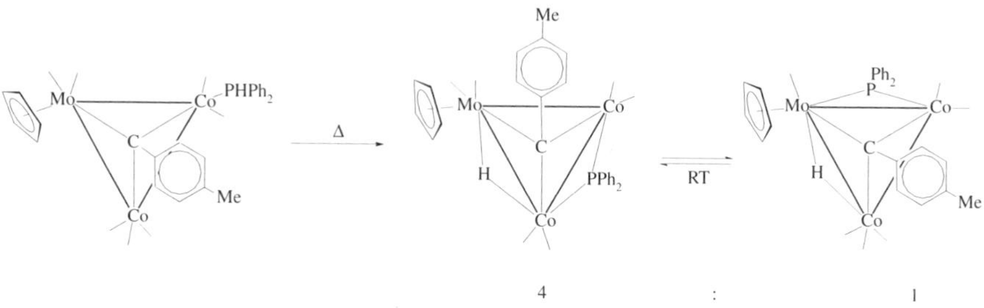

FIG. 31. P—H activation at $[MoCo_2(\mu_3\text{-}CC_6H_4Me\text{-}4)(CO)_7(\eta^5\text{-}C_5H_5)]$.

$\{\mu\text{-CH(OH)-4-}C_6H_4CHOH\}]$, $[MoRuCo(\mu_3\text{-Se})(CO)_8(\eta^5\text{-}C_5H_4CHMeOH\}]$, and $[MFeNi(\mu_3\text{-S})(CO)_5(\eta^5\text{-}C_5H_5)(\eta^5\text{-}C_5H_4CHROH)]$, respectively;[92–95] the secondary alcohol derivatives $[MFeNi(\mu_3\text{-S})(CO)_5(\eta^5\text{-}C_5H_5)(\eta^5\text{-}C_5H_4CHMeOH)]$ were then alkylated by Et_3OBF_4 to afford ether derivatives $[MFeNi(\mu_3\text{-S})(CO)_5(\eta^5\text{-}C_5H_5)\{\eta^5\text{-}C_5H_4CH(OEt)Me\}]$.[95] The ketone functional group in $[MFeNi(\mu_3\text{-S})(CO)_5(\eta^5\text{-}C_5H_5)\{\eta^5\text{-}C_5H_4C(O)Me\}]$ (M = Mo, W) reacted with 2,4-dinitrophenylhydrazine to afford the expected phenylhydrazone derivatives $[MFeNi(\mu_3\text{-S})(CO)_5(\eta^5\text{-}C_5H_5)\{\eta^5\text{-}C_5H_4C(Me){=}NNH\text{-}2,4\text{-}C_6H_3(NO_2)_2\}]$.[95]

2. *Other Ligands*

Very few reports concerning transformations of ligands with other donor atoms exist (Table V). P—H activation at secondary phosphines is the most common motif, with the metal–metal bonds at the heterometallic faces stabilizing the resulting fragments in each case (Figs. 31, 32, 33).[20,70,97] In the formation of both

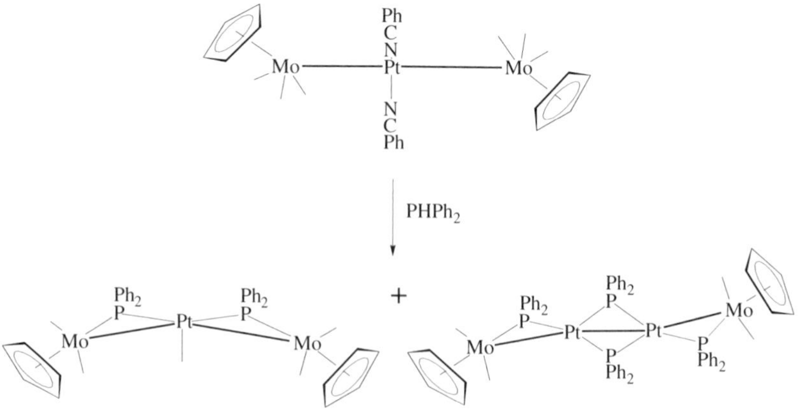

FIG. 32. P—H activation at a dimolybdenum–platinum cluster.

TABLE V

OTHER LIGAND TRANSFORMATIONS AT VERY MIXED-METAL CLUSTERS

Transformation	Product Cluster	Face/Bond/Metal Coordinating Transformed Ligand	Ref.
P−H activation: $PHPh_2 \rightarrow PPh_2 + H$	$[MoCo_2(\mu\text{-}H)(\mu_3\text{-}CC_6H_4Me\text{-}4)(\mu\text{-}PPh_2)(CO)_6(\eta^5\text{-}C_5H_5)]$	Mo−Co, Co−Co bonds/Mo−Co bonds	70
P−H activation: $PHPh_2 \rightarrow PPh_2 + H$	$[Mo_2Pt(\mu\text{-}PPh_2)_2(CO)_5(\eta^5\text{-}C_5H_5)_2]$ $[Mo_2Pt_2(\mu\text{-}PPh_2)_4(CO)_4(\eta^5\text{-}C_5H_5)_2]$	Mo−Pt bonds/Mo−Pt, Pt−Pt bonds	97
P−H activation: $PHR_2 \rightarrow PR_2 + H$ (R = Et, Ph)	$[WCo_2(\mu\text{-}PR_2)_3(CO)_5(\eta^5\text{-}C_5H_5)]$	W−Co, Co−Co bonds	20
P−H activation: $PH_2Ph \rightarrow PPh + H_2$	$[Mo_2Co_2(\mu_3\text{-}PPh)(\mu_3\text{-}S)_3(CO)(PH_2Ph)(\eta^5\text{-}C_5H_4Me)_2]$ $[Mo_2Co_2(\mu_3\text{-}PPh)(\mu_3\text{-}S)_3(CO)_2(\eta^5\text{-}C_5H_4Me)_2]$	$MoCo_2$ face	52, 98
P−C, C−H activation, C−H formation: $PPh_3 \rightarrow PPhC_6H_4\text{-}2 + C_6H_6$	$[WIr_3\{\mu_3\text{-}\eta^2\text{-}PPh(C_6H_4)\}(\mu\text{-}CO)_2(CO)_7(\eta^5\text{-}C_5H_5)]$ $[WIr_3\{\mu_3\text{-}\eta^2\text{-}PPh(C_6H_4)\}(\mu\text{-}CO)_2(CO)_6(PPh_3)(\eta^5\text{-}C_5H_5)]$	Ir_3 face	99
Ar−S activation: Ar−S → Ar + S	$[Mo_2Co_2(\mu_3\text{-}S)_4(CO)_2(\eta^5\text{-}L)_2]^-$ ($L = C_5H_4Me$, C_5Me_4Et) $[Mo_2Co_2(\mu_4\text{-}S)(\mu_3\text{-}S)_2(CO)_3(SR)(\eta^5\text{-}C_5H_4Me)_2]$	Mo−Co bonds	100, 101
S−R, S−H activation: RSH → S + RH	$[Mo_2Co_2(\mu_3\text{-}S)_4(CO)_2(\eta^5\text{-}C_5H_4Me)_2]$	Mo−Co bonds	52
P−C, C−H formation: $PPh_2 + H + RC{\equiv}CR \rightarrow PPh_2(cis\text{-}CR{=}CHR)$	$[WCo_2\{\mu_3\text{-}C(C_6H_4Me\text{-}4)CRCRCO\}(\mu\text{-}CO)(CO)_4\{PPh_2(cis\text{-}CR{=}CHR)\}(\eta^5\text{-}C_5H_5)]$ (R = Me, Et)	WCo_2 face	20
C−C activation, P−C formation $CMeCO + PPh_2 \rightarrow CMePPh_2 + CO$	$[W_3Rh_2(\mu_3\text{-}CMe)(\mu\text{-}CMe)(\mu\text{-}CMePPh_2)(\mu\text{-}CO)_3(\mu\text{-}PPh_2)(CO)_2(\eta^5\text{-}C_5H_5)_3]$	W−Rh bond	82, 102

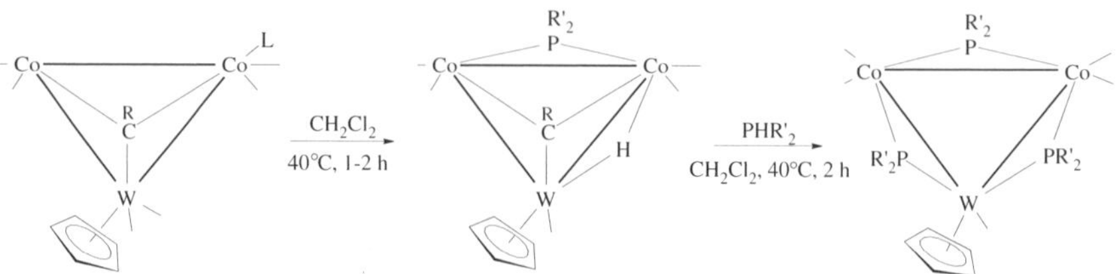

FIG. 33. P—H activation at a tungsten–dicobalt cluster: R = Me, C_6H_4Me-4; R′ = Ph, Et; L = $PHPh_2$, $PHEt_2$.

$[MoCo_2(\mu\text{-}H)(\mu_3\text{-}CC_6H_4Me\text{-}4)(\mu\text{-}PPh_2)(CO)_6(\eta^5\text{-}C_5H_5)]$[70] and $[WCo_2(\mu\text{-}PR_2)_3(CO)_5(\eta^5\text{-}C_5H_5)]$ (R = Et, Ph),[20] coordination of phosphine at the later transition metal was observed to precede P—H activation and cleavage. Not surprisingly, double P—H activation at primary phosphines is also possible (Fig. 34).[52,98]

More interesting are examples of C-heteroatom cleavage. Orthometallation of coordinated PPh_3 and elimination of benzene was observed on heating $[WIr_3(CO)_{11\text{-}x}(PPh_3)_x(\eta^5\text{-}C_5H_5)]$ (x = 1−3);[99] unlike other examples in this section, the ligated residue is coordinated at the homometallic face. The product distribution suggests that the M-P bond was cleaved under the reaction conditions (Fig. 35). C—S cleavage has been reported at mixed molybdenum-cobalt clusters,[33,52,58,63,68,100,101,103,104] which contain a metal combination active for the industrially important hydrodesulfurization of liquid fuels. At low temperatures, coordination of thiophenoxide occurred at a cobalt, but on warming the thiophenoxide bridged a heterometallic linkage and the C—S bond eventually cleaved (Fig. 36). At higher temperatures, the same molybdenum-cobalt cluster desulfurized a range of sulfur-containing organic compounds (Fig. 37). The C—S bonds in propylene sulfide were cleaved by $[RePt_3(\mu\text{-}dppm)_3(CO)_3]^+$ with evolution of propene; the product clusters $[RePt_3(\mu_3\text{-}S)_2(\mu\text{-}dppm)_3(CO)_3]^+$ and $[RePt_3(\mu\text{-}dppm)_3(CO)_3(S)]^+$ contained the sulfido residues.[105,106]

Examples of "very mixed"-metal cluster-assisted C-heteroatom bond formation are still rare, with both literature extant examples involving coupling of bridging phosphido ligand with a C-ligand. Phosphido, hydrido, and alkyne were assembled stereospecifically to afford $PPh_2(cis\text{-}CR{=}CHR)$ (Fig. 29),[20] while an unusual

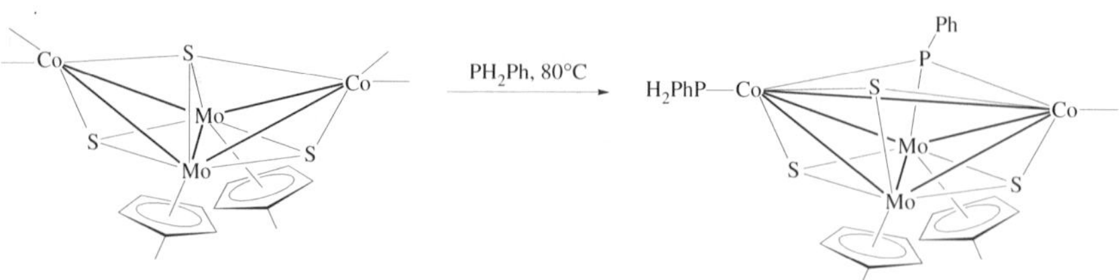

FIG. 34. Double P—H activation at $[Mo_2Co_2(\mu_4\text{-}S)(\mu_3\text{-}S)_2(CO)_4(\eta^5\text{-}C_5H_4Me)_2]$.

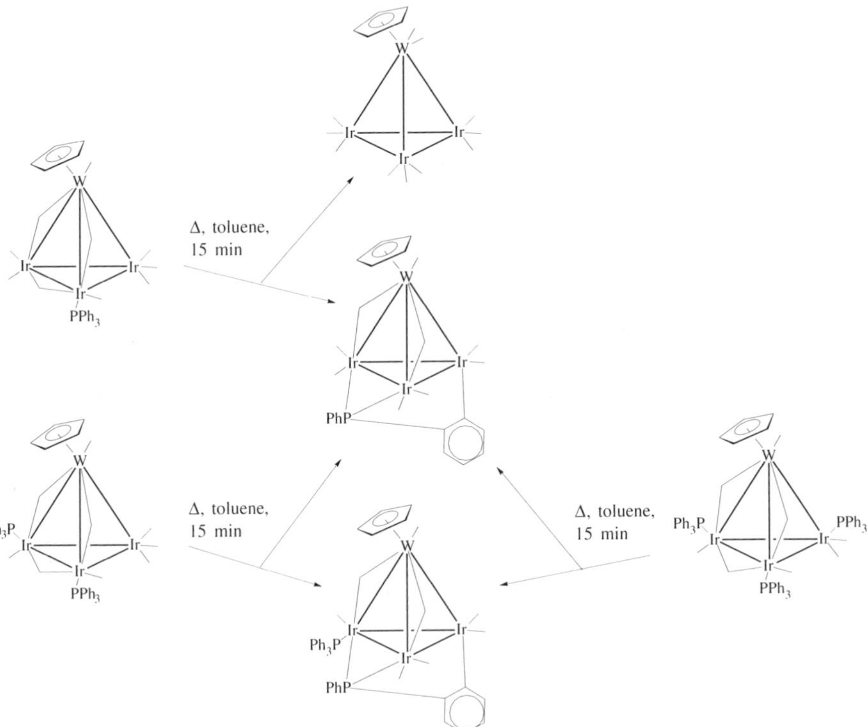

FIG. 35. P–C activation at tungsten–triiridium clusters.

isomerization replaced the bridging phosphido at a W—Rh linkage with a bridging carbonyl, affording the $Ph_2P{=}CMe$ ligand (Fig. 38).[102]

E. *Core Transformations*

The premise of this review is that synthetic procedures for "very mixed"-metal clusters are comparatively well understood, but that reactivity and physical properties are less well studied. Metal core transformations (modifications of a pre-existing cluster) fall into both the synthesis and reactivity categories. A summary is presented here, but as they have been reviewed elsewhere (see Refs. 4, 107–109), the account below is necessarily brief. Section II.E.1. considers core transformations where the cluster core nuclearity is preserved, whereas Section II.E.2. summarizes reactions involving a change in core size.

1. *Metal Exchange*

Efficient routes into "very mixed"-metal clusters by metal exchange reactions have been developed, principally by Vahrenkamp and co-workers. Metal exchange

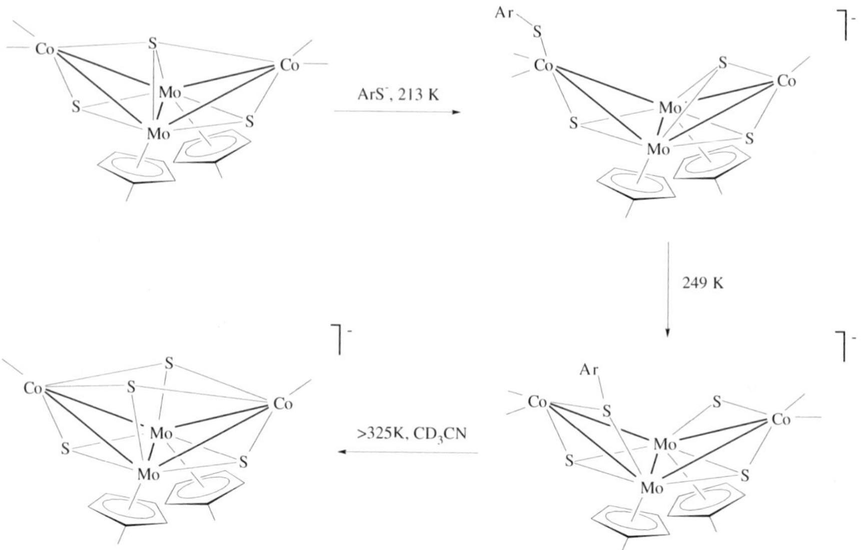

FIG. 36. Ar—S activation at $[Mo_2Co_2(\mu_4\text{-}S)(\mu_3\text{-}S)_2(CO)_4(\eta^5\text{-}C_5H_4Me)_2]$.

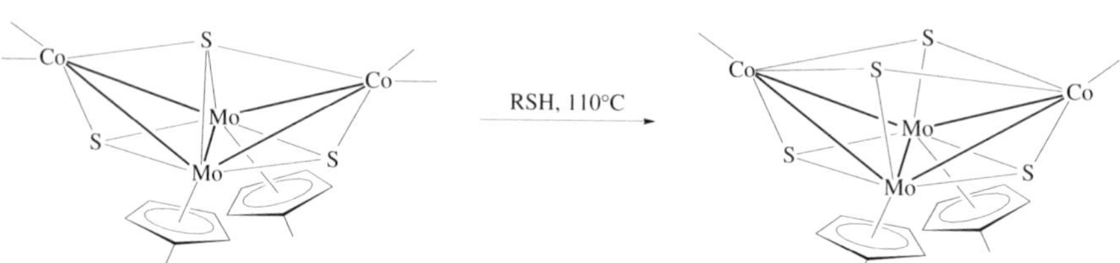

FIG. 37. C—S activation at $[Mo_2Co_2(\mu_4\text{-}S)(\mu_3\text{-}S)_2(CO)_4(\eta^5\text{-}C_5H_4Me)_2]$.

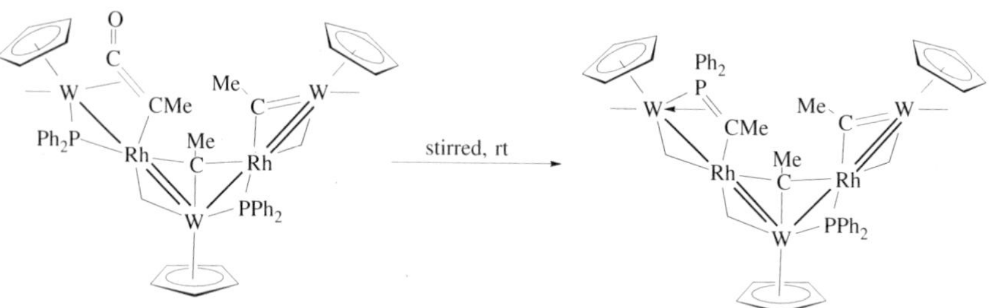

FIG. 38. P—C formation at $[W_3Rh_2(\mu\text{-}CMe)(\mu\text{-}CMeCO)(\mu\text{-}CO)_2(\mu\text{-}PPh_2)_2(CO)_2(\eta^5\text{-}C_5H_5)_3]$.

reactions are those in which one or more metal-ligand groups of a cluster are replaced by a different metal-ligand group to afford a new cluster containing the same total number of metal atoms.[1] Although metal exchange reactions could be expected to be quite complex, many reactions proceed by one-step processes, with others occurring by multistep addition and substitution reactions involving systematic addition and incorporation of organometallic units.[4] Specific classes of reagents have been utilized to introduce the heterometal vertices by metal exchange reactions:

A. Metal carbonyl anions/cyclopentadienylmetal carbonyl anions
B. Metal carbonyls/cyclopentadienylmetal carbonyls/nickelocene/ $[Pt(\eta^2\text{-}C_2H_4)(PPh_3)_2]$
C. Cyclopentadienylmetal carbonyl arsenides
D. Cyclopentadienylmetal carbonyl hydrides/chlorides

and Table VI contains a classification by each reaction type. The first two classes of reagents have been by far the most popular; the arsenide and hydride/chloride routes are considerably more limited in scope, with examples thus far confined to the group 6 metals. The arsenide route has been defined mechanistically (Fig. 39).

The trinuclear clusters $[Co_3(\mu_3\text{-}CR)(CO)_9]$ are the most common precursors utilized in the "very mixed"-metal exchange reactions reported thus far (Fig. 40),

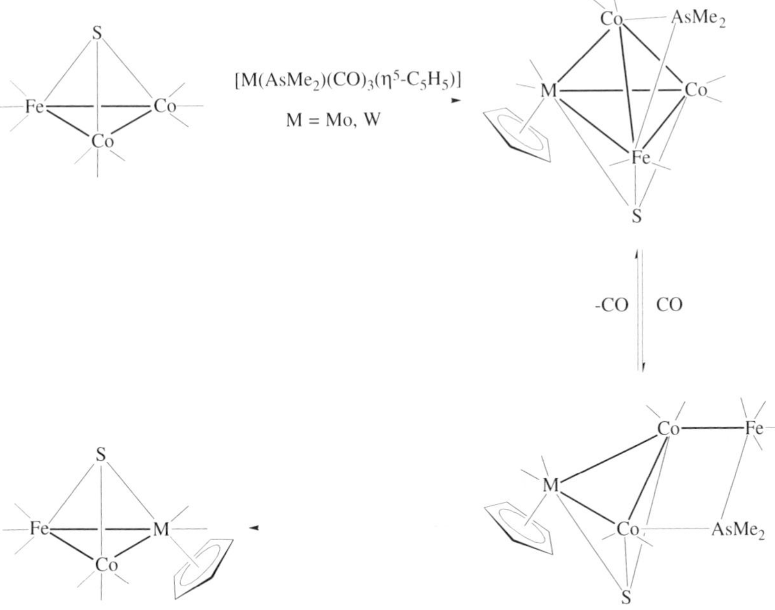

FIG. 39. Mixed-metal cluster formation employing $[M(AsMe_2)(CO)_3(\eta^5\text{-}C_5H_5)]$.

TABLE VI

METAL EXCHANGE AT VERY MIXED-METAL CLUSTERS

Product Cluster	Precursor Cluster, Reagent	Reaction Type[a]	Ref.
$[MCo_2(\mu_3\text{-}CR)(CO)_8(\eta^5\text{-}C_5H_5)]$ (M = Cr, Mo, W; R = H, Me, Ph, Cl)	$[Co_3(\mu_3\text{-}CR)(CO)_9]^-$, $[M(CO)_3(\eta^5\text{-}C_5H_5)]^-$ (R = H, Me, Ph); $[Co_3(\mu_3\text{-}CCl)(CO)_9]$, [ppn]$[Mo(CO)_3(\eta^5\text{-}C_5H_5)]$	A	96, 110
$[MCo_2(\mu_3\text{-}CR)(CO)_8(\eta^5\text{-}C_5H_4R')]$ [M = Mo, W, R = CO_2Et, Ph, R′ = Me; M = Mo, W, R = Ph, R′ = C(O)H, C(O)Me, CO_2Et; M = Mo, R = $SiEt_3$, H, Cl, R′ = H; M = W, R = Me, $SiEt_3$, H, R′ = H)	$[Co_3(\mu_3\text{-}CR)(CO)_9]$, $[M(CO)_3(\eta^5\text{-}C_5H_4R')]^-$	A	89, 111–115
$[WCo_2(\mu_3\text{-}CCO_2Pr^i)(CO)_5(\eta^5\text{-}L)]$ (L = C_5H_4Me, L = C_5Me_5)	$[Co_3(\mu_3\text{-}CCO_2Pr^i)(CO)_9]$, $[W(CO)_3(\eta^5\text{-}L)]^-$	A	116
$[M_2Co(\mu_3\text{-}CR)(CO)_7(\eta^5\text{-}C_5H_4R')]$ (M = Cr, Mo, W; R = R′ = H; M = Mo, W; R = Ph, CO_2Et, R′ = Me; M = Mo, R = Cl, R′ = H)	$[Co_3(\mu_3\text{-}CCl)(CO)_9]$, $[M(CO)_3(\eta^5\text{-}C_5H_5)]^-$ (K^+, R = H); $[Co_3(\mu_3\text{-}CR)(CO)_9]$, $[M(CO)_3(\eta^5\text{-}C_5H_4Me)]^-$; $[Co_3(\mu_3\text{-}CCl)(CO)_9]$, $[Mo(CO)_3(\eta^5\text{-}C_5H_5)]^-[ppn^+$, R = Cl]	A	96,111,117
$[MoWCo(\mu_3\text{-}CCl)(CO)_7(\eta^5\text{-}C_5H_5)_2]$	$[MoCo_2(\mu_3\text{-}CCl)(CO)_8(\eta^5\text{-}C_5H_5)]$, $[W(CO)_3(\eta^5\text{-}C_5H_5)]^-$	A	96
$[MoFeCo(H)(\mu_3\text{-}CR')(CO)_8(\eta^5\text{-}C_5H_4R)]$ [R′ = Ph, R = C(O)H, C(O)Me, CO_2Et; R′ = CO_2menthyl, R = H; R′ = CO_2Et, R = Me]	$[MoCo_2(\mu_3\text{-}CR')(CO)_8(\eta^5\text{-}C_5H_4R)]$, $[Fe(CO)_4]^{2-}$, H^+	A	29, 89, 112
$[MM'Co(H)(\mu_3\text{-}CR)(CO)_8(\eta^5\text{-}C_5H_5)]$ (M = Mo, W; R = H, Me, Ph; M′ = Ru, Os)	$[MCo_2(\mu_3\text{-}CR)(CO)_8(\eta^5\text{-}C_5H_5)]$, $[M'(CO)_4]^{2-}$	A	24, 30
$[MoFe_2(\mu\text{-}H)(\mu_3\text{-}S)(CO)_8(\eta^5\text{-}C_5H_5)]$	$[MoFeCo(\mu_3\text{-}S)(CO)_8(\eta^5\text{-}C_5H_5)]$, $[Fe(CO)_4]^{2-}$, H_2O	A	118
$[MoCo_2(\mu_3\text{-}S)\{\mu\text{-}(R)\text{-}CMeNCHMeCy\}(CO)_6(\eta^5\text{-}C_5H_5)]$	$[Co_3(\mu_3\text{-}S)(CO)_7\{\mu\text{-}(R)\text{-}CMeNCHMeCy\}]$, $[Mo(CO)_3(\eta^5\text{-}C_5H_5)]^-$	A	119
$[MM'Fe(\mu_3\text{-}S)(CO)_7(\eta^5\text{-}C_5H_4R)(\eta^5\text{-}C_5H_4R')]$ (M = M′ = Mo, W; R = R′ = H, CO_2Me, CO_2Et, C(O)Me)	$[MFeCo(\mu_3\text{-}S)(CO)_8(\eta^5\text{-}C_5H_4R)]$, $[M'(CO)_3(\eta^5\text{-}C_5H_4R')]^-$	A	120
$[MFeCo(\mu_3\text{-}E)(CO)_8(\eta^5\text{-}C_5H_4R)]$ [M = Mo, E = S, R = C(O)H, C(O)Me, CO_2Et; M = Mo, W, E = S, Se, $PNEt_2$, R = H]	$[FeCo_2(\mu_3\text{-}E)(CO)_9]$, $[M(CO)_3(\eta^5\text{-}C_5H_4R)]^-$	A	92, 121

$[\{MFeCo(\mu_3\text{-}S)(CO)_8(\eta^5\text{-}C_5H_4)\}_2(\mu\text{-}R)]$ $[M = Mo, W, R = 4\text{-}C(O)C_6H_4C(O),$ $2\text{-}CH_2C_6H_4CH_2; M = Mo, R = Me_2SiOSiMe_2]$	$[FeCo_2(\mu_3\text{-}S)(CO)_9]$, $[\{M(CO)_3(\eta^5\text{-}C_5H_4)\}_2(\mu\text{-}R)]^{2-}$	A	93
$[\{MRuCo(\mu_3\text{-}E)(CO)_8(\eta^5\text{-}C_5H_4C(O)\}_2(\mu\text{-}4\text{-}C_6H_4)]$ $(M = Mo, W, E = S; M = Mo, E = Se)$	$[RuCo_2(\mu_3\text{-}E)(CO)_9]$, $[\{M(CO)_3(\eta^5\text{-}C_5H_4)\}_2\{\mu\text{-}C(O)\text{-}4\text{-}C_6H_4C(O)\}]^{2-}$	A	94, 122, 123
$[MRuCo(\mu_3\text{-}E)(CO)_8(\eta^5\text{-}C_5H_4R)]$ $[M = Mo, W, E = S; M = Mo, E = Se; R = C(O)H,$ $C(O)Me, C(O)Ph, C(O)C_6H_4CO_2Me]$	$[RuCo_2(\mu_3\text{-}E)(CO)_9]$, $[M(CO)_3(\eta^5\text{-}C_5H_4R)]^-$	A	94, 123, 124
$[MoFeRu(H)(\mu_3\text{-}Se)(CO)_8\{\eta^5\text{-}C_5H_4C(O)Me\}]$	$[MoRuCo(\mu_3\text{-}Se)(CO)_8\{\eta^5\text{-}C_5H_4C(O)Me\}]$, $[Fe(CO)_4]^{2-}$, H^+	A	94
$[MoRuCo(\mu_3\text{-}\eta^2\text{-}RC_2R')(CO)_8(\eta^5\text{-}C_5H_5)]$ $(R = R' = Ph, Me)$	$[RuCo_2(\mu_3\text{-}\eta^2\text{-}RC_2R')(CO)_9]$, $[Mo(CO)_3(\eta^5\text{-}C_5H_5)]^-$	A	125
$[WRuCo(\mu_3\text{-}PPh)(CO)_8(\eta^5\text{-}C_5H_5)]$	$[RuCo_2(\mu_3\text{-}PPh)(CO)_9]$, $[W(CO)_3(\eta^5\text{-}C_5H_5)]^-$	A	24
$[L_nMRe_2Pt(\mu\text{-}H)_2(CO)_9]$ $(L_nM = [W(CO)_3$ $(\eta^5\text{-}C_5H_5)]^-$, $[Mn(CO)_5]^-$, $[HRe_2(CO)_9]^-$, $[Co(CO)_4]^-)$	$[Re_3Pt(\mu\text{-}H)_3(CO)_{14}]^-$, ML_n	A	73
$[MCo_2(\mu_3\text{-}CR)(CO)_8(\eta^5\text{-}L)]$ $(M = Mo, R = CO_2Pr^i,$ $L = C_5H_5, C_5H_4Me, C_5Me_5; M = Mo,$ $R = CO_2$ menthyl, $L = C_5H_5, C_5H_4Pr^i, C_9H_7;$ $M = Mo, W, R = Me, L = C_5H_5; M = W,$ $R = CO_2Pr^i, L = C_5H_5.$	$[Co_3(\mu_3\text{-}CR)(CO)_9]$, $[M(CO)_3(\eta^5\text{-}L)]_2$	B	16, 17, 24, 28, 56, 116, 126
$[MoCo_2(\mu_3\text{-}As)(CO)_3(\eta^5\text{-}C_5H_5)_3]$	$[Mo_3(\mu_3\text{-}As)(CO)_6(\eta^5\text{-}C_5H_5)_3]$, $[Co(CO)_2(\eta^5\text{-}C_5H_5)]$	B	127
$[MoPt_2(\mu_3\text{-}As)(CO)_4(PPh_3)_2]$	$[Mo_3(\mu_3\text{-}As)(CO)_6(\eta^5\text{-}C_5H_5)_3]$, $[Pt(\eta^2\text{-}C_2H_4)(PPh_3)_2]$	B	128
$[Mo_2Ni(\mu_3\text{-}As)(CO)_4(\eta^5\text{-}C_5H_5)_3]$	$[Mo_3(\mu_3\text{-}As)(CO)_6(\eta^5\text{-}C_5H_5)_3]$, $[Ni(CO)(\eta^5\text{-}C_5H_5)_2]$	B	128
$[MM'Ru(\mu_3\text{-}S)(CO)_7(\eta^5\text{-}C_5H_5)_2]$ $(M, M' = Mo, W)$	$[MRuCo(\mu_3\text{-}S)(CO)_8(\eta^5\text{-}C_5H_5)]$, $[M'(CO)_3(\eta^5\text{-}C_5H_5)]_2$	B	129
$[MRuCo(\mu_3\text{-}S)(CO)_8(\eta^5\text{-}C_5H_5)]$ $(M = Mo, W)$	$[RuCo_2(\mu_3\text{-}S)(CO)_9]$, $[M(CO)_3(\eta^5\text{-}C_5H_5)]_2$	B	130
$[MoRuRh(\mu_3\text{-}\eta^2\text{-}MeC_2Me)(CO)_8(\eta^5\text{-}C_5H_5)]$	$[MoFeRu(\mu\text{-}H)(\mu_3\text{-}\eta^2\text{-}MeC_2Me)(CO)_8(\eta^5\text{-}C_5H_5)]$, base, $[Rh(CO)_2Cl]_2$	B	131
$[MFeNi(\mu_3\text{-}S)(CO)_5(\eta^5\text{-}C_5H_5)(\eta^5\text{-}C_5H_4R)]$ $[M = Mo, W;$ $R = CHO, C(O)Me, CO_2Me]$	$[MFeCo(\mu_3\text{-}S)(CO)_8(\eta^5\text{-}C_5H_4R)]$, $[Ni(\eta^5\text{-}C_5H_5)_2]$	B	95
$[MoRuNi(\mu_3\text{-}\eta^2\text{-}MeC_2Me)(CO)_5(\eta^5\text{-}C_5H_5)_2]$	$[MoRuCo(\mu_3\text{-}\eta^2\text{-}MeC_2Me)(CO)_8(\eta^5\text{-}C_5H_5)]$, $[Ni(CO)(\eta^5\text{-}C_5H_5)]_2$	B	125

(Continued)

TABLE VI (*Continued*)

Product Cluster	Precursor Cluster, Reagent	Reaction Type[a]	Ref.
$[MoCoNi(\mu_3\text{-}CR)(CO)_5(\eta^5\text{-}C_5H_5)_2]$	$[MoCo_2(\mu_3\text{-}CR)(CO)_8(\eta^5\text{-}C_5H_5)]$ (R = Me, CO_2menthyl, CO_2Pr^i), $[Ni(CO)(\eta^5\text{-}C_5H_5)]_2$; $[Co_2Ni(\mu_3\text{-}CMe)(CO)_6(\eta^5\text{-}C_5H_5)]$, $[Mo(CO)_3(\eta^5\text{-}C_5H_5)]_2$	B	28, 29, 56, 126
$[MoNi_2(\mu_3\text{-}CCO_2Pr^i)(CO)_2(\eta^5\text{-}C_5H_5)_3]$	$[MoCo_2(\mu_3\text{-}CCO_2Pr^i)(CO)_8(\eta^5\text{-}C_5H_5)]$, $[Ni(\eta^5\text{-}C_5H_5)_2]$	B	56
$[MoCo_3(\mu\text{-}CO)_3(CO)_8(\eta^5\text{-}C_5H_5)]$	$[Co_4(\mu\text{-}CO)_3(CO)_6(\eta^6\text{-mesitylene})]$, $[Mo(CO)_3(\eta^5\text{-}C_5H_5)]_2$	B	132
$[Mo_2Co_2(\mu\text{-}CO)_3(CO)_7(\eta^5\text{-}C_5H_5)_2]$	$[Co_4(\mu\text{-}CO)_3(CO)_6(\eta^6\text{-mesitylene})]$, $[Mo(CO)_3(\eta^5\text{-}C_5H_5)]_2$	B	132
$[Mo_4Ni_2(\mu_4\text{-}As)(\mu_3\text{-}As)(CO)_7(\eta^5\text{-}C_5H_5)_6]$	$[Mo_3(\mu_3\text{-}As)(CO)_6(\eta^5\text{-}C_5H_5)_3]$, $[Ni(CO)(\eta^5\text{-}C_5H_5)]_2$	B	128
$[MCo_2(\mu_3\text{-}CR)(CO)_8(\eta^5\text{-}C_5H_5)]$ (M = Mo, R = CO_2menthyl, H; M = W, R = H)	$[Co_3(\mu_3\text{-}CR)(CO)_9]$, $[M(AsMe_2)(CO)_3(\eta^5\text{-}C_5H_5)]$	C	29, 115
$[MCo_2(\mu_3\text{-}GeR)(CO)_8(\eta^5\text{-}C_5H_5)]$ (M = Mo, R = Me, Ph, Bu^t; M = W, R = Bu^t)	$[Co_3(\mu_3\text{-}GeR)(CO)_9]$, $[M(AsMe_2)(CO)_3(\eta^5\text{-}C_5H_5)]$	C	133
$[MoCo_2(\mu_3\text{-}PPh)(\mu\text{-}AsMe_2)(CO)_6(\eta^5\text{-}C_5H_5)]$	$[Co_3(\mu_3\text{-}PPh)(CO)_9]$, $[Mo(AsMe_2)(CO)_3(\eta^5\text{-}C_5H_5)]$	C	134
$[MoFeCo(\mu_3\text{-}PPh)(CO)_8(\eta^5\text{-}C_5H_5)]$	$[FeCo_2(\mu_3\text{-}PPh)(CO)_9]$, $[Mo(AsMe_2)(CO)_3(\eta^5\text{-}C_5H_5)]$	C	134
$[MM'Fe(\mu_3\text{-}S)(CO)_7(\eta^5\text{-}C_5H_5)]$ (M, M′ = Mo, W)	$[MFeCo(\mu_3\text{-}S)(CO)_8(\eta^5\text{-}C_5H_5)]$, $[M'(AsMe_2)(CO)_3(\eta^5\text{-}C_5H_5)]$	C	129
$[MFeCo(\mu_3\text{-}S)(CO)_8(\eta^5\text{-}C_5H_5)]$ (M = Mo, W)	$[FeCo_2(\mu_3\text{-}S)(CO)_9]$, $[M(AsMe_2)(CO)_3(\eta^5\text{-}C_5H_5)]$	C	135
$[MRuCo(\mu_3\text{-}S)(CO)_8(\eta^5\text{-}C_5H_5)]$ (M = Cr, Mo, W)	$[RuCo_2(\mu_3\text{-}S)(CO)_8\{AsMe_2M(CO)_3(\eta^5\text{-}C_5H_5)\}]$	C	130
$[MCo_2(\mu_3\text{-}CR)(CO)_8(\eta^5\text{-}C_5H_5)]$ (M = Cr, Mo, W, R = H, Me, Ph; M = Mo, W, R = C_6H_4Me-4)	$[Co_3(\mu_3\text{-}CR)\{AsMe_2M(CO)_3(\eta^5\text{-}C_5H_5)\}(CO)_8]$	C	136
$[MoCo_2(\mu_3\text{-}CR)(CO)_8(\eta^5\text{-}C_5H_5)]$ (R = Me, H, Cl)	$[Co_3(\mu_3\text{-}CR)(CO)_9]$; $[MoCl(CO)_3(\eta^5\text{-}C_5H_5)]$ (R = Me), $[MoH(CO)_3(\eta^5\text{-}C_5H_5)]$ (R = Me, H, Cl)	D	115
$[MoFeCo(\mu_3\text{-}S)(CO)_8(\eta^5\text{-}C_5H_4Me)]$	$[Fe_2Co(\mu\text{-}H)(\mu_3\text{-}S)(CO)_9]$, $[MoCl(CO)_3(\eta^5\text{-}C_5H_4Me)]$	D	137
$[MoRuCo(\mu_3\text{-}\eta^2\text{-}RC_2H)(CO)_8(\eta^5\text{-}C_5H_5)]$ (R = Me, Ph)	$[RuCo_2(\mu_3\text{-}\eta^2\text{-}RC_2H)(CO)_9]$, $[MoCl(CO)_3(\eta^5\text{-}C_5H_5)]$	D	138
$[MCo_3(\mu\text{-}CO)_3(CO)_8(\eta^5\text{-}C_5H_5)]$ (M = Mo, W)	$[Co_4(CO)_{12}]$, $[MH(CO)_3(\eta^5\text{-}C_5H_5)]$	D	139
$[CrWPd_2(\mu_3\text{-}CO)_2(\mu\text{-}CO)_4(PPh_3)_2(\eta^5\text{-}C_5H_5)_2]$	$[Cr_2Pd_2(\mu_3\text{-}CO)_2(\mu\text{-}CO)_4(PPh_3)_2(\eta^5\text{-}C_5H_5)_2]$, $[WH(CO)_3(\eta^5\text{-}C_5H_5)]$	D	140
$[L_nMRe_2Pt(\mu\text{-}H)_2(CO)_9]$($L_nM$ = $[ReH(CO)_4(PPh_3)]$, $[ReH(CO)_3(PPh_3)_2]$)	$[Re_3Pt(\mu\text{-}H)_3(CO)_{14}]^-$, ML_n	D	73

[a]Reaction types: A, metal carbonyl anions/cyclopentadienyl metal carbonyl anions; B, metal carbonyls/cyclopentadienylmetal carbonyls/nickelocene/[Pt(η^2-

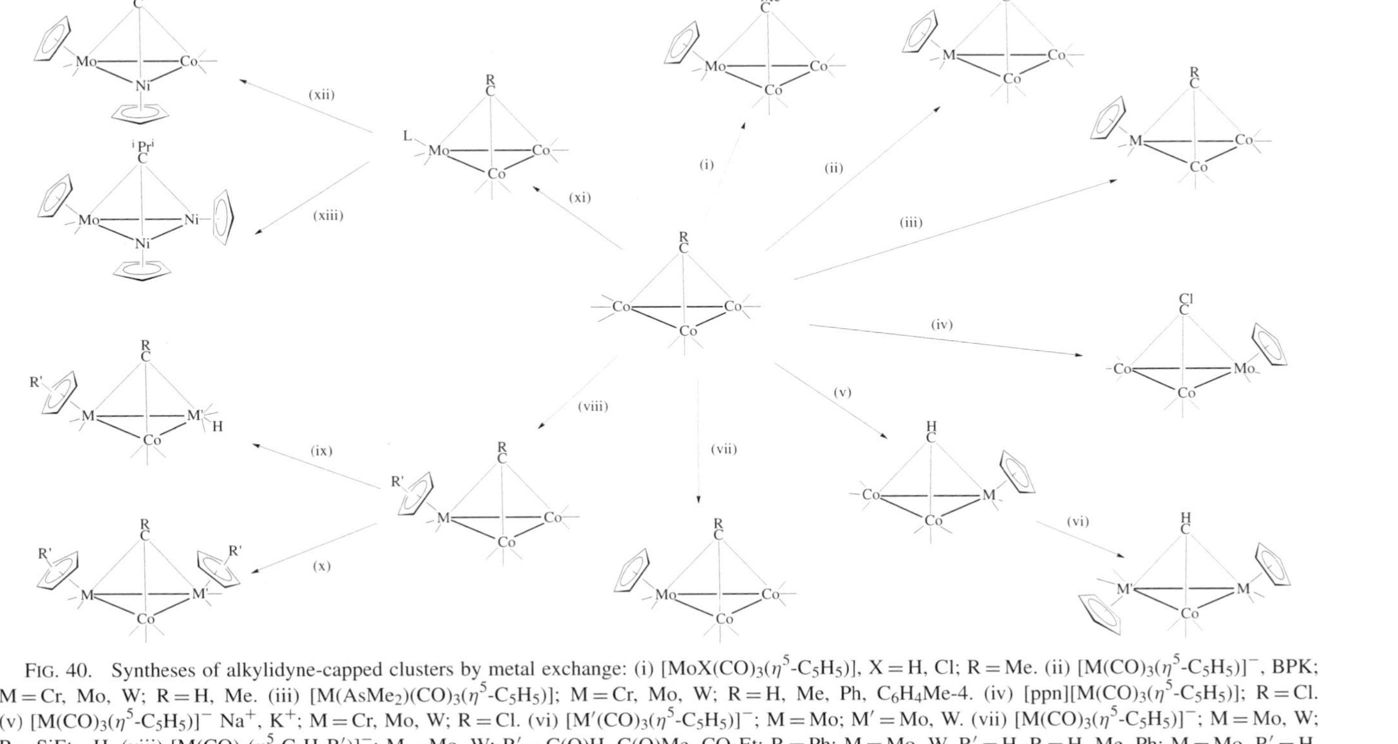

FIG. 40. Syntheses of alkylidyne-capped clusters by metal exchange: (i) $[MoX(CO)_3(\eta^5\text{-}C_5H_5)]$, X = H, Cl; R = Me. (ii) $[M(CO)_3(\eta^5\text{-}C_5H_5)]^-$, BPK; M = Cr, Mo, W; R = H, Me. (iii) $[M(AsMe_2)(CO)_3(\eta^5\text{-}C_5H_5)]$; M = Cr, Mo, W; R = H, Me, Ph, C_6H_4Me-4. (iv) $[ppn][M(CO)_3(\eta^5\text{-}C_5H_5)]$; R = Cl. (v) $[M(CO)_3(\eta^5\text{-}C_5H_5)]^-$ Na^+, K^+; M = Cr, Mo, W; R = Cl. (vi) $[M'(CO)_3(\eta^5\text{-}C_5H_5)]^-$; M = Mo; M′ = Mo, W. (vii) $[M(CO)_3(\eta^5\text{-}C_5H_5)]^-$; M = Mo, W; R = $SiEt_3$, H. (viii) $[M(CO)_3(\eta^5\text{-}C_5H_4R')]^-$; M = Mo, W; R′ = C(O)H, C(O)Me, CO_2Et; R = Ph; M = Mo, W, R′ = H, R = H, Me, Ph; M = Mo, R′ = H, R = CO_2menthyl; M = Mo, W, R′ = CO_2Et, R = Me; M = Mo, W; R′ = Me, R = Ph, CO_2Et. (ix) $[Fe(CO)_4]^{2-}/H^+$; R′ = C(O)H, C(O)Me, CO_2Et; R = Ph, M = Mo, M′ = Fe; R′ = H, R = CO_2 menthyl; M = Mo, M′ = Fe, R′ = Me, R = CO_2Et; M = Mo, W, M′ = Fe. (x) $[M'(CO)_4]^{2-}/H^+$; R = H, Me, Ph; M = Mo, W; M′ = Ru, Os; M, M′ = Mo, W, R = Ph, CO_2Et; R′ = Me. (xi) $[M(CO)_3(\eta^5\text{-}L)]_2$; M = W, L = C_5H_5, R = Me; M = Mo, L = $C_5H_4Pr^i$, C_5H_5, C_9H_7, R = CO_2 menthyl; M = Mo, L = C_5H_5, C_5H_4Me, C_5Me_5, R = Pr^i. (xii) $[Ni(CO)(\eta^5\text{-}C_5H_5)]_2$; M = Mo, L = $\eta^5\text{-}C_5H_5$; R = Pr^i, Me. (xiii) $[Ni(\eta^5\text{-}C_5H_5)_2]$; M = Mo, L = $\eta^5\text{-}C_5H_5$; R = Pr^i.

as the $Co(CO)_3$ unit is the most readily displaced group in metal exchange reactions.[16,17,28,30,56,110,112,116,117,134] Early studies in this area have been reviewed;[4] more recently, extensions to the metal exchange procedure have been explored using different μ_3-CR groups, and varying coligands on the metal exchange reagents. For example, the reactions between $[Co_3(\mu_3\text{-CR})(CO)_9]$ (R = $SiEt_3$, H, Cl, Br) and $[M(CO)_3(\eta^5\text{-}C_5H_5)]^-$ (M = Mo, W) afforded $[MCo_2(\mu_3\text{-CR})(CO)_8(\eta^5\text{-}C_5H_5)]$, the reactivity of the μ_3-CR apical substituent resulting in low yields of the products.[115] The reactions of $[Co_3(\mu_3\text{-CPh})(CO)_9]$ and $[M(CO)_3(\eta^5\text{-}C_5H_4R)]^-$ [M = Mo, W; R = C(O)H, C(O)Me, CO_2Et] afforded $[MCo_2(\mu_3\text{-CR})(CO)_8(\eta^5\text{-}C_5H_4R)]$; incorporation of an electron-withdrawing group on the cyclopentadienyl ligand reduced the activity of the exchange reagent.[112–114]

An alternative route to $[MCo_2(\mu_3\text{-CR})(CO)_8(\eta^5\text{-}C_5H_5)]$ (M = Cr, Mo, W; R = H, Me, Ph) involving electron transfer catalysis has been demonstrated.[110] Reaction of $[Co_3(\mu_3\text{-CR})(CO)_9]$, the metal exchange reagent, and a catalytic amount of sodium diphenylketyl under varying conditions afforded the desired clusters in good yield. This reaction proceeded rapidly at room temperature, in contrast to the classical metal exchange reaction which required heating in benzene for 3 days.[28,134]

Many examples of sequential metal exchange at face-capped trinuclear clusters have been reported. Two successive metal exchange reactions can afford tetrahedral clusters which are chiral by virtue of having four different core constituents (Fig. 40). Metal exchange can proceed by a further step; the cluster $[MoNi_2(\mu_3\text{-CPr}^i)(CO)_2(\eta^5\text{-}C_5H_5)_2]$[56] is an example of metal exchange in which all three $Co(CO)_3$ units of the precursor cluster have been replaced with other moieties.

While the apical substituent in these clusters has usually been an alkylidyne group, other face capping ligands have also been successfully employed. For example, the clusters $[MFeCo(\mu_3\text{-X})(CO)_8(\eta^5\text{-}C_5H_5)]$ (M = Mo, W; X = S, Se, $PNEt_2$),[121 129] $[MoFeCo(\mu_3\text{-S})(CO)_8(\eta^5\text{-}C_5H_4R)]$ [R = C(O)H, C(O)Me, CO_2 Et],[92] $[MoRuCo(\mu_3\text{-S})(CO)_8(\eta^5\text{-}C_5H_5)]$,[129] and $[\{MFeCo(\mu_3\text{-S})(CO)_8(\eta^5\text{-}C_5H_4)\}_2(\mu\text{-R})]$ [M = Mo, W; R = 2-$CH_2C_6H_4CH_2$, 4-$C(O)C_6H_4C(O)$][93] were formed by metal exchange reactions with $[M'Co_2(\mu_3\text{-X})(CO)_9]$ (M′ = Fe, Ru), itself prepared by metal exchange from the precursor cluster $[Co_3(\mu_3\text{-X})(CO)_9]$ (Fig. 41). The metal exchange products $[MFeCo(\mu_3\text{-S})(CO)_8(\eta^5\text{-}C_5H_4R)]$ (M = Mo, W; R = CHO, OMe, CO_2Me) reacted with nickelocene in refluxing THF to afford $[MFeNi(\mu_3\text{-S})(CO)_5(\eta^5\text{-}C_5H_5)(\eta^5\text{-}C_5H_4R)]$.[95] $[Fe_2Co(\mu\text{-H})(\mu_3\text{-S})(CO)_9]$, prepared from $[FeCo_2(\mu_3\text{-S})(CO)_9]$, reacted with $[MoCl(CO)_3(\eta^5\text{-}C_5H_4Me)]$ to afford $[MoFeCo(\mu_3\text{-S})(CO)_8(\eta^5\text{-}C_5H_4Me)]$ and $[Mo_2Fe(\mu_3\text{-S})(CO)_7(\eta^5\text{-}C_5H_4Me)_2]$.[137] Similarly, reaction between $[RuCo_2(\mu_3\text{-S})(CO)_9]$ and $Na_2[(CO)_3M\{\eta^5\text{-}C_5H_4C(O)\text{-}4\text{-}C_6H_4C(O)\text{-}\eta^5\text{-}C_5H_4\}M(CO)_3]$ (M = Mo, W) has afforded the clusters $[MRuCo(\mu_3\text{-S})(CO)_8\{\eta^5\text{-}C_5H_4C(O)\text{-}4\text{-}C_6H_4C(O)\text{-}\eta^5\text{-}C_5H_4\}MRuCo(\mu_3\text{-S})(CO)_8]$ in which the exchange reagent links two cluster cores (Fig. 42),[122,123] and $[MRuCo(\mu_3\text{-S})(CO)_8(\eta^5\text{-}C_5H_4R)]$ [M = Mo, W; R = C(O)H, C(O)Me, C(O)Ph, 4-$C(O)C_6H_4$

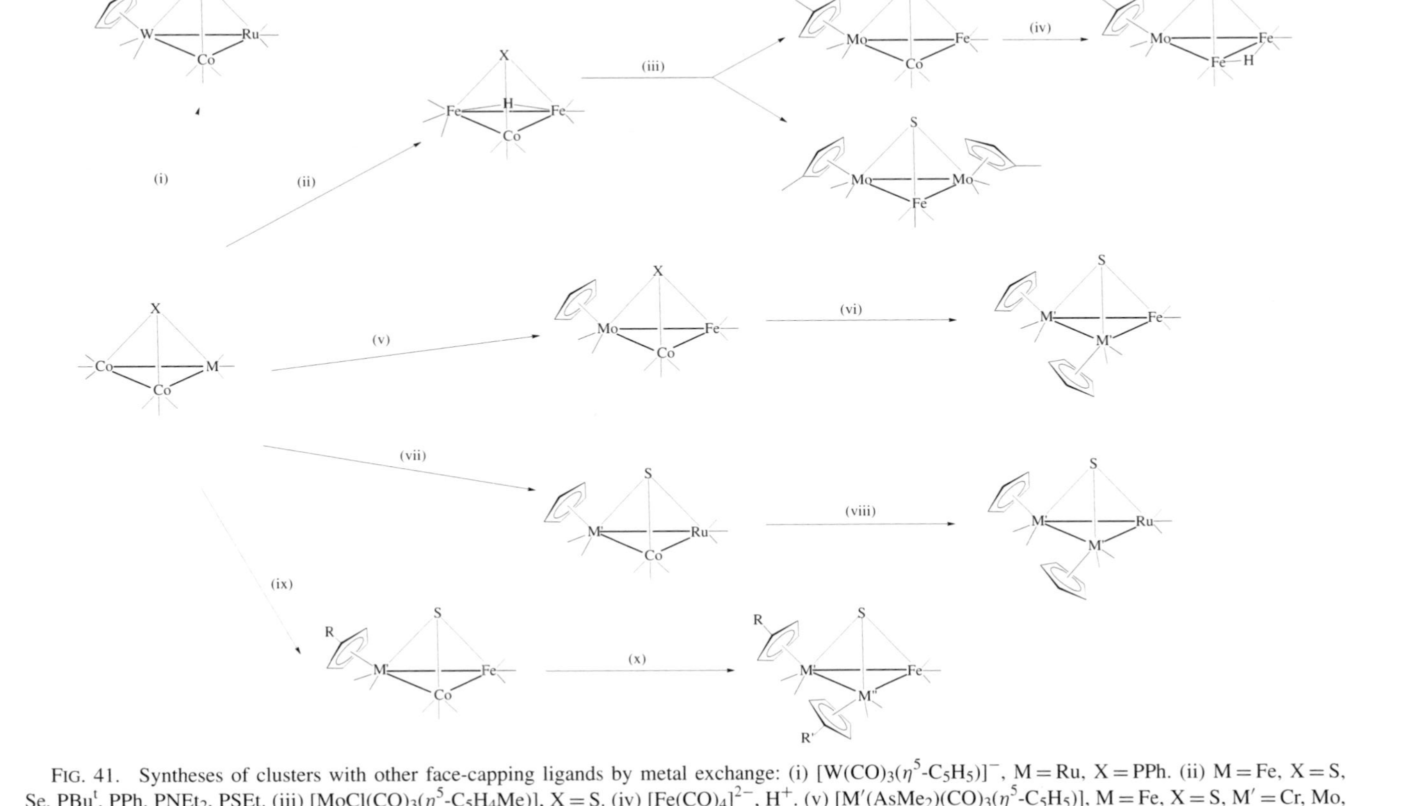

FIG. 41. Syntheses of clusters with other face-capping ligands by metal exchange: (i) $[W(CO)_3(\eta^5\text{-}C_5H_5)]^-$, M = Ru, X = PPh. (ii) M = Fe, X = S, Se, PBu^t, PPh, $PNEt_2$, PSEt. (iii) $[MoCl(CO)_3(\eta^5\text{-}C_5H_4Me)]$, X = S. (iv) $[Fe(CO)_4]^{2-}$, H^+. (v) $[M'(AsMe_2)(CO)_3(\eta^5\text{-}C_5H_5)]$, M = Fe, X = S, M′ = Cr, Mo, W, M = Ru, X = S; $[M'(CO)_3(\eta^5\text{-}C_5H_5)]^-$, M = Fe, M′ = Mo, W, X = S, Se, $PNEt_2$. (vi) $[M'(AsMe_2)(CO)_3(\eta^5\text{-}C_5H_5)]$, X = S. (vii) $[M'(CO)_3(\eta^5\text{-}C_5H_5)]_2$, M′ = Mo, W; $[M'(AsMe_2)(CO)_3(\eta^5\text{-}C_5H_5)]$, M′ = Cr, Mo, W, M = Ru, X = S. (viii) $[M'(CO)_3(\eta^5\text{-}C_5H_5)]_2$, M′ = Mo, W. (ix) $[M'(CO)_3(\eta^5\text{-}C_5H_4R)]^-$; M = Mo, W, R = C(O)H, C(O)Me, CO_2Et, CO_2Me, M = Fe, X = S. (x) $[M''(CO)_3(\eta^5\text{-}C_5H_4R')]^-$, M″ = Mo, W, R′ = H, C(O)Me, CO_2Me, Me, R = CO_2Et, CO_2Me, M = Mo, W.

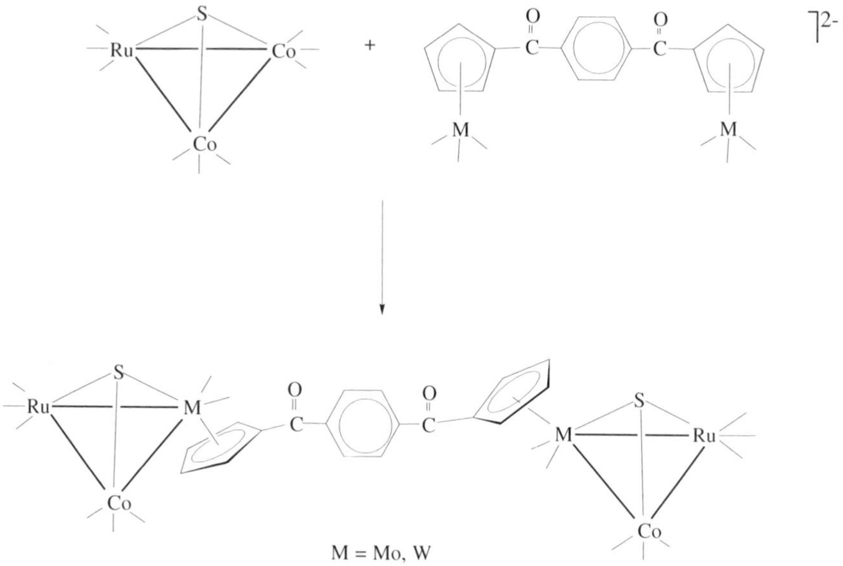

FIG. 42. Syntheses of linked clusters by metal exchange.

CO_2Me] were obtained by metal exchange from the same ruthenium-dicobalt precursor and analogous functionalized (cyclopentadienyl)metal carbonyl anions.[123,124] Related reactions of the selenido-containing cluster [$RuCo_2(\mu_3$-Se)$(CO)_9$] with the functionalized (cyclopentadienyl)molybdenum carbonyl anion reagents afforded the analogous [MoRuCo(μ_3-Se)$(CO)_8(\eta^5$-$C_5H_4R)$], and with the linked dianionic dimolybdenum reagent afforded [MoRuCo(μ_3-Se)$(CO)_8(\eta^5$-$C_5H_4C(O)$-4-$C_6H_4C(O)$-η^5-C_5H_4}MoRuCo(μ_3-Se)$(CO)_8$].[94] A capping arsenic atom has been employed; one or two $Mo(CO)_2(\eta^5$-$C_5H_5)$ vertices in the trimolybdenum precursor [$Mo_3(\mu_3$-As)$(CO)_6(\eta^5$-$C_5H_5)_3$] were replaced (Fig. 43).[127,128] The flexibility of this procedure has been further demonstrated utilizing clusters with capping germylidyne groups (Fig. 44).[133]

Metal exchange methods have also been utilized with alkyne-bridged clusters. For example, [$RuCo_2(\mu_3$-η^2-$RC_2R')(CO)_9$] was the precursor to [MoRuCo(μ_3-η^2-$RC_2R')(CO)_8(\eta^5$-$C_5H_5)$] (R, R′ = Ph, Me), [MoRuNi(μ_3-η^2-$MeC_2Me)(CO)_5(\eta^5$-$C_5H_5)_2$], and [MoRuRh(μ_3-η^2-$MeC_2Me)(CO)_8(\eta^5$-$C_5H_5)$] (Fig. 45).[125]

Although metal exchange procedures to afford "very mixed"-metal clusters have been utilized most frequently with trinuclear clusters, reactions involving tetranuclear clusters have also been successful. The reaction of [$Co_4(\mu$-$CO)_3(CO)_9$] with [$MH(CO)_3(\eta^5$-$C_5H_5)$] (M = Mo, W) afforded the tetrahedral cluster [$MCo_3(\mu$-$CO)_3(CO)_8(\eta^5$-$C_5H_5)$].[139] A similar reaction between [$Co_4(\mu$-$CO)_3(CO)_6(\eta^6$-1,3,5-$C_6H_3Me_3)$] and [$Mo(CO)_3(\eta^5$-$C_5H_5)]_2$ likewise afforded [$MCo_3(\mu$-$CO)_3(CO)_8$

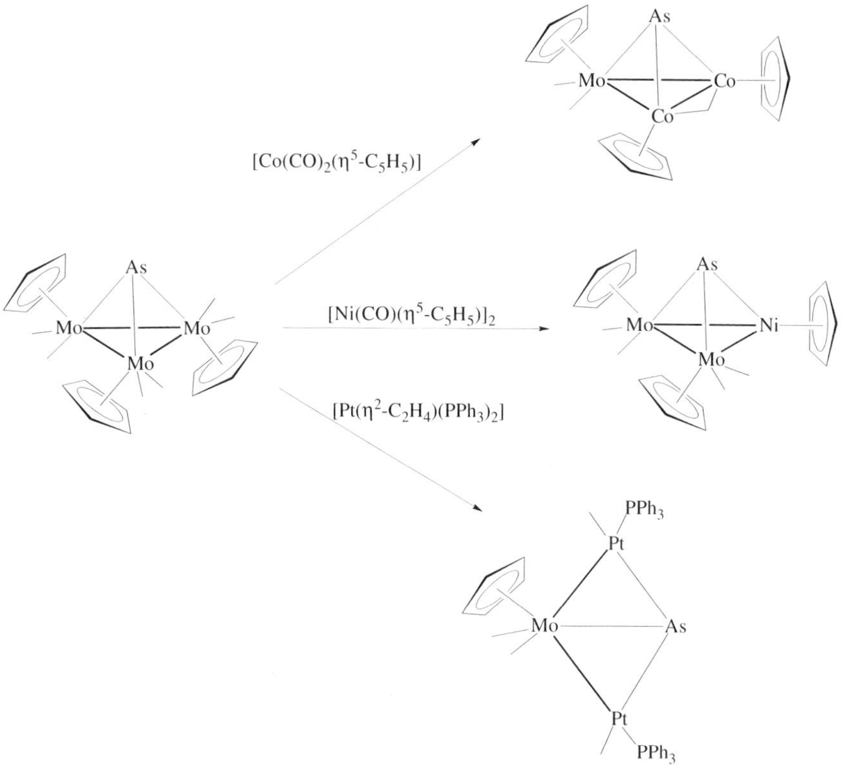

FIG. 43. Syntheses of arsenido-capped clusters by metal exchange.

(η^5-C_5H_5)], together with [$Mo_2Co_2(\mu$-$CO)_3(CO)_7(\eta^5$-$C_5H_5)_2$] (Fig. 46).[132] The butterfly cluster [$Cr_2Pd_2(\mu_3$-$CO)_2(\mu$-$CO)_4(PPh_3)_2(\eta^5$-$C_5H_5)_2$] reacted with [$WH(CO)_3$ (η^5-C_5H_5)] to afford [$CrWPd_2(\mu_3$-$CO)_2(\mu$-$CO)_4(PPh_3)_2(\eta^5$-$C_5H_5)_2$] (Fig. 47).[140] The "spiked" triangular cluster [$Re_3Pt(\mu$-$H)_3(CO)_{14}$] reacted with a range of metal carbonyl anions ([$HRe_2(CO)_9$]$^-$, [$Mn(CO)_5$]$^-$, [$W(CO)_3(\eta^5$-$C_5H_5)$]$^-$, [$Co(CO)_4$]$^-$) and metal carbonyl hydrides ([$ReH(CO)_4(PPh_3)$], [$ReH(CO)_3(PPh_3)_2$]) to afford

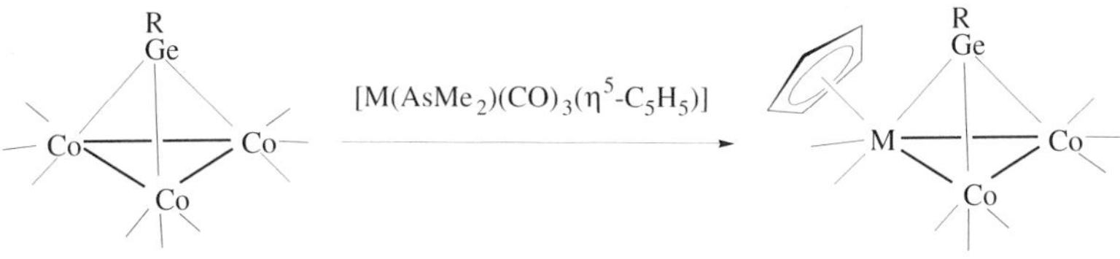

FIG. 44. Syntheses of germylidyne-capped clusters by metal exchange: M = Mo, R = Me, Ph, Bu^t; M = W, R = Bu^t.

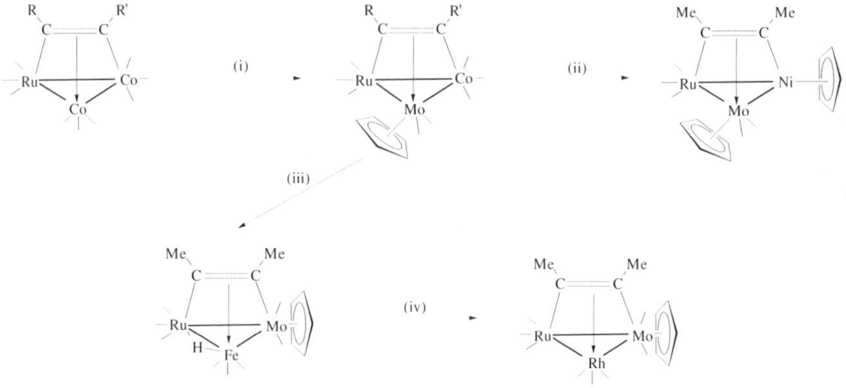

FIG. 45. Syntheses of alkyne-capped clusters by metal exchange: (i) $[Mo(CO)_3(\eta^5\text{-}C_5H_5)]^-$, R′, R = Me, Ph; $[MoCl(CO)_3(\eta^5\text{-}C_5H_5)]$, R = Me, Ph, R′ = H; NEt_3, CuI. (ii) $[Ni(CO)(\eta^5\text{-}C_5H_5)]_2$, R = R′ = Me. (iii) $[Fe_2(CO)_8]^{2-}$, R = R′ = Me. (iv) DBU, $[Rh(CO)_2Cl]_2$.

"spiked" triangular cluster products. Metal exchange at higher nuclearity "very mixed"-metal clusters may also be possible, but has yet to be exploited.

2. *Core Expansion or Core Contraction*

Reactions of clusters with mononuclear or dinuclear metal complexes frequently provide a method of expanding the metal core nuclearity under controlled conditions. The majority of medium- and high-nuclearity homometallic clusters has been prepared from lower-nuclearity cluster precursors by thermolyses ("heat-it-and-hope") reactions. This is less true of the heterometallic clusters in this

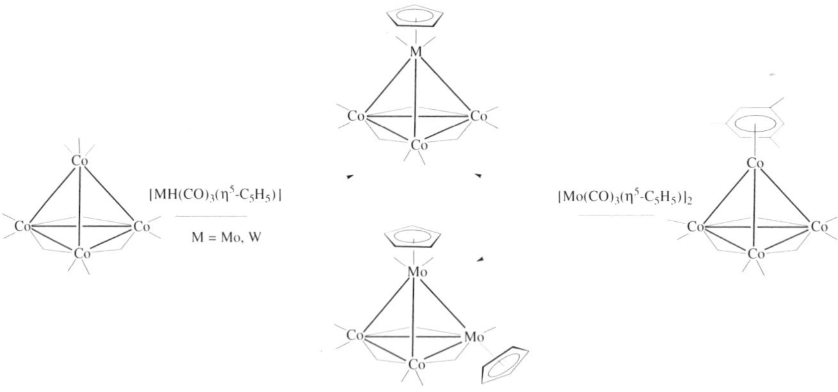

FIG. 46. Syntheses of group 6–group 9 tetranuclear clusters by metal exchange.

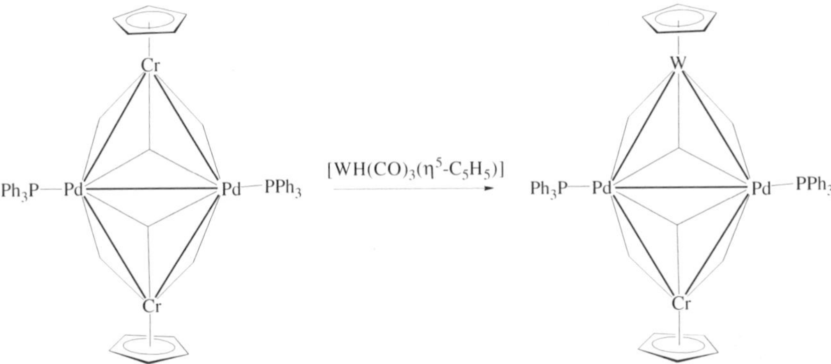

FIG. 47. Syntheses of group 6–group 10 tetranuclear clusters by metal exchange.

section, with three synthetic procedures employed extensively to afford a controlled means of obtaining clusters of different nuclearity. Nucleophilic substitution by a metal complex nucleophile on a cluster, or a cluster nucleophile on a metal complex ("redox condensation"), bridge-assisted procedures employing a variety of flexible ligands, and Stone's systematic stepwise assembly of medium- and high-nuclearity cluster chains and stars have all proved highly successful. Table VII collects "very mixed"-metal clusters synthesized by expansion or contraction of pre-existing cluster complexes.

The reaction of $[M_4(\mu\text{-CO})_3(CO)_9]$ ($M = Co$, Rh) with $[Cr(CO)_3(\eta^6\text{-PhC}_2Ph)]$ (which has a pendant acetylene linkage) afforded products dependent on the precursor cluster; whereas the lighter cluster fragmented to afford $[Co_2(\mu_3, \eta^2{:}\eta^6\text{-PhC}_2Ph)(CO)_6Cr(CO)_3]$, the heavier cluster $[Rh_4(\mu\text{-CO})_3(CO)_9]$ reacted by expansion of the cluster core, yielding the crystallographically characterized $[CrRh_4(\mu_5, \eta^2{:}\eta^6\text{-PhC}_2Ph)(\mu\text{-CO})_2(CO)_{10}]$, with a spiked butterfly coordination geometry (Fig. 48).[155]

Sulfido bridging ligands can facilitate cluster expansion. Photolysis of the triangular clusters $[Cr_2Co(\mu_3\text{-S})_2(\mu\text{-SBu}^t)(CO)_2(\eta^5\text{-C}_5H_4Me)_2]$ or $[Cr_2Rh(\mu_3\text{-S})_2(\mu\text{-SBu}^t)(\eta^4\text{-cod})(\eta^5\text{-C}_5H_5)_2]$ in the presence of $[Fe_2(\mu\text{-S})_2(CO)_6]$ afforded the clusters $[Cr_2Fe_2M(\mu_3\text{-S})_4(\mu\text{-SBu}^t)(CO)_6(\eta^5\text{-C}_5H_4R)_2]$ ($M = Co$, $R = Me$; $M = Rh$, $R = H$) (Fig. 49).[157] Heating $[Cr_2Co(\mu_3\text{-S})_2(\mu\text{-SBu}^t)(CO)_2(\eta^5\text{-C}_5H_5)_2]$ in the presence of diphenylacetylene afforded the tetrahedral cluster $[Cr_3Co(\mu_3\text{-S})_4(CO)(\eta^5\text{-C}_5H_5)_3]$, rather than the expected substitution product.[144] Sulfido bridging ligands can also maintain some metal–metal connectivity upon cluster fragmentation. Reaction of the "bow-tie" cluster $[Cr_4Ni(\mu_3\text{-S})_4(\mu\text{-SBu}^t)_2(\eta^5\text{-C}_5H_5)_4]$ and $[Co_2(CO)_8]$ afforded a lower nuclearity cluster $[Cr_2Co_2(\mu_4\text{-S})(\mu_3\text{-S})_2(CO)_4(\eta^5\text{-C}_5H_5)(\eta^5\text{-C}_5H_4Bu^t)]$ with a butterfly coordination geometry; this complex reaction also involves an unusual transfer of a tertiary butyl group from the μ-SBu^t ligand to one of the η^5-cyclopentadienyl ligands (Fig. 50).[145]

TABLE VII

CORE EXPANSION/CONTRACTION AT VERY MIXED-METAL CLUSTERS

M_n	Product Cluster	Precursor Cluster, Reagent	Ref.
n = 3	$[MCo_2(\mu_3\text{-}CR)(CO)_8(\eta^5\text{-}C_5H_5)]$ (M = Mo, W; R = Me, Ph)	$[MCo_3(\mu_3\text{-}CR)(\mu\text{-}AsMe_2)(CO)_8(\eta^5\text{-}C_5H_5)]$, CO	141
	$[MFeCo(\mu_3\text{-}PR)(CO)_8(\eta^5\text{-}C_5H_5)]$ (M = Mo, R = Ph, Bu^t; M = W, R = Bu^t)	$[MFeCo_2(\mu_3\text{-}PR)(\mu\text{-}AsMe_2)(CO)_8(\eta^5\text{-}C_5H_5)]$, CO	141
	$[MM'Co(\mu_3\text{-}S)(CO)_8(\eta^5\text{-}C_5H_5)]$ (M = Mo, W; M′ = Fe, Ru)	$[MM'Co_2(\mu_3\text{-}S)(\mu\text{-}AsMe_2)(CO)_8(\eta^5\text{-}C_5H_5)]$, CO	141
n = 4	$[M_2M'Pd(\mu\text{-}dppm)_2(CO)_5(\eta^5\text{-}C_5H_5)_2]$ (M = Mo, W; M′ = Pd, Pt)	$[MM'Pd(\mu_3\text{-}CO)_2(\mu\text{-}dppm)_2(Cl)(\eta^5\text{-}C_5H_5)]$, $[M(CO)_3(\eta^5\text{-}C_5H_5)]^-$	55
	$[MoCoMPd(\mu_3\text{-}CO)_2(\mu\text{-}dppm)_2(CO)_4(\eta^5\text{-}C_5H_5)]$ (M = Pd, Pt)	$[CoMPd(\mu_3\text{-}CO)_2(\mu\text{-}dppm)_2(Cl)(CO)]$, $[Mo(CO)_3(\eta^5\text{-}C_5H_5)]^-$	142
	$[MM'CoPd(\mu_3\text{-}CO)_2(\mu\text{-}dppm)_2(CO)_4(\eta^5\text{-}C_5H_5)]$ (M = Pd, Pt; M′ = Mo, W)	$[MM'Pd(\mu_3\text{-}CO)_2(\mu\text{-}dppm)_2(Cl)(CO)]$, $[Co(CO)_4]^-$	142
	$[MnCoMPd(\mu\text{-}CO)_2(\mu\text{-}dppm)_2(CO)_7]$ (M = Pd, Pt; X = Cl, I)	$[CoMPd(\mu_3\text{-}CO)_2(\mu\text{-}dppm)_2(X)(CO)]$, $[Mn(CO)_5]^-$	142, 143
	$[MnCoPdPt(\mu\text{-}dppm)_2(CO)_8]$	$[CoPdPtCl_2(\mu\text{-}dppm)_2]$, $[Mn(CO)_5]^-$	143
	$[Cr_3Co(\mu_3\text{-}S)_4(CO)(\eta^5\text{-}C_5H_5)_3]$	$[Cr_2Co(\mu_3\text{-}S)_2(\mu\text{-}SBu^t)(CO)_2(\eta^5\text{-}C_5H_5)_2]$, PhC_2Ph	144
	$[Cr_2Co_2(\mu_4\text{-}S)(\mu_3\text{-}S)_2(CO)_4(\eta^5\text{-}C_5H_5)(\eta^5\text{-}C_5H_4Bu^t)]$	$[Cr_4Ni(\mu_3\text{-}S)_4(\mu\text{-}SBu^t)_2(\eta^5\text{-}C_5H_5)_4]$, $[Co_2(CO)_8]$	145
	$[M_2Co_2(\mu_4\text{-}\eta^2\text{-}MeC_2Me)(CO)_8(\eta^5\text{-}C_5H_5)_2]$ (M = Mo, W; M′ = Ru, Os)	$[MCo_2(\mu_3\text{-}CMe)(CO)_8(\eta^5\text{-}C_5H_5)]$, $[M'(CO)_4]^{2-}$	30
	$[MoCo_3(\mu_3\text{-}As)(\mu_3\text{-}CO)_2(\eta^5\text{-}C_5H_5)_3(\eta^5\text{-}L)]$ (L = C_5H_5, C_5H_4Me)	$[Mo_3(\mu_3\text{-}As)(CO)_6(\eta^5\text{-}L)_3]$, $[Co(CO)_2(\eta^5\text{-}C_5H_5)]$	127
	$[Mn_2Pd_2(\mu\text{-}dppm)_2(CO)_9]$	$[MnPd_2(\mu\text{-}dppm)_2(Cl)(CO)_4]$, $[Mn(CO)_5]^-$	142
	$[MoRh_3(\mu_4\text{-}As)(\mu_3\text{-}CO)_2(\eta^5\text{-}C_5H_5)_4Rh(CO)(\eta^5\text{-}C_5H_5)]$	$[Mo_3(\mu_3\text{-}As)(CO)_6(\eta^5\text{-}C_5H_5)_3]$, $[Rh(CO)_2(\eta^5\text{-}C_5H_5)]$	65
	$[MoNi_3(\mu_3\text{-}CO)_3(\eta^5\text{-}C_5H_5)_4]$	$[Ni_3(\mu_3\text{-}CO)_2(\eta^5\text{-}C_5H_5)_3]$, $[Mo(CO)_3(\eta^5\text{-}C_5H_5)]_2$	146

$[Mn_2Pd_2(\mu\text{-}\eta^2\text{-CN})_2(CO)_6(\eta^5\text{-}C_5H_4Me)_2]_2$	$[Pd_4(\mu\text{-OAc})_4(\mu\text{-CO})_4]^{2+}$, $[Mn(CN)(CO)_2(\eta^5\text{-}C_5H_4Me)]^-$	147
$[M_2Pd_2(CO)_6(L)_2(\eta^5\text{-}C_5H_5)_2]$ (M = Cr, Mo, W; L = PMe_3, PEt_3, PBu^n_3, PMe_2Ph, PPh_3)	$[M_2Pd(CO)_6(NCPh)_2(\eta^5\text{-}C_5H_5)_2]$, L	49
$[M_2Pt_2(\mu_3\text{-CO})_2(\mu\text{-CO})_4(L)_2(\eta^5\text{-}C_5H_4R)_2]$ [M = Mo, W; R = H, Me; L = PCy_3, PPr^i_3, $PCyPh_2$, PBu^i_3, $P(C_6H_4Me\text{-}4)_3$, $P(C_6H_4Me\text{-}3)_3$, $Ph_2PCH_2PPh_2Mn(CO)_2(\eta^5\text{-}C_5H_4Me)$]	$[M_2Pt(CO)_6(NCPh)_2(\eta^5\text{-}C_5H_4R)_2]$, L	48, 51
$[W_2Pd_2(\mu_3\text{-CO})_2(\mu\text{-CO})_4(PPh_3)_2(\eta^5\text{-}C_5H_5)_2]$	$[WPd_2(\mu_3\text{-CO})(\mu\text{-CO})_2(\mu\text{-OH})(Ph)_2(PPh_3)_2(\eta^5\text{-}C_5H_5)]$, $[WH(CO)_3(\eta^5\text{-}C_5H_5)]$	140
$[M_2Pd_2(\mu_3\text{-CO})_2(\mu\text{-CO})_4(PR_3)_2(\eta^5\text{-}C_5H_5)_2]$ (M = Cr, Mo, W; R = Pr^i, Et)	$[MPd_2(\mu_3\text{-CO})(\mu\text{-Cl})(\mu\text{-CO})_2(PR_3)_2(\eta^5\text{-}C_5H_5)]$, $[M(CO)_3(\eta^5\text{-}C_5H_5)]^-$	59
$[Mo_2Pt_2(CO)_6(dppe)(\eta^5\text{-}C_5H_5)_2]$	$[Mo_2Pt(CO)_6(NCPh)_2(\eta^5\text{-}C_5H_5)_2]$, dppe	148
$[M_2Pt_2(\mu_3\text{-CR})(\mu\text{-CR})(CO)_4(\eta^4\text{-cod})(\eta^5\text{-L})_2]$ (M = Mo, W; R = C_6H_4Me-4, L = C_5H_5, C_5Me_5; R = Me, L = C_5Me_5)	$[M_2Pt(\mu\text{-CR})_2(CO)_4(\eta^5\text{-L})_2]$, $[Pt(\eta^4\text{-cod})_2]$	41, 76, 149, 150
$[MPt_3(\mu\text{-dppm})_3(CO)_3]^+$ (M = Mn, Re)	$[Pt_3(\mu_3\text{-CO})(\mu\text{-dppm})_3]^{2+}$, $[M(CO)_5]^-$	151
$[Re_3Pt(\mu\text{-H})(CO)_{13}]^{2-}$	$[\{Pt_3(\mu\text{-CO})_3(CO)_3\}_n]^{2-}$ ($n \approx 10$), $[Re_2(\mu\text{-H})(H)_2(CO)_8]^-$	152
$[Re_2Pt(\mu\text{-H})_2(CO)_9(\eta^4\text{-cod})\{M'\}]$ ([M′] = $[Mn(CO)_5]^-$, $[Re(CO)_5]^-$, $[W(CO)_3(\eta^5\text{-}C_5H_5)]^-$, $[Co(CO)_4]^-$; [M′]H = $[ReH(CO)_4(PPh_3)]$, $[ReH(CO)_3(PPh_3)_2]$)	$[Re_2Pt(\mu\text{-H})_2(CO)_8(\eta^4\text{-cod})]$, [M′], [M′]H	72, 73
$[M_2M'Pt(\mu_3\text{-CR})(\mu\text{-CR})(CO)_4(\eta^4\text{-cod})(L)_2]$ (M = Mo, W; M′ = Ni, Pt; L = $\eta^5\text{-}C_5H_5$, $\eta^5\text{-}C_5Me_5$; R = C_6H_4Me-4, Me)	$[M_2M'(\mu\text{-CR})_2(CO)_4(L)_2]$, $[Pt(\eta^4\text{-cod})_2]$	149, 150
$[MoWPt_2(\mu_3\text{-}\sigma\text{:}\eta^6\text{-7,9-}Me_2\text{-7,9-}C_2B_{10}H_8\text{-12-OC})(\mu_3\text{-}CC_6H_4Me\text{-4})(CO)_5(PMe_3)_2(\eta^5\text{-}C_5H_5)]$	$[MoW_2Pt_2(\mu_3\text{-}\sigma\text{:}\eta^6\text{-7,9-}Me_2\text{-7,9-}C_2B_{10}H_8\text{-12-OC})(\mu_3\text{-}CC_6H_4Me\text{-4})(\mu\text{-}CC_6H_4Me\text{-4})(\mu\text{-CO})_2(CO)_5(\eta^5\text{-}C_5H_5)_2]$, PMe_3	153
$[MoFeCo_2(\mu_3\text{-}CCO_2Pr^i)(CO)_{11}(\eta^5\text{-}C_5H_5)]$	$[MoCo_2(\mu_3\text{-}CCO_2Pr^i)(CO)_8(\eta^5\text{-}C_5H_5)]$, $[Fe_2(\mu\text{-CO})_3(CO)_6]$	56

(Continued)

TABLE VII (*Continued*)

M_n	Product Cluster	Precursor Cluster, Reagent	Ref.
	$MoFeCo_2(\mu_4\text{-}CCO_2Pr^i)(CO)_5(\eta^5\text{-}C_5H_5)_3]$	$[MoCo_2(\mu_3\text{-}CCO_2Pr^i)(\mu\text{-}CO)(CO)_2(\eta^5\text{-}C_5H_5)_3]$, $[Fe_2(\mu\text{-}CO)_3(CO)_6]$	56
	$[MoFeRuCo(\mu_3\text{-}\eta^2\text{-}MeC_2Me)(CO)_{10}(\eta^5\text{-}C_5H_5)]$	$[MoRuCo(\mu_3\text{-}\eta^2\text{-}MeC_2Me)(CO)_8(\eta^5\text{-}C_5H_5)]$, $[Fe_2(\mu\text{-}CO)_3(CO)_6]$	154
	$[MRuCoRh(\mu_3\text{-}\eta^2\text{-}MeC_2Me)(CO)_7(\eta^5\text{-}C_5H_5)_2]$ (M = Mo, W)	$[MRuCo(\mu_3\text{-}\eta^2\text{-}MeC_2Me)(CO)_8(\eta^5\text{-}C_5H_5)]$, $[Rh(CO)_2(\eta^5\text{-}C_5H_5)]$	154
	$[MoFeNi_2(\mu_3\text{-}CCO_2Pr^i)(CO)_5(\eta^5\text{-}C_5H_5)_3]$	$[MoNi_2(\mu_3\text{-}CCO_2Pr^i)(CO)_2(\eta^5\text{-}C_5H_5)_3]$, $[Fe_2(\mu\text{-}CO)_3(CO)_6]$	56
n = 5	$[CrRh_4(\mu_5\text{-}\eta^2\text{:}\eta^6\text{-}PhC_2Ph)(\mu\text{-}CO)_2(CO)_{10}]$	$[Rh_4(CO)_{12}]$, $[Cr(CO)_3(\eta^6\text{-}PhC_2Ph)]$	155
	$[W_3Rh_2(\mu_3\text{-}CMe)(\mu\text{-}CMe)\{\mu\text{-}C(Me)C(O)\}(\mu\text{-}CO)_2(\mu\text{-}PPh_2)_2(CO)_2(\eta^5\text{-}C_5H_5)_3]$	$[W_2Rh_2(\mu_3\text{-}CMe)\{\mu\text{-}C(Me)C(O)\}(\mu\text{-}CO)(\mu\text{-}PPh_2)_2(CO)_2(\eta^5\text{-}C_5H_5)_2]$, $[W(\equiv CMe)(CO)_2(\eta^5\text{-}C_5H_5)]$	82, 102
	$[Mo_2Pt_3(\mu_3\text{-}CC_6H_4Me\text{-}4)_2(CO)_4(\eta^4\text{-cod})_2(\eta^5\text{-}C_5H_5)_2]$	$[Mo_2Pt(\mu\text{-}CC_6H_4Me\text{-}4)_2(CO)_4(\eta^5\text{-}C_5H_5)_2]$, $2[Pt(\eta^4\text{-cod})_2]$	156
	$[M_2Pt_3(\mu_3\text{-}CR)_2(CO)_4(\eta^4\text{-cod})_2(\eta^5\text{-}L)_2]$ (M = Mo, W; R = C_6H_4Me-4, Me, L = C_5H_5, C_5Me_5)	$[M_2Pt(\mu\text{-}CR)_2(CO)_4(\eta^5\text{-}L)_2]$, $2.5[Pt(\eta^4\text{-cod})_2]$	76, 149
	$[W_3Pt_2(\mu_3\text{-}CR)(\mu\text{-}CR)_2(CO)_6(L)_3]$ (R = C_6H_4Me-4, Me, L = $\eta^5\text{-}C_5H_5$, $\eta^5\text{-}C_5Me_5$)	$[W_2Pt_2(\mu_3\text{-}CR)(\mu\text{-}CR)(CO)_4(\eta^4\text{-cod})(L)_2]$, $[W(\equiv CR)(CO)_2(\eta^5\text{-}C_5H_5)]$	41, 149
	$[W_3Pt_2(\mu_3\text{-}CR)(\mu\text{-}CR')(\mu\text{-}CR'')(CO)_6(\eta^5\text{-}L)(\eta^5\text{-}L')_2]$ (R, R′, R″ = Me, C_6H_4Me-4; L, L′ = C_5H_5, C_5Me_5)	$[W_3Pt_2(\mu_3\text{-}CR)(\mu\text{-}CR')(CO)_4(\eta^4\text{-cod})(\eta^5\text{-}L)_2]$, $[W(\equiv CR)(CO)_2(\eta^5\text{-}L')]$	149
	$[W_3Pt_2(\mu_3\text{-}CR)_2(\mu\text{-}CR')(\mu\text{-}CR'')(CO)_6(\eta^5\text{-}L)(\eta^5\text{-}L')_2]$ (R, R′, R″ = Me, C_6H_4Me-4; L, L′ = C_5H_5, C_5Me_5)	$[W_2Pt_2(\mu_3\text{-}CR)(\mu\text{-}CR')_2(CO)_4(\eta^4\text{-cod})(\eta^5\text{-}L)_2]$, $[W(\equiv CR'')(CO)_2(\eta^5\text{-}L)$	149
	$[Re_2Pt(\mu\text{-}H)_2(CO)_9\{Re_2H(CO)_9\}]^-$	$[Re_2Pt(\mu\text{-}H)_2(CO)_8(\eta^4\text{-cod})]$, $[Re_2H(CO)_9]^-$	72
	$[Re_4Pt(\mu\text{-}H)_6(CO)_{16}]$	$[Re_2Pt(\mu\text{-}H)_2(CO)_8(\eta^4\text{-cod})]$, $[Re_2(\mu\text{-}H)_2(CO)_8]$	71
	$[Re_4Pt(CO)_{17}]^{2-}$	$[\{Pt_3(\mu\text{-}CO)_3(CO)_3\}_n]^{2-}$ ($n \approx 10$), $[Re(CO)_5]^-$	152
	$[Cr_2Fe_2M(\mu_3\text{-}S)_4(\mu\text{-}SBu^t)(CO)_6(\eta^5\text{-}L)_2]$	$[Fe_2(\mu\text{-}S)_2(CO)_6]$, $[Cr_2M(\mu_3\text{-}S)_2(\mu\text{-}SBu^t)(L)(\eta^5\text{-}L')_2]$ [M = Co, L = $(CO)_2$, L′ = C_5H_4Me; M = Rh, L = η^4-cod, L′ = C_5H_5]	157

	$[MoW_2Pt_2(\mu_3\text{-}\sigma{:}\eta^6\text{-}7,9\text{-}Me_2\text{-}7,9\text{-}C_2B_{10}H_8\text{-}12\text{-}OC)(\mu_3\text{-}CC_6H_4Me\text{-}4)(\mu\text{-}CC_6H_4Me\text{-}4)(\mu\text{-}CO)_2(CO)_5(\eta^5\text{-}C_5H_5)_2]$	$[MoPt_2(\mu_3\text{-}\sigma\text{-}\sigma'{:}\eta^6\text{-}7,9\text{-}Me_2\text{-}7,9\text{-}C_2B_{10}H_8)(\mu\text{-}CO)(CO)_2(\eta^4\text{-}cod)_2]$, $[W(\equiv CC_6H_4Me\text{-}4)(CO)_2(\eta^5\text{-}C_5H_5)]$	153
	$[Mo_2NiPt_2(\mu_3\text{-}CC_6H_4Me\text{-}4)_2(CO)_4(\eta^4\text{-}cod)_2(\eta^5\text{-}C_5H_5)_2]$	$[Mo_2NiPt(\mu_3\text{-}CC_6H_4Me\text{-}4)(\mu\text{-}CC_6H_4Me\text{-}4)(CO)_4(\eta^4\text{-}cod)(\eta^5\text{-}C_5H_5)_2]$, $[Pt(\eta^4\text{-}cod)_2]$	150
n = 6	$[W_3Pt_3(\mu_3\text{-}CMe)_2(\mu\text{-}CMe)(CO)_6(\eta^4\text{-}cod)(\eta^5\text{-}C_5Me_5)_3]$	$[W_3Pt_2(\mu_3\text{-}CMe)(\mu\text{-}CMe)_2(CO)_6(\eta^5\text{-}C_5Me_5)_3]$, $[Pt(\eta^4\text{-}cod)_2]$,$C_2H_4$	156
	$[RePt_5(\mu\text{-}O)_3(\mu\text{-}OH)(CO)_5(PCy_3)_4]^+$	$[Pt_3(\mu\text{-}CO)_3(PCy_3)_3]$, $[Re_2O_7]$	158
	$[W_2Mn_2Pt_2(\mu_3\text{-}CO)_2(\mu\text{-}CO)_4(CO)_4(\mu\text{-}dppm)_2(\eta^5\text{-}C_5H_4Me)_2]$	$[Mn(\eta^1\text{-}dppm)(CO)_2(\eta^5\text{-}C_5H_4Me)]$, $[W_2Pt(CO)_6(NCPh)_2(\eta^5\text{-}C_5H_4Me)_2]$	48
n = 7	$[W_3Ir_4(\mu\text{-}H)(CO)_{12}(\eta^5\text{-}C_5H_5)_3]$	$[W_2Ir_2(CO)_{10}(\eta^5\text{-}C_5H_5)_2]$	159
	$[W_4Ni_3(\mu_3\text{-}CMe)_2(\mu\text{-}CMe)_2(CO)_8(\eta^5\text{-}C_5Me_5)_4]$	$[W_2Ni(\mu\text{-}CMe)_2(CO)_4(\eta^5\text{-}C_5Me_5)_2]$, $[Ni(\eta^4\text{-}cod)_2]$	156
	$[W_4Pt_3(\mu_3\text{-}CC_6H_4Me\text{-}4)_2(\mu\text{-}CC_6H_4Me\text{-}4)_2(CO)_8(\eta^5\text{-}C_5H_5)_4]$	$[W_2Pt_3(\mu_3\text{-}CC_6H_4Me\text{-}4)_2(CO)_4(\eta^4\text{-}cod)_2(\eta^5\text{-}C_5H_5)_2]$, $[W(\equiv CC_6H_4Me\text{-}4)(CO)_2(\eta^5\text{-}C_5H_5)]$	156, 160
	$[W_4Pt_3(\mu_3\text{-}CMe)_2(\mu\text{-}CMe)_2(CO)_8(\eta^5\text{-}L)(\eta^5\text{-}C_5Me_5)_3](L = C_5H_5, C_5Me_5)$	$[W_3Pt_3(\mu_3\text{-}CMe)_2(\mu\text{-}CMe)(CO)_6(\eta^4\text{-}cod)(\eta^5\text{-}C_5Me_5)_3]$, $[W(\equiv CMe)(CO)_2(\eta^5\text{-}L)]$ (3 atm C_2H_4)	156
	$[Mo_2WPt_3(\mu_3\text{-}CC_6H_4Me\text{-}4)_2(\mu\text{-}CMe)(CO)_6(\eta^4\text{-}cod)(\eta^5\text{-}C_5H_5)_3]$	$[Mo_2Pt_3(\mu_3\text{-}CC_6H_4Me\text{-}4)_2(CO)_4(\eta^4\text{-}cod)_2(\eta^5\text{-}C_5H_5)_2]$, $[W(\equiv CMe)(CO)_2(\eta^5\text{-}C_5H_5)]$ (3 atm C_2H_4)	156
	$[Mo_2WPt_4(\mu_3\text{-}CC_6H_4Me\text{-}4)_2(\mu_3\text{-}CMe)(CO)_6(\eta^4\text{-}cod)_2(\eta^5\text{-}C_5H_5)_3]$	$[Mo_2WPt_3(\mu_3\text{-}CC_6H_4Me\text{-}4)_2(\mu\text{-}CMe)(CO)_6(\eta^4\text{-}cod)(\eta^5\text{-}C_5H_5)_3]$, 1.5 $[Pt(\eta^4\text{-}cod)_2]$	45, 46
	$[Mo_2W_2Pt_3(\mu_3\text{-}CC_6H_4Me\text{-}4)_2(\mu\text{-}CMe)_2(CO)_8(\eta^5\text{-}C_5H_5)(\eta^5\text{-}L)_2](L = C_5H_5, C_5Me_5)$	$[Mo_2Pt_3(\mu_3\text{-}CC_6H_4Me\text{-}4)_2(CO)_4(\eta^4\text{-}cod)_2(\eta^5\text{-}C_5H_5)_2]$, $[W(\equiv CMe)(CO)_2(\eta^5\text{-}L)]$ (3 atm C_2H_4)	156
	$[W_4NiPt_2(\mu_3\text{-}CMe)_2(\mu\text{-}CMe)_2(CO)_8(\eta^5\text{-}C_5Me_5)_4]$	$[W_2Pt(\mu\text{-}CMe)_2(CO)_4(\eta^5\text{-}C_5Me_5)_2]$, $[Ni(\eta^4\text{-}cod)_2]$	156
	$[Mo_2W_2NiPt_2(\mu_3\text{-}CC_6H_4Me\text{-}4)_2(\mu\text{-}CR)_2(CO)_8(\eta^5\text{-}C_5H_5)_2(\eta^5\text{-}L)_2]$	$[Mo_2NiPt_2(\mu_3\text{-}CC_6H_4Me\text{-}4)_2(CO)_4(\eta^4\text{-}cod)_2(\eta^5\text{-}C_5H_5)_2]$, $[W(\equiv CPh)(CO)_2(\eta^5\text{-}C_5H_5)]$ (2 atm C_2H_4) (R = Ph, L = C_5H_5); $[MoWPt(\mu\text{-}CC_6H_4Me\text{-}4)(\mu\text{-}CMe)(CO)_4(\eta^5\text{-}C_5H_5)(\eta^5\text{-}C_5Me_5)]$, $[Ni(\eta^4\text{-}cod)_2]$ (R = Me, L = C_5Me_5)	150

(Continued)

TABLE VII (*Continued*)

M_n	Product Cluster	Precursor Cluster, Reagent	Ref.
n = 8	$[Re_7Pd(\mu_6\text{-}C)(CO)_{21}(L)]^{2-}$ ($L = \eta^3$-1-phenylallyl, $\eta^3\text{-}C_3H_5$)	$[Re_7(\mu_6\text{-}C)(CO)_{21}]^{3-}$, $[Pd(L)Cl]_2$	161, 162
	$[Re_7Pt(\mu_6\text{-}C)(CO)_{21}(L)]^{2-}$ ($L = \eta^3$-1-methylallyl, Me_3)	$[Re_7(\mu_6\text{-}C)(CO)_{21}]^{3-}$, $[Pt(\eta^3\text{-1-methylallyl})Cl]_2$ or $[PtMe_3I]_4$	161
	$[Mo_4Pd_4(\mu_3\text{-}CO)_4(\mu\text{-}CO)_8(\eta^5\text{-}C_5H_5)_4]^{2-}$	$[Pd_4(\mu\text{-}OAc)_4(\mu\text{-}CO)_4]$, $[Mo(CO)_3(\eta^5\text{-}C_5H_5)]^-$	163
	$[W_4Ni_2Pt_2(\mu_3\text{-}CPh)_4(CO)_8(\eta^5\text{-}C_5H_5)_4]$, $[W_4Ni_2Pt_2(\mu_3\text{-}CPh)_3(\mu\text{-}CPh)(\mu\text{-}CO)(CO)_7(\eta^5\text{-}C_5H_5)_4]$	$[W_2Pt(\mu\text{-}CPh)(CO)_4(\eta^5\text{-}C_5H_5)_2]$, $[Ni(\eta^4\text{-cod})_2]$	164, 165
	$[M_4Ni_2Pt_2(\mu_3\text{-}CC_6H_4Me\text{-}4)_3(\mu\text{-}CC_6H_4Me\text{-}4)(\mu\text{-}CO)(CO)_7(\eta^5\text{-}C_5H_5)_4]$(M = Mo, W)	$[M_2Pt(\mu\text{-}CC_6H_4Me\text{-}4)_2(CO)_4(\eta^5\text{-}C_5H_5)_2]$, $[Ni(\eta^4\text{-cod})_2]$	150, 164, 165
	$[W_4Pt_4(\mu_3\text{-}CC_6H_4Me\text{-}4)_3(\mu\text{-}CC_6H_4Me\text{-}4)(\mu\text{-}CO)(CO)_7(\eta^5\text{-}C_5H_5)_4]$	$[W_4Pt_3(\mu_3\text{-}CC_6H_4Me\text{-}4)_2(\mu\text{-}CC_6H_4Me\text{-}4)_2(CO)_8(\eta^5\text{-}C_5H_5)_4]$, $[Pt(\eta^2\text{-}C_2H_4)_3]$ or $[Pt(\eta^4\text{-cod})_2]$; $[W_2Pt_2(\mu_3\text{-}CC_6H_4Me\text{-}4)(\mu\text{-}CC_6H_4Me\text{-}4)(CO)_4(\eta^5\text{-}C_5H_5)_2]$, C_2H_4	160, 165
	$[W_4NiPt_3(\mu_3\text{-}CC_6H_4Me\text{-}4)_3(\mu\text{-}CC_6H_4Me\text{-}4)(\mu\text{-}CO)(CO)_7(\eta^5\text{-}C_5H_5)_4]$	$[W_4Pt_3(\mu_3\text{-}CC_6H_4Me\text{-}4)_2(\mu\text{-}CC_6H_4Me\text{-}4)_2(CO)_8(\eta^5\text{-}C_5H_5)_4]$, $[Ni(\eta^4\text{-cod})_2]$	165
	$[M_2W_2Ni_2Pt_2(\mu_3\text{-}CR)_2(\mu_3\text{-}CR')_2(CO)_8(\eta^5\text{-}C_5H_5)_4]$ (M = Mo, W; R = Ph, C_6H_4Me-4; R′ = Me, C_6H_4Me-4)	$[MWPt(\mu\text{-}CR)(\mu\text{-}CR')(CO)_4(\eta^5\text{-}C_5H_5)_2]$, $[Ni(\eta^4\text{-cod})_2]$	150, 165
n = 9	$[Mo_2W_3Pt_4(\mu_3\text{-}CC_6H_4Me\text{-}4)_2(\mu_3\text{-}CMe)(\mu\text{-}CR)_2(CO)_{10}(\eta^5\text{-}C_5H_5)_5]$ (R = Me, C_6H_4Me-4)	$[Mo_2WPt_4(\mu_3\text{-}CC_6H_4Me\text{-}4)_2(\mu_3\text{-}CMe)(CO)_6(\eta^4\text{-cod})_2(\eta^5\text{-}C_5H_5)_3]$, 2 $[W(\equiv CR)(CO)_2(\eta^5\text{-}C_5H_5)]$	45, 46
n = 10	$[Mo_2W_3Pt_5(\mu_3\text{-}CC_6H_4Me\text{-}4)_3(\mu_3\text{-}CMe)(\mu\text{-}CC_6H_4Me\text{-}4)(CO)_{10}(\eta^4\text{-cod})(\eta^5\text{-}C_5H_5)_5]$	$[Mo_2W_3Pt_4(\mu_3\text{-}CC_6H_4Me\text{-}4)_2(\mu_3\text{-}CMe)(\mu\text{-}CC_6H_4Me\text{-}4)_2(CO)_{10}(\eta^5\text{-}C_5H_5)_5]$, 2.4 $[Pt(\eta^4\text{-cod})_2]$	46
n = 11	$[Mo_2W_3Pt_6(\mu_3\text{-}CC_6H_4Me\text{-}4)_4(\mu_3\text{-}CMe)(CO)_{10}(\eta^4\text{-cod})_2(\eta^5\text{-}C_5H_5)_5]$	$[Mo_2W_3Pt_4(\mu_3\text{-}CC_6H_4Me\text{-}4)_2(\mu_3\text{-}CMe)(\mu\text{-}CC_6H_4Me\text{-}4)_2(CO)_{10}(\eta^5\text{-}C_5H_5)_5]$, 2.4 $[Pt(\eta^4\text{-cod})_2]$	46
	$[Mo_2W_3Pt_6(\mu_3\text{-}CC_6H_4Me\text{-}4)_2(\mu_3\text{-}CMe)_3(CO)_{10}(\eta^4\text{-cod})_2(\eta^5\text{-}C_5H_5)_5]$	$[Mo_2W_3Pt_4(\mu_3\text{-}CC_6H_4Me\text{-}4)_2(\mu_3\text{-}CMe)(\mu\text{-}CMe)_2(CO)_{10}(\eta^5\text{-}C_5H_5)_5]$, 2.4 $[Pt(\eta^4\text{-cod})_2]$	45, 46

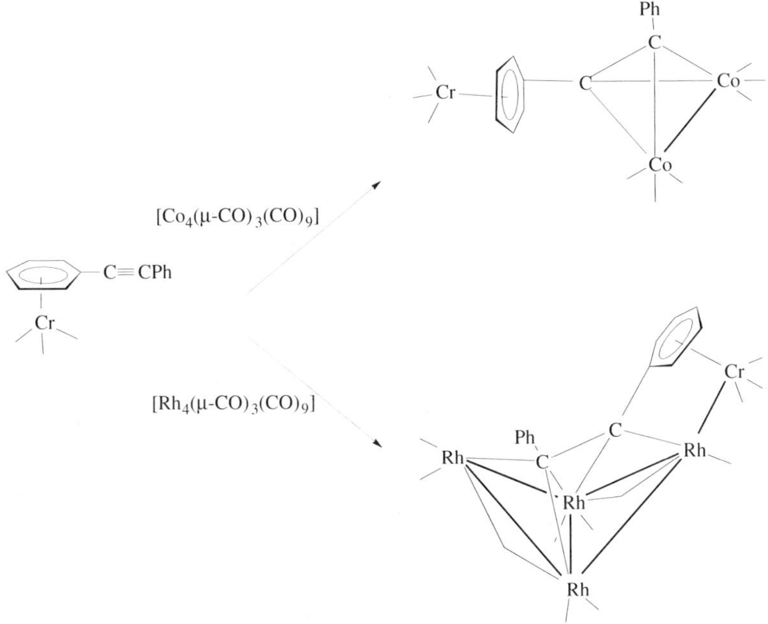

FIG. 48. Core expansion using a pendant acetylene group.

Expansion of the capped octahedral $[Re_7(\mu_6\text{-}C)(CO)_{21}]^{3-}$ with platinum group electrophiles has afforded the clusters $[Re_7M(\mu_6\text{-}C)(CO)_{21}(L)]^{2-}$ (M = Pd, L = C_3H_5;[161] M = Pd, L = η^3-1-phenylallyl;[162] M = Pt, L = η^3-1-methylallyl;[161] M = Pt, L = Me_3[161]), with 1,4-bicapped octahedral coordination geometries (Fig. 51).

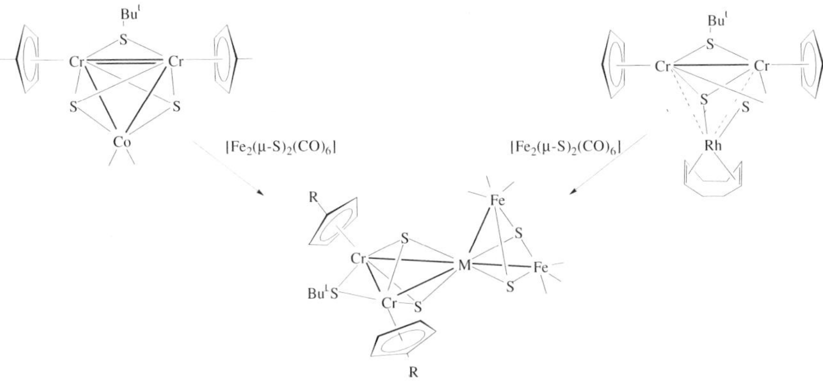

FIG. 49. Core expansion using $[Fe_2(\mu\text{-}S)_2(CO)_6]$: M = Co, R = Me; M = Rh, R = H.

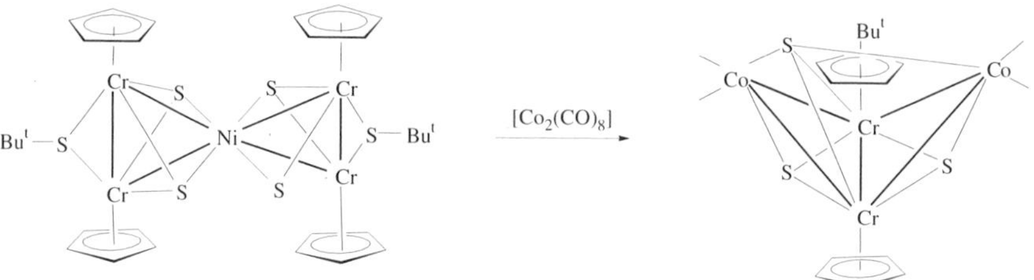

FIG. 50. Reaction of $[Cr_4Ni(\mu_3\text{-}S)_4(\mu\text{-}SBu^t)_2(\eta^5\text{-}C_5H_5)_4]$ with $[Co_2(CO)_8]$.

A number of mononuclear and binuclear rhenium complexes have been utilized in the formation and expansion of mixed platinum-rhenium clusters. Reaction of the coordinatively unsaturated cluster $[Pt_3(\mu_3\text{-}CO)(\mu\text{-}dppm)_3]^{2+}$ with $[M(CO)_5]^-$ (M = Re, Mn) afforded the tetrahedral clusters $[MPt_3(\mu\text{-}dppm)_3(CO)_3]^+$, unusually short metal–metal bond distances being noted for the rhenium example (Fig. 52).[151] The Re—Pt clusters formed from the columnar clusters $[\{Pt_3(\mu\text{-}CO)_3(CO)_3\}_n]^{2-}$ ($n \approx 10$) are dependent on the nature of the rhenium complexes utilized. For example, reaction of $[\{Pt_3(\mu\text{-}CO)_3(CO)_3\}_n]^{2-}$ with $[Re(CO)_5]^-$ afforded $[Re_4Pt(CO)_{17}]^{2-}$, with an edge-bridged butterfly ("swallow") core geometry,[152] whereas reaction with $[Re_2(\mu\text{-}H)(H)_2(CO)_8]^-$ afforded $[Re_3Pt(\mu\text{-}H)(CO)_{13}]^{2-}$ with a butterfly core geometry (Fig. 53).[152] $[Re_2Pt(\mu\text{-}H)_2(CO)_8(\eta_4\text{-}cod)]$ was prepared by reaction of $[Pt(\eta^4\text{-}cod)_2]$ and $[Re_2(\mu\text{-}H)_2(CO)_8]$; it reacted with $[Re_2(\mu\text{-}H)_2(CO)_8]$ in the presence of H_2 to afford the structurally characterized $[Re_4Pt(\mu\text{-}H)_6(CO)_{16}]$, with a bow-tie core geometry (Fig. 54).[71] In a related series of reactions, reaction of $[Re_2Pt(\mu\text{-}H)_2(CO)_8(\eta^4\text{-}cod)]$ with metal hydrides or metal carbonyl anions caused displacement of the η^4-cod ligand and expansion of the metal core nuclearity, affording the spiked triangular clusters $[Re_2Pt(\mu\text{-}H)_2(CO)_8\{M'\}]$ $[\{M'\} = Mn(CO)_5$, $Re(CO)_4(PPh_3)$, $Re(CO)_3(PPh_3)_2$, $W(CO)_3(\eta^5\text{-}C_5H_5)$, $Co(CO)_4]$ (Fig. 55).[72,73] High oxidation state rhenium precursors can also be successfully employed; reaction of $[Pt_3(\mu\text{-}CO)_3(PCy_3)_3]$ with

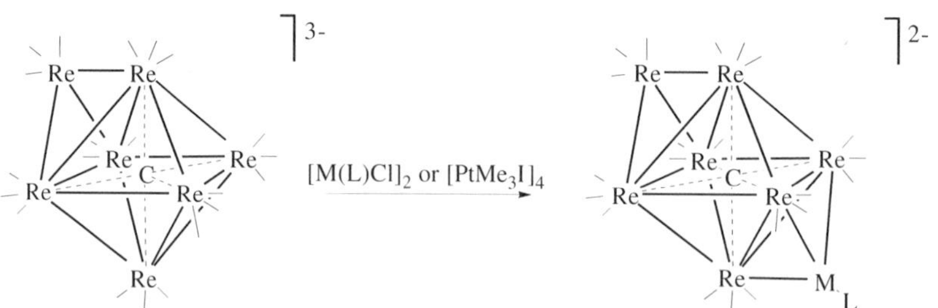

FIG. 51. Core expansion of $[Re_7(\mu_6\text{-}C)(CO)_{21}]^{3-}$ using platinum group electrophiles: M = Pd, L = (η^3-1-phenylallyl), M = Pt, L = (η^3-2-methylallyl), M = Pt, L = Me_3.

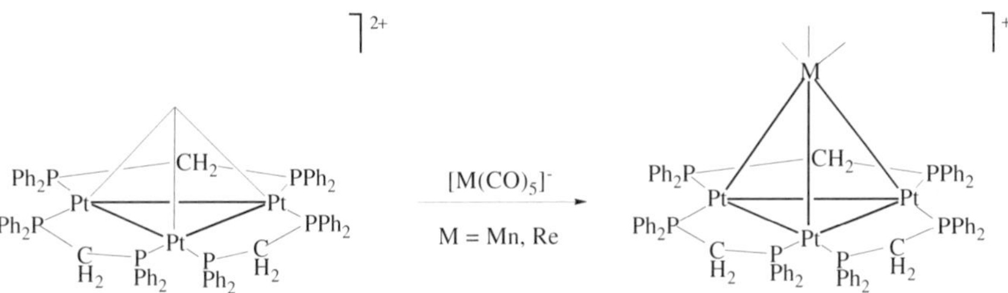

FIG. 52. Core expansion of $[Pt_3(\mu_3\text{-CO})(\mu\text{-dppm})_3]^{2+}$.

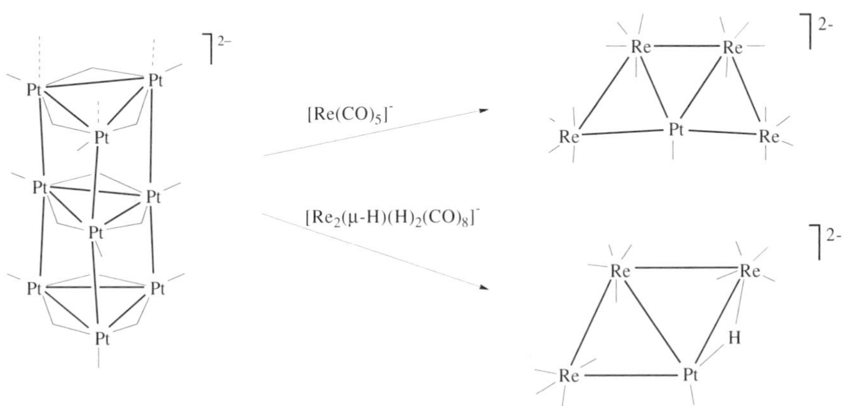

FIG. 53. Mixed rhenium-platinum cluster syntheses utilizing $[\{Pt(\mu\text{-CO})_3(CO)_{3n}\}]^{2-}$.

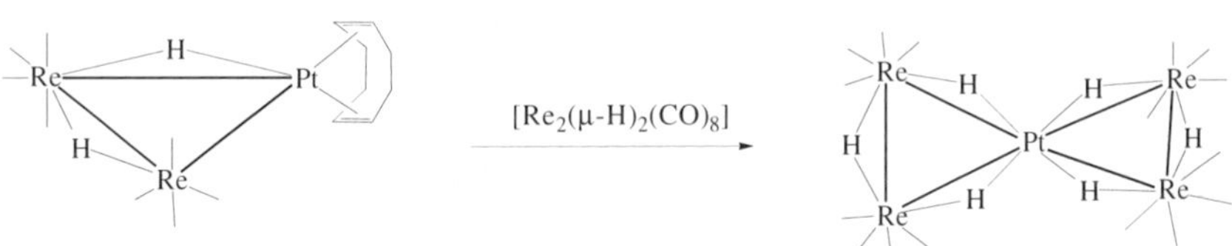

FIG. 54. Core expansion of $[Re_2Pt(\mu\text{-H})_2(CO)_8(\eta^4\text{-cod})]$.

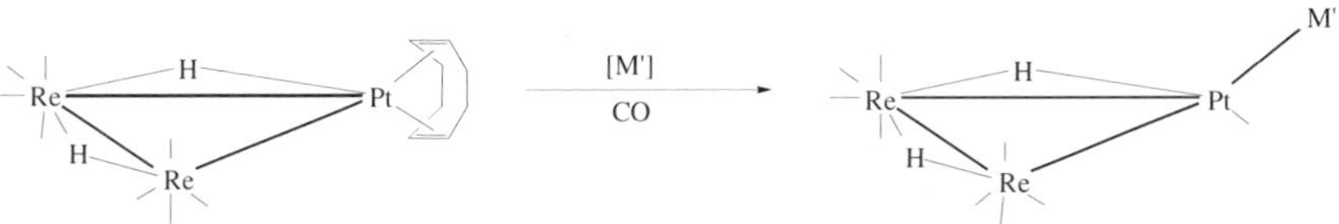

FIG. 55. Core expansion utilizing $[Re_2Pt(\mu\text{-H})_2(CO)_8(\eta^4\text{-cod})]$: $M' = [Mn(CO)_5]^-$, $[ReH(CO)_4(PPh_3)]$, $[ReH(CO)_3(PPh_3)_2]$, $[W(CO)_3(\eta^5\text{-}C_5H_5)]^-$, $[Co(CO)_4]^-$.

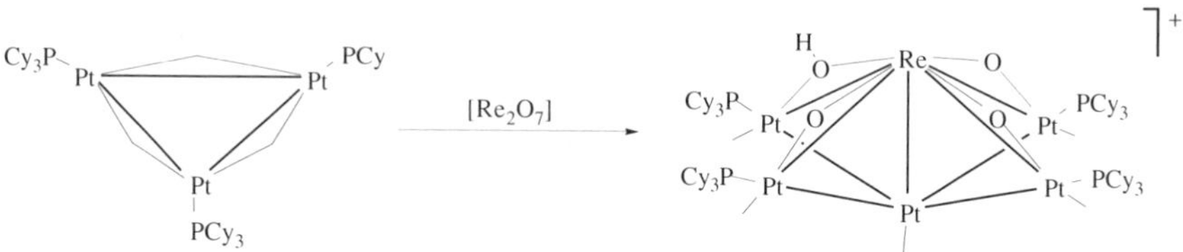

FIG. 56. Reaction of $[Pt(\mu\text{-}CO)_3(PCy_3)_3]$ with $[Re_2(CO)_7]$.

$[Re_2O_7]$ gave $[RePt_5(\mu\text{-}O)_3(\mu\text{-}OH)(CO)_5(PCy_3)_4]^+$, structurally characterized with the unusual bi-butterfly core geometry shown in Fig. 56.[158]

The rectangular cluster $[Pd_4(\mu\text{-}OAc)_4(\mu\text{-}CO)_4]^{2+}$ has acetate groups which are readily displaced; reaction with $[Mo(CO)_3(\eta^5\text{-}C_5H_5)]^-$ afforded $[Mo_4Pd_4(\mu_3\text{-}CO)_4(\mu\text{-}CO)_8(\eta^5\text{-}C_5H_5)_4]^{2-}$ with a tetra-edge-bridged square core geometry, and the palladium atoms in the unusual $+1/2$ oxidation state,[163,166] whereas reaction with $[Mn(CN)(CO)_2(\eta^5\text{-}C_5H_4Me)]^-$ gave $[Mn_4Pd_4(\mu\text{-}NC)_4(CO)_{12}(\eta^5\text{-}C_5H_4Me)_4]$,[147] which consists of two Mn_2Pd_2 chains linked together by bridging CN-moieties (Fig. 57). The V-shaped cluster $[WPd_2(\mu_3\text{-}CO)(\mu\text{-}CO)_2(\mu\text{-}OH)(Ph)_2(PPh_3)_2(\eta^5\text{-}C_5H_5)]$ reacted with $[WH(CO)_3(\eta^5\text{-}C_5H_5)]$ to afford the butterfly cluster $[W_2Pd_2(\mu_3\text{-}CO)_2(\mu\text{-}CO)_4(PPh_3)_2(\eta^5\text{-}C_5H_5)_2]$.[140] The closely related clusters $[M_2Pd_2(\mu_3\text{-}CO)_2(\mu\text{-}CO)_4(PR_3)_2(\eta^5\text{-}C_5H_5)_2]$ (M = Cr, Mo, W; R = Pr^i, Et) were

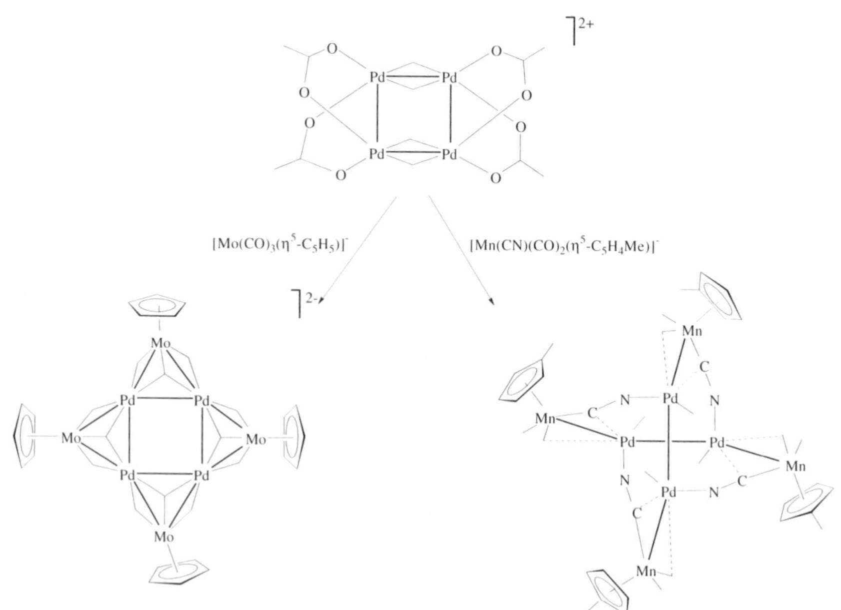

FIG. 57. Core expansion utilizing $[Pd_4(\mu\text{-}CO)_4(\mu\text{-}OAc)_4]^{2+}$.

formed by nucleophilic displacement of chloride by $[M(CO)_3(\eta^5\text{-}C_5H_5)]^-$ at $[MPd_2(\mu_3\text{-}CO)(\mu\text{-}Cl)(\mu\text{-}CO)_2(PR_3)_2(\eta^5\text{-}C_5H_5)]$.[59]

As was mentioned in the introduction to this section, purely thermolytic procedures have been seldom employed to afford higher nuclearity "very mixed"-metal clusters. Heating $[W_2Ir_2(CO)_{10}(\eta^5\text{-}C_5H_5)_2]$ in refluxing tetrahydrofuran afforded the heptanuclear cluster $[W_3Ir_4(\mu\text{-}H)(CO)_{12}(\eta^5\text{-}C_5H_5)_3]$ in a very poor (1%) yield.[159] On thermolysis, the clusters $[Mo_3(\mu_3\text{-}As)(CO)_6(\eta^5\text{-}C_5H_4R)_3]$ (R = H, Me) provide a source of $MoAs(\eta^5\text{-}C_5H_4R)$ fragments which can combine with suitable organometallic reagents; thus, reaction with $[Co(CO)_2(\eta^5\text{-}C_5H_5)]$ afforded $[MoCo_3(\mu_3\text{-}As)(\mu_3\text{-}CO)_2(\eta^5\text{-}C_5H_5)_3(\eta^5\text{-}C_5H_4R)]$,[127] whereas reaction with $[Rh(CO)_2(\eta^5\text{-}C_5H_5)]$ (R = H) gave $[MoRh_4(\mu_3\text{-}As)(\mu_3\text{-}CO)_2(CO)(\eta^5\text{-}C_5H_5)_5]$ (Fig. 58).[65] In contrast, photolysis of $[Mo_3(\mu_3\text{-}As)(CO)_6(\eta^5\text{-}C_5H_5)_3]$ in the

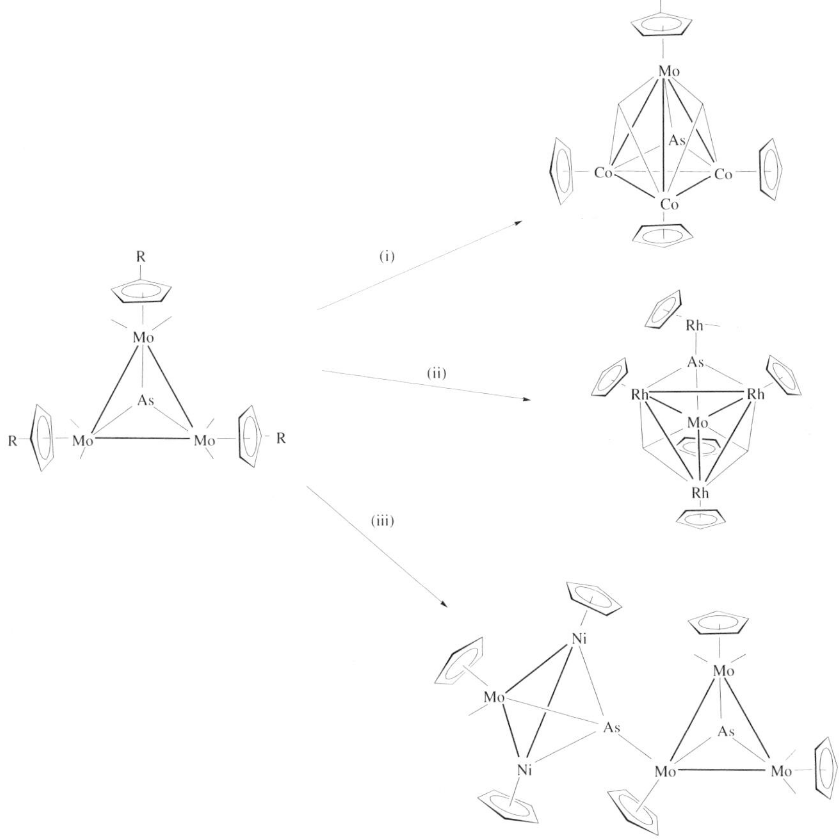

FIG. 58. Core expansion utilizing $[Mo_3(\mu_3\text{-}As)(CO)_6(\eta^5\text{-}C_5H_4R)_3]$: (i) Δ, $[Co(CO)_2(\eta^5\text{-}C_5H_5)]$, R = H, Me. (ii) Δ, $[Rh(CO)_2(\eta^5\text{-}C_5H_5)]$, R = H. (iii) $h\nu$, $[Ni(CO)(\eta^5\text{-}C_5H_5)]_2$, R = H.

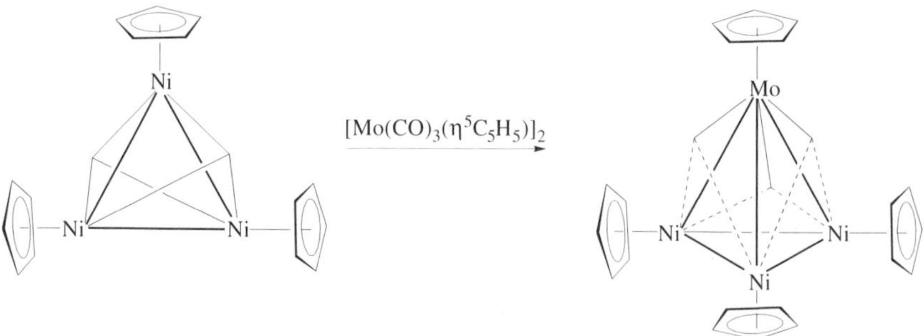

FIG. 59. Core expansion of $[Ni_3(\mu_3\text{-}CO)_2(\eta^5\text{-}C_5H_5)_3]$.

presence of $[Ni(CO)(\eta^5\text{-}C_5H_5)]_2$ afforded the metal exchange product $[MoNi_2(\mu_3\text{-}As)(CO)_4(\eta^5\text{-}C_5H_5)_3]$ together with $[Mo_4Ni_2(\mu_4\text{-}As)(\mu_3\text{-}As)(CO)_7(\eta^5\text{-}C_5H_5)_6]$, shown crystallographically to consist of Mo_3 and $MoNi_2$ triangles linked by a μ_4-As bridge.[128] The triangular cluster $[Ni_3(\mu_3\text{-}CO)_2(\eta^5\text{-}C_5H_5)_3]$ reacted with $[Mo(CO)_3(\eta^5\text{-}C_5H_5)]_2$ by core expansion rather than metal exchange, affording the paramagnetic $[MoNi_3(\mu_3\text{-}CO)_3(\eta^5\text{-}C_5H_5)_4]$ (Fig. 59); structural characterization of the analogue $[MoNi_3(\mu_3\text{-}CO)_3(\eta^5\text{-}C_5H_5)_3(\eta^5\text{-}C_5H_4Me)]$ (prepared by an alternative route from dinuclear precursors) confirmed the tetrahedral core geometry.[146]

Braunstein and co-workers have demonstrated the utility of carbonylmetalate anions and dppm-stabilized triangular mixed-metal clusters as precursors to clusters with spiked triangular core geometries, with the products retaining the face-capping and bridging ligands of the mixed-metal precursor (Fig. 60).[55,142,143] The exocyclic metal–metal bonds can be cleaved in some instances, carbon monoxide and halide reacting with the "spiked" triangular clusters by loss of the ligated metal spike.[142] Reaction of $[Mo_2Pt(CO)_6(NCPh)_2(\eta^5\text{-}C_5H_5)_2]$, with one equivalent of dppe proceeded with cluster core expansion to afford the spiked triangular cluster $[Mo_2Pt_2(CO)_6(dppe)(\eta^5\text{-}C_5H_5)_2]$, with the bidentate ligand chelating at a platinum

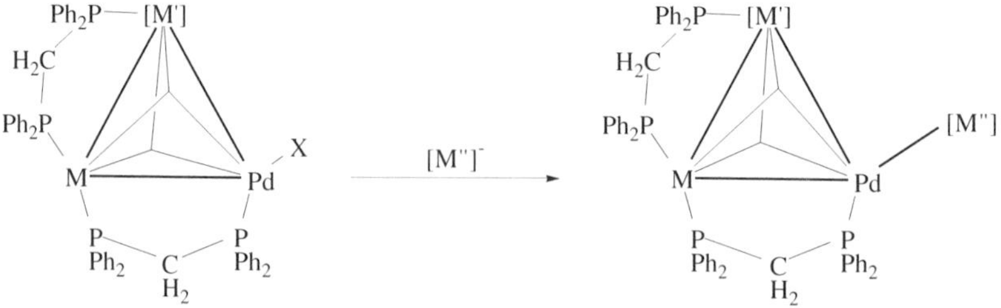

FIG. 60. Mixed-metal clusters from carbonylmetalate anions and dppm-stabilized triangular clusters: M = Pd, Pt, [M'] = $Mo(\eta^5\text{-}C_5H_5)$, $W(\eta^5\text{-}C_5H_5)$, $Mn(CO)_2$, $Co(CO)$, [M''] = $[Mo(CO)_3(\eta^5\text{-}C_5H_5)]$, $[W(CO)_3(\eta^5\text{-}C_5H_5)]$, $[Mn(CO)_5]$, $[Co(CO)_4]$, X = Cl, I.

atom.[148] Similarly, the palladium-containing analogues $[M_2Pd(CO)_6(NCPh)_2(\eta^5\text{-}C_5H_5)_2]$ (M = Cr, Mo, W) reacted with phosphines to give the isostructural $[M_2Pd_2(CO)_6(PR_3)_2(\eta^5\text{-}C_5H_5)_2]$.[49] In contrast, the platinum-containing chain complexes $[M_2Pt(CO)_6(NCPh)_2(\eta^5\text{-}C_5H_4R)_2]$ (M = Mo, W; R = H, Me) reacted with phosphines to afford the butterfly clusters $[M_2Pt_2(\mu_3\text{-}CO)(\mu\text{-}CO)_4(L)_2(\eta^5\text{-}C_5H_4R)_2]$ [L = phosphines, $Ph_2PCH_2PPh_2Mn(CO)_2(\eta^5\text{-}C_5H_4Me)$].[48,51]

As can be seen in Fig. 40, a significant number of mixed-metal clusters have been derived from the triangular cluster $[Co_3(\mu_3\text{-}CR)(CO)_9]$. Whereas further treatment with molybdenum, tungsten, or nickel complexes usually induces additional metal exchange, treatment of $[MoCo_2(\mu_3\text{-}CCO_2Pr^i)(CO)_8(\eta^5\text{-}C_5H_5)]$, $[MoNi_2(\mu_3\text{-}CCO_2Pr^i)(CO)_2(\eta^5\text{-}C_5H_5)_3]$, or $[MoCo_2(\mu_3\text{-}CCO_2Pr^i)(\mu\text{-}CO)(CO)_2(\eta^5\text{-}C_5H_5)_3]$ with $[Fe_2(\mu\text{-}CO)_3(CO)_6]$ resulted in cluster expansion forming $[MoFeCo_2(\mu_4\text{-}CCO_2Pr^i)(CO)_{11}(\eta^5\text{-}C_5H_5)]$, $[MoFeNi_2(\mu_3\text{-}CCO_2Pr^i)(CO)_5(\eta^5\text{-}C_5H_5)_3]$, and $[MoFeCo_2(\mu_3\text{-}CCO_2Pr^i)(CO)_5(\eta^5\text{-}C_5H_5)_3]$, respectively, with butterfly coordination geometries (Fig. 61).[56] Similarly, photolysis of $[MRuCo(\mu_3\text{-}\eta^2\text{-}MeC_2Me)(CO)_8(\eta^5\text{-}C_5H_5)]$ (M = Mo) in the presence of $[Fe_2(\mu\text{-}CO)_3(CO)_6]$ afforded $[MoFeRuCo(\mu_3\text{-}\eta^2\text{-}MeC_2Me)(CO)_{10}(\eta^5\text{-}C_5H_5)]$, and both the molybdenum and tungsten-containing clusters reacted with $[Rh(CO)_2(\eta^5\text{-}C_5H_5)]$ to form

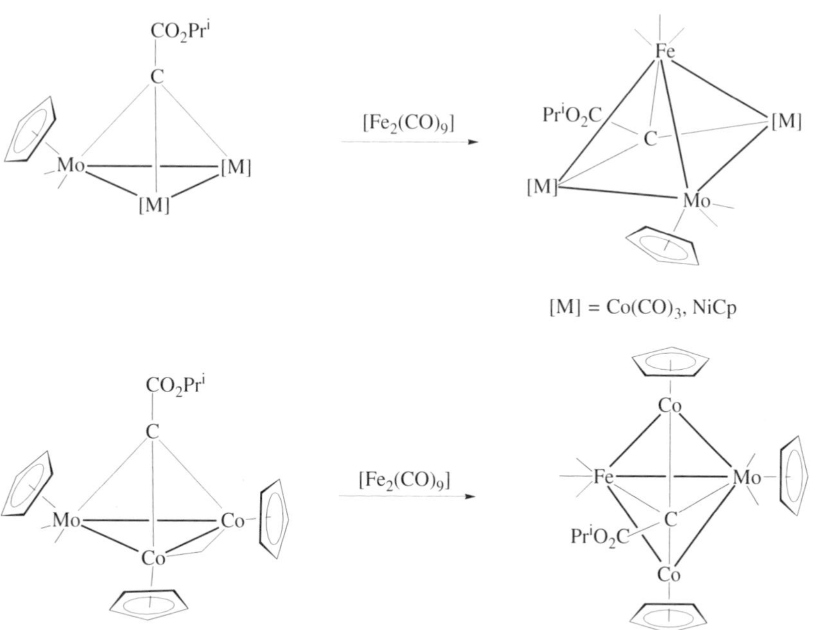

FIG. 61. Core expansion of alkylidyne-capped clusters.

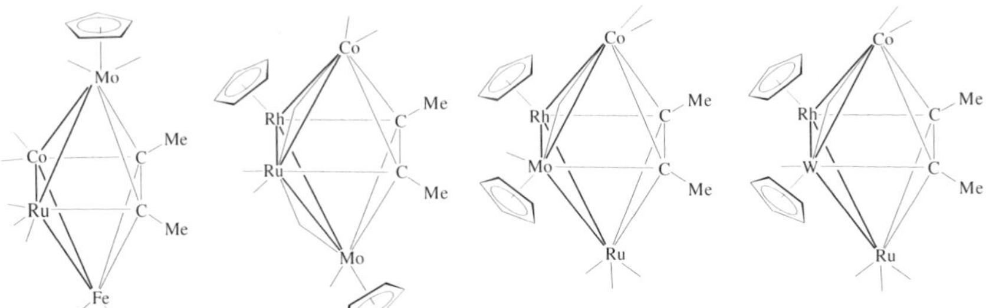

FIG. 62. Core expansion of alkyne-capped clusters.

$[MRuCoRh(\mu_3\text{-}\eta^2\text{-}MeC_2Me)(CO)_7(\eta^5\text{-}C_5H_5)_2]$ (M = Mo, W) (Fig. 62).[154] In a similar fashion, metal exchange with $[Ru(CO)_4]^{2-}$ and $[Os(CO)_4]^{2-}$ at $[MCo_2(\mu_3\text{-}CMe)(CO)_8(\eta^5\text{-}C_5H_5)]$ (M = Mo, W) afforded as by-products the core-expanded products $[M_2Co_2(\mu_4\text{-}\eta^2\text{-}MeC_2Me)(CO)_8(\eta^5\text{-}C_5H_5)_2]$.[30] The tetranuclear clusters $[MM'Co_2(\mu_3\text{-}E)(\mu\text{-}AsMe_2)(CO)_8(\eta^5\text{-}C_5H_5)]$ (M = Mo, W; M′ = Fe, Ru, Co; E = S, PR, CR; R = Me, Ph, Bu^t; not all combinations) and $[MM'M''Co(\mu_3\text{-}S)(\mu\text{-}AsMe_2)(CO)_7(\eta^5\text{-}C_5H_5)_2]$ (M, M′ = Mo, W; M″ = Fe, Ru; not all combinations) reacted with CO under slightly elevated pressures with core fragmentation and elimination of a Co—As unit to afford trinuclear products; at normal pressure, tetranuclear intermediates corresponding to addition of CO and rupture of two metal–metal bonds are observed.[141]

Stone and co-workers have reported the systematic expansion of group 6–group 10 clusters using $[M(\eta^2\text{-cod})_2]$ (M = Ni, Pt) and $[LM'(CO)_2(\equiv CR)]$ (M′ = Mo, W; L = η^5-C_5H_5, η^5-C_5Me_5; R = Me, Ph, C_6H_4Me-4). This work has been reviewed elsewhere[3] and will be summarized only briefly here. Figure 63 reveals the logical formation of chain complexes of three, four, five, six, seven, nine, and eleven metal atoms under mild conditions,[41,45,46,149,150,153,156,160,165] and this methodology has been extended to embrace other metal-ligand systems (e.g. $[Ru(CO)_4(\eta^2\text{-}C_2H_4)]$, a source of the $Ru(CO)_4$ fragment).[40] More recently, this work has been expanded to include metallacarborane cluster compounds with the preparation of $[MoW_2Pt_2(\mu_3\text{-}\eta^7\text{-}7,9\text{-}Me_2\text{-}7,9\text{-}C_2B_{10}H_8\text{-}12\text{-}OC)(\mu_3\text{-}CC_6H_4Me\text{-}4)(\mu\text{-}CC_6H_4Me\text{-}4)(\mu\text{-}CO)_2(CO)_5(\eta^5\text{-}C_5H_5)_2]$ (Fig. 64),[153] and tungsten-rhodium examples, with the preparation of $[W_3Rh_2(\mu_3\text{-}CMe)(\mu\text{-}CMe)\{\mu\text{-}C(Me)C(O)\}(\mu\text{-}CO)_2(\mu\text{-}PPh_2)_2(CO)_2(\eta^5\text{-}C_5H_5)_3]$ from $[W_2Rh_2(\mu_3\text{-}CMe)\{\mu\text{-}C(Me)C(O)\}(\mu\text{-}CO)(\mu\text{-}PPh_2)_2(CO)_2(\eta^5\text{-}C_5H_5)_2]$.[82,102] Attempts to extend the lengths of chains beyond seven metal atoms resulted in chain cyclization to form macrocyclic "star" clusters in instances where the heptametallic precursor has terminal tungsten carbyne groups (Fig. 65),[150,160,164,165] but the chains can be successfully extended if the heptametallic intermediate has terminal $[Pt(\eta^4\text{-cod})]$ groups (Fig. 63), a result attributed to differing chain conformations.

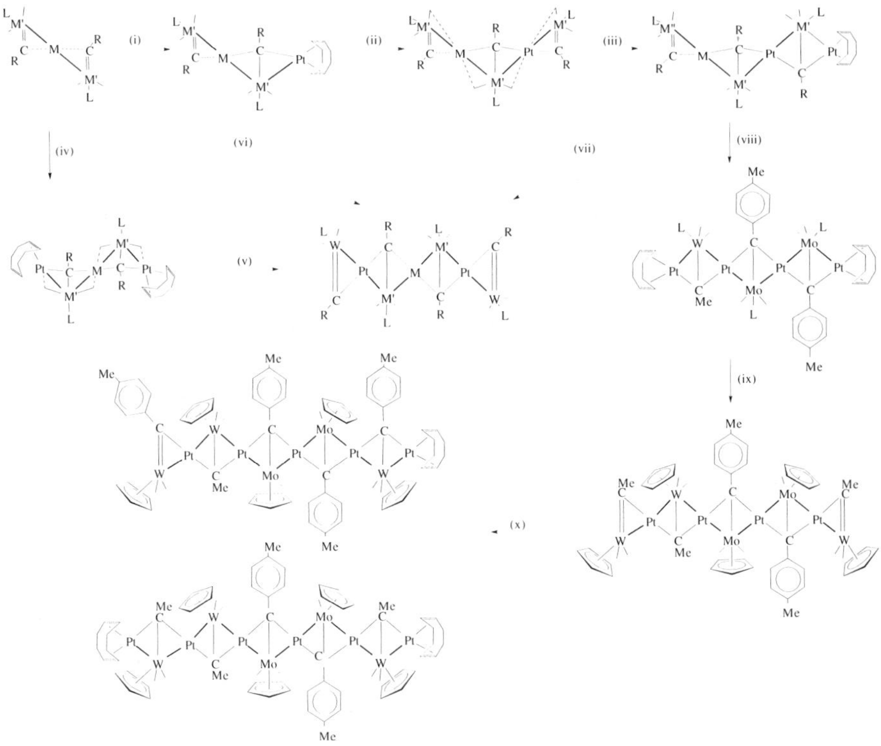

FIG. 63. Formation of chain clusters: (i) $[Pt(\eta^4\text{-cod})_2]$; M = Ni, Pt; M′ = Mo, W; R = Me, Ph, C_6H_4Me-4; L = η^5-C_5H_5, η^5-C_5Me_5. (ii) $[W(\equiv CR)(CO)_2(L)]$; M = Ni, Pt; M′ = Mo, W; R = C_6H_4Me-4, Me; L = η^5-C_5H_5, η^5-C_5Me_5; not all combinations. (iii) $[Pt(\eta^4\text{-cod})_2]$; M = Ni, Pt, M′ = Mo, W; R = Me , L = η^5-C_5Me_5. (iv) 2-2.5$[Pt(\eta^4\text{-cod})_2]$; M = Ni, Pt; M′ = Mo, W; R = C_6H_4Me-4, Me; L = η^5-C_5H_5, η^5-C_5Me_5. (v) $[W(\equiv CR)(CO)_2(L)]$, 3 atm C_2H_4; M = Ni, Pt; M′ = Mo, W; R = C_6H_4Me-4, Me; L = η^5-C_5H_5, η^5-C_5Me_5; not all combinations. (vi) $[Ni(\eta^4\text{-cod})_2]$; M = Ni, Pt; M′ = W; R = Me, L = η^5-C_5Me_5. (vii) $[W(\equiv CR)(CO)_2(L)]$, 3 atm C_2H_4; M = Pt; M′ = W; R = Me; L = η^5-C_5H_5, η^5-C_5Me_5. (viii) 1.5$[Pt(\eta^4\text{-cod})_2]$; M = Pt; M′ = Mo; R = C_6H_4Me-4, Me; L = η^5-C_5H_5, η^5-C_5Me_5. (ix) $2[W(\equiv CMe)(CO)_2(\eta^5\text{-}C_5H_5)]$; M′ = Mo, M″ = W; L = η^5-C_5H_5. (x) 2.4$[Pt(\eta^4\text{-cod})_2]$, 3 atm C_2H_4.

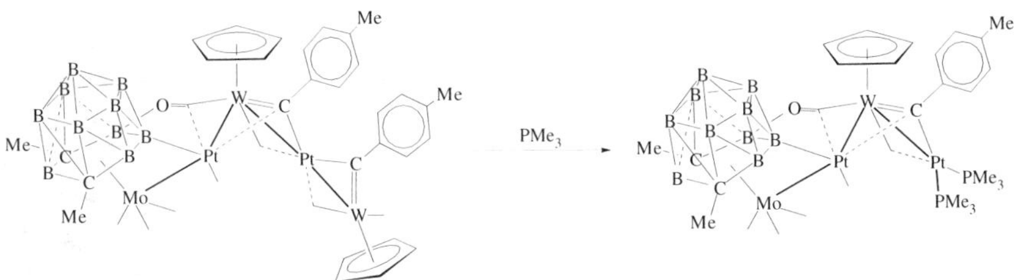

FIG. 64. Metallocarborane-containing mixed-metal cluster core contraction: B-bound H′s omitted for clarity.

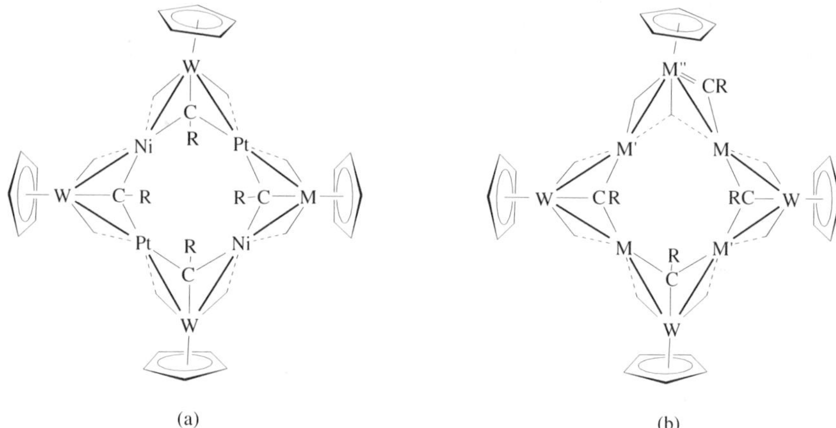

FIG. 65. "Star" clusters: (a) M = Mo, R = C_6H_4Me-4; M = W, R = Ph. (b) M = Pt, M′ = Ni, M″ = W, R = C_6H_4Me-4, Ph; M = Ni, M′ = Pt, M″ = W, R = C_6H_4Me-4, Ph; M = Ni, M′ = Pt, M″ = Mo, R = C_6H_4Me-4.

F. *Catalytic Studies*

Research into cluster catalysis has been driven by both intrinsic interest and utilitarian potential. Catalysis involving "very mixed"-metal clusters is of particular interest as many established heterogeneously catalyzed processes couple mid and late transition metals (e.g., hydrodesulfurization and petroleum reforming). Attempts to model catalytic transformations are summarized in Section II.F.1., while the use of "very mixed"-metal clusters as homogeneous and heterogeneous catalysis precursors are discussed in Sections II.F.2. and II.F.3., respectively. The general area of mixed-metal cluster catalysis has been summarized in excellent reviews by Braunstein and Rosé;[167,168] while the tabulated results are intended to be comprehensive in scope, the discussion below focuses on the more recent results.

1. *Modelling Catalysis*

Reactivity studies of organic ligands with mixed-metal clusters have been utilized in an attempt to shed light on the fundamental steps that occur in heterogeneous catalysis (Table VIII), although the correspondence between cluster chemistry and surface-adsorbate interactions is often poor.[169] While some of these studies have been mentioned in Section II.D., it is useful to revisit them in the context of the catalytic process for which they are models. Shapley and co-workers have examined the solution chemistry of tungsten-iridium clusters in an effort to understand hydrogenolysis of butane. The reaction of excess diphenylacetylene with

TABLE VIII

VERY MIXED-METAL CLUSTERS MODELING HETEROGENEOUS CATALYSIS

Metals	Precursor Cluster	Product(s)	Transformation Modeled	Ref.
Mo–Co	$[Mo_2Co_2(\mu_4\text{-}S)(\mu_3\text{-}S)_2(CO)_4(\eta^5\text{-}C_5H_4Me)_2]$ + RSH (R = Bu^t, $Me_2CHCH_2CH_2$, Ph)	$[Mo_2Co_2(\mu_3\text{-}S)_4(CO)_2(\eta^5\text{-}C_5H_4Me)_2]$ + RH	Desulfurization	68, 103, 104
	$[Mo_2Co_2(\mu_4\text{-}S)(\mu_3\text{-}S)_2(CO)_4(\eta^5\text{-}C_5H_4Me)_2]$ + *cis*-2,3-dimethylthiirane	$[Mo_2Co_2(\mu_3\text{-}S)_4(CO)_2(\eta^5\text{-}C_5H_4Me)_2]$ + *cis*-2-butene	Desulfurization	68, 103
	$[Mo_2Co_2(\mu_4\text{-}S)(\mu_3\text{-}S)_2(CO)_4(\eta^5\text{-}C_5H_4Me)_2]$ + diphenylthietane (*cis* or *trans*)	$[Mo_2Co_2(\mu_3\text{-}S)_4(S)(\eta^5\text{-}C_5H_4Me)_2]$ + *cis*-1,3-diphenylpropene	Desulfurization	103
	$[Mo_2Co_2(\mu_4\text{-}S)(\mu_3\text{-}S)_2(CO)_4(\eta^5\text{-}C_5H_4Me)_2]$ + $SCNBu^t$	$[Mo_2Co_2(\mu_4\text{-}S)(\mu_3\text{-}S)_2(CO)_3(CNR)(\eta^5\text{-}C_5H_4Me)_2]$, $[Mo_2Co_2(\mu_3\text{-}S)_4(CNBu^t)(CO)(\eta^5\text{-}C_5H_4Me)_2]$	Desulfurization	68, 103
	$[Mo_2Co_2(\mu_4\text{-}S)(\mu_3\text{-}S)_2(CO)_4(\eta^5\text{-}C_5H_4Me)_2]$ + thiophenes	Saturated and unsaturated C_1–C_4 hydrocarbons	Desulfurization	68, 103
	$[Mo_2Co_2(\mu_3\text{-}S)_4(CO)_2(\eta^5\text{-}C_5H_4Et)_2]$ + PhSSPh	$[Mo_2Co_2(\mu_3\text{-}S)_4(SPh)_2(\eta^5\text{-}C_5H_4Et)_2]$	Desulfurization	63, 64
	$[Mo_2Co_2(\mu_3\text{-}S)_4(CO)_2(\eta^5\text{-}C_5H_4Et)_2]$ + PhSH	$[Mo_2Co_2(\mu_3\text{-}S)_4(SPh)_2(\eta^5\text{-}C_5H_4Et)_2]$	Desulfurization	63, 64
	$[Mo_2Co_2(\mu_3\text{-}S)_4(CO)_2(\eta^5\text{-}C_5H_4Et)_2]$ + PhSH + CO	$[Mo_2Co_2(\mu_3\text{-}S)_4(SPh)_2(\eta^5\text{-}C_5H_4Et)_2]$, PhSSPh, PhS(CO)Ph	Desulfurization	63, 64
	$[Mo_2Co_2(\mu_3\text{-}S)_4(SPh)_2(\eta^5\text{-}C_5H_4Et)_2]$ + H_2 + CO	PhSSPh	Desulfurization	63, 64
W–Ir	$[WIr_3(CO)_{11}(\eta^5\text{-}C_5H_5)]$ + PhC≡CPh	$[WIr_3(\mu_3\text{-}CPh)(\mu\text{-}CPh)(\mu\text{-}\eta^4\text{-}CPhCPhCPhCPh)(CO)_5(\eta^5\text{-}C_5H_5)]$	Hydrogenolysis	62
	$[W_2Ir_2(CO)_{10}(\eta^5\text{-}C_5H_5)_2]$ + PhC≡CPh	$[W_2Ir_2(\mu_3\text{-}CPh)(\mu_3\text{-}\eta^2\text{-}CPhCPhCPh)(\mu\text{-}CO)_2(CO)_4(\eta^5\text{-}C_5H_5)_2]$	Hydrogenolysis	78
Re–Pt	$[RePt_3(\mu\text{-}dppm)_3(CO)_3]^+$ + propylene sulfide	$[RePt_3(\mu\text{-}dppm)_3(CO)_3(S)]^+$ + propene	Desulfurization	105, 106
	$[RePt_3(\mu\text{-}dppm)_3(CO)_3]^+$ + excess propylene sulfide	$[RePt_3(\mu_3\text{-}S)_2(\mu\text{-}dppm)_3(CO)_3]^+$ + propene	Desulfurization	105, 106
	$[RePt_3(\mu\text{-}dppm)_3(O)_3]^+$ + excess propylene sulfide	$[RePt_3(\mu_3\text{-}S)_2(\mu\text{-}dppm)_3(O)_3]^+$ + propene	Desulfurization	106

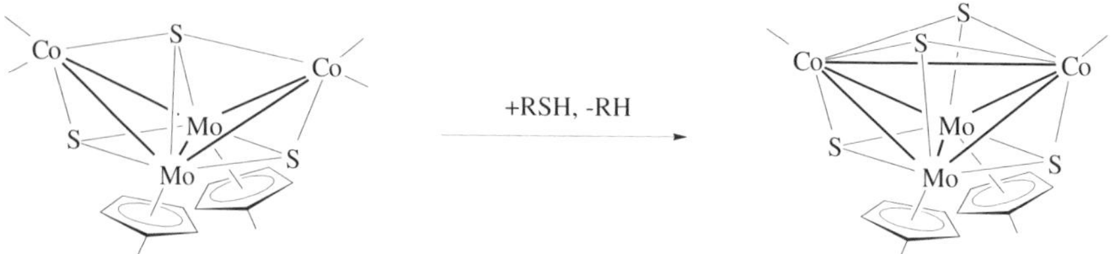

FIG. 66. Desulfurization of thiols by $[Mo_2Co_2(\mu_4\text{-}S)(\mu_3\text{-}S)_2(CO)_4(\eta^5\text{-}C_5H_4Me)_2]$.

$[WIr_3(CO)_{11}(\eta^5\text{-}C_5H_5)]$ afforded two products, one of which, $[WIr_3(\mu_3\text{-}CPh)(\mu\text{-}CPh)(\mu\text{-}\eta^4\text{-}CPhCPhCPhCPh)(CO)_5(\eta^5\text{-}C_5H_5)]$, has an open (butterfly) core with the tungsten atom at the "hinge" position; the alkyne ligands have undergone either C≡C scission to give μ_2- and μ_3-coordinated alkylidyne moieties or dimerization via C—C coupling to give an iridacyclopentadienyl system (see Fig. 27).[62] The related cluster $[W_2Ir_2(CO)_{10}(\eta^5\text{-}C_5H_5)_2]$ has also been reacted with diphenylacetylene and yielded two products, one of which has allylic and phenylmethylidyne fragments resulting from C≡C scission and C—C bond formation.[78] These transformations are believed to illustrate processes involved in butane hydrogenolysis over the cluster-derived catalysts, as experimental studies suggest C—C cleavage of an adsorbed dehydrogenated intermediate in ethane hydrogenolysis.[62]

Curtis and co-workers have studied the reaction of organosulfur compounds with $[Mo_2Co_2(\mu_4\text{-}S)(\mu_3\text{-}S)_2(CO)_4(\eta^5\text{-}C_5H_4Me)_2]$ and $[Mo_2Co_2(\mu_3\text{-}S)_4(CO)_2(\eta^5\text{-}C_5H_4Et)_2]$ to understand the desulfurization process. The reaction pathways that lead to C—S scission have been compared to the commercial hydrodesulfurization reactions with alumina-supported "MoCoS" catalysts. The bimetallic cluster $[Mo_2Co_2(\mu_4\text{-}S)(\mu_3\text{-}S)_2(CO)_4(\eta^5\text{-}C_5H_4Me)_2]$ desulfurized thiols to parent hydrocarbons and the cubane cluster $[Mo_2Co_2(\mu_3\text{-}S)_4(CO)_2(\eta^5\text{-}C_5H_4Me)_2]$ (Fig. 66); thiophene desulfurization products were saturated and unsaturated C_1—C_4 hydrocarbons, and isothiocyanates were desulfurized to RNC, which replaced a carbonyl ligand affording isocyanide-coordinated clusters $[Mo_2Co_2(\mu_4\text{-}S)(\mu_3\text{-}S)_2(CO)_3(CNR)(\eta^5\text{-}C_5H_4Me)_2]$ and $[Mo_2Co_2(\mu_3\text{-}S)_4(CNR)(CO)(\eta^5\text{-}C_5H_4Me)_2]$.[68,103] Kinetic and mechanistic details of the thiol desulfurization reactions have also been investigated.[104]

Reaction of thiophenol with $[Mo_2Co_2(\mu_3\text{-}S)_4(CO)_2(\eta^5\text{-}C_5H_4Et)_2]$ formed the electron-deficient, paramagnetic cluster $[Mo_2Co_2(\mu_3\text{-}S)_4(SPh)_2(\eta^5\text{-}C_5H_4Et)_2]$, which, in the presence of CO, regenerated the original carbonyl cluster and afforded PhSSPh. The combination of these two reactions constitutes the basis of a catalytic cycle (Fig. 67).[63,64]

Supported bimetallic Re—Pt catalysts are important in selective reforming of petroleum. It is believed that sulfiding the catalyst before use gives ReS units which act as inert diluents to reduce the size of a local ensemble of platinum atoms. Selectivity for desirable dehydrocyclization and isomerization reactions

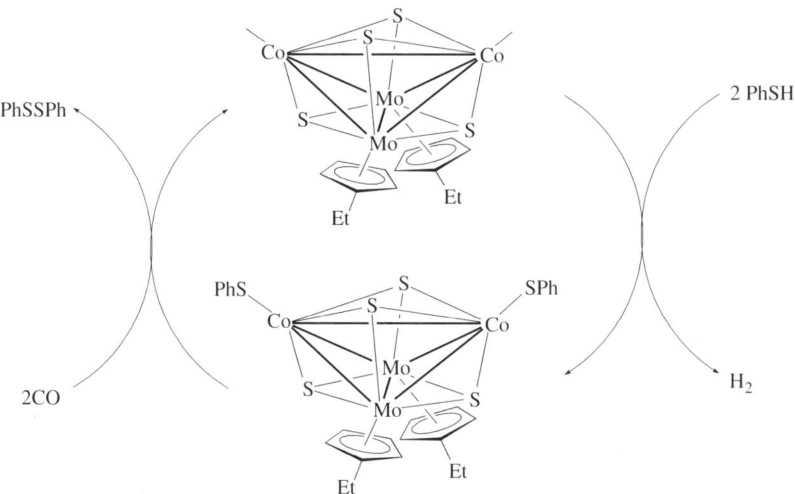

FIG. 67. Conversion of thiophenol to diphenyl disulfide by $[Mo_2Co_2(\mu_3\text{-}S)_4(CO)_2(\eta^5\text{-}C_5H_4Et)_2]$.

is thereby increased, and hydrogenolysis of alkanes, which requires several adjacent platinum atoms, is diminished. Metalloselective sulfidation of rhenium in Re—Pt clusters has been studied as evidence for these hypotheses. On reaction with propylene sulfide, one or two sulfur atoms were added to the tetrahedral cluster cation $[RePt_3(\mu\text{-dppm})_3(CO)_3]^+$, to give $[RePt_3(\mu\text{-dppm})_3(S)_n(CO)_3]^+$ (n = 1 or 2). Reaction under mild conditions of one equivalent of propylene sulfide gave a cluster with a rhenium-coordinated sulfur atom and propene; a second propylene sulfide was then desulfurized to give a cluster with two face-capping sulfur atoms bridging $RePt_2$ faces, and another propene molecule.[105,106] A similar sulfidation of the oxo clusters $[RePt_3(\mu_3\text{-}O)(\mu\text{-dppm})_3(CO)_3]^+$ and $[RePt_3(\mu_3\text{-}O)_2(\mu\text{-dppm})_3(CO)_3]^+$ with propylene sulfide has recently been studied; an equivalent of propene was generated and the sulfur coordinated in a μ_3-fashion with a decrease in metal–metal bonding in the cluster complexes (Fig. 68).[106]

2. *Homogeneous Catalysis*

Homogeneous catalysis by transition metal clusters has been reviewed from the perspective of the specific transformations.[170] Examples of very mixed-metal clusters catalyzing processes homogeneously are collected in Table IX. As is generally the case with homogeneous catalysis, the catalytic precursor is well defined, but the nature of the active catalyst is unclear.

Sulfided bimetallic clusters which mimic the metal composition of commercial hydrodesulfurization (HDS) catalysts have been prepared and their homogeneous catalytic behavior studied. Reaction of thiophenol with $[Mo_2Co_2(\mu_4\text{-}S)$

TABLE IX

HOMOGENEOUS CATALYSIS EMPLOYING VERY MIXED-METAL CLUSTER PRECURSORS

Catalyst	Precursor Cluster	Ref. (Cluster Syntheses)	Catalyzed Reaction	Ref. (Catalysis Studies)
Mo–Co	$[MoCo_2(\mu_3\text{-}CMe)(CO)_7(L)(\eta^5\text{-}C_5H_5)]$ $[L = CO, P(OMe)_3]$	28, 171	Hydroformylation of 1-pentene and styrene	182
			Hydrosilation of acetophenone	183
	$[MoCo_2(\mu_3\text{-}S)(CO)_8(\eta^5\text{-}C_5H_5)]$	5	Hydroformylation of 1-pentene and styrene	182
	$[MoCo_2(\mu_3\text{-}S)(\mu\text{-}\eta^2\text{-}MeCNPh)(CO)_6(\eta^5\text{-}C_5H_5)]$	119	Hydroformylation of styrene	119
	$[Mo_2Co_2(\mu_4\text{-}S)(\mu_3\text{-}S)_2(CO)_4(\eta^5\text{-}C_5H_4Me)_2]$	172, 173	Hydrodesulfurization of thiophenol	58
	$[Mo_2Co_2(\mu_3\text{-}S)_4(CO)_2(\eta^5\text{-}C_5H_4Et)_2]$	174	Hydrodesulfurization of thiophenol	63, 64
W–Co	$[WCo_2(\mu_3\text{-}CH)(CO)_8(\eta^5\text{-}C_5H_5)]$	136	Hydrosilation of acetophenone	183
Cr–Pd	$[Cr_2Pd_2(\mu_3\text{-}CO)_2(\mu\text{-}CO)_4(PEt_3)_2(\eta^5\text{-}C_5H_5)_2]$	175	Hydrogenation of cod	184
Mo–Pd	$[Mo_2Pd_2(\mu_3\text{-}CO)_2(\mu\text{-}CO)_4(L)_2(\eta^5\text{-}C_5H_5)_2]$ $(L = PPh_3, PEt_3)$	49	Hydrogenation of cod and 1-hexene	184
			Hydrogenation of phenylacetylene and 1-hexyne	184
			Hydroformylation of 1-pentene	184
			Hydrosilation of 1-pentene	184
	$[Mo_4Pd_4(\mu_3\text{-}CO)_4(\mu\text{-}CO)_8(\eta^5\text{-}C_5H_5)_4]$	163	Dehydration of alcohols	166
Mo–Pt	$[Mo_2Pt(CNCy)_2(CO)_6(\eta^5\text{-}C_5H_5)_2]$	176	Hydrogenation of alkenes	185
			Hydrogenation of alkynes	185
	$[Mo_2Pt(3\text{-}MeC_5H_4N)_2(CO)_6(\eta^5\text{-}C_5H_5)_2]$	177, 178	Hydrogenation of alkynes	185

	$[Mo_2Pt_2(\mu_3\text{-}CO)_2(\mu\text{-}CO)_4(L)_2(\eta^5\text{-}C_5H_5)_2]$ ($L = PPh_3, PEt_3$)	50, 179	Hydrogenation of cod and 1-hexene	184
			Hydrogenation of phenylacetylene	184
			Hydroformylation of 1-pentene	184
			Hydrosilation of 1-pentene	184
W–Pd	$[W_2Pd_2(\mu_3\text{-}CO)_2(\mu\text{-}CO)_4(L)_2(\eta^5\text{-}C_5H_5)_2]$ ($L = PPh_3, PEt_3$)	49	Hydrogenation of cod	184
			Hydrogenation of phenylacetylene	184
			Hydrosilation of 1-pentene	184
W–Pt	$[W_2Pt_2(\mu_3\text{-}CO)_2(\mu\text{-}CO)_4(L)_2(\eta^5\text{-}C_5H_5)_2]$ ($L = PPh_3, PEt_3$)	50	Hydrogenation of cod	184
			Hydrogenation of phenylacetylene	184
Mo–Fe–Co	$[MoFeCo(H)(\mu_3\text{-}CMe)(CO)_8(\eta^5\text{-}C_5H_5)]$	23, 180	Hydrogenation of styrene	186
			Hydrosilation of acetophenone	183
Mo–Ru–Co	$[MoRuCo(\mu_3\text{-}S)(CO)_8(\eta^5\text{-}C_5H_5)]$	130	Hydrogenation of styrene	186
Mo–Co–Ni	$[MoCoNi(\mu_3\text{-}CMe)(CO)_5(\eta^5\text{-}C_5H_5)_2]$	126	Hydrogenation of styrene	186
			Hydroformylation of 1-pentene and styrene	182
			Hydrosilation of acetophenone	183
W–Fe–Co	$[WFeCo(\mu_3\text{-}PMe)(CO)_8(\eta^5\text{-}C_5H_5)]$	181	Hydrogenation of styrene	186

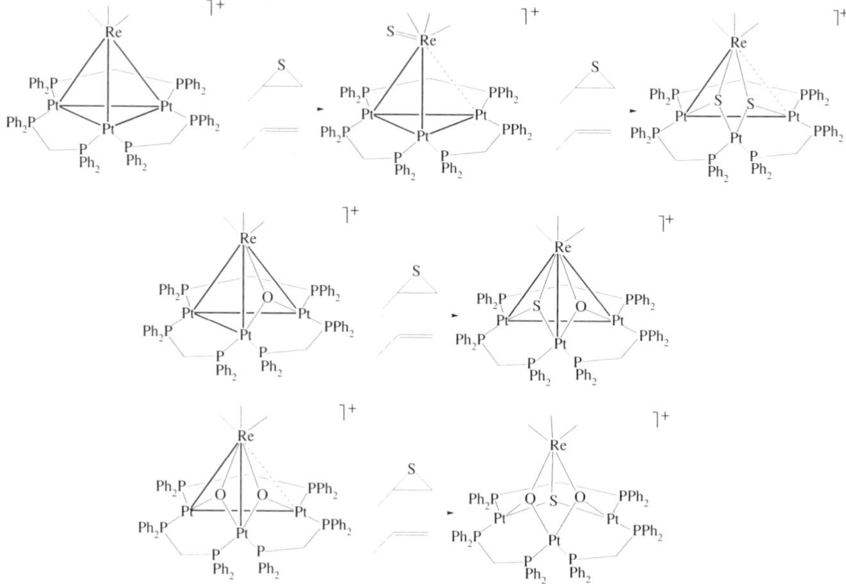

FIG. 68. Desulfurization of propylene sulfide by $[RePt_3(\mu\text{-dppm})_3(\mu\text{-CO})_3]^+$.

$(\mu_3\text{-S})_2(CO)_4(\eta^5\text{-}C_5H_4Me)_2]$ under catalytic conditions (1000 psi CO, 150°C) afforded only a stoichiometric amount of HDS products; PhSSPh is formed catalytically. The cluster regeneration step involved desulfidation by CO to give COS (Fig. 14).[58] Under the same conditions, $[Mo_2Co_2(\mu_3\text{-S})_4(CO)_2(\eta^5\text{-}C_5H_4Et)_2]$ was reacted with thiophenol to yield PhSSPh (171%, based on cluster) and PhS(CO)Ph (161%). The residual organometallic product was recrystallized to give $[Mo_2Co_2(\mu_3\text{-S})_4(SPh)_2(\eta^5\text{-}C_5H_4Et)_2]$ (57%).[63,64]

The octanuclear cluster $Na_2[Mo_4Pd_4(\mu_3\text{-CO})_4(\mu\text{-CO})_8(\eta^5\text{-}C_5H_5)_4]$ was found to catalyze dehydration of aliphatic (MeOH, EtOH, Pr^nOH, Bu^tCH_2OH) and arylaliphatic ($PhCH_2OH$, Ph_2CHOH) alcohols. Dehydration proceeded slowly under mild conditions (60–80°C) in alcohol solutions containing the catalyst under an argon atmosphere; an acidic medium was not required. Alcohols having no hydrogen atom in the α-position were unable to undergo elimination. All the data available suggest that the reaction proceeded via oxidative addition of the alcohol across the Mo—Pd bond and proton transfer to give an intermediate carbene species (Fig. 69). The same cluster was inefficient in the hydrogenolysis of alcohols.[166]

3. *Heterogeneous Catalysis*

Heterogeneous catalysis by metals has been of long-standing interest,[187–189] with bimetallic catalysts a particular focus.[190] Transition metal carbonyls have

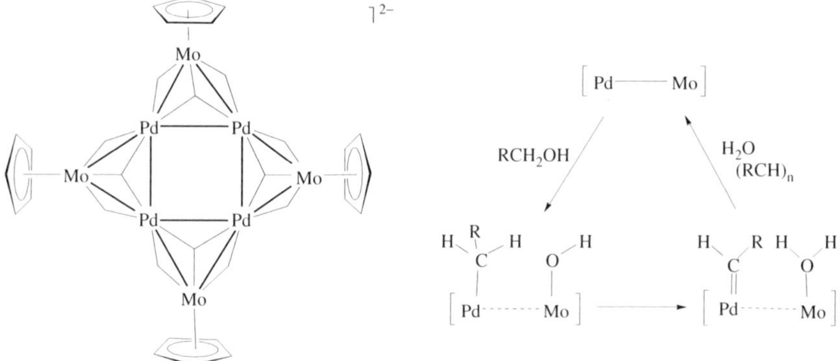

FIG. 69. Dehydration of alcohols by $[Mo_4Pd_4(\mu_3\text{-}CO)_4(CO)_8(\eta^5\text{-}C_5H_5)_4]^{2-}$.

attracted attention as precursors to active catalysts,[191] with heterometallic clusters being examined as precursors to bimetallic particles of controlled stoichiometry. Examples of heterogeneous catalysis by cluster-derived species are collected in Table X.

A MgO-supported W–Pt catalyst has been prepared from $[W_2Pt(CO)_6(NCPh)_2(\eta^5\text{-}C_5H_5)_2]$ (Fig. 70), reduced under a H_2 stream at 400°C, and characterized by IR, EXAFS, TEM and chemisorption of H_2, CO, and O_2. Activity in toluene hydrogenation at 1 atm and 60°C was more than an order of magnitude less for the bimetallic cluster-derived catalyst, than for a catalyst prepared from the two monometallic precursors.[196]

A silica-supported catalyst was prepared by anaerobic impregnation of $[Mo_2Rh(\mu\text{-}CO)(CO)_4(\eta^5\text{-}C_5H_5)_3]$ (Fig. 70) from CH_2Cl_2 solution, followed by evacuation at room temperature. Decomposition processes were observed at the

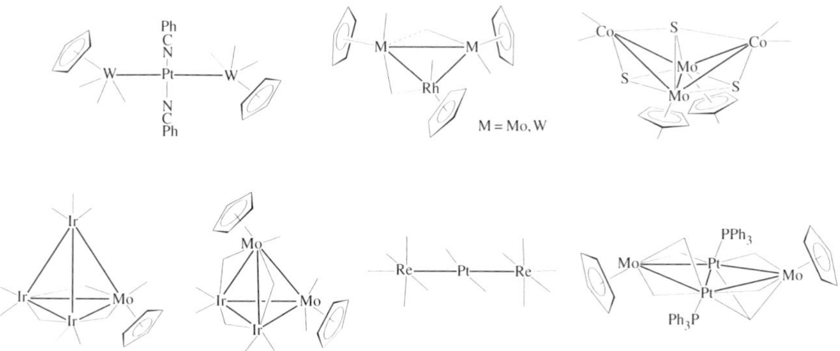

FIG. 70. Cluster precursors to heterogeneous catalysts.

TABLE X

HETEROGENEOUS CATALYSIS EMPLOYING VERY MIXED-METAL CLUSTER PRECURSORS

Catalyst	Precursor Cluster	Ref. (Cluster Syntheses)	Support	Catalyzed Reaction and/or Characterization	Ref. (Catalysis Studies)
Mo–Co	$[Mo_2Co_2(\mu_4\text{-}S)(\mu_3\text{-}S)_2(CO)_4(\eta^5\text{-}C_5H_4Me)_2]$	192	γ-Al_2O_3, SiO_2, TiO_2, MgO	Hydrodesulfurization and characterization of catalyst (TPD)	198
Mo–Rh	$[Mo_2Rh(\mu\text{-}CO)(CO)_4(\eta^5\text{-}C_5H_5)_3]$	193	SiO_2	Hydrogenation of CO	199
	$[Mo_2Rh(\mu\text{-}CO)(CO)_4(\eta^5\text{-}C_5H_5)_3]$ (no pretreatment)	193	SiO_2	Hydrogenation of CO and characterization of catalyst (FTIR, CO chemisorption, TPR, EPR)	193
Mo–Ir	$[MoIr_3(\mu\text{-}CO)_3(CO)_8(\eta^5\text{-}C_5H_5)]$	139	γ-Al_2O_3	Butane hydrogenolysis and catalyst characterization (EXAFS, XANES)	200
	$[Mo_2Ir_2(\mu\text{-}CO)_3(CO)_7(\eta^5\text{-}C_5H_5)_2]$	139, 194	γ-Al_2O_3	Butane hydrogenolysis and catalyst characterization (EXAFS, XANES)	200
W–Rh	$[W_2Rh(\mu\text{-}CO)(CO)_4(\eta^5\text{-}C_5H_5)_3]$	193	SiO_2	Hydrogenation of CO and characterization of catalyst (FTIR, CO chemisorption, TPR, EPR)	199
W–Ir	$[WIr_3(CO)_{11}(\eta^5\text{-}C_5H_5)]$	195	γ-Al_2O_3	Butane hydrogenolysis	195
	$[W_2Ir_2(CO)_{10}(\eta^5\text{-}C_5H_5)_2]$	195	γ-Al_2O_3	Butane hydrogenolysis	195
Mo–Pd	$[Mo_2Pd_2(\mu_3\text{-}CO)_2(\mu\text{-}CO)_4(PPh_3)_2(\eta^5\text{-}C_5H_5)_2]$	49	γ-Al_2O_3	Carbonylation of $ArNO_2$	201
			MgO	Catalyst characterization (EXAFS, IR, TEM, TPD, chemisorption)	202
W–Pt	$[W_2Pt(CO)_6(NCPh)_2(\eta^5\text{-}C_5H_5)_2]$	196	MgO	Hydrogenation of toluene	196
Re–Pt	$[Re_2Pt(CO)_{12}]$	197	γ-Al_2O_3	Catalyst characterization (IR, XPS, TPR, chemisorption)	203
				Catalyst characterization (EXAFS, chemisorption) and methylcyclohexane dehydrogenation	204

beginning of the CO hydrogenation at 1 MPa in the flow reactor. At 523 K the decomposition was complete and the catalyst exhibited slightly smaller activity (ca. 50% conversion) compared with conventionally prepared samples, but higher selectivity toward oxygenated products.[193] This has been extended to include the tungsten-containing homologue. Hydrogenation of carbon monoxide on both molybdenum- and tungsten-promoted silica-supported rhodium catalysts has also been compared employing both monometallic and cluster catalyst precursors. The silica-adsorbed clusters $[M_2Rh(\mu\text{-}CO)(CO)_4(\eta^5\text{-}C_5H_5)_3]$ (M = Mo, W) were activated at 473 K in H_2 and the catalyst characterized by FTIR, CO chemisorption, temperature-programmed reduction (TPR), and EPR. A significant increase in activity was observed with the presence of a group 6 metal as a promoter, relative to that observed for a monometallic-derived rhodium catalyst. The use of these heterometallic clusters as catalyst precursors prevented the formation of active sites responsible for CH_4 production, increasing the proportion of oxygen-containing products.[199] Curtis *et al.* have supported $[Mo_2Co_2(\mu_4\text{-}S)(\mu_3\text{-}S)_2(CO)_4(\eta^5\text{-}C_5H_4Me)_2]$ (Fig. 70) on the refractory oxides Al_2O_3, SiO_2, TiO_2, and MgO, and subjected them to temperature-programmed decomposition (TPD) under H_2. Evolution of CO commenced near 100°C, followed by evolution of cyclopentadienyl ligands from 180 to 400°C, together with small amounts of CO_2, CH_4, and Me_2S. The alumina-supported catalyst was used in the hydrodesulfurization of thiophene, and its activity compared with the commercial Catalco catalyst. Product distributions and activities of both catalysts were nearly identical; the cluster-derived catalyst showed a greater increase in activity upon presulfiding, while cracking activity was enhanced in the reduced form.[198]

The tetrahedral heterometallic clusters $[MoIr_3(\mu\text{-}CO)_3(CO)_8(\eta^5\text{-}C_5H_5)]$ and $[Mo_2Ir_2(\mu\text{-}CO)_3(CO)_7(\eta^5\text{-}C_5H_5)_2]$ (Fig. 70) have been deposited on alumina, and methane evolution profiles observed during activation in H_2. Materials with comparable metal compositions were prepared from stoichiometric mixtures of $[Ir_4(CO)_{12}]$ and $[Mo_2(CO)_6(\eta^5\text{-}C_5H_5)_2]$. All the MoIr catalysts were active for the hydrogenolysis of *n*-butane at 215°C; the $[MoIr_3]$ catalyst exhibited enhanced activity (5–10 times) over the $[Ir_4 + 2Mo_2]$ sample, but selectivity toward ethane production (70–75%) was the same for both. In contrast, the $[Mo_2Ir_2]$ catalyst showed greater selectivity for C_1 and C_3 production (ca. 50%) but activity comparable to $[Ir_4 + 2Mo_2]$. Metal–metal interactions in the activated materials were characterized with the use of Mo K edge X-ray absorption spectra.[200]

A catalyst supported on $\gamma\text{-}Al_2O_3$ was prepared from $[Re_2Pt(CO)_{12}]$ (Fig. 70) and characterized by IR, X-ray photoelectron spectroscopy (XPS), and TPR. The chemisorbed cluster was treated with H_2 at about 150°C resulting in fragmentation and formation of rhenium subcarbonyls; at 400°C the sample was completely decarbonylated. A catalyst prepared from a mixture of $[Re_3(\mu\text{-}H)_3(CO)_{12}]$ and $[PtMe_2(\eta^4\text{-}cod)]$ and treated under equivalent conditions showed the rhenium to be present in a high valent cationic form and the platinum in a metallic form. The platinum is believed to facilitate the reduction of rhenium and is likely to be near the rhenium in the sample prepared from $[Re_2Pt(CO)_{12}]$.[203]

Dehydration of methylcyclohexane in the presence of a $[Re_2Pt(CO)_{12}]$-derived catalyst and catalysts derived from $[Pt(NH_3)_2](NO_3)_2$ and $[Pt(NH_3)_2](NO_3)_2 + NH_4[ReO_4]$ has also been examined. The alumina-supported cluster was decomposed under flowing H_2 at 400°C for 4 h. The catalysts were similar to each other in their selectivities for the dehydrogenation of methylcyclohexane and were characterized by high initial conversion (>90%) to toluene. The catalyst made from $[Re_2Pt(CO)_{12}]$ was found to be more resistant to deactivation than catalysts prepared conventionally from rhenium and platinum salt precursors. Characterization by EXAFS revealed the cluster-derived catalyst to be more highly dispersed than the others; its resistance to deactivation was attributed to the role of rhenium in stabilizing the dispersion of the platinum.[204]

MgO-supported model Mo—Pd catalysts have been prepared from the bimetallic cluster $[Mo_2Pd_2(\mu_3\text{-}CO)_2(\mu\text{-}CO)_4(PPh_3)_2(\eta^5\text{-}C_5H_5)_2]$ (Fig. 70) and monometallic precursors. Each supported sample was treated in H_2 at various temperatures to form metallic palladium, and characterized by chemisorption of H_2, CO, and O_2, transmission electron microscopy, TPD of adsorbed CO, and EXAFS. The data showed that the presence of molybdenum in the bimetallic precursor helped to maintain the palladium in a highly dispersed form. In contrast, the sample prepared from the monometallic precursors was characterized by larger palladium particles and by weaker Mo—Pd interactions.[202]

III

PHYSICAL MEASUREMENTS

A. *General Comments*

The focus of research on "very mixed"-metal clusters has been on their synthesis and structure, and the limited physical measurements of these clusters have thus far been largely restricted to fluxionality and electrochemical investigations. Studies of ligand fluxionality are summarized in Section III.A. and reports of electrochemical investigations are reviewed in Section III.B. The few reports of the magnetic behavior of these clusters are discussed in Section III.C.1., and theoretical studies are summarized in Section III.C.2.

B. *Fluxionality*

Ligand fluxionality on metal clusters has been the subject of many studies, the majority of reports focusing on carbonyl migration on homometallic tri- and tetranuclear clusters.[1,205–212] Extending fluxionality studies to mixed-metal clusters affords a significant advantage; mechanistic details may be more accessible due to the decreased symmetry and effective labeling provided by introduction of the heterometal. Additionally, though, activation energies for ligand scrambling in mixed-metal clusters may differ markedly from those of their homometallic analogues, and it would be expected that this difference will be accentuated utilizing "very mixed"-metal clusters. A combination of decreased symmetry, labeled core nuclei, and differing energetics for processes involving metals with disparate electronegativities may permit discrimination of fluxional pathways not possible in homometallic clusters. Despite these significant advantages, comparatively few reports of fluxionality at "very mixed"-metal clusters have appeared.

Three major classes of fluxional behavior are commonly observed in clusters: (A) Metal localized scrambling, usually seen in $M(CO)_3$ or $M(CO)_2L$ groups. (B) Intermetallic ligand migrations, usually involving CO or hydrides (although examples with phosphines and alkynes are extant) and proceeding via edge-bridged or face-capped intermediates. (C) Metal framework rearrangements. Examples of all three classes of fluxional behavior have been observed at "very mixed"-metal clusters. It should be emphasized that absolute atomic motions are not accessible; considering ligands rotating around a fixed metal core or a metal core rotating within a fixed ligand polytope are equally valid viewpoints (the latter has been documented in the solid state: see, for example, Ref. 205). The former description has been utilized far more often in literature reports and, following this convention, this review summarizes specific cases in terms of ligands exchanging by rotating around a metal core where possible.

Table XI summarizes reports of ligand fluxionality at "very mixed"-metal clusters. A number of studies (e.g., Refs. 28, 81, 116, 213, 214) have reported that both metal-localized and global carbonyl fluxionality occur but give little mechanistic detail. The following discussion focuses on the examples for which detailed studies have been undertaken and/or those for which mechanistic speculation is available.

A number of examples of metal-localized ligand scrambling have been documented. Rotation of the $Ru(CO)_2(\eta^5\text{-}C_5H_5)$ units about the Zr—Ru bonds in the V-shaped $[ZrRu_2(CO)_4(\eta^5\text{-}C_5H_5)_4]$ resulted in equivalence of the carbonyl ligands (Fig. 71);[213] replacing the cyclopentadienyl ligands by $CH_2(CH_2NSiMe_3)_2$ reduced significantly the barrier to rotation in the ruthenium-containing analogue and its iron-containing homologue.[215] A similar rotation of the $W(CO)_2(\eta^5\text{-}C_5Me_5)$ group in $[WCo_2(\mu_3\text{-}CC_6H_4Me\text{-}4)(CO)_8(\eta^5\text{-}C_5Me_5)]$ has also been noted.[216] A detailed study of the rotation of the $M(CO)_2(\eta^5\text{-}C_5R_4R')$ vertices (M = Mo, W; R, R′ = H, Me) in $[MCo_2(\mu_3\text{-}CCO_2Pr^i)(CO)_6(L)(\eta^5\text{-}C_5R_4R')]$ (L = 2CO, $Ph_2PCH_2CH_2PPh_2$, $Ph_2AsCH_2CH_2PPh_2$)[17,116] revealed that the barriers to rotation of the group 6 metal-containing vertices in these clusters (37–42 kJ mol^{-1}) are similar to those of the

TABLE XI

FLUXIONALITY AT VERY MIXED-METAL CLUSTERS

Cluster	Exchange Type[a]	Fluxional Process (Exchange Barrier)	Ref.
$[ZrRu_2(CO)_4(\eta^5\text{-}C_5H_5)_4]$	A	$CpRu(CO)_2$ rotation ($G^{\ddagger}_{324} = 56.5 \pm 2.1$ kJ mol^{-1})	213
$[ZrM_2\{CH_2(CH_2NSiMe_3)_2\}(CO)_4(\eta^5\text{-}C_5H_5)_2]$ (M = Fe, Ru)	A	$CpM(CO)_2$ rotation	215
$[MCo_2(\mu_3\text{-}CR'')(CO)_6(L)(\eta^5\text{-}C_5R_4R')]$[M = Mo, W;	A	$Co(CO)_3$ tripodal rotation	17, 116, 216
R = R' = H, Me; R'' = CO_2Pr^i, C_6H_4Me-4; L = 2CO, $Ph_2ECH_2CH_2PPh_2$(E = P, As)]	A	$CpM(CO)_2$ rotation ($G^{\ddagger} = 39 \pm 2.5$ kJ mol^{-1}, M = Mo, R'' = CO_2Pr^i, L = $Ph_2PCH_2CH_2PPh_2$, R = R' = Me)	
	B	Co–Mo intramolecular CO exchange ($G^{\ddagger}_{333} = 69.0 \pm 2.5$ kJ mol^{-1}, M = Mo, R'' = CO_2Pr^i, L = $Ph_2PCH_2CH_2PPh_2$, R = R' = Me)	
$[WCo_2(\mu\text{-}EtC_2Et)(\mu\text{-}\eta^4\text{-}CEtCEtCEtCEt)(CO)_8]$	A	$Co(CO)_3$ tripodal rotation	21
	C	ligated cobalt "twisting"	
$[Re_2Pt(\mu\text{-}H)_2(CO)_8(L)]$ [L = η^4-cod, $(PPh_3)_2$]	A	$Re(CO)_3$ tripodal rotation at $Re(CO)_3(PPh_3)$	217–219
	A	$Re(CO)_2(PPh_3)$ tripodal rotation at $Re(CO)_3(PPh_3)$($G^{\ddagger} = 74 \pm 1.7$ kJ mol^{-1} [L = $(PPh_3)_2$])	
	B	hydride exchange across Re–Pt vectors	
$[RePt_3(\mu\text{-}dppm)_3(CO)_3(L)]^+$ [L = $P(OMe)_3$, $P(OPh)_3$]	A	$Re(CO)_3(L)$ tetrapodal rotation	42
$[Re_7Pt(\mu_6\text{-}C)(L)(CO)_{21}]^{2-}$ (L = Me_3, η^3-2-methylallyl)	A	tripodal rotation at $Re(CO)_3$	161
$[MoFeNi(\mu_3\text{-}\eta^2\text{-}PhC_2CO_2Pr^i)(CO)_6(\eta^5\text{-}C_5H_5)_2]$	B	acetylene migration ($G^{\ddagger} = 69 \pm 4$ kJ mol^{-1})	220
$[MW_2Co(CO)_9(\eta^5\text{-}C_5H_4Me)_3]$ (M = Mo, W)	B	global MW_2-localized CO exchange	221
$[Mo_2Co_2(\mu_3\text{-}S)_2(\mu\text{-}S)(\mu\text{-}SC_6H_4Me\text{-}4)(CO)_4(\eta^5\text{-}C_5H_4Me)_2]$	B	migration of thiolate between Mo–Co bonds [$G^{\ddagger} = 42 \pm 4$ kJ mol^{-1}]	100

Compound	Pathway[a]	Process	Ref.
$[Mo_2Co_2(\mu_3\text{-}\eta^2\text{-}PBu^tC_6H_4\text{-}2PBu^t)(\mu\text{-}CO)(CO)_6(\eta^5\text{-}C_5H_5)_2]$	B	migration of phosphido from μ-(Mo–Co)$_2$ to μ-(Mo–Co, Co–Co) [$G^{\ddagger}_{298} = 120.9 \pm 9.2$ kJ mol^{-1}]	222
$[Re_3Pt(\mu\text{-}H)_3(L)(CO)_{13}]$ [L = CO, $(CO_2Me)^-$]	B	hydride exchange across Re–Pt vectors [$G^{\ddagger}_{180} = 43.0$ kJ mol^{-1}, L = CO; $G^{\ddagger}_{243} = 70 \pm 1$ kJ mol^{-1}, L = $(CO_2Me)^-$]	72, 214
	C	intermolecular exchange of $ReH(CO)_5$ ($G^{\ddagger}_{180} = 41.9$ kJ mol^{-1}, L = CO)	
$[RePt_3(\mu\text{-}dppm)_3(O)_3\{P(OMe)_3\}]^+$	B	migration of phosphite around core	81
$[Re_4Pt(\mu\text{-}H)_3(CO)_{18}]^-$	B	hydride exchange across Re–Pt vectors [$G^{\ddagger}_{273} = 74 \pm 7$ kJ mol^{-1}]	72
$[MM'Rh(\mu\text{-}CO)_2(CO)_5(\eta^5\text{-}C_5Me_5)_2]$ (M = Cr, Mo, W; M′ = Co, Rh)	C	$M(CO)_5$ rotation	223, 224
$[MPd_2(\mu\text{-}Cl)(CO)_3(L)_2(\eta^5\text{-}C_5H_5)]$ (M = Cr, Mo, W; L = dimethylbenzylamine-C^2, *N* (dmba), 2-(dimethylamino)toluene-*C*, *N* (dmat))	C	dissociation of $[M(CO)_3(\eta^5\text{-}C_5H_5)]^-$ and reassociation [$G^{\ddagger} \sim 70$ kJ mol^{-1} (M = Mo, W, L = dmba); $G^{\ddagger} \sim 50$ kJ mol^{-1} (M = Mo, W, L = dmat)]	225
$[W_2Re_2Pt_2(\mu\text{-}C_6H_4Me\text{-}4)_2(CO)_{18}]$	C	rotation about midpoint of μ-C-Pt vectors	84

[a]Exchange pathways: A, metal localized ligand scrambling; B, intermetallic ligand migrations; C, metal framework rearrangements.

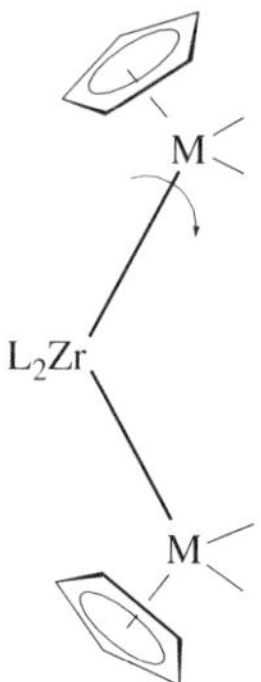

FIG. 71. Rotation of the $M(CO)_2(\eta^5-C_5H_5)$ units about the Zr—M bonds: M = Fe, $L_2 = \eta^2$-$(NSiMe_3CH_2)_2CH_2$; M = Ru, $L_2 = (\eta^5-C_5H_5)_2$, η^2-$(NSiMe_3CH_2)_2CH_2$.

$M(CO)_2(\eta^5-C_5H_{5-n}Me_n)$ vertices in tetrahedral $[M_2(\mu_3-\eta^2-RC_2R)(CO)_4(\eta^5-C_5H_{5-n}Me_n)_2]$.[17] Tripodal $Co(CO)_3$ rotation was sufficiently rapid in $[MoCo_2(\mu_3-CCO_2Pr^i)(CO)_6(L)(\eta^5-C_5R_4R')]$ that it could not be frozen out in solution NMR studies. A similarly rapid turnstile motion was observed at one cobalt atom in $[WCo_2(\mu-EtC_2Et)(\mu-\eta^4-CEtCEtCEtCEt)(CO)_8]$, but $Co(CO)_3$ rotation at the other cobalt was retarded by adjacent ethyl groups.[21]

Two localized exchange processes were observed at the triangular $[Re_2Pt(\mu-H)_2(CO)_8(PPh_3)_2]$; the lower energy process was a scrambling of the three carbonyl ligands in the $Re(CO)_3(PPh_3)$ unit, while the higher energy process corresponded to a trigonal twist at the same metal, but involving phosphine as well as carbonyls.[218] Three-fold exchange of this type at $M(CO)_4$ groups has been documented previously in homometallic clusters, e.g. $[Os_3(\mu-H)(\mu_3-CR)(CO)_{10}]$ (R = H, OMe, Ph).[226] Tripodal $Re(CO)_3$ rotation in the bicapped octahedral $[Re_7Pt(\mu_6-C)(Me)_3(CO)_{21}]^{2-}$ is facile at room temperature.[161] Upon cooling a sample of the cluster, rotation at the three rheniums at the base of the $PtMe_3$ cap was frozen first, followed by cessation at the three rheniums at the base of the $Re(CO)_3$ cap. Investigation of the isostructural $[Re_7Pt(\mu_6-C)(CO)_{21}(\eta^3$-2-methylallyl$)]^{2-}$ revealed analogous tripodal rotation; decoalescence of resonances was observed in the reverse order to that in the $PtMe_3$-containing cluster, but it was not clear if this results from a reversal of relative activation energies or from chemical shift order. Tetrapodal rotation of the $[Re(CO)_3(L)]$ vertex was observed in $[RePt_3(\mu-dppm)_3(CO)_3(L)]^+$ [L = $P(OMe)_3$, $P(OPh)_3$]; the trimethylphosphite-containing cluster had a lower barrier to rotation, a result ascribed to steric considerations.[42]

Several examples of intermetallic hydride migration in rhenium-platinum clusters have been investigated. Hydride hopping between the two Re—Pt edges in triangular $[Re_2Pt(\mu-H)_2(CO)_8(L)]$ [L = η^4-cod, $(PPh_3)_2$] occurred at a lower energy than rotation of the Pt(H)(L) moiety.[219] A similar hydride migration between the two Re—Pt edges was observed in the spiked-triangular $[Re_3Pt(\mu-H)_3(CO)_{14}]^{214}$ and

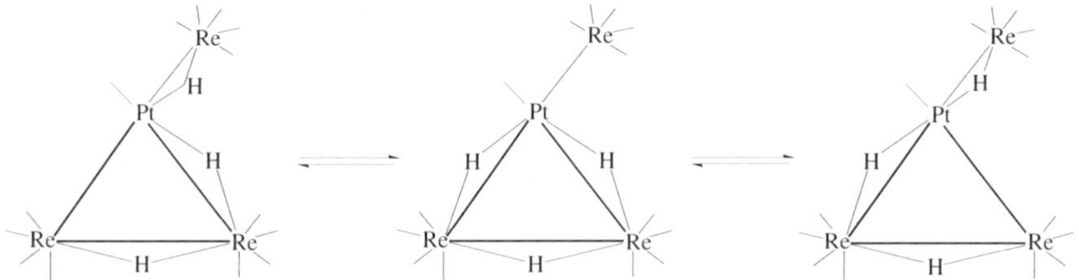

FIG. 72. Hydride migration at $[Re_3Pt(\mu\text{-}H)_3(CO)_{14}]$.

$[Re_3Pt(\mu\text{-}H)_3(CO_2Me)(CO)_{13}]^-$,[72] and the "scorpion" $[Re_4Pt(\mu\text{-}H)_3(CO)_{18}]^-$;[72] an intermediate in which both exchanging hydrides are bridging Re—Pt linkages in the triangular fragment of the cluster was proposed (Fig. 72). Similarly, the hydrides bridging the Re—Pt bonds in the "bow-tie" cluster $[Re_4Pt(\mu\text{-}H)_5(CO)_{16}]^-$ exchange rapidly, with only one ^1H NMR signal for these hydrides observed down to 193 K.[71]

Alkyne migration at $[MoFeNi(\mu_3\text{-}\eta^2\text{-}PhC_2CO_2Pr^i)(CO)_6(\eta^5\text{-}C_5H_5)_2]$ has been examined;[220] a formal rotation of the alkyne relative to the metal triangle proceeding by a "modified windscreen wiper" mechanism was proposed (Fig. 73).

Carbonyl exchange between metals has been the most frequently studied intermetallic ligand migration in homometallic clusters, but the few examples which have been examined in the "very mixed"-metal domain have afforded inconclusive results. At the triangular cluster $[MoCo_2(\mu_3\text{-}CCO_2Pr^i)(\mu\text{-}Ph_2ECH_2CH_2PPh_2)(CO)_6(\eta^5\text{-}C_5Me_5)]$, the mechanism interconverting molybdenum- and cobalt-bound carbonyls was not conclusively established, but rotation of the $[Mo(CO)_2(\eta^5\text{-}$

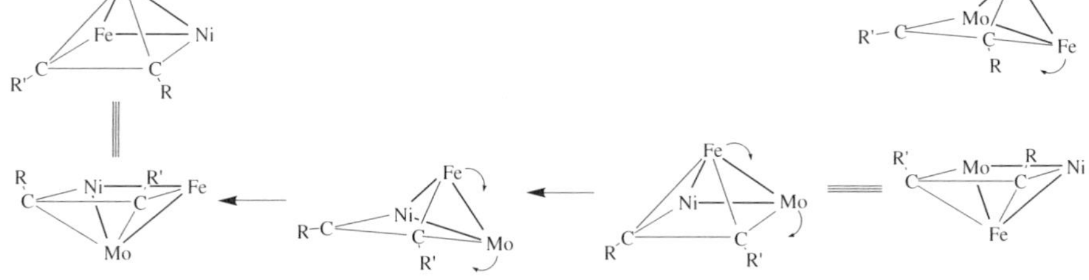

FIG. 73. Alkyne migration at $[MoFeNi(\mu_3\text{-}PhC_2CO_2Pr^i)(CO)_6(\eta^5\text{-}C_5H_5)_2]$: other ligands omitted for clarity.

$C_5Me_5)]$ unit was a critical requirement.[17] Global MW_2-localized CO exchange was noted at the tetrahedral cluster $[MW_2Co(CO)_9(\eta^5\text{-}C_5H_4Me)_3]$,[221] although again the fluxional process or processes were not conclusively established.

Platinum-bound phosphine and rhenium-ligated carbonyl irreversibly exchange on warming a solution of $[Re_2Pt(\mu\text{-}H)_2(CO)_8(PPh_3)_2]$ above 273 K;[217,218] a mixture of two isomers is formed which interconvert by a restricted trigonal twist at a $[Re(CO)_3(PPh_3)]$ unit (see above). Intermetallic phosphite migration has been observed at $[RePt_3(\mu\text{-dppm})_3(O)_3\{P(OMe)_3\}]^+$; [81] a fluxional process similar to that observed earlier in the homometallic cluster $[Pt_3(\mu_3\text{-}CO)(\mu\text{-dppm})_3\{P(OMe)_3\}]^{2+}$ was proposed, involving the intermediacy of a phosphite ligand triply bridging the Pt_3 face. Thiolate migration between Co—Mo bonds in the butterfly cluster anion $[Mo_2Co_2(\mu_3\text{-}S)_2(\mu\text{-}S)(\mu\text{-}SC_6H_4Me\text{-}4)(CO)_4(\eta^5\text{-}C_5H_4Me)_2]^-$ occurs with the thiolate crossing the Co—Co axis and a barrier to this motion of 42 ± 4 kJ mol^{-1}.[100] The *o*-phenylenebis(μ-*tert*-butylphosphido) ligand in $[Mo_2Co_2(\mu_3\text{-}\eta^2\text{-}PBu^tC_6H_4\text{-}2\text{-}PBu^t)(\mu\text{-}CO)(CO)_6(\eta^5\text{-}C_5H_5)_2]$ bridges the Mo—Co bonds; when heated at 75–80°C, it undergoes an irreversible transformation to afford an isomer with the bis-phosphido ligand spanning Mo—Co and Co—Co vectors.[222]

Few examples of metal framework rearrangements have been observed by NMR methods at "very mixed"-metal clusters. A study of the V-shaped cluster $[WCo_2(\mu\text{-}EtC_2Et)(\mu\text{-}\eta^4\text{-}CEtCEtCEtCEt)(CO)_8]$ revealed that the fluxionality on the NMR timescale was not alkyne ligand rotation; rather, a twisting of one ligated cobalt with respect to the other introduced a molecular mirror plane (Fig. 74). The triangular clusters $[MM'Rh(\mu\text{-}CO)_2(CO)_5(\eta^5\text{-}C_5Me_5)_2]$ (M = Cr, Mo, W; M′ = Co, Rh) underwent a formal rotation of the $M(CO)_5$ unit about an axis through M and the midpoint of the M′-Rh vector (Fig. 75).[223,224] The cluster is an isolobal analogue of a d^6-ML_5 alkene complex. In this context, it is interesting that the barrier to rotation in $[MoCoRh(\mu\text{-}CO)_2(CO)_5(\eta^5\text{-}C_5Me_5)_2]$ ($\Delta G^{\ddagger}_{258} = 52.3 \pm 1.3$ kJ mol^{-1}) is in the range observed for hindered rotation at d^6-ML_5 organic

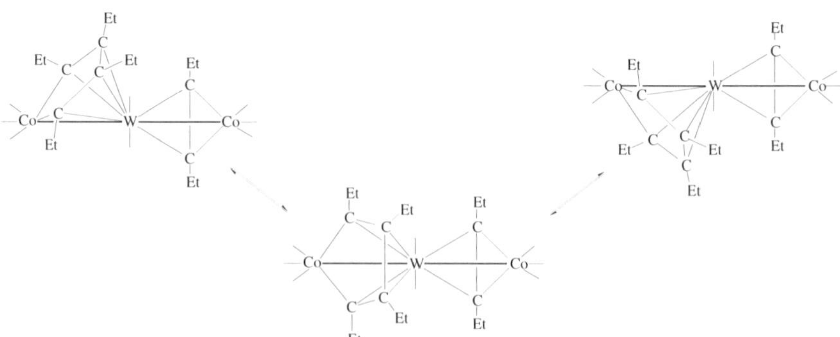

FIG. 74. Ligated cobalt twisting at $[WCo_2(\mu\text{-}CEtCEt)(\mu\text{-}\eta^4\text{-}CEtCEtCEtCEt)(CO)_8]$.

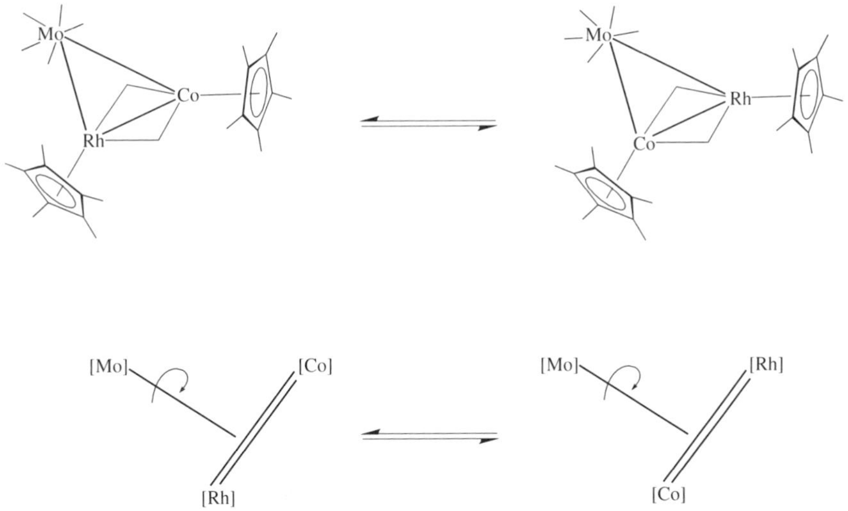

FIG. 75. $M(CO)_5$ rotation at $[MoCoRh(\mu\text{-}CO)_2(CO)_5(\eta^5\text{-}C_5Me_5)_2]$.

alkene complexes ($\Delta G^{\ddagger} = 35 - 80$ kJ mol^{-1}). A similar rotation, about an axis through the tungsten and midpoint of the μ-C-Pt vector, was proposed to explain the equivalence of the radial carbonyls in $[W_2Re_2Pt(\mu\text{-}CC_6H_4Me\text{-}4)_2(CO)_{18}]$.[84] The clusters $[MFeCo_2(\mu_3\text{-}E)(\mu\text{-}AsMe_2)(CO)_8(\eta^5\text{-}C_5H_5)]$ (M = Mo, W; E = S, PMe) exist as isomers corresponding to differing positions of the iron and cobalt atoms; detailed equilibration and isomerization studies showed that isomer interconversion does not occur under rigorously clean conditions, but instead requires the presence of impurities, suggesting that radical type metal–metal bond cleavages initiated the isomerization process.[26]

Examples of intermolecular ligated metal exchange are also extant. The V-shaped clusters $[MPd_2(\mu\text{-}Cl)(CO)_3(L)_2(\eta^5\text{-}C_5H_5)]$ (M = Cr, Mo, W; L = dmba, dmat) underwent exchange by dissociation and reassociation of the carbonylmetallate anion $[M(CO)_3(\eta^5\text{-}C_5H_5)]^-$ (Fig. 76).[225] The spiked triangular cluster $[Re_3Pt(\mu\text{-}H)_3(CO)_{14}]$ exchanged the "spike" fragment $[ReH(CO)_5]$ in solution (Fig. 77), in a process which is a formal substitution at a square planar platinum.[214] Such reactions are usually associative; accordingly, a strong dependence on the concentration of added $[ReH(CO)_5]$ was noted.

"Very mixed"-metal clusters offer significant advantages in fluxionality studies, but these advantages remain to be fully exploited. Studies of tetrahedral rhodium and mixed rhodium-iridium clusters have revealed intermetallic carbonyl ligand site exchange proceeding by way of merry-go-round and change-of-basal-face processes.[206] Substitution of Rh by Ir in the apical site slowed down the former

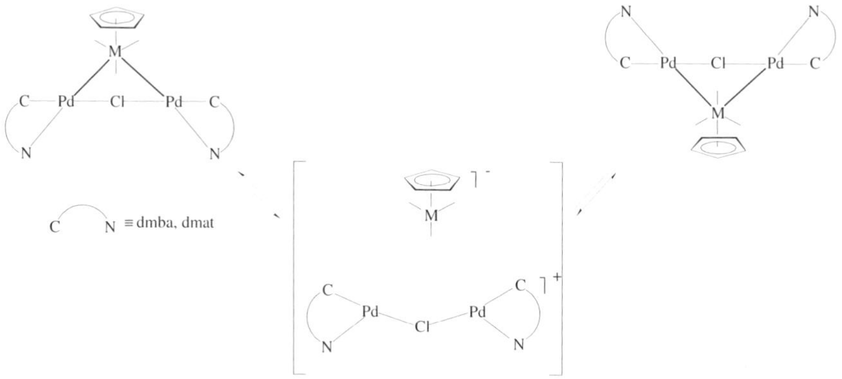

FIG. 76. Isomerization at $[MPd_2(\mu\text{-Cl})(CO)_3(L)_2(\eta^5\text{-}C_5H_5)]$: M = Cr, Mo, W; L = dmba, dmat.

process, rationalized as resulting from the more electronegative Rh inducing a shift of electron density to the basal face and stabilizing the bridging CO-containing ground state. Tetrahedral iridium and mixed rhodium-iridium clusters underwent tripodal rotation at the apical metal. Substitution of Ir by Rh in the basal plane accelerated this process, rationalized as the more electronegative Rh removing electron density from the apical 5d orbitals between the apical Ir—CO bonds. Both of these examples involved an electronic effect from a remote metal affecting relative energies of activation. Replacing the vertices in homometallic clusters to afford "very mixed"-metal clusters should accentuate these effects, but this remains to be demonstrated. One problem for fluxional processes involving the heterometal is that isolobal equivalence of ligated metal fragments involves introducing a (formally) higher coordination number metal, and possibly increased steric effects [e.g., replacing $Ir(CO)_3$ by $W(CO)_2(\eta^5\text{-}C_5H_5)$ in proceeding from $[Ir_4(CO)_{12}]$ to isostructural $[WIr_3(CO)_{11}(\eta^5\text{-}C_5H_5)]$ introduces a (formally) eight coordinate tungsten, with a sterically demanding cyclopentadienyl ligand]. However, for fluxional processes remote from the heterometal (such as with the rhodium-iridium examples above), this problem may not be important.

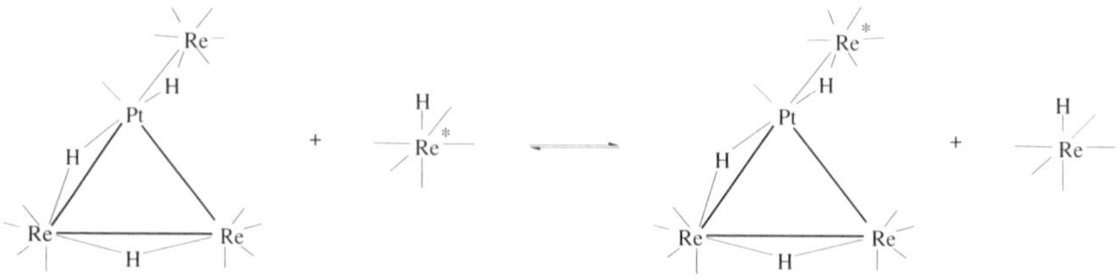

FIG. 77. Intermolecular $ReH(CO)_5$ exchange at $[Re_3Pt(\mu\text{-H})_3(CO)_{14}]$.

C. *Electrochemistry*

The electrochemical behavior of heterometallic clusters has been reviewed elsewhere.[227] The interest in examining clusters stems from their potential to act as "electron sinks;" in principle, an aggregate of several metal atoms may be capable of multiple redox state changes. The incorporation of heterometals provides the opportunity to tune the electrochemical response, effects which should be maximized in "very mixed"-metal clusters. Few "very mixed"-metal clusters have been subjected to detailed electrochemical studies; the majority of reports deal with cyclic voltammetry only. Table XII contains a summary of electrochemical investigations of "very mixed"-metal clusters.

Not surprisingly, given the success of metal exchange procedures in generating a wide range of systematically varied "very mixed"-metal clusters (Section II.E.1.), derivatives of $[Co_3(\mu_3\text{-E})(CO)_9]$ have been studied intensively, especially capped triangular clusters of general formulas $[MCo_2(\mu_3\text{-E})(CO)_8(\eta^5\text{-}C_5H_5)]$ (M = Cr, Mo, W; E = CH, CMe, CPh, PTol, GeMe, GePh) and $[MM'Co(\mu_3\text{-E})(CO)_8(\eta^5\text{-}C_5H_5)]$ (M = Mo, W; M′ = Fe, Ru; E = CMe, CBu^t, CPh, PMe, S, Se) where the capping group may either be electrochemically active or provide structural rigidity important in subsequent oxidation state changes. Jensen *et al.*[110] have described an electrosynthesis of some alkylidyne examples ($CrCo_2$, $MoCo_2$, WCo_2), with higher yields than the classical thermally activated route (Fig. 78). The direct reaction between $[Co_3(\mu_3\text{-CPh})(CO)_9]$ and $K[Mo(CO)_3(\eta^5\text{-}C_5H_5)]$ can be initiated by electrolysis at the potential for the couple $[Co_3(\mu_3\text{-CPh})(CO)_9]^{0/-1}$.

Two reduction processes were observed for the $[MCo_2(\mu_3\text{-CR})(CO)_8(\eta^5\text{-}C_5H_5)]$ (M = Cr, Mo, W; R = Me, Ph) clusters. The first, an electrochemically and chemically irreversible process, was followed by an electrochemically and chemically reversible step; both steps involved cleavage of a M—Co bond accompanied by a carbonyl moving from a terminal position to bridge these metals. The peak separation decreased in the order Cr > Mo ≈ W, with a dependence on apical substituent. EPR data has been used in the characterization of the electrogenerated species, with experiments carried out at several temperatures and under argon and CO atmospheres. These clusters are chemically more reactive than their homonuclear tricobalt counterparts, with the $[MoCo_2(\mu_3\text{-CR})(CO)_8(\eta^5\text{-}C_5H_5)]$ clusters the most

$$[Co_3(\mu_3\text{-CPh})(CO)_9] + [Mo(CO)_3(\eta5\text{-}C5H5)]^- \xrightarrow[\text{fast}]{-0.40\ \text{V, THF}} \underset{\text{good yield}}{[MoCo_2(\mu_3\text{-CPh})(CO)_8(\eta^5\text{-}C_5H_5)]} + [Co(CO)_4]^-$$

$$[Co_3(\mu_3\text{-CPh})(CO)_9] + {}^1/_2[Mo(CO)_3(\eta^5\text{-}C_5H_5)]_2 \xrightarrow[\text{3 days}]{\text{heat, benzene}} \underset{\text{low yield}}{[MoCo_2(\mu_3\text{-CPh})(CO)_8(\eta^5\text{-}C_5H5]} + \text{'}[Co(CO)_4]\text{'}$$

FIG. 78. Electrochemical and thermal routes to $[MoCo_2(\mu_3\text{-CPh})(CO)_8(\eta^5\text{-}C_5H_5)]$.

TABLE XII

ELECTROCHEMICAL STUDIES OF VERY MIXED-METAL CLUSTERS

Cluster	Technique(s)[a]	Ref.
$[MCo_2(\mu_3\text{-}CH)(CO)_8(\eta^5\text{-}C_5H_5)]$ (M = Cr, Mo, W)	ETCS	110
$[MCo_2(\mu_3\text{-}CMe)(CO)_8(\eta^5\text{-}C_5H_5)]$ (M = Cr, W)	ETCS, CV, DCP, DPP	110, 228, 229
$[MCo_2(\mu_3\text{-}CR)(CO)_8(\eta^5\text{-}C_5H_5)]$ (M = Cr, Mo, W, R = Ph: M = Mo, R = Me)	ETCS, CV, DCP, DPP, EPR	110, 228, 229
$[MoCo_2(\mu_3\text{-}CC_6H_4Me\text{-}4)(\mu\text{-}PPh_2)(CO)_6(\eta^5\text{-}C_5H_5)]^-$	CV	70
$[MoCo_2(\mu_3\text{-}GeR)(CO)_8(\eta^5\text{-}C_5H_5)]$ (R = Me, Ph)	CV, DCP, EPR	228
$[MoFeCo(\mu_3\text{-}PBu^t)(CO)_8(\eta^5\text{-}C_5H_5)]$	CV, DCP, DPP	228
$[MoFeCo(\mu_3\text{-}PR)(CO)_8(\eta^5\text{-}C_5H_5)]$ (R = Me, Ph)	CV, DCP, DPP, EPR	228, 229
$[MM'Co(\mu_3\text{-}E)(CO)_8(\eta^5\text{-}C_5H_5)]$ (M = Mo, M′ = Fe, E = S, Se; M = Mo, M′ = Ru, E = S; M = W, M′ = Fe, E = S)	CV	229
$[WFeCo(\mu_3\text{-}PR)(CO)_8(\eta^5\text{-}C_5H_5)]$ (R = Me, Ph)	CV	229
$[Cr_3Co(\mu_3\text{-}S)_4(CO)(\eta^5\text{-}C_5H_4Me)_3]$	CV	230
$[MoCo_3(\mu_3\text{-}\eta^2\text{-}PBu^tC_6H_4\text{-}2\text{-}PBu^t)(\mu\text{-}CO)(CO)_7(\eta^5\text{-}C_5H_5)]$	CV	74
$[Mo_2Co_2(\mu_3\text{-}\eta^2\text{-}PBu^tC_6H_4\text{-}2\text{-}PBu^t)(\mu\text{-}CO)(CO)_6(\eta^5\text{-}C_5H_5)_2]$	CV	74
$[Mo_2Co_2(\mu_3\text{-}S)_4(CO)_2(\eta^5\text{-}C_5H_4Et)_2]$	CV	231
$[Mo_2Co_2(\mu_4\text{-}S)(\mu_3\text{-}S)_2(CO)_4(\eta^5\text{-}C_5H_4R)_2]$ (R = Me, Et)	CV	172, 231
$[W_2Co_2(\mu_3\text{-}S)_4(NO)_2(\eta^5\text{-}C_5H_4Et)_2]$	CV	231
$[W_2Co_2(\mu_3\text{-}S)_3(CO)_5(\eta^5\text{-}C_5H_4Et)_2]$	CV	231
$[W_2Co_2(\mu_3\text{-}S)_4(NO)_2(\eta^5\text{-}C_5Me_5)_2]$	CV	231
$[Cr_2M_2(\mu_3\text{-}CO)_2(\mu\text{-}CO)_4(PR_3)_2(\eta^5\text{-}C_5H_5)_2]$ (M = Pd, Pt; R = Ph, Me, Et, Bu^n)	CV	232, 233
$[Mo_2Ni_2(\mu_3\text{-}S)_4(CO)_2(\eta^5\text{-}C_5H_4Me)_2]$	CV	172
$[Mo_2M_2(\mu_3\text{-}CO)_2(\mu\text{-}CO)_4(PPh_3)_2(\eta^5\text{-}C_5H_5)_2]$ (M = Pd, Pt; R = Ph, Me, Et, Bu^n)	CV	232, 233
$[W_2Pt_2(\mu_3\text{-}CO)_2(\mu\text{-}CO)_4[PPh_2CH_2PPh_2\{Mn(CO)_2(\eta^5\text{-}C_5H_4Me\}]_2(\eta^5\text{-}C_5H_4Me)_2]$	CV, C	48
$[W_2M_2(\mu_3\text{-}CO)_2(\mu\text{-}CO)_4(PPh_3)_2(\eta^5\text{-}C_5H_5)_2]$ (M = Pd, Pt; R = Ph, Me, Et, Bu^n)	CV, EPR	232, 233
$[Re_7M(\mu_6\text{-}C)(CO)_{21}(L)]^{2-}$ [M = Pd, L = η^3-C_3H_5; M = Pt, L = η^3-2-methylallyl, (Me)$_3$]	CV	161

[a]Techniques utilized: ETCS, electron transfer-catalyzed synthesis; CV, cyclic voltammetry; DCP, d.c. polarography; DPP, differential pulse polarography; EPR, electron paramagnetic resonance spectroscopy; C, coulometry.

chemically reversible.[228,229] The d.c. polarograms and cyclic voltammograms of the related germanium-capped $[MoCo_2(\mu_3\text{-}GeR)(CO)_8(\eta^5\text{-}C_5H_5)]$ (R = Me, Ph) also showed two reduction waves. The first process (reversible) corresponded to formation of the radical anion $[MoCo_2(\mu_3\text{-}GeR)(CO)_8(\eta^5\text{-}C_5H_5)]^{\cdot-}$, an assignment supported by EPR data; the second process was chemically irreversible and had the characteristics of an ECE mechanism.[228]

The anion $[MoCo_2(\mu_3\text{-}CC_6H_4Me\text{-}4)(\mu\text{-}PPh_2)(CO)_6(\eta^5\text{-}C_5H_5)]^-$ underwent two reversible one-electron oxidations to give radical and cationic species sequentially (Fig. 79). The structure of the radical was established by a single-crystal X-ray diffraction study, which confirmed that the μ-PPh_2 ligand bridges a Mo—Co bond. Low-temperature protonation, thermally induced migration of the phosphido ligand to the Co—Co bond, and subsequent deprotonation gave an isomer of $[MoCo_2(\mu_3\text{-}CC_6H_4Me\text{-}4)(\mu\text{-}PPh_2)(CO)_6(\eta^5\text{-}C_5H_5)]^-$. The CV of the deprotonated form of this isomer showed two oxidation waves; the first wave was fully reversible and corresponded to the formation of the neutral radical species. Multiple scanning through the second wave, however, showed that rapid isomerization occurred via μ-PPh_2 bridge migration to give the Mo—Co bridging form.[70]

Reduction of the trimetallic clusters $[MoFeCo(\mu_3\text{-}PR)(CO)_8(\eta^5\text{-}C_5H_5)]$ (R = Me, Ph, Bu^t) was chemically and electrochemically irreversible, even at low temperatures, and is believed to arise from a fast fragmentation reaction following radical anion formation. Introduction of the third heterometal made reduction more difficult than reduction of its phosphinidene-capped iron-dicobalt cluster precursor.[228,229] The analogous tungsten-containing clusters $[WFeCo(\mu_3\text{-}PR)(CO)_8(\eta^5\text{-}C_5H_5)]$ (R = Me, Ph) were irreversibly reduced at more negative potentials.[229]

The chalcogen-capped clusters $[M_1M_2Co(\mu_3\text{-}S)(CO)_8(\eta^5\text{-}C_5H_5)]$ (M_1M_2 = MoFe, MoRu, WFe) and $[MoFeCo(\mu_3\text{-}Se)(CO)_8(\eta^5\text{-}C_5H_5)]$ underwent a one-electron, quasi-reversible reduction. Addition of an electron proceeded more readily for the clusters with the lighter metals and for the selenium capped cluster relative to its sulfur analogue.[229]

While trimetallic clusters have been the most intensively studied, clusters of higher nuclearity have also been investigated. Cyclic voltammetry of the tetra-capped cluster $[Cr_3Co(\mu_3\text{-}S)_4(CO)(\eta^5\text{-}C_5H_4Me)_3]$ in DMF revealed four oxidation waves, three (−1.07, −0.23, +0.23 V) reversible one-electron transfer steps, and an irreversible two-electron process (+0.80 V). Another tetra-capped cluster which has been electrochemically studied is $[Mo_2Co_2(\mu_3\text{-}S)_4(CO)_2(\eta^5\text{-}C_5H_4Et)_2]$. Recent work of Mansour *et al.* showed that this compound undergoes a single, reversible, one-electron reduction at −1.04 V, a significantly more positive reduction potential than that of its electron-rich 62 CVE nitrosyl analogue $[Mo_2Co_2(\mu_3\text{-}S)_4(NO)_2(\eta^5\text{-}C_5H_4Et)_2]$ (−1.53 V).[231] The electrochemistry of the butterfly cluster $[Mo_2Co_2(\mu_4\text{-}S)(\mu_3\text{-}S)_2(CO)_4(\eta^5\text{-}C_5H_4Me)_2]$ was complicated by an electrode-surface reaction

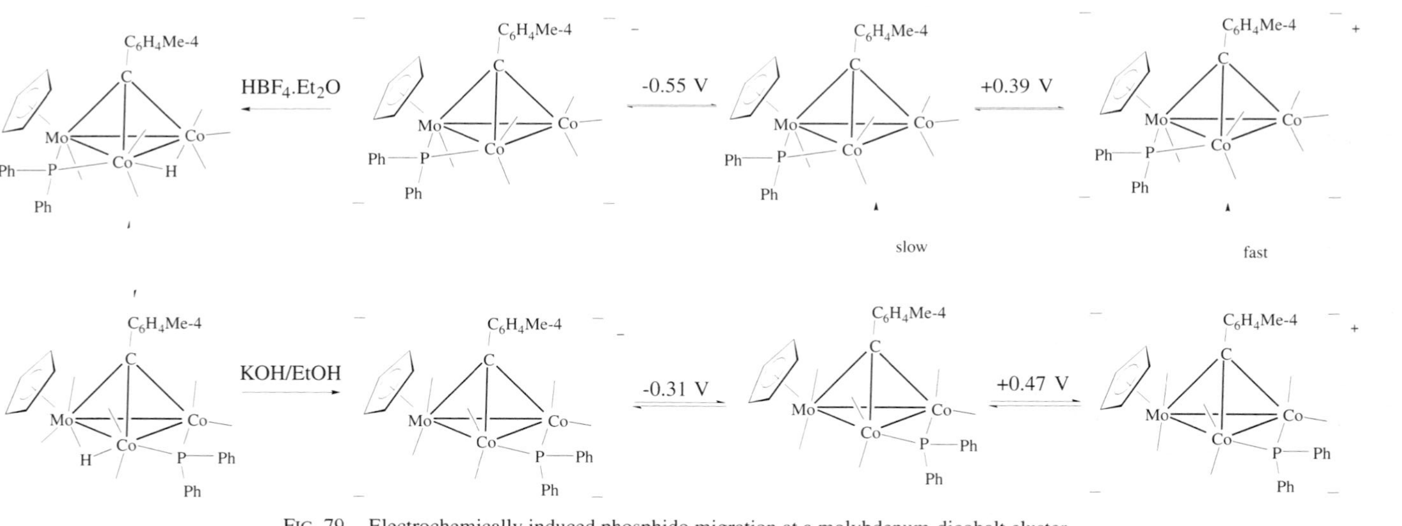

FIG. 79. Electrochemically induced phosphido migration at a molybdenum-dicobalt cluster.

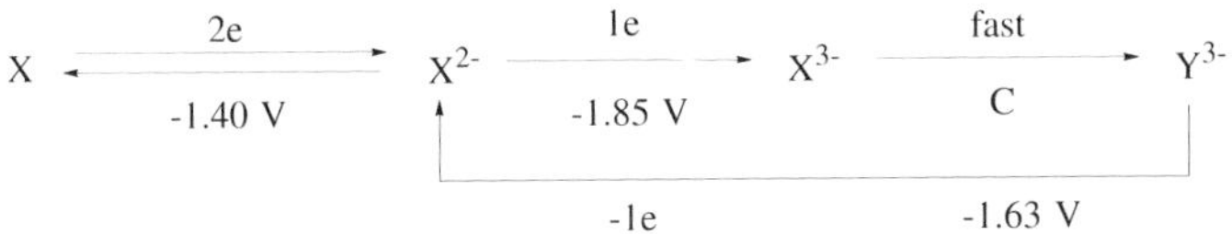

FIG. 80. Electrochemical behavior of a sulfido dimolybdenum-dicobalt cluster: X = $[Mo_2Co_2(\mu_4\text{-}S)(\mu_3\text{-}S)_2(CO)_4(\eta^5\text{-}C_5H_4Me)_2]$

that led to small spurious peaks; starting the reduction sweep at −0.70 V eliminated this problem and produced a well-defined two-electron reduction at −1.40 V. This is followed by a further cathodic wave at −1.85 V that is coupled to a reverse anodic wave at −1.63 V, behavior that can be explained by an EEC-type mechanism (Fig. 80).[172]

The clusters $[MoCo_3(\mu_3\text{-}\eta^2\text{-}PBu^tC_6H_4\text{-}2\text{-}PBu^t)(\mu\text{-}CO)(CO)_7(\eta^5\text{-}C_5H_5)]$ and $[Mo_2Co_2(\mu_3\text{-}\eta^2\text{-}PBu^tC_6H_4\text{-}2\text{-}PBu^t)(\mu\text{-}CO)(CO)_6(\eta^5\text{-}C_5H_5)_2]$, with tetrahedral cores, have contrasting electrochemical behavior; the latter exhibited a quasi-reversible oxidation wave at 0.65 V (E_p = 180 mV), while no oxidation wave was observed for the former (up to 0.75 V).[74]

The sulfur-capped tetrahedral cluster $[Mo_2Ni_2(\mu_3\text{-}S)_4(CO)_2(\eta^5\text{-}C_5H_4Me)_2]$ displayed fairly simple electrochemical behavior. Reversible, one-electron reduction waves were observed at −1.58 and −2.11 V, while oxidation processes at −0.11 and 0.09 V were irreversible.[172] An uncomplicated reduction cyclic voltammogram of $[W_2Co_2(\mu_3\text{-}S)_3(CO)_5(\eta^5\text{-}C_5H_4Et)_2]$ showed three facile electron-transfer processes. The first two reversible reductions (−1.11, −1.71 V) were followed by a third, "quasi-reversible" reduction at −2.23 V (reverse reoxidation at −2.04 V). The oxidation behavior for this cluster is more complex, though. An initial irreversible oxidation at 0.16 V was followed by four successive oxidations thought to be due to ECE processes.[231] The lower oxidation potential of the tungsten-containing clusters relative to their molybdenum analogues is consistent with tungsten-containing clusters being more electron rich. This trend was confirmed by the electrochemical studies on the systematically varied tetrametallic clusters $[M_2M'_2(\mu_3\text{-}CO)_2(\mu\text{-}CO)_4(PR_3)_2(\eta^5\text{-}C_5H_5)_2]$ (M = Cr, Mo, W; M′ = Pd, Pt; R = Ph, Me, Et, Bu^t). All combinations underwent an irreversible two-electron reduction corresponding to degradation of the cluster into identified fragments; the M′(I) centers were formally electroreduced to M′(0) in one step, accompanied by formation of $[M(CO)_3(\eta^5\text{-}C_5H_5)]^-$ (confirmed by IR). Oxidation occurred in two distinct one-electron transfers, with the second accompanied by some chemical decomposition. The radical cation generated by the first oxidation was reasonably stable and has been studied by EPR as a 106 K frozen solution; results suggested that the HOMO is located on the metals in the cluster. The stability of the dications was dependent on the solvent, with the most stable clusters being $[M_2Pt_2(\mu_3\text{-}CO)_2(\mu\text{-}CO)_4(PPh_3)_2(\eta^5\text{-}C_5H_5)_2]$ (M = Mo or W).

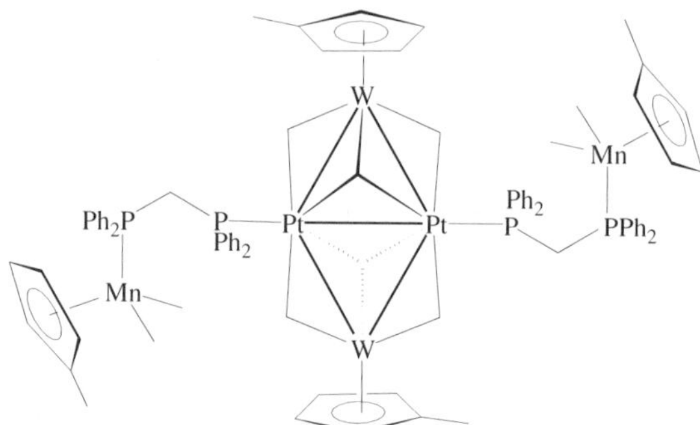

FIG. 81. $[W_2Pt_2(\mu_3\text{-}CO)_2(\mu\text{-}CO)_4\{Ph_2PCH_2PPh_2Mn(CO)_2(\eta^5\text{-}C_5H_4Me)\}_2(\eta^5\text{-}C_5H_4Me)_2]$

While the reduction potentials were sensitive to R substituent effects, the oxidation potentials were reasonably insensitive. This could indicate that the phosphine only plays a limited role in the HOMO, consistent with the EPR results.[232,233] The related $[W_2Pt_2(\mu_3\text{-}CO)_2(\mu\text{-}CO)_4[PPh_2CH_2PPh_2\{Mn(CO)_2(\eta^5\text{-}C_5H_4Me\}]_2(\eta^5\text{-}C_5H_4Me)_2]$ (Fig. 81), with pendant manganese metal centers, displayed some similarities to its bis(triphenylphosphine) analogue. In particular, a single irreversible reduction wave was seen at −1.45 V (c.f. −1.44 V); an associated reverse oxidation appeared at −0.18 V. Oxidation of the cluster occurred at 0.86 V and was followed by a reduction peak at −0.67 V in the reverse scan, thought to originate from an ECE process.[48]

The anodic scans of the three related octanuclear anionic clusters $[Re_7Pd(\mu_6\text{-}C)(CO)_{21}(\eta^3\text{-}C_3H_5)]^{2-}$, $[Re_7Pt(\mu_6\text{-}C)(CO)_{21}(Me)_3]^{2-}$ and $[Re_7Pt(\mu_6\text{-}C)(CO)_{21}(\eta^3\text{-}2\text{-methylallyl})]^{2-}$ showed two oxidation waves, the first quasi-reversible and the second almost completely irreversible. Oxidation potentials for the first process increased in the order $[Re_7Pt(\mu_6\text{-}C)(CO)_{21}(\eta^3\text{-}2\text{-methylallyl})]^{2-}$ (0.42 V) < $[Re_7Pd(\mu_6\text{-}C)(CO)_{21}(\eta^3\text{-}C_3H_5)]^{2-}$ (0.48 V) < $[Re_7Pt(\mu_6\text{-}C)(CO)_{21}(Me)_3]^{2-}$ (0.64 V). The large difference between the latter two can perhaps be explained by the oxidation state of platinum in these compounds; in the cluster with the highest oxidation potential, the capping metal is Pt(IV), while in the other clusters the capping metal is M(II).[161]

D. *Other Physical Methods*

The electronic, optical, and magnetic properties of metal clusters are of great current interest,[2] but these properties have been little studied with "very mixed"-metal clusters. This is to some extent a reflection of the difficulty of preparing high-nuclearity examples; many of these interesting properties become important upon increasing cluster size. The limited magnetic studies to date are

TABLE XIII

MAGNETISM STUDIES OF VERY MIXED-METAL CLUSTERS

Cluster	Measurements	Ref.
$[Cr_2Co(\mu_3\text{-}S)_2(\mu\text{-}SBu^t)(CO)_2(\eta^5\text{-}C_5H_4R)_2]$ (R = H, Me)	μ_{eff} vs T	234, 235
$[Cr_2M(\mu_3\text{-}S)_2(\mu\text{-}SBu^t)(CO)(PPh_3)(\eta^5\text{-}C_5H_5)_2]$ (M = Rh, Ir)	μ_{eff} vs T	236
$[Cr_3Co(\mu_3\text{-}S)_4(CO)(\eta^5\text{-}C_5R_5]$ (R = H, Me)	χ_m	144, 230
$[MNi_3(\mu_3\text{-}CO)_3(\eta^5\text{-}C_5H_5)_3(\eta^5\text{-}C_5H_4Me)]$ (M = Mo, W)	μ_{eff} vs T	146

summarized in Section III.D.1., with complementary MO studies in Section III.D.2. Few other studies have been reported, one example being the XPS binding energies of $[RePt_3(\mu\text{-}dppm)_3(CO)_3]^+$, $[RePt_3(\mu_3\text{-}S)_2(\mu\text{-}dppm)_3(CO)_3]^+$ and $[RePt_3(\mu\text{-}dppm)_3(CO)_3(S)]^+$ which reveal an increase along the series in Re $4f(^7/_2)$ binding energies. Sulfidation of $[RePt_3(\mu\text{-}dppm)_3(CO)_3]^+$ affords successively $[RePt_3(\mu\text{-}dppm)_3(CO)_3(S)]^+$ and $[RePt_3(\mu_3\text{-}S)_2(\mu\text{-}dppm)_3(CO)_3]^+$, and the XPS data are consistent with oxidation at Re as the Re=S group is formed, but oxidation at platinum on subsequent sulfidation when the μ_3-S groups are formed.[105]

1. *Magnetic Measurements of "Very Mixed"-Metal Clusters*

The limited magnetic measurements of "very mixed"-metal clusters are summarized in Table XIII. The magnetic behavior of some anti-ferromagnetic "very mixed"-metal carbonyl clusters (Fig. 82) has been studied by Pasynskii and coworkers. Temperature dependences of the magnetic susceptibilities χ_m of $[Cr_2Co(\mu_3\text{-}S)_2(\mu\text{-}SBu^t)(CO)_2(\eta^5\text{-}C_5H_4R)_2]$ (R = H, Me) have been determined using the Faraday method.[234,235] From χ_m, the effective magnetic moments μ_{eff} were calculated using the formula $\mu_{eff} = \sqrt{8\chi_m T}$, and plots of μ_{eff} vs T prepared. The plots fitted the Heisenberg-Dirac-Van Vleck (HDVV) model for two exchange-coupled ions [Cr(III), spin 3/2; Co(I) is diamagnetic] in the absence of orbital degeneracy of the complexes in the ground state. The related phosphine-substituted clusters $[Cr_2M(\mu_3\text{-}S)_2(\mu\text{-}SBu^t)(CO)(PPh_3)(\eta^5\text{-}C_5H_5)_2]$ (M = Rh, Ir) also exhibited a decrease in effective magnetic moment with temperature, which agreed

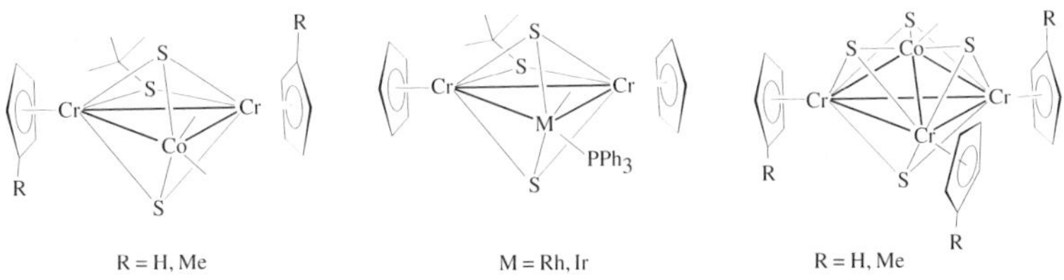

FIG. 82. Clusters examined for their magnetic behavior.

with the HDVV model. In contrast to the clusters above, the tetrametallic clusters $[Cr_3Co(\mu_3\text{-}S)_4(CO)(\eta^5\text{-}C_5H_4R)_3]$ (R = H, Me) were found to be diamagnetic (i.e., $\chi_m < 0$), consistent with the 18-electron saturation of each metal atom.[144,230] The magnetic susceptibilities of $[MNi_3(\mu_3\text{-}CO)_3(\eta^5\text{-}C_5H_5)_3(\eta^5\text{-}C_5H_4Me)]$ (M = Mo, W) in solution have been measured over a range of temperatures.[146] At 40°C, the magnetic moment of the molybdenum-nickel cluster corresponds to an average value of less than two unpaired electrons per molecule (1.72 μ_B), monotonically decreasing to 1.39 μ_B at −60°C. The tungsten-nickel cluster has magnetic moments of 1.19 μ_B at 20°C and 0.93 μ_B at −40°C. These data, in combination with the fact that these clusters are EPR-silent, suggest temperature-dependent singlet-triplet equilibria.

2. *Molecular Orbital Studies of "Very Mixed"-Metal Clusters*

The need to rationalize bonding, structural data, and magnetic and electrochemical behavior has encouraged MO calculations: a list of "very mixed"-metal clusters studied by MO methods is presented in Table XIV. In order to determine which chromium-based orbitals participated in the metal–metal and metal–ligand interactions of $[Cr_2Co(\mu_3\text{-}S)_2(\mu\text{-}SBu^t)(CO)_2(\eta^5\text{-}C_5H_4R)_2]$ and $[Cr_2Co_2(\mu_4\text{-}S)(\mu_3\text{-}S)_2(CO)_4(\eta^5\text{-}C_5H_4R)_2]$ (R = H, Me) (Fig. 83), the electronic structure of the fragment $[Cr_2(\mu_3\text{-}S)_2(\mu\text{-}SBu^t)(\eta^5\text{-}C_5H_5)]^-$ was calculated by the Extended Hückel (EH) method.[237] Interaction of the lower lying p-orbitals of the bridging sulfur atoms with the chromium d-orbitals resulted in the destabilization of the d-orbitals;

TABLE XIV

MOLECULAR ORBITAL STUDIES OF VERY MIXED-METAL CLUSTERS

Cluster	Method	Ref.
$[Cr_2Co_2(\mu_4\text{-}S)(\mu_3\text{-}S)_2(CO)_4(\eta^5\text{-}C_5H_4R)_2]$ (R = H, Me)	Extended-Hückel	237
$[Cr_2Ni_2(\mu_4\text{-}S)(\mu_3\text{-}S)_2(\eta^5\text{-}C_5H_5)_4]$	Extended-Hückel	237
$[Cr_2Co(\mu_3\text{-}S)_2(\mu\text{-}SBu^t)(CO)_2(\eta^5\text{-}C_5H_4R)_2]$ (R = H, Me)	Extended-Hückel	237
$[Cr_2Co(\mu_3\text{-}\eta^6\text{-}B_4H_7)(\mu\text{-}CO)(CO)_2(\eta^5\text{-}C_5H_5)_2]$	Fenske-Hall	238
$[MoCo_2(\mu_3\text{-}\eta^2\text{-}CCH_2)(CO)_8(\eta^5\text{-}C_5H_5)]^+$	Extended-Hückel	239
$[M_2Rh(CO)_5(\eta^5\text{-}C_5H_5)_3]$ (M = Mo, W)	Extended-Hückel	240
$[Cr_2Co_2(\mu_4\text{-}S)(\mu_3\text{-}S)_2(CO)_4(\eta^5\text{-}C_5H_4R)_2]$ (M = Cr, R = H, Me; M = Mo, R = H)	Extended-Hückel	173, 237
$[Mo_2Co_2(\mu_3\text{-}S)_4(CO)_2(\eta^5\text{-}C_5H_4Me)_2]$	Extended-Hückel	231
$[Mo_2Ni_2(\mu_4\text{-}CO)(\mu_3\text{-}S)_2(\eta^5\text{-}C_5H_5)_4]$	Extended-Hückel	241
$[RePt_3(\mu_3\text{-}I)(\mu\text{-}dppm)_3(CO)_3]$	Extended-Hückel	79
$[RePt_3(\mu\text{-}dppm)_3(CO)_3]^+$	Extended-Hückel	42, 83, 151
$[RePt_3(\mu\text{-}dppm)_3(CO)_4]^+$	Extended-Hückel	42
$[RePt_5(\mu\text{-}O)_3(\mu\text{-}OH)(CO)_5(OReO_3)(PCy_3)_4]^+$	Extended-Hückel	158

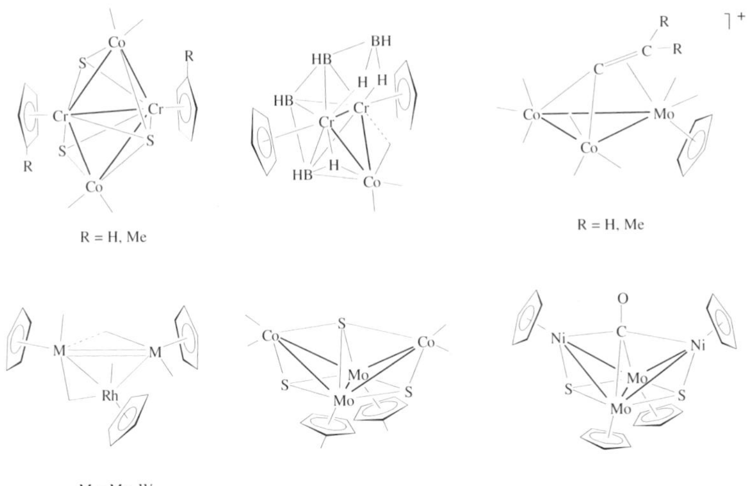

FIG. 83. Clusters examined by molecular orbital methods.

the unusual ordering of energy levels [δ^*(d–d) < δ(d–d)] resulted from the importance of the Cr—S interactions. The use of four half-occupied orbitals of the dichromium fragment for forming Cr—Co bonds with two $Co(CO)_2$ fragments was thought to explain the diamagnetic behavior of [$Cr_2Co_2(\mu_4\text{-}S)(\mu_3\text{-}S)_2(CO)_4(\eta^5\text{-}C_5H_4R)_2$], in contrast to [$Cr_2Co(\mu_3\text{-}S)_2(\mu\text{-}SBu^t)(CO)_2(\eta^5\text{-}C_5H_4R)_2$] which possess antiferromagnetic properties because of half-occupied orbitals. Bonding of all four (s, p_x, p_y, and p_z) orbitals of the sulfide bridges with the chromium atoms corresponds to participation of formally unshared electron pairs of the sulfur atoms in "double-bonding;" this is consistent with the decrease in Cr—S distances to 2.25–2.30 Å, significantly less than the sum of covalent radii of Cr (1.46 Å) and S (1.04 Å) atoms.

Fenske-Hall MO calculations were carried out on [$Cr_2Co(\mu_3\text{-}\eta^6\text{-}B_4H_7)(\mu\text{-}CO)(CO)_2(\eta^5\text{-}C_5H_5)_2$] (Fig. 83) in order to determine the nature of the interactions between the odd electron $Co(CO)_3$ and $Cr_2(B_4H_7)(\eta^5\text{-}C_5H_5)_2$ fragments, as the latter is effectively acting as a cluster "ligand" to the former. The principal changes between the ground state molecule [$Cr_2(B_4H_8)(\eta^5\text{-}C_5H_5)_2$] and the $Cr_2(B_4H_7)(\eta^5\text{-}C_5H_5)_2$ fragment in the cluster resulted in a shorter Cr—Cr distance and longer Cr—B distances and involved increases in energy for those fragment orbitals which are Cr—Cr antibonding and Cr—B bonding. Interaction of an empty σ orbital of the $Cr(CO)_3$ fragment with a filled B—H—Cr orbital caused the hydrogen atom to become Cr—B—Co face-bridging, rather than edge-bridging.[238]

In order to explain why the bridging vinylidene group of [$MoCo_2(\mu_3\text{-}\eta^2\text{-}CCH_2)(CO)_8(\eta^5\text{-}C_5H_5)]^+$ (Fig. 83) leans toward the molybdenum atom rather than a cobalt atom, EHMO calculations have been employed. Results showed that the vinylidene capping unit was particularly well stabilized by direct interaction with the molybdenum atom, where the positive charge is better tolerated. Antarafacial migration of the C=CH_2 group, possessing a large transition energy barrier, was shown to be severely disfavored. This is in accord with NMR experimental data, where the dimethylvinylidene analogue exhibited a sharp methyl singlet over the range 183–273 K resulting from the higher symmetry molybdenum-bonded vinylidene complex.[239]

The electronic structure of $[M_2Rh(\mu\text{-}CO)(CO)_4(\eta^5\text{-}C_5H_5)_3]$ (M=Mo, W) (Fig. 83) has been studied utilizing EHMO calculations. The nature of the metal–metal bonds in the clusters have been analyzed in terms of the bonding between the $[M_2(CO)_4(\eta^5\text{-}C_5H_5)_2]$ and $[Rh(CO)(\eta^5\text{-}C_5H_5)]$ fragments. Multiple M=M bonding in $[M_2(CO)_4(\eta^5\text{-}C_5H_5)_2]$ was realized by σ interactions between the orbitals $5d_{z^2}$ and $6p_z$, and π interactions via $5d_{yz}$ and $5d_{xz}$ orbitals. The energetic consequences of the step by step transformation of a terminal CO group, coordinated to M, into a carbonyl ligand bridging one of the M—Rh bonds has been considered; terminal coordination of all carbonyl groups led to an energetic destabilization of about 11 eV, while the experimentally determined geometry was calculated to have the greatest binding energy.[240]

The calculated EHMO energy levels for $M_2M'_2S_4$ type cubane clusters have been applied to $[Mo_2Co_2(\mu_4\text{-}S)(\mu_3\text{-}S)_2(CO)_4(\eta^5\text{-}C_5H_4Me)_2]$ (Fig. 83) in order to explain bonding behavior.[242] It was suggested that the HOMO would have σ(Mo—Mo) character and that the LUMO was most likely σ^*(Mo—Mo); addition of electrons to the cluster is believed to result in cleavage of the Mo—Mo, a proposal supported by electrochemical studies (see Section III.C.). Bonding of the sulfido ligands in $[Mo_2Co_2(\mu_4\text{-}S)(\mu_3\text{-}S)_2(CO)_4(\eta^5\text{-}C_5H_5)_2]$ has also been studied using EHMO methods.[173] The total M—S overlap population to the μ_4-S atom is 1.55 compared to 1.57 for the M—μ_3-S bonds (suggesting that they donate the same number of electrons). The same overlap population spread over four M—S bonds in the μ_4-ligand rather than concentrated in three bonds in the μ_3-ligand provides the rationale for the longer μ_4-S to metal bonds. It is primarily the p-orbitals of the μ_4-sulfur that overlap with the metal d-orbitals [overlap distributions Mo—S (40%), Co—S (60%)].

The butterfly metal core of $[Mo_2Ni_2(\mu_4\text{-}CO)(\mu_3\text{-}S)_2(\eta^5\text{-}C_5H_5)_4]$ is bridged by a crystallographically confirmed carbonyl with μ_4-η^1-coordination (Fig. 83). MO calculations have been used to support this assignment.[241] Orbital interactions of the μ_4-CO with the Mo atoms were both σ-bonding and multicenter bonding with the CO π-system. The Ni—CO bonding was best described as a dative bond from Ni to the π^* orbitals of the carbonyl. Low C—O overlap population reflected the additional electron donation into the C—O antibonding orbitals and was consistent with a low ν_{CO}.

The bonding in $[RePt_3(\mu\text{-dppm})_3(CO)_3]^+$ can be understood in terms of the donation of electron density from three filled Pt—Pt bonding orbitals of a Pt_3

(μ-dppm)$_3$ fragment to the three vacant acceptor orbitals of a $[Re(CO)_3]^+$ fragment (or $[ReO_3]^+$). In the MO treatment, the fragment orbitals combine to form bonding and antibonding MOs each with $a_1 + e$ symmetry, and there are six cluster electrons which just fill the bonding MOs.[42,66,83,151] Reaction of this coordinatively unsaturated cluster (54 CVE) with CO gives $[RePt_3(\mu\text{-dppm})_3(CO)_4]^+$ by selective addition to rhenium.[42] The bonding may still be considered in terms of a $Pt_3(\mu\text{-dppm})_3$ fragment donating to $[Re(CO)_4]^+$, but the latter has only two acceptor orbitals, limiting the interaction to two donor-acceptor bonds. When compared to $[RePt_3(\mu\text{-dppm})_3(CO)_3]^+$, weaker rhenium-platinum bonding is expected. Halide ions add at the Pt_3 face of $[RePt_3(\mu\text{-dppm})_3(CO)_3]^+$ to yield $[RePt_3(\mu_3\text{-I})(\mu\text{-dppm})_3(CO)_3]$. Analysis of the interaction of I^- with the model cluster $[RePt_3(\mu\text{-H}_2PCH_2PH_2)_3(CO)_3]^+$ to give the simplified analogue $[RePt_3(\mu_3\text{-I})(\mu\text{-H}_2PCH_2PH_2)_3(CO)_3]^+$ has been made using the EHMO method.[79] The filled *p*-orbitals of I^- overlap with unoccupied platinum p_z orbitals. The calculations suggest that, although the net bonding is weak, the $Pt_3(\mu_3\text{-I})$ interaction is covalent in nature and that iodide can act as a weak six-electron donor by using all its filled *p*-orbitals in bonding.

In order to study bonding in $[RePt_5(\mu\text{-O})_3(\mu\text{-OH})(CO)_5(OReO_3)(PCy_3)_4]^+$, EH calculations have been carried out on the model cluster $[RePt_5(\mu\text{-O})_3(\mu\text{-OH})(CO)_5(O)(PH_3)_4]^+$.[158] Calculations indicate at least a partial triple-bond character in the central Re—Pt bond, with complications due to the presence of peripheral metal–metal bonds. Structural data show a very short Re—Pt distance and the very unusual structure suggests a degree of multiple metal–metal bonding in $[RePt_5(\mu\text{-O})_3(\mu\text{-OH})(CO)_5(OReO_3)(PCy_3)_4]^+$, consistent with the theoretical results.

IV

CONCLUSION AND OUTLOOK

With flexible routes into "very mixed"-metal clusters in hand, the time is ripe for systematic investigations of their reactivity and physical properties. Ligand substitution generally proceeds with metalloselectivity, but the reactions are frequently not site-selective; many of these complexes are highly fluxional, and more than one configuration is often sufficiently stable so that mixtures of isomers are obtained. Characterization of these configurations has largely been crystallographic, which generally results in one configuration only being identified; comprehensive spectroscopic studies are needed to identify all configurations. A number of examples of C-ligand transformation have been reported and, in almost all instances, this involves a heterometallic bond or face. This likely indicates the importance of heterometallic centers in effecting the transformations, but mechanistic studies to confirm this remain to be carried out; it is always possible, though unlikely in all instances, that the transformation occurs elsewhere on the cluster, and that the heterometallic unit is the favorable center to stabilize the resulting fragments. Investigations of "very mixed"-metal cluster catalysis have been dominated by clusters with metal combinations effective in heterogeneous catalysis (e.g., Mo—Co in HDS; Re—Pt in petroleum reforming); clearly, there is great scope for exploring the potential of clusters with other metal combinations. The physical properties of "very mixed"-metal clusters have been intensively studied only recently. The significance of heterometals in controlling ligand mobility and tuning electrochemical response has yet to be fully exploited, and other physical properties (e.g., magnetism, nonlinear optical properties) are almost untouched. These are likely to be of increasing importance in the future.

V
APPENDIX: ABBREVIATIONS

arphos	1-(diphenylarsino)-2-(diphenylphosphino)ethane
C	coulometry
cod	1,5-cyclooctadiene
CV	cyclic voltammetry
CVE	cluster valence electrons
DCP	d.c. polarography
dmat	2-(dimethylethylamino)toluene-*C,N*
dmba	dimethyl benzylamine-C^2-*N*
dmpe	1,2-bis(dimethylphosphino)ethane
DPP	differential pulse polarography
dppa	bis(diphenylphosphino)acetylene
dppe	1,2-bis(diphenylphosphino)ethane
dppm	bis(diphenylphosphino)methane
ECE	electrochemical-chemical-electrochemical
EEC	electrochemical-electrochemical-chemical
ETCS	electron transfer-catalyzed synthesis
EXAFS	extended x-ray absorption fine structure
HDS	hydrodesulfurization
pz	pyrazolyl
TEM	transmission electron microscopy
THF	tetrahydrofuran
TPD	temperature-programmed desorption/decomposition
TPR	temperature-programmed reduction
XANES	x-ray absorption near-edge spectroscopy
Xy	xylyl

ACKNOWLEDGMENTS

We thank the Australian Research Council for support. N.T.L. is the recipient of an Australian Postgraduate Award, and M.G.H. acknowledges an ARC Australian Research Fellowship and an ARC Senior Research Fellowship.

REFERENCES

(1) Shriver, D. F.; Kaesz, H. D.; Adams, R. D., Eds. *The Chemistry of Metal Cluster Complexes;* V.C.H.: New York, 1990.
(2) de Jongh, L. L., Ed. *Physics and Chemistry of Metal Cluster Complexes;* Kluwer: Amsterdam, 1994.
(3) Stone, F. G. A. *Angew. Chem., Int. Ed. Engl.* **1984,** *23,* 89.
(4) Vahrenkamp, H. *Comments Inorg. Chem.* **1985,** *4,* 253.
(5) Vahrenkamp, H. *Adv. Organomet. Chem.* **1983,** *22,* 169.
(6) Mingos, D. M. P. *Acc. Chem. Res.* **1984,** *17,* 311.
(7) Hoffmann, R. *Angew. Chem., Int. Ed. Engl.* **1982,** *21,* 1711.
(8) Gladfelter, W. L.; Geoffroy, G. L. *Adv. Organomet. Chem.* **1981,** *18,* 207.
(9) Farrugia, L. J. In *Comprehensive Organometallic Chemistry II;* Abel, E. W.; Stone, F. G. A.; Wilkinson, G., Eds.; Pergamon Press: Oxford, 1994, Vol. 10, Ch. 4.
(10) Chi, Y. In *Comprehensive Organometallic Chemistry II;* Abel, E. W.; Stone, F. G. A.; Wilkinson, G., Eds.; Pergamon Press: Oxford, 1994, Vol. 6, Ch. 3.
(11) Bruce, M. I. *J. Organomet. Chem.* **1983,** *257,* 417.
(12) Roberts, D. A.; Geoffroy, G. L. In *Comprehensive Organometallic Chemistry;* Abel, E. W.; Stone, F. G. A.; Wilkinson, G. Eds.; Pergamon Press: Oxford, 1982, Vol. 6, pp. 763.
(13) Comstock, M. C.; Shapley, J. R. *Coord. Chem. Rev.* **1995,** *143,* 501.
(14) Sappa, E.; Tiripicchio, A.; Braunstein, P. *Coord. Chem. Rev.* **1985,** *65,* 219.
(15) Vahrenkamp, H. *Pure Appl. Chem.* **1989,** *61,* 1777.
(16) Clark, D. T.; Sutin, K. A.; McGlinchey, M. J. *Organometallics* **1989,** *8,* 155.
(17) Sutin, K. A.; Kolis, J. W.; Mlekuz, M.; Bougeard, P.; Sayer, B. G.; Quilliam, M. A.; Faggiani, R.; Lock, C. J. L.; McGlinchey, M. J.; Jaouen, G. *Organometallics* **1987,** *6,* 439.
(18) Jeffery, J. C.; Lawrence-Smith, J. G. *J. Organomet. Chem.* **1985,** *280,* C34.
(19) Chetcuti, M. J.; Chetcuti, P. A. M.; Jeffery, J. C.; Mills, R. M.; Mitrprachachon, P.; Pickering, S. J.; Stone, F. G. A.; Woodward, P. *J. Chem. Soc., Dalton Trans.* **1982,** 699.
(20) Dunn, P.; Jeffery, J. C.; Sherwood, P. *J. Organomet. Chem.* **1986,** *311,* C55.
(21) Scott, I. D.; Smith, D. O.; Went, M. J. *J. Chem. Soc., Dalton Trans.* **1989,** 1375.
(22) Jeffery, J. C.; Ruiz, M. A.; Stone, F. G. A. *J. Chem. Soc., Dalton Trans.* **1988,** 1131.
(23) Richter, F.; Vahrenkamp, H. *Angew. Chem., Int. Ed. Engl.* **1980,** *19,* 65.
(24) Mani, D.; Vahrenkamp, H. *Chem. Ber.* **1986,** *119,* 3639.
(25) Wang, Q.-L.; Chen, S.-N.; Wang, X.; Sun, W.-H.; Wang, H.-Q.; Yang, S.-Y. *Polyhedron* **1996,** *15,* 2613.
(26) Müller, M.; Schacht, H.-T.; Fischer, K.; Ensling, J.; Gütlich, P.; Vahrenkamp, H. *Inorg. Chem.* **1986,** *25,* 4032.
(27) Planalp, R. P.; Vahrenkamp, H. *Organometallics* **1987,** *6,* 492.
(28) Beurich, H.; Blumhofer, R.; Vahrenkamp, H. *Chem. Ber.* **1982,** *115,* 2409.
(29) Blumhofer, R.; Vahrenkamp, H. *Chem. Ber.* **1986,** *119,* 683.
(30) Schacht, H.-T.; Vahrenkamp, H. *Chem. Ber.* **1989,** *122,* 2239.
(31) Mani, D.; Schacht, H.-T.; Powell, A.; Vahrenkamp, H. *Organometallics* **1987,** *6,* 1360.
(32) Curnow, O. J.; Kampf, J. W.; Curtis, M. D.; Shen, J.-K.; Basolo, F. *J. Am. Chem. Soc.* **1994,** *116,* 224.
(33) Curnow, O. J.; Curtis, M. D.; Kampf, J. W. *Organometallics* **1997,** *16,* 2523.

(34) Waterman, S. M.; Humphrey, M. G.; Tolhurst, V.-A.; Skelton, B. W.; White, A. H.; Hockless, D. C. R. *Organometallics* **1996,** *15,* 934.
(35) Lucas, N. T.; Whittall, I. R.; Humphrey, M. G.; Hockless, D. C. R.; Perera, M. P. S.; Williams, M. L. *J. Organomet. Chem.* **1997,** *540,* 147.
(36) Waterman, S. M.; Humphrey, M. G.; Hockless, D. C. R. *J. Organomet. Chem.* **1998,** *555,* 25.
(37) Waterman, S. M.; Humphrey, M. G.; Hockless, D. C. R. *J. Organomet. Chem.* **1998,** *565,* 81.
(38) Lee, J.; Humphrey, M. G.; Hockless, D. C. R.; Skelton, B. W.; White, A. H. *Organometallics* **1993,** *12,* 3468.
(39) Lucas, N. T.; Humphrey, M. G.; Healy, P. C.; Williams, M. L. *J. Organomet. Chem.* **1997,** *545–546,* 519.
(40) Davies, S. J.; Howard, J. A. K.; Pilotti, M. U.; Stone, F. G. A. *J. Chem. Soc., Dalton Trans.* **1989,** 2289.
(41) Awang, M. R.; Carriedo, G. A.; Howard, J. A. K.; Mead, K. A.; Moore, I.; Nunn, C. M.; Stone, F. G. A. *J. Chem. Soc., Chem. Commun.* **1983,** 964.
(42) Xiao, J.; Hao, L.; Puddephatt, R. J.; Manojlovic-Muir, L.; Muir, K. W.; Torabi, A. A. *Organometallics* **1995,** *14,* 4183.
(43) Muir, K. W.; Manojlovic-Muir, L.; Ashgar Torabi, A. *J. Organomet. Chem.* **1997,** *536–537,* 319.
(44) Xiao, J.; Hao, L.; Puddephatt, R. J.; Manojlovic-Muir, L.; Muir, K. W. *J. Am. Chem. Soc.* **1995,** *117,* 6316.
(45) Davies, S. J.; Howard, J. A. K.; Musgrove, R. J.; Stone, F. G. A. *Angew. Chem., Int. Ed. Engl.* **1989,** *28,* 624.
(46) Davies, S. J.; Howard, J. A. K.; Musgrove, R. J.; Stone, F. G. A. *J. Chem. Soc., Dalton Trans.* **1989,** 2269.
(47) Curnow, O. J.; Kampf, J. W.; Curtis, M. D.; Mueller, B. L. *Organometallics* **1992,** *11,* 1984.
(48) Braunstein, P.; Knorr, M.; Strampfer, M.; Dusausoy, Y.; Bayeul, D.; DeCian, A.; Fischer, J.; Zanello, P. *J. Chem. Soc., Dalton Trans.* **1994,** 1533.
(49) Bender, R.; Braunstein, P.; Jud, J.-M.; Dusausoy, Y. *Inorg. Chem.* **1983,** *22,* 3394.
(50) Bender, R.; Braunstein, P.; Jud, J.-M.; Dusausoy, Y. *Inorg. Chem.* **1984,** *23,* 4489.
(51) Braunstein, P.; de Meric de Bellefon, C.; Bouaoud, S.-E.; Grandjean, D.; Halet, J.-F.; Saillard, J.-Y. *J. Am. Chem. Soc.* **1991,** *113,* 5282.
(52) Curtis, M. D.; Riaz, U.; Curnow, O. J.; Kampf, J. W.; Rheingold, A. L.; Haggerty, B. S. *Organometallics* **1995,** *14,* 5337.
(53) Bars, O.; Braunstein, P.; Jud, J.-M. *Nouv. J. Chem.* **1984,** *8,* 771.
(54) Pasynskii, A. A.; Eremenko, I. L.; Zalmanovich, V. R.; Kaverin, V. V.; Orazsakhatov, B.; Novotortsev, V. M.; Ellert, O. G.; Yanovsky, A. I.; Struchkov, Y. T. *J. Organomet. Chem.* **1988,** *356,* 57.
(55) Braunstein, P.; Ries, M.; de Méric de Bellefon, C.; Dusausoy, Y.; Mangeot, J.-P. *J. Organomet. Chem.* **1988,** *355,* 533.
(56) Mlekuz, M.; Bougeard, P.; Sayer, B. G.; Faggiani, R.; Lock, C. J. L.; McGlinchey, M. J.; Jaouen, G. *Organometallics* **1985,** *4,* 2046.
(57) Curtis, M. D.; Curnow, O. J. *Organometallics* **1994,** *13,* 2489.
(58) Riaz, U.; Curnow, O.; Curtis, M. D. *J. Am. Chem. Soc.* **1991,** *113,* 1416.
(59) Werner, H.; Thometzek, P.; Krüger, C.; Kraus, H.-J. *Chem. Ber.* **1986,** *119,* 2777.
(60) Riaz, U.; Curtis, M. D.; Rheingold, A.; Haggerty, B. S. *Organometallics* **1990,** *9,* 2647.
(61) Green, M.; Howard, J. A. K.; James, A. P.; Nunn, C. M.; Stone, F. G. A. *J. Chem. Soc., Dalton Trans.* **1986,** 187.
(62) Shapley, J. R.; Humphrey, M. G.; McAteer, C. H. In *Selectivity in Catalysis;* Davis, M. E.; Suib, S. L., Eds.; American Chemical Society: Washington, DC, 1993; p. 127.
(63) Mansour, M. A.; Curtis, M. D.; Kampf, J. W. *Organometallics* **1995,** *14,* 5460.
(64) Mansour, M. A.; Curtis, M. D.; Kampf, J. W. *Organometallics* **1997,** *16,* 3363.

(65) Neumann, H.-P.; Ziegler, M. L. *J. Organomet. Chem.* **1989,** *377,* 255.
(66) Xiao, J.; Puddephatt, R. J. *J. Am. Chem. Soc.* **1994,** *116,* 1129.
(67) Hao, L.; Xiao, J.; Vittal, J. J.; Puddephatt, R. J. *Angew. Chem., Int. Ed. Engl.* **1995,** *34,* 346.
(68) Riaz, U.; Curnow, O. J.; Curtis, M. D. *J. Am. Chem. Soc.* **1994,** *116,* 4357.
(69) Braunstein, P.; de Jesús, E.; Tiripicchio, A.; Ugozzoli, F. *Inorg. Chem.* **1992,** *31,* 411.
(70) Bradford, M. R.; Connelly, N. G.; Harrison, N. C.; Jeffery, J. C. *Organometallics* **1989,** *8,* 1829.
(71) Ciani, G.; Moret, M.; Sironi, A.; Antognazza, P.; Beringhelli, T.; D'Alfonso, G.; Della Pergola, R.; Minoja, A. *J. Chem. Soc., Chem. Commun.* **1991,** 1255.
(72) Bergamo, M.; Beringhelli, T.; D'Alfonso, G.; Ciani, G.; Moret, M.; Sironi, A. *Organometallics* **1996,** *15,* 1637.
(73) Bergamo, M.; Beringhelli, T.; Ciani, G.; D'Alfonso, G.; Moret, M.; Sironi, A. *Inorg. Chim. Acta* **1997,** *259,* 291.
(74) Kyba, E. P.; Kerby, M. C.; Kashyap, R. P.; Mountzouris, J. A.; Davis, R. E. *J. Am. Chem. Soc.* **1990,** *112,* 905.
(75) Davies, S. J.; Stone, F. G. A. *J. Chem. Soc., Dalton Trans.* **1989,** 1865.
(76) Chetcuti, M. J.; Fanwick, P. E.; Gordon, J. C. *Inorg. Chem.* **1991,** *30,* 4710.
(77) Churchill, M. R.; Biondi, L. V. *J. Organomet. Chem.* **1989,** *366,* 265.
(78) Shapley, J. R.; McAteer, C. H.; Churchill, M. R.; Biondi, L. V. *Organometallics* **1984,** *3,* 1595.
(79) Xiao, J.; Hao, L.; Puddephatt, R. J.; Manojlovic-Muir, L.; Muir, K. W.; Torabi, A. A. *Organometallics* **1995,** *14,* 2194.
(80) Xiao, J.; Hao, L.; Puddephatt, R. J.; Manojlovic-Muir, L.; Muir, K. W.; Torabi, A. A. *J. Chem. Soc., Chem. Commun.* **1994,** 2221.
(81) Hao, L.; Xiao, J.; Vittal, J.; Puddephatt, R. J.; Manojlovic-Muir, L.; Muir, K. W.; Torabi, A. A. *Inorg. Chem.* **1996,** *35,* 658.
(82) Davies, S. J.; Howard, J. A. K.; Pilotti, M. U.; Stone, F. G. A. *J. Chem. Soc., Chem. Commun.* **1989,** 190.
(83) Xiao, J.; Vittal, J. J.; Puddephatt, R. J. *J. Am. Chem. Soc.* **1993,** *115,* 7882.
(84) Jeffery, J. C.; Lewis, D. B.; Lewis, G. E.; Parrott, M. J.; Stone, F. G. A. *J. Chem. Soc., Dalton Trans.* **1986,** 1717.
(85) Bernhardt, W.; Schacht, H.-T.; Vahrenkamp, H. *Z. Naturforsch.* **1989,** *44b,* 1060.
(86) Bernhardt, W.; von Schering, C.; Vahrenkamp, H. *Angew. Chem., Int. Ed. Engl.* **1986,** *25,* 279.
(87) Pasynskii, A. A.; Eremenko, I. L.; Nefedov, S. E.; Kolobkov, B. I.; Shaposhnikova, A. D.; Stadnitchenko, R. A.; Drab, M. V.; Struchkov, Y. T.; Yanovsky, A. I. *New. J. Chem.* **1994,** *18,* 69.
(88) Adams, R. D.; Belinski, J. A. *Organometallics* **1991,** *10,* 2114.
(89) Wu, H.-P.; Zhao, Z.-Y.; Liu, S.-M.; Ding, E.-R.; Yin, Y.-Q. *Polyhedron* **1996,** *15,* 4117.
(90) Churchill, M. R.; Biondi, L. V.; Shapley, J. R.; McAteer, C. H. *J. Organomet. Chem.* **1985,** *280,* C63.
(91) Churchill, M. R.; Biondi, L. V. *J. Organomet. Chem.* **1988,** *353,* 73.
(92) Heping, W.; Yuanqi, Y.; Qingchuan, Y. *Polyhedron* **1996,** *15,* 43.
(93) Song, L.-C.; Shen, J.-Y.; Wu, X.-X.; Hu, Q.-M. *Polyhedron* **1998,** *17,* 35.
(94) Ding, E.-R.; Yin, Y.-Q.; Sun, J. *J. Organomet. Chem.* **1998,** *559,* 157.
(95) Song, L.-C.; Dong, Y.-B.; Hu, Q.-M.; Li, Y.-K.; Sun, J. *Polyhedron* **1998,** *17,* 1579.
(96) Duffy, D. N.; Kassis, M. M.; Rae, A. D. *J. Organomet. Chem.* **1993,** *460,* 97.
(97) Blum, T.; Braunstein, P.; Tiripicchio, A.; Tiripicchio Camellini, M. *Organometallics* **1989,** *8,* 2504.
(98) Curnow, O. J.; Kampf, J. W.; Curtis, M. D. *Organometallics* **1991,** *10,* 2546.
(99) Waterman, S. M.; Tolhurst, V.-A.; Humphrey, M. G.; Skelton, B. W.; White, A. H. *J. Organomet. Chem.* **1996,** *515,* 89.
(100) Druker, S. H.; Curtis, M. D. *J. Am. Chem. Soc.* **1995,** *117,* 6366.

(101) Curtis, M. D.; Druker, S. H.; Goossen, L.; Kampf, J. W. *Organometallics* **1997,** *16,* 231.
(102) Davies, S. J.; Howard, J. A. K.; Pilotti, M. U.; Stone, F. G. A. *J. Chem. Soc., Dalton Trans.* **1989,** 1855.
(103) Curtis, M. D. *J. Cluster Sci.* **1996,** *7,* 247.
(104) Curtis, M. D.; Druker, S. H. *J. Am. Chem. Soc.* **1997,** *119,* 1027.
(105) Hao, L.; Xiao, J.; Vittal, J. J.; Puddephatt, R. J. *J. Chem. Soc., Chem. Commun.* **1994,** 2183.
(106) Hao, L.; Xiao, J.; Vittal, J. J.; Puddephatt, R. J. *Organometallics* **1997,** *16,* 2165.
(107) Vahrenkamp, H. *Phil. Trans. R. Soc. Lond. A* **1982,** *308,* 17.
(108) Stone, F. G. A. *Phil. Trans. R. Soc. Lond. A* **1982,** *308,* 87.
(109) Stone, F. G. A. *Pure Appl. Chem.* **1986,** *58,* 529.
(110) Jensen, S.; Robinson, B. H.; Simpson, J. *J. Chem. Soc., Chem. Commun.* **1983,** 1081.
(111) Heping, W.; Zhuanyun, Z.; Yuanqi, Y.; Daosen, J.; Xiaoying, H. *Polyhedron* **1995,** *14,* 1543.
(112) Wu, H.-P.; Yin, Y.-Q.; Huang, X.-Y.; Yu, K.-B. *J. Organomet. Chem.* **1995,** *498,* 119.
(113) Wu, H.-P.; Yin, Y.-Q.; Huang, X.-Y. *Polyhedron* **1995,** *14,* 2993.
(114) Wu, H.-P.; Yin, Y.-Q.; Huang, X.-Y. *Inorg. Chim. Acta* **1997,** *255,* 167.
(115) Schacht, H. T.; Vahrenkamp, H. *J. Organomet. Chem.* **1990,** *381,* 261.
(116) Sutin, K. A.; Li, L.; Frampton, C. S.; Sayer, B. G.; McGlinchey, M. J. *Organometallics* **1991,** *10,* 2362.
(117) Duffy, D. N.; Kassis, M. M.; Rae, A. D. *Acta Cryst.* **1991,** *C47,* 2343.
(118) Fischer, K.; Deck, W.; Schwarz, M.; Vahrenkamp, H. *Chem. Ber.* **1985,** *118,* 4946.
(119) Mahe, C.; Patin, H.; Le Marouille, J.-Y.; Benoit, A. *Organometallics* **1983,** *2,* 1051.
(120) Song, L.-C.; Shen, J.-Y.; Hu, Q.-M.; Qin, X.-D. *Polyhedron* **1995,** *14,* 2079.
(121) Honrath, U.; Vahrenkamp, H. *Z. Naturforsch.* **1984,** *39b,* 559.
(122) Ding, E.-R.; Yin, Y.-Q.; Sun, J. *Polyhedron* **1997,** *16,* 3067.
(123) Ding, E.-R.; Wu, S.-L.; Xia, C.-G.; Yin, Y.-Q. *J. Organomet. Chem.* **1998,** *568,* 157.
(124) Ding, E.-R.; Li, Q.-S.; Yin, Y.-Q.; Sun, J. *J. Chem. Res. (S)* **1998,** 624.
(125) Bantel, H.; Powell, A. K.; Vahrenkamp, H. *Chem. Ber.* **1990,** *123,* 1607.
(126) Beurich, H.; Vahrenkamp, H. *Angew. Chem., Int. Ed. Engl.* **1981,** *20,* 98.
(127) Ziegler, M. L.; Neumann, H.-P. *Chem. Ber.* **1989,** *122,* 25.
(128) Gorzellik, M.; Nuber, B.; Bohn, T.; Ziegler, M. L. *J. Organomet. Chem.* **1992,** *429,* 173.
(129) Richter, F.; Roland, E.; Vahrenkamp, H. *Chem. Ber.* **1984,** *117,* 2429.
(130) Roland, E.; Vahrenkamp, H. *Chem. Ber* **1984,** *117,* 1039.
(131) Albiez, T.; Bantel, H.; Vahrenkamp, H. *Chem. Ber.* **1990,** *123,* 1805.
(132) Kaganovich, V. S.; Slovokhotov, Y. L.; Mironov, A. V.; Struchkov, Y. T.; Rybinskaya, M. I. *J. Organomet. Chem.* **1989,** *372,* 339.
(133) Gusbeth, P.; Vahrenkamp, H. *Chem. Ber.* **1985,** *118,* 1758.
(134) Richter, F.; Beurich, H.; Vahrenkamp, H. *J. Organomet. Chem.* **1979,** *166,* C5.
(135) Richter, F.; Vahrenkamp, H. *Organometallics* **1982,** *1,* 756.
(136) Beurich, H.; Vahrenkamp, H. *Angew. Chem., Int. Ed. Engl.* **1978,** *17,* 863.
(137) Sun, W.-H.; Wang, H.-Q.; Yang, S.-Y.; Zhou, Q.-F. *Polyhedron* **1994,** *13,* 389.
(138) Bernhardt, W.; Vahrenkamp, H. *J. Organomet. Chem.* **1988,** *355,* 427.
(139) Churchill, M. R.; Li, Y.-J.; Shapley, J. R.; Foose, D. S.; Uchiyama, W. S. *J. Organomet. Chem.* **1986,** *312,* 121.
(140) Kuznetsov, V. F.; Bensimon, C.; Facey, G. A.; Grushin, V. V.; Alper, H. *Organometallics* **1997,** *16,* 97.
(141) Richter, F.; Müller, M.; Gärtner, N.; Vahrenkamp, H. *Chem. Ber.* **1984,** *117,* 2438.
(142) Braunstein, P.; de Méric de Bellefon, C.; Ries, M. *Inorg. Chem.* **1990,** *29,* 1181.
(143) Braunstein, P.; de Méric de Bellefon, C.; Ries, M. *J. Organomet. Chem.* **1984,** *262,* C14.
(144) Pasynskii, A. A.; Eremenko, I. L.; Orazsakhatov, B.; Kalinnikov, V. T.; Aleksandrov, G. G.; Struchkov, Y. T. *J. Organomet. Chem.* **1981,** *214,* 367.

(145) Eremenko, I. L.; Pasynskii, A. A.; Katugin, A. S.; Orazsakhatov, B.; Shirokii, V. L.; Shklover, V. E.; Struchkov, Y. T. *J. Organomet. Chem.* **1988,** *345,* 177.
(146) Chetcuti, M. J.; Huffman, J. C.; McDonald, S. R. *Inorg. Chem.* **1989,** *28,* 238.
(147) Braunstein, P.; Oswald, B.; Tiripicchio, A.; Tiripicchio-Camellini, M. *Angew. Chem., Int. Ed. Engl.* **1990,** *29,* 1140.
(148) Braunstein, P.; Jud, J.-M.; Dusausoy, Y.; Fischer, J. *Organometallics* **1983,** *2,* 180.
(149) Elliot, G. P.; Howard, J. A. K.; Mise, T.; Moore, I.; Nunn, C. M.; Stone, F. G. A. *J. Chem. Soc., Dalton Trans.* **1986,** 2091.
(150) Davies, S. J.; Stone, F. G. A. *J. Chem. Soc., Dalton Trans.* **1989,** 785.
(151) Xiao, J.; Kristof, E.; Vittal, J. J.; Puddephatt, R. J. *J. Organomet. Chem.* **1995,** *490,* 1.
(152) Beringhelli, T.; Ceriotti, A.; Ciani, G.; D'Alfonso, G.; Garlaschelli, L.; Della Pergola, R.; Moret, M.; Sironi, A. *J. Chem. Soc., Dalton Trans.* **1993,** 199.
(153) Carr, N.; Mullica, D. F.; Sappenfield, E. L.; Stone, F. G. A.; Went, M. J. *Organometallics* **1993,** *12,* 4350.
(154) Bantel, H.; Powell, A. K.; Vahrenkamp, H. *Chem. Ber.* **1990,** *123,* 677.
(155) Tunik, S. P.; Shipil, P. N.; Vlasov, A. V.; Denisov, V. R.; Nikols'kii, A. B.; Dolgushin, F. M.; Yanovsky, A. I.; Struchkov, Y. T. *J. Organomet. Chem.* **1996,** *515,* 11.
(156) Davies, S. J.; Elliot, G. P.; Howard, J. A. K.; Nunn, C. M.; Stone, F. G. A. *J. Chem. Soc., Dalton Trans.* **1987,** 2177.
(157) Eremenko, I. L.; Pasynskii, A. A.; Katugin, A. S.; Zalmanovitch, V. R.; Orazsakhatov, B.; Sleptsova, S. A.; Nekhaev, A. I.; Kaverin, V. V.; Ellert, O. G.; Novotortsev, V. M.; Yanovsky, A. I.; Shklover, V. E.; Struchkov, Y. T. *J. Organomet. Chem.* **1989,** *365,* 325.
(158) Hao, L.; Vittal, J. J.; Xiao, J.; Puddephatt, R. J. *J. Am. Chem. Soc.* **1995,** *117,* 8035.
(159) Waterman, S. M.; Humphrey, M. G.; Hockless, D. C. R. *Organometallics* **1996,** *15,* 1745.
(160) Elliot, G. P.; Howard, J. A. K.; Nunn, C. M.; Stone, F. G. A. *J. Chem. Soc., Chem. Commun.* **1986,** 431.
(161) Henly, T. J.; Shapley, J. R.; Rheingold, A. L.; Gelb, S. J. *Organometallics* **1988,** *7,* 441.
(162) Henly, T. J.; Wilson, S. R.; Shapley, J. R. *Inorg. Chem.* **1988,** *27,* 2551.
(163) Stromnova, T. A.; Busygina, I. N.; Katser, S. B.; Antsyshkina, A. S.; Porai-Koshits, M. A.; Moiseev, I. I. *J. Chem. Soc., Chem. Commun.* **1988,** 114.
(164) Elliot, G. P.; Howard, J. A. K.; Mise, T.; Nunn, C. M.; Stone, F. G. A. *Angew. Chem., Int. Ed. Engl.* **1986,** *25,* 190.
(165) Elliot, G. P.; Howard, J. A. K.; Mise, T.; Nunn, C. M.; Stone, F. G. A. *J. Chem. Soc., Dalton Trans.* **1987,** 2189.
(166) Moiseev, I. I.; Stromnova, T. A.; Vargaftik, M. N. *J. Mol. Catal.* **1994,** *86,* 71.
(167) Braunstein, P.; Rosé, J., Eds. *Heterometallic Clusters in Catalysis;* Elsevier: Amsterdam, 1989; Vol. 3.
(168) Braunstein, P.; Rosé, In *J. Comprehensive Organometallic Chemistry II;* Abel, E. W.; Stone, F. G. A.; Wilkinson, G. Eds.; Pergamon Press: Oxford, **1994;** Vol. 10, Ch. 7.
(169) Johnson, B. F. G.; Gallup, M.; Roberts, Y. V. *J. Mol. Catal.* **1994,** *86,* 51.
(170) Süss-Fink, G.; Meister, G. *Adv. Organomet. Chem.* **1993,** *35,* 41.
(171) Beurich, H.; Vahrenkamp, H. *Chem. Ber.* **1982,** *115,* 2385.
(172) Curtis, M. D.; Williams, P. D.; Butler, W. M. *Inorg. Chem.* **1988,** *27,* 2853.
(173) Li, P.; Curtis, D. *Inorg. Chem.* **1990,** *29,* 1242.
(174) Brunner, H.; Wachter, J. *J. Organomet. Chem.* **1982,** *240,* C41.
(175) Bender, R.; Braunstein, P.; Dusausoy, Y.; Protas, J. *Angew. Chem., Int. Ed. Engl.* **1978,** *17,* 596.
(176) Barbier, J.-P.; Braunstein, P. *J. Chem. Res. (S)* **1978,** 412.
(177) Braunstein, P.; Dehand, J. *J. Organomet. Chem.* **1970,** *24,* 497.
(178) Braunstein, P.; Dehand, J. *J. Chem. Soc., Chem. Commun.* **1972,** 164.
(179) Bender, R.; Braunstein, P.; Dusausoy, Y.; Protas, J. *J. Organomet. Chem.* **1979,** *172,* C51.

(180) Richter, F.; Vahrenkamp, H. *Angew. Chem., Int. Ed. Engl.* **1978,** *17,* 864.
(181) Müller, M.; Vahrenkamp, H. *Chem. Ber.* **1983,** *116,* 2748.
(182) Richmond, M. G.; Absi-Halbi, M.; Pittman, C. U. *J. Mol. Catal.* **1984,** *22,* 367.
(183) Pittman, C. U.; Richmond, M. G.; Absi-Halabi, M.; Beurich, H.; Richter, F.; Vahrenkamp, H. *Angew. Chem., Int. Ed. Engl.* **1982,** *21,* 786.
(184) Pittman, C. U.; Honnick, W.; Absi-Halabi, M.; Richmond, M. G.; Bender, R.; Braunstein, P. *J. Mol. Catal.* **1985,** *32,* 177.
(185) Fusi, A.; Ugo, R.; Psaro, R.; Braunstein, P.; Dehand, J. *J. Mol. Catal.* **1982,** *16,* 217.
(186) Mani, D.; Vahrenkamp, H. *J. Mol. Catal.* **1985,** *29,* 305.
(187) Sinfelt, J. H. *Progr. Solid-State Chem.* **1975,** *10,* 55.
(188) Boudart, M. *J. Mol. Catal.* **1965,** *30,* 27.
(189) Biswas, J.; Bickle, G. M.; Gray, P. G.; Do, D. D.; Barbier, J. *Catal. Rev. Sci. Eng.* **1988,** *30,* 161.
(190) Sinfelt, J. H. *Sci. Am.* **1985,** *253,* 96.
(191) Bailey, D. C.; Langer, S. H. *Chem. Rev.* **1981,** *81,* 109.
(192) Curtis, M. D.; Williams, P. D. *Inorg. Chem.* **1983,** *22,* 2661.
(193) Walthur, B.; Scheer, M.; Böttcher, H.-C.; Trunschke, A.; Ewald, H.; Gutschick, D.; Miessner, H.; Skupin, M.; Vorbeck, G. *Inorg. Chim. Acta* **1989,** *156,* 285.
(194) Lucas, N. T.; Humphrey, M. G.; Hockless, D. C. R. *J. Organomet. Chem.* **1997,** *535,* 175.
(195) Shapley, J. R.; Hardwick, S. J.; Foose, D. S.; Stucky, G. D. *J. Am. Chem. Soc.* **1981,** *103,* 7383.
(196) Alexeev, O.; Shelef, M.; Gates, B. C. *J. Catal.* **1996,** *164,* 1.
(197) Urbancic, M. A.; Wilson, S. R.; Shapley, J. R. *Inorg. Chem.* **1984,** *23,* 2954.
(198) Curtis, M. D.; Penner-Hahn, J. E.; Schwank, J.; Baralt, O.; McCabe, D. J.; Thompson, L.; Waldo, G. *Polyhedron* **1988,** *7,* 2411.
(199) Trunschke, A.; Ewald, H.; Gutschick, D.; Miessner, H.; Skupin, M.; Walthur, B.; Böttcher, H.-C. *J. Mol. Catal.* **1989,** *56,* 95.
(200) Shapley, J. R.; Uchiyama, W. S.; Scott, R. A. *J. Phys. Chem.* **1990,** *94,* 1190.
(201) Braunstein, P.; Bender, R.; Kervennal, J. *Organometallics* **1982,** *1,* 1236.
(202) Kawi, S.; Alexeev, O.; Shelef, M.; Gates, B. C. *J. Phys. Chem.* **1995,** *99,* 6296.
(203) Fung, A. S.; McDevitt, M. R.; Tooley, P. A.; Kelley, M. J.; Koningsberger, D. C.; Gates, B. C. *J. Mol. Catal.* **1993,** *140,* 190.
(204) Fung, A. S.; Kelley, M. J.; Koningsberger, D. C.; Gates, B. C. *J. Am. Chem. Soc.* **1997,** *119,* 5877.
(205) Farrugia, L. J. *J. Chem. Soc., Dalton Trans.* **1997,** 1783.
(206) Roulet, R. *Chimia* **1996,** *50,* 629.
(207) Johnson, B. F. G.; Roberts, Y. V.; Parisini, E.; Benfield, R. E. *J. Organomet. Chem.* **1994,** *478,* 21.
(208) Orrell, K.; Sik, V. *Ann. Rep. NMR Spectroscopy* **1993,** *27,* 103.
(209) Orrell, K.; Sik, V. *Ann. Rep. NMR Spectroscopy* **1987,** *19,* 79.
(210) Mann, B. E. *Ann. Rep. NMR Spectroscopy* **1982,** *12,* 263.
(211) Band, E.; Muetterties, E. L. *Chem. Rev.* **1978,** *78,* 639.
(212) Johnson, B. F. G.; Benfield, R. E. *J. Chem. Soc., Dalton Trans.* **1978,** 1554.
(213) Casey, C. P.; Jordan, R. F.; Rheingold, A. L. *Organometallics* **1984,** *3,* 504.
(214) Beringhelli, T.; D'Alfonso, G.; Minoja, A. P. *Organometallics* **1994,** *13,* 663.
(215) Friedrich, S.; Gade, L. H.; Scowan, I. J.; McPartlin, M. *Angew. Chem., Int. Ed. Engl.* **1996,** *35,* 1338.
(216) Delgado, E.; Hein, J.; Jeffrey, J. C.; Ratermann, A. L.; Stone, F. G. A.; Farrugia, L. J. *J. Chem. Soc., Dalton Trans.* **1987,** 1191.
(217) Beringhelli, T.; Ceriotti, A.; D'Alfonso, G.; Della Pergola, R. *Organometallics* **1990,** *9,* 1053.
(218) Beringhelli, T.; D'Alfonso, G.; Minoja, A. P. *Organometallics* **1991,** *10,* 394.

(219) Antognazza, P.; Beringhelli, T.; D'Alfonso, G.; Minoja, A.; Ciani, G.; Moret, M.; Sironi, A. *Organometallics* **1992,** *11,* 1777.
(220) Mlekuz, M.; Bougeard, P.; Sayer, B. G.; Peng, S.; McGlinchey, M. J.; Marinetti, A.; Saillard, J.-Y.; Naceur, J. B.; Mentzen, B.; Jaouen, G. *Organometallics* **1985,** *4,* 1123.
(221) Chetcuti, M. J.; Gordon, J. C.; Fanwick, P. E. *Inorg. Chem.* **1990,** *29,* 3781.
(222) Kyba, E. P.; Kerby, M. C.; Kashyap, R. P.; Mountzouris, J. A.; Davis, R. E. *Organometallics* **1989,** *8,* 852.
(223) Barr, R. D.; Green, M.; Howard, J. A. K.; Marder, T. B.; Stone, F. G. A. *J. Chem. Soc., Chem. Commun.* **1983,** 759.
(224) Barr, R. D.; Green, M.; Marsden, K.; Stone, F. G. A.; Woodward, P. *J. Chem. Soc., Dalton Trans.* **1983,** 507.
(225) Pfeffer, M.; Fischer, J.; Mitschler, A. *Organometallics* **1984,** *3,* 1531.
(226) Yeh, W.-Y.; Shapley, J. R. *Organometallics* **1985,** *4,* 767.
(227) Zanello, P. *Struct. Bonding (Berlin)* **1992,** *79,* 101.
(228) Lindsay, P. N.; Peake, B. M.; Robinson, B. H.; Simpson, J.; Honrath, U.; Vahrenkamp, H.; Bond, A. M. *Organometallics* **1984,** *3,* 413.
(229) Honrath, U.; Vahrenkamp, H. *Z. Naturforsch.* **1984,** *39b,* 545.
(230) Pasynskii, A. A.; Eremenko, I. L.; Katugin, A. S.; Gasanov, G. S.; Turchanova, E. A.; Ellert, O. G.; Struchkov, Y. T.; Shklover, V. E.; Berberova, N. T.; Sogomonova, A. G.; Okhlobystin, O. Y. *J. Organomet. Chem.* **1988,** *344,* 195.
(231) Mansour, M. A.; Curtis, M. D.; Kampf, J. W. *Organometallics* **1997,** *16,* 275.
(232) Jund, R.; Lemoine, P.; Gross, M.; Bender, R.; Braunstein, P. *J. Chem. Soc., Chem. Commun.* **1983,** 86.
(233) Jund, R.; Lemoine, P.; Gross, M.; Bender, R.; Braunstein, P. *J. Chem. Soc., Dalton Trans.* **1985,** 711.
(234) Pasynskii, A. A.; Eremenko, I. L.; Orazsakhatov, B.; Rakitin, Y. V.; Novotortsev, V. M.; Ellert, O. G.; Kalinnikov, V. T.; Aleksandrov, G. G.; Struchkov, Y. T. *J. Organomet. Chem.* **1981,** *214,* 351.
(235) Pasynskii, A. A.; Eremenko, I. L.; Orazsakhatov, B.; Gasanov, G. S.; Novotortsev, V. M.; Ellert, O. G.; Seifulina, Z. M.; Shklover, V. E.; Struchkov, Y. T. *J. Organomet. Chem.* **1984,** *270,* 53.
(236) Pasynskii, A. A.; Eremenko, I. L.; Zalmanovitch, V. R.; Kaverin, V. V.; Orazsakhatov, B.; Nefedov, S. E.; Ellert, O. G.; Novotortsev, V. M.; Yanovsky, A. I.; Struchkov, Y. T. *J. Organomet. Chem.* **1991,** *414,* 55.
(237) Eremenko, I. L.; Pasynskii, A. A.; Orazsakhatov, B.; Shestakov, A. F.; Gasanov, G. S.; Katugin, A. S.; Struchkov, Y. T.; Shklover, V. E. *J. Organomet. Chem.* **1988,** *338,* 369.
(238) Aldridge, S.; Hashimoto, H.; Kawamura, K.; Shang, M.; Fehlner, T. P. *Inorg. Chem.* **1998,** *37,* 928.
(239) D'Agostino, M. F.; Frampton, C. S.; McGlinchey, M. J. *J. Organomet. Chem.* **1990,** *394,* 145.
(240) Winter, G.; Schultz, B.; Trunschke, A.; Miessner, H.; Böttcher, H.-C.; Walther, B. *Inorg. Chim. Acta* **1991,** *184,* 27.
(241) Li, P.; Curtis, M. D. *J. Am. Chem. Soc.* **1989,** *111,* 8279.
(242) Harris, S. *Polyhedron* **1989,** *8,* 2843.

ADVANCES IN ORGANOMETALLIC CHEMISTRY, VOL. 46

Friedel–Crafts Alkylations with Silicon Compounds

IL NAM JUNG and BOK RYUL YOO

Organosilicon Chemistry Laboratory
Korea Institute of Science & Technology
Seoul 130-650 Korea

I

INTRODUCTION

Friedel-Crafts alkylations catalyzed by Lewis acids[1–6] have been studied for quite some time and much work has been done in this field since the reaction of benzene with amyl chloride to produce amylbenzene was first reported by Friedel and Crafts in 1877.[2,3] Friedel–Crafts alkylations with organic compounds such as organic halides, alkenes, and alkynes are now widely used in the laboratory and the petrochemical industry as valuable and established routes to introduce alkyl substituents onto aromatic rings.[1] Alkylation with silicon compounds was first described by Wagner *et al.* in 1953, for the preparation of (phenylethyl)trichlorosilane, a monomer that was hydrolyzed to polysiloxanes.[7] In the following year, Petrov reported the Friedel–Crafts alkylation of benzene derivatives with (chloroalkyl)silanes.[8] At that time, the silicon industry was growing rapidly and supplying monomers on a large scale by the direct synthesis of organochlorosilanes, a process discovered by Rochow.[9] Alkylations with silicon compounds such as alkenylchlorosilanes[10–13] and (chloroalkyl)silanes[14–20] were expected to be a way of preparing useful monomers for polysiloxane products, and a few articles on this topic were published by 1968.[10–20] Soon after, interest in Friedel–Crafts alkylations with silicon compounds faded due to the difficulties of

0065-3055/01 $35.00

the syntheses and separations of alkenylchlorosilanes[21,22] and (chloroalkyl)chlorosilanes.[23,24]

In 1993, a successful direct synthesis[25] of allyldichlorosilane from silicon metal and allyl chloride gave us a motive for exploring the alkylation of benzene derivatives using allylchlorosilanes.[26] We have reinvestigated the Friedel–Crafts alkylation of various aromatic compounds such as substituted benzenes, biphenyls, and naphthalenes with allylchlorosilanes[26] in the presence of Lewis acid catalysts. This work has also been extended to alkylation with vinylchlorosilanes[27,28] and (chloroalkyl)chlorosilanes.[29] Substantial progress has been made in the development of the chemistry of this interesting class of organosilicon compounds. This review will describe the chemistry of the Friedel–Crafts alkylation of aromatic compounds with silicon compounds, in particular the aluminum chloride catalyzed alkylation reactions.

II

CHARACTERISTICS OF FRIEDEL–CRAFTS ALKYLATIONS WITH SILICON COMPOUNDS

In this section, the reactivities of organosilicon compounds for the Friedel–Crafts alkylation of aromatic compounds in the presence of aluminum chloride catalyst and the mechanism of the alkylation reactions will be discussed, along with the orientation and isomer distribution in the products and associated problems such as the decomposition of chloroalkylsilanes to chlorosilanes. Side reactions such as transalkylation and reorientation of alkylated products will also be mentioned, and the insertion reaction of allylsilylation and other related reactions will be explained.

A. *Reactivities and Mechanism*

The reactivities of alkenylsilanes in the presence of a Lewis acid vary depending upon the nature of substituents on silicon as shown in Table I.

In the case of alkylation using allylsilanes in the presence of aluminum chloride as a catalyst, allylsilanes containing one or more chlorine substituents on the silicon react with aromatic compounds at room temperature or below 0°C to give alkylated products, 2-aryl-1-silylpropanes,[26] while allyltrimethylsilane did not give the alkylated product but instead dimerized to give the allylsilylation product, 5-(trimethylsilyl)-4-(trimethylsilylmethyl)-1-pentene (Eq. (1)).[30] In the alkylation reaction, the reactivity of allylsilanes increased as the number of chlorine

TABLE I

REACTIVITY OF ALKENYLSILANES $CH_2{=}CH(CH_2)_nSiR_3$ FOR THE $AlCl_3$-CATALYZED ALKYLATION

Alkenylsilanes $[CH_2{=}CH(CH_2)_nSiR_3]$		Reactivity for Alkylation to Benzene (Ferrocene)		Other Reaction
n	R_3			
0	Cl_3	High	(No)	
0	$MeCl_2$	High	(No)	
0	Me_2Cl	Low	(No)	
0	Me_3	No	(No)	
1	Cl_3	High	(No)	
1	HCl_2	High	(No)	
1	$MeCl_2$	Middle	(Middle)	
1	Me_2Cl	No	(High)	Decomposition[a]
1	Me_3	No	(No)	Allylsilylation[b]

[a] Propene and dimethyldichlorosilane were produced by the decomposition of allyl(dimethyl)chlorosilane.

[b] Allylsilylation of allylsilane to carbon–carbon multiple bonds.

atoms on the silicon increased, but decreased as the number of methyl groups increased.

(1)

Alkylation with vinylchlorosilanes requires a relatively higher reaction temperature and prolonged reaction time, likely due to the lower stability of the carbocation intermediates.

It is well known in the literature that aluminum chloride, a strong Lewis acid, is a very effective catalyst in Friedel–Crafts alkylations with silicon compounds.[7,8,10–20,26–28] In the Friedel–Crafts alkylations of arenes with aluminum chloride as catalyst, a small amount of hydrogen chloride resulting from the reaction of anhydrous aluminum chloride with moisture inevitably present in the reactants initiates the reaction.[31] The proton from hydrogen chloride interacts with the π-bond of alkenylsilanes to give the carbenium ion intermediate on the carbon β to silicon. This occurs because the intermediate silylethyl cation is stabilized by the electron-donating silyl group through σ–π conjugation known as β-stabilization[32–34] and the more stable secondary carbenium ion is generated through protonation of the terminal carbon of the allyl group.[35] Electrophilic attack of this carbenium ion on the π-bond of the aromatic ring generates a cation on the aromatic ring,[1] which is followed by deprotonation to give silylalkylated aromatic compounds along with the regeneration of a proton. This proton initiates the catalytic cycle of the Friedel–Crafts alkylation with alkenylsilanes.

In the case of allylsilanes, the protonated intermediate 1-silylpropyl cation can be stabilized by the electron-donating silyl group through β-stabilization as described above. This can be more effective for allyltrimethylsilane than for allylchlorosilanes because of the electron-donating methyl groups on silicon.[30] This facilitates the protodesilylation[30,36,37] of allyltrimethylsilane by hydrogen chloride in the presence of aluminum chloride which gives propene and a Me_3SiCl-$AlCl_3$ complex, which catalyzes the allylsilylation of alkenes.[37–40] In contrast, the collapse of the β-silyl cation intermediates for allylchlorosilanes is largely retarded due to less effective σ–π conjugation due to the presence of the electronegative chlorine atom(s) on silicon. The cation intermediates undergo alkylation faster than protodesilylation, which is why allylchlorosilanes show higher reactivity and give higher yields in the alkylation compared to allyltrimethylsilane.[26]

The substituent effect of vinylsilanes is similar to that of allylsilanes. The reactivity of vinylsilanes increased as the number of chlorine atoms on the silicon increased, but decreased as the number of methyl groups increased. However, vinyltrimethylsilane does not react with benzene to give alkylated products.[41] In the aluminum chloride-catalyzed alkylation of arenes with allylsilanes or vinylsilanes, one or more chlorine substituents on the silicon atom of silanes are required.

The reactivity of (ω-chloroalkyl)chlorosilanes for the alkylation of arenes varies depending upon the length of the ω-chloroalkyl group and the substituents on the silicon. Generally, the reactivity increases as the length of the alkyl group on the silicon of (ω-chloroalkyl)chlorosilanes increases from methylene to propylene.[19,42] That may be mainly attributed to the steric hindrance between the incoming silanes and the aromatic ring, and partly to the electronic nature of the silyl group. The rearrangement of alkylating agent under Friedel–Crafts reaction conditions and chloride exchange between aluminum chloride and alkyl chloride are well known.[1] These indicate that the alkylation proceeds via the transitory existence of carbocations resulting from complexation between the alkylating agent and Lewis acid catalyst.

$$AlCl_3 + RCl \longrightarrow R^+AlCl_4^-$$

Brown and Grayson reported that the rate of alkylation reactions with benzyl chloride was third order overall: first order in aromatic component, first order in $AlCl_3$, and first order in benzyl chloride. This indicates that a rate determining nucleophilic attack by the aromatic component on a polar alkyl chloride-aluminum chloride adduct is involved in the alkylation.[43] If the reaction proceeded by an ionization mechanism, the rate of alkylation should be determined solely by the rate of ionization of the alkyl chloride and should be independent of the concentration or nucleophilic properties of the aromatic compound undergoing the alkylation.

B. *Orientation and Isomer Distributions*

The alkylation of monosubstituted benzenes gives an isomeric mixture of *ortho, meta,* and *para*-products.[1] The ratio of isomeric products varies depending upon the electronic nature and steric bulk of the substituents on the benzene ring. Different isomeric distributions of the alkylation products are obtained at different reaction temperatures, probably due to the temperature dependence of isomer formation and/or isomerization reactions of the products with the Lewis acid catalysts. The ratio of *ortho* to the *meta* and *para* products derived from the alkylation of monosubstituted benzenes with allylsilanes decreases as the size of the substituents on the benzene ring increases.[44,45] No *ortho*-alkylation products are obtained in the case of isopropylbenzene due to the steric interaction between the isopropyl group and the incoming allyl groups. Steric hindrance arising from the size of the alkyl groups at the *ortho* position may be the principal reason for the differences in the alkylation rates of the substituted benzenes. The yield of the *ortho* isomer decreases and that of the *meta* isomer increases as the reaction proceeds for long periods or at higher temperatures in the presence of aluminum chloride. This indicates that the *ortho* and *para* adducts are the kinetically controlled products and isomerization or alkylation-dealkylation of the resulting alkylated benzenes favors the thermodynamically more stable *meta* product.

The alkylation of halogen-substituted benzenes with allylsilanes gives *ortho* and *para* adducts predominantly. Considering the high electronegativities of halogen atoms, the *meta* isomer was expected to be the major product. The predominance of *ortho* and *para* isomers in the products indicates that the resonance effect of halogen substituents to the benzene ring should be considered in addition to their electronic effects.[46] The isomerization of the products from the alkylation of alkylbenzenes is faster than that of the products obtained from the reaction of halobenzenes.

C. *Transalkylations and Reorientations*

Alkylation of benzene does not stop with the introduction of one alkyl group onto the ring. Di, tri, and higher alkylation products are also produced. The production of higher alkylation compounds can be controlled to some extent by using a molar excess of benzene and by variations in other experimental parameters. The alkylation of benzene with excess vinylmethyldichlorosilane in the presence of aluminum chloride at room temperature for 4 h gave peralkylated product, hexakis[2-(methyldichlorosilyl)ethyl]benzene,[27] along with less alkylated products. However, the alkylation with allylsilanes gave no higher alkylation products than the 1,3,5-trialkylated compound due to higher steric interactions among the silylalkyl groups.

It is also well known that alkyl groups can be transferred intramolecularly from one position to another on the same ring and intermolecularly from one aromatic ring to another through dealkylation reactions catalyzed by Lewis acid. The intramolecular alkyl-transfer is called reorientation or isomerization and the intermolecular alkyl transfer is referred to as disproportionation. Reorientation processes are normally faster than disproportionation.

III

FRIEDEL–CRAFTS ALKYLATION OF ARENES WITH ALKENYLCHLOROSILANES

This section will describe the Friedel–Crafts alkylation reactions of aromatic hydrocarbons with alkenylchlorosilanes containing short chain alkenyl groups such as allyl and vinyl. The reaction will be discussed in terms of the substituent effect on silicon and the arene rings.

A. *Alkylation with Allylchlorosilanes*

1. *Aromatic Hydrocarbons*

Nametkin and co-workers first reported the alkylation of benzene derivatives with allylchlorosilanes in the presence of aluminum chloride as catalyst.[12] 2-(Aryl)propylsilanes were obtained from the alkylation of substituted benzenes (Ph—X; X = H, Cl, Br) with allylsilanes such as allyldichlorosilane and allyltrichlorosilane.[12,26] The yields ranged from 34 to 66% depending upon the substituents on the benzene ring, but information concerning reaction rates and product isomer distribution was not reported.

After succeeding in the direct synthesis of allyldichlorosilane by reacting elemental silicon with a mixture of allyl chloride and hydrogen chloride in 1993,[25] Jung *et al.* reinvestigated the Friedel–Crafts reactions of benzene derivatives with allyldichlorosilanes in detail (Eq. (2)).

$$CH_2{=}CHCH_2SiCl_2X + C_6H_4R^1R^2 \xrightarrow{AlCl_3} R^1R^2C_6H_3CH(CH_3)CH_2SiCl_2X \quad (2)$$

X = H, Me, Cl
R^1 = H, Me
R^2 = halides, alkyl, Ph, OPh, etc

The results of these alkylation reactions with allyldichlorosilane (**1**) in the presence of aluminum chloride catalyst are summarized in Table II.

As shown in Table II, the alkylation of halobenzenes with **1** at room temperature for 50 min afforded the monoadducts in relatively good yields ranging from 60 to 66%. The reaction with alkylbenzenes took shorter times at lower temperatures and the yields obtained within 20 min at 0°C ranged from 70 to 78%. The results indicated that electron-withdrawing groups such as halogens deactivated the alkylation, while electron donating groups such as alkyl groups on benzene generally facilitated the reaction.[44,46] This is consistent with the electrophilic nature of the alkylation reaction.

The reactivities of allylsilanes for the alkylation of benzene derivatives varied depending on the substituents on silicon.[12] Allylchlorosilanes having more than two chlorine atoms on silicon, such as allyldichloromethylsilane and allyltrichlorosilane, readily reacted with substituted benzenes to give 2-(aryl)propylchlorosilanes in the presence of aluminum chloride, but allyltrimethylsilane did not react.[12] Among the allylchlorosilanes, allyltrichlorosilanes gave the highest yield while allylchlorodimethylsilane gave the lowest,[47] thus indicating higher reactivity for the polychlorine substituted allylchlorosilanes.

Monoalkylation products, 3-aryl-1,1-dichloro-1-silabutanes, were obtained from the alkylation of aromatic compounds with **1** in the presence of aluminum chloride catalyst in good isolated yields (60–80%) along with small amounts of higher alkylation products. Dialkylation products were obtained in yields ranging from 2 to 8% when a 5-fold excess of the aromatic compounds with respect to **1** was used. The amount of dialkylated products can be further reduced by using a greater excess of the aromatic compounds.

In the alkylation of toluene at various temperatures ranging from −45°C to room temperature, different product distributions (*o*-: *m*-: *p*-) were observed ranging from

TABLE II
ALKYLATION CONDITIONS[a] AND ISOMER DISTRIBUTION OF PRODUCTS

Substituents		Reaction Temp.	Monoalkylated Products	
R^1	R^2	(°C)	Yield (%)	Isomer Distribn. (*o*: *m*: *p*)[b]
H	H	rt	72	
H	F	rt	69	31:2:67
H	Cl	rt	68	27:6:67
H	Br	rt	60	29:9:62
H	Me	0	70	3:66:31
H	Et	0	78	2:64:34
H	*i*-Pr	0	71	-:33:67
H	Ph[c]	rt	83	-:37:63
H	PhO[c]	rt	74	47:3:50
Me	*o*-Me	0	81	
Me	*m*-Me	0	73	ac:bb = 62:38
Me	*p*-Me	0	85	ab:bc = 2:98
Me	*p*-Cl	0	66	ab:ba = 80:20

[a] The reactions were carried out at 0°C or rt for 20 min neat except for 50 min for halobenzenes (Ph–X; X = F, Cl, Br) and using the 1:5:0.1 mol ratio of **1** to aromatic compound to aluminum chloride.

[b] **a–c** represent *ortho-*, *meta-*, and *para-*substituted products, respectively.

[c] The reaction was carried out in hexane solution for biphenyl and diphenyl ether.

53:9:38 at −45°C to 17:20:63 at 0°C and to 3:65:32 at room temperature. These results indicate that the alkylation of the benzene ring with **1** produces kinetically controlled *ortho-* and *para-*products, due to the *ortho-* and *para-*directing nature of the methyl group,[48] at an early stage of the reaction. Then at a later stage, the isomerization or alkylation–dealkylation of the alkylated benzenes favors the thermodynamically more stable *meta-*product, as had been observed in the reaction of toluene with *n*-butene.[49] Aluminum chloride in Friedel–Crafts reactions was reported to be an effective catalyst for intra- and intermolecular isomerizations[28,50] of the alkylated products.

The ratio of *ortho-* to the *meta-* and *para-*products of monoalkylbenzenes with **1** decreased as the size of the substituents on benzene ring increased. No *ortho-*alkylation product was found in the case of *i*-propylbenzene due to the steric interaction between *i*-propyl and the incoming allyl groups. Steric hindrance arising from the size of the alkyl groups at *ortho* positions of the substituted benzenes appeared to be the principal cause of the differences in isomer product ratio.[44,45]

The yield of *ortho*-isomer decreased and *meta*-isomer increased as the reaction proceeded for longer periods or at higher temperatures in the presence of aluminum chloride.

The alkylation of halogen (R = F, Cl, Br) substituted benzenes with **1** results in predominantly *ortho*- and *para*-adducts (91–98%), indicating that the halogen atoms are *ortho*- and *para*-directing groups. Considering the high electronegativities of halogen atoms, the *meta* isomer was expected to be the major product. Higher resonance effects compared to electronic effects might be responsible for the *ortho*- and *para*-directing properties of the halogen atoms. The results indicate that the alkylation of electron donating group-substituted benzenes with **1** was faster than the reaction of electron withdrawing group-substituted benzenes.[51,52] Isomerization of the products obtained from alkylbenzenes was also faster than those of the products obtained from halobenzenes.

Substituent effects of the benzene ring in the alkylation of benzene derivatives with **1** were studied by comparison with benzene itself. These results are summarized in Table III.

TABLE III

RELATIVE ALKYLATION RATES (R^1–C_6H_4–R^2/C_6H_6)[a] AND SUBSTITUENT CONSTANTS (σ)[b] FOR SUBSTITUTED BENZENES (R^1–C_6H_4–R^2)

Substituents		Relative Rates		Substituent Constants[c]		
R^1	R^2	k_R/k_H	log k_R/k_H	σ_R	σ_I	σ
H	H	1.00	0.00	0.00	0.00	0.00
H	F	0.081	−1.04	0.74	−0.60	0.14
H	Cl	0.016	−1.44	0.72	−0.24	0.48
H	Br	0.011	−1.96	0.72	−2.52	0.54
H	Me	3.89	0.59	−0.01	−0.41	−0.42
H	Et	4.10	0.61	−0.02	−0.44	−0.46
H	*i*-Pr	5.65	0.72	—	—	—
H	Ph	10.27	1.01	0.25	−0.37	−0.12
H	PhO	9.42	0.97	0.76	−1.29	−0.53
Me	*p*-Cl	—	−1.22	—	—	—
Me	*o*-Me	—	0.13	—	—	—
Me	*m*-Me	—	0.46	—	—	—
Me	*p*-Me	—	0.05	—	—	—

[a] Reaction mole ratio of **1**:benzene:substituted benzene: aluminum chloride = 1:20:20:0.1.

[b] See Ref. 46.

[c] Definitions: σ_R, resonance constant; σ_I, field constant; $\sigma = \sigma_R + \sigma_I$.

The reactivity of substituted benzenes (Ph—R) decreased in the following order: R = Ph > PhO > *i*-Pr > Et > Me > H > F > Cl > Br.[26] The alkylation of alkyl benzenes was faster than that of halogen substituted benzenes. Among the halogen substituted benzenes, the fastest rate was observed for fluorobenzene and the slowest for bromobenzenes, which was not consistent with the electronegativities of the substituents.[46,51,52] The results indicate that the resonance effect of halogen substituents to the benzene ring should be considered in addition to the electronic effect in order to explain the reaction rates.[46] The relative reaction rates for the alkylation of substituted benzenes with respect to benzene (log k_R/k_H) were plotted against substituent coefficients (σ)[46] for alkyl, aryl, and halogen groups. According to the Hammet equation:[53] $\log k_R/k_H = \rho\sigma$, where ρ is a Hammet constant, ρ was found to be -3.1 from the relationship between the substituent coefficients (σ) and the relative reaction rates for the alkylation of substituted benzenes with respect to benzene (log k_R/k_H). In the case of disubstituted benzenes, the reactivity for alkylation decreased as follows: *meta*-xylene > *ortho*-xylene > *para*-xylene > *para*-chlorotoluene.

In order to study the polyalkylation reaction of benzene derivatives with allylchlorosilanes, allyldichlorosilane was reacted with benzene derivatives. The reactions were carried out under various reaction conditions. The polyalkylated products increased as the mole ratio of allyldichlorosilane to benzene derivatives increased, but the polyalkylated products obtained from the one step reaction were difficult to purify by distillation due to their high boiling points. Thus, the reaction was carried out by adding allyldichlorosilane to benzene derivatives in the presence of aluminum chloride catalyst to obtain dialkylated products as the major product. The reaction using a 2:1 mixture of allyldichlorosilane and benzene derivatives afforded dialkylated products (32–41%) along with trialkylated products (23–28%) (Eq. (3)). In the reaction using a 3:1 mixture of allyldichlorosilane and benzene derivatives, trialkylated compounds were obtained as the major product.

1 + R-benzene ($AlCl_3$, RT) → ; R = H, Me, Et, iPr

$1/AlCl_3$, RT → ; n = 1 or 2 (3)

R = H; n = 1 (41%), n = 2 (26%)
R = Me; n = 1 (32%), n = 2 (28%)
R = Et; n = 1 (34%), n = 2 (26%)
R = iPr; n = 1 (38%), n = 2 (23%)

In the reaction using a 1:4 mixture of allyldichlorosilane and benzene derivatives, higher alkylated compounds than trialkylation products were not obtained. This is apparently due to the steric hindrance between substituents on the benzene ring and the incoming allyldichlorosilane. Alternatively, step by step alkylation gives the corresponding products in higher yield than that of the one step reaction. In the case of benzene with allyldichlorosilane, a mixture of dialkylated products, *m*- and *p*-bis[(1-dichlorosilylmethyl)ethyl]benzenes, were obtained in 59% yield from the reaction of monoalkylated product of benzene with another mole of allyldichlorosilane. Purification was also easier.

Allylchlorosilanes reacted with naphthalene to give isomeric mixtures of polyalkylated products. However, it was difficult to distill and purify the products for characterization from the reaction mixture due to the high boiling points of the products and the presence of many isomeric compounds. The alkylation of anthracene with allylchlorosilanes failed due to deactivation by complex formation with anthracene and the self-polymerization of anthracene to solid char.

2. *Ferrocene*

Ferrocene behaves in many respects like an aromatic electron-rich organic compound which is activated toward electrophilic reactions.[54] In Friedel–Crafts type acylation of aromatic compounds with acyl halides, ferrocene is 10^6 times more reactive than benzene and gives yields over 80%.[55] However, ferrocene is different from benzene in respect to reactivity and yields in the Friedel–Crafts alkylation with alkyl halides or olefins. The yields of ferrocene alkylation are often very low, and the separations of the polysubstituted byproducts are tedious.[56,57]

Jung *et al.* reported the Friedel–Crafts type alkylation of ferrocene with allylchlorosilanes.[58] The reaction of ferrocene with allylchlorosilanes in the presence of Lewis acid afforded regiospecific alkylated ferrocenes bearing chlorosilyl groups at the β-carbon to the ferrocene ring (Eq. (4)).

To optimize the alkylation conditions, ferrocene was reacted with allyldimethylchlorosilane (**2**) in the presence of various Lewis acids such as aluminum halides and Group 10 metal chlorides. Saturated hydrocarbons and polychloromethanes such as hexane and methylene chloride or chloroform were used as solvents because of the stability of the compounds in the Lewis acid catalyzed Friedel–Crafts reactions. The results obtained from various reaction conditions are summarized in Table IV.

As shown in Table IV, the highest catalytic activity of metal halides used as Lewis acid for the alkylation reaction of ferrocene with **2** was observed in methylene chloride solvent. Among Lewis acids such as aluminum chloride, aluminum bromide, and Group 4 transition metal chlorides ($TiCl_4$, $ZrCl_4$, $HfCl_4$), catalytic efficiency for the alkylation decreases in the following order: hafnium chloride > zirconium chloride > aluminum chloride > aluminum bromide. Titanium chloride

Lewis acid

Lewis acid (4)

R^1 = H, Me, Cl
R^2 = alkyl groups (Me, ^{t}Bu, $^nC_6H_{13}$)
n = 0, 1, 2
m = 0, 1, 2
n + m = 1 or 2

showed no catalytic activity for the alkylation. It is interesting to note that mild Lewis acids such as $HfCl_4$ are more effective than aluminum chloride, a strong Lewis acid.

When a catalytic amount of aluminum chloride was used, the catalytic activity decreased and the color changed to green–blue as the reaction proceeded. In the case of higher reaction temperatures, the alkylation reaction eventually stopped with the mixture turning a deep blue color. It seems likely that aluminum chloride becomes deactivated by forming a charge transfer complex with ferrocene. Generally, both the reactivity of ferrocene toward alkylation and the Lewis acidity of aluminum chloride decrease as a ferrocene/aluminum chloride complex is formed. It is well documented that ferrocene can be oxidatively converted to ferrocenium cation species by most common electrophiles, and as a result is more reluctant to undergo electrophilic substitutions.[59] The yields of alkylated ferrocenes are not related to the acidity order of Lewis acids used as catalysts. This suggests that ferrocene undergoes two competing reactions: Friedel–Crafts alkylation with allylchlorosilanes and the complexation with Lewis acid catalyst, which deactivates the catalyst.[59] To find the optimum amount of catalyst for the alkylation, hafnium chloride-catalyzed alkylation was carried out using various mol% of the catalyst with respect to allylsilane used. The alkylation rate increased as the concentration of catalyst increased from 5 to 80 mol%, but the reaction was too slow with 5 mol% catalyst. In these experiments, it was found that hafnium chloride was the best catalyst for the alkylation of ferrocene with allylchlorosilane.

Among methylene chloride, chloroform, carbon tetrachloride, and hexane, fast reaction rate was observed in methylene chloride or chloroform solvents, but slow

TABLE IV

ALKYLATION CONDITIONS OF FERROCENE WITH **2** AND PRODUCT YIELDS

Reaction Conditions[a]				Products (%)[b]	
Lewis Acid (mol%)[c]	Solvent	Temp. (°C)	Time (h)	Mono-Alkylated	Di-Alkylated
$AlCl_3$ (10)	CH_2Cl_2	0	1.5	77	6
$AlBr_3$ (10)	CH_2Cl_2	0	1.5	74	7
$TiCl_4$ (10)	CH_2Cl_2	0	1.5	0	0
$ZrCl_4$ (10)	CH_2Cl_2	0	1.5	81	8
$HfCl_4$ (10)	CH_2Cl_2	0	1.5	89	8
$HfCl_4$ (5)	CH_2Cl_2	0	2	30[d]	—
$HfCl_4$ (20)	CH_2Cl_2	0	1	81	9
$HfCl_4$ (40)	CH_2Cl_2	0	1	78	15
$HfCl_4$ (80)	CH_2Cl_2	0	0.75	67	6
$HfCl_4$ (10)	Hexane	Reflex	6	40[e]	—
$HfCl_4$ (10)	CCl_4	rt	6	44[f]	—
$HfCl_4$ (10)	$CHCl_3$	0	3.0	86	6

[a] 1:4 Mixture of **2** and ferrocene was used.
[b] Yields were determined based on allylsilane used.
[c] Mol% of Lewis acid was based on allylsilane used.
[d] 70% of unreacted **2.**
[e] 60% of unreacted **2.**
[f] 56% of unreacted **2.**

rates were observed in non-polar solvents such as hexane and carbon tetrachloride. Methylene chloride was found to be the best solvent. The substituent effects on the silicon of allylsilanes were studied in the presence of hafnium chloride as catalyst. The results obtained from the alkylation with several different allylsilanes are summarized in Table V.

The reactivity of allylchlorosilanes for the alkylation of ferrocene depends on the substituents on the silicon atom. The reactivity increases as the number of alkyl groups on the silicon of allylsilanes increases. Allyl(dialkyl)chlorosilanes having two alkyl groups and one chlorine reacted with ferrocene in the presence of 10 mol% $HfCl_4$ under mild reaction conditions (0°C, 1.5 h) to give alkylated products, (2-silyl-1-methylethyl)ferrocenes, in good yields (89–96%), but allyl(alkyl)dichlorosilanes required 50 mol% $HfCl_4$ as a catalyst and reflux temperatures to give alkylation products. Allyltrialkylsilanes undergo decomposition in the presence of hafnium chloride catalyst and allyltrichlorosilane and allyldichlorosilane showed little reactivity. The reactivity of allylsilanes for the alkylation of ferrocene decreased in the following order: allyldialkylchlorosilane >

TABLE V

$HfCl_4$-CATALYZED ALKYLATION OF FERROCENE WITH ALLYLSILANE ($X_3SiCH_2CH{=}CH_2$) AND PRODUCT YIELDS

Substits. on Si (X_3)	Reaction Conditions[a]			Products (%)[b]	
	$HfCl_4$ (mol%)[c]	Temp. (°C)	Time (h)	Mono-Alkylated	Di-Alkylated
Me_3	0.4	Reflux	2	Decomposition[d]	
Me_2Cl	0.1	0	1.5	89	8
$MeCl_2$	0.4	Reflux	7	90	0
${}^iPrCl_2$	0.4	Reflux	7	91	5
${}^nC_6H_{13}Cl_2$	0.4	Reflux	6	96	0
HCl_2[e]	0.8	Reflux	16	8	0
Cl_3[f]	0.8	Reflux	16	4	0

[a] 1:4 Reaction of allylsilane and ferrocene was carried out in the presence of $HfCl_4$.

[b] Yields were determined based on allylsilane used.

[c] Mol% of $HfCl_4$ was based on allylsilane used.

[d] Me_3SiCl and propene were obtained as major products.

[e] Polymeric materials were obtained as major products.

[f] 47% of unreacted **2**.

allyl(alkyl)dichlorosilane ≫ allyldichlorosilane ≈ allyltrichlorosilane. This reactivity order of allylchlorosilanes for the alkylation of ferrocene is opposite to that of allylchlorosilanes for the alkylation of substituted benzenes. Considering that the alkylations of both substituted benzenes and ferrocene are electrophilic substitution reactions, these results are unexpected and surprising. Although the cause of the reactivity differences in the alkylations of benzene derivatives and ferrocene with allylchlorosilanes is not clear, it is probably related to the complexation of ferrocene with the Lewis acid catalysts.[59]

B. *Alkylation with Vinylchlorosilanes*

Vinylchlorosilanes react with aromatic compounds in the presence of Lewis acid to give the alkylation products 2-(chlorosilyl)ethylarenes.[7,10–12] In the Friedel–Crafts alkylation of aromatic compounds, the reactivity of vinylchlorosilanes is slightly lower than that of allylchlorosilanes.[10,11–26] Friedel–Crafts alkylation of benzene derivatives with vinylsilanes to give 2-(chlorosilyl)ethylarenes was first reported by the Andrianov group (Eq. (5)).[11] The reactivity of vinylsilanes in the alkylation to aromatic compounds varies depending upon the substituents on the silicon of vinylsilanes.

$$\text{C}_6\text{H}_6 + \text{CH}_2{=}\text{CH–Si}X^1X^2X^3 \xrightarrow[\text{reflux, 4 h}]{\text{AlCl}_3} \text{C}_6\text{H}_5\text{CH}_2\text{CH}_2\text{Si}X^1X^2X^3 \tag{5}$$

$X^1 = X^2 = X^3 = Cl$ (67%)
$X^1 = Me$, $X^2 = X^3 = Cl$ (59%)
$X^1 = X^2 = Me$, $X^3 = Cl$ (25%)

Vinyldialkylsilanes and vinyltrimethylsilane having no chlorine atoms do not undergo alkylation with benzene derivatives in the presence of aluminum chloride[41] but vinylchlorosilanes react with benzene to give the alkylation products. The reactivities of vinylchlorosilanes decrease in the following order: vinyl(methyl)dichlorosilane > vinyltrichlorosilane > vinyl(dimethyl)chlorosilane.

The alkylation of benzene derivatives with methyl(vinyl)dichlorosilane (**3**) will be described in detail. Alkylation of monosubstituted benzenes such as toluene, chlorobenzene, and biphenyl at 75–80°C for 2 h afforded the corresponding alkylated products in 50–63% yields.[10]

1. *Polyalkylation of Benzene with Excess 3*

Benzene reacted with excess **3** in the presence of aluminum chloride at room temperature for 4 h to give peralkylated product,[27] hexakis[2-(methyldichlorosilyl)ethyl]benzene (**4a**) and other alkylated products: pentakis[2-(methyldichlorosilyl)ethyl]benzene (**4b**), tetrakis[2-(methyldichlorosilyl)ethyl]benzene (**4c**), tris [2-(methyldichlorosilyl)ethyl]benzene (**4d**), and bis[2-(methyldichlorosilyl)ethyl]-benzene (**4e**) (Eq. (6)). The product distributions were plotted against mole ratios of **3**/benzene in Fig. 1.

As shown in Fig.1, product distributions varied depending upon the mole ratio of **3** to benzene. When one equivalent or a two-fold excess of benzene was used with respect to **3,** the yields of monoalkylated and dialkylated benzenes remained about the same, around 50 and 30%, respectively. As the proportion of **3** to benzene increased, the mono- and dialkylation products decreased and were formed in less than 5% yield, while **4a** increased rapidly, becoming the major product (56–58%) at a 6-fold excess or more of **3** to benzene. The yield of **4b** increased to 26% at the maximum and then decreased smoothly to 18%. Products **4c** and **4d,** however, were produced in yields near 10% regardless of the mole ratios used of **3**/benzene.

To optimize the reaction time for the synthesis of **4a,** the alkylation was carried out using a 1:6 mole ratio of benzene to **3** at room temperature. Product distributions

+ $CH_2=CH-SiMeCl_2$ (3) $\xrightarrow{AlCl_3}$ mono-alylated product ($PhCH_2CH_2SiMeCl_2$)

$\xrightarrow{3/AlCl_3}$ $C_6H_{6-n}(CH_2CH_2SiMeCl_2)_n$

n = 5 (**4b**) n = 3 (**4d**)

n = 4 (**4c**) n = 2 (**4e**)

$\xrightarrow{3/AlCl_3}$ $C_6(CH_2CH_2SiMeCl_2)_6$ **4a** (6)

were determined at various time intervals. Product distributions are plotted against reaction time in Fig. 2.

As shown in Fig. 2, monoalkylated compound was the major product at the initial stage of alkylation, but decreased drastically and almost disappeared after 4 h. However, **4a** increased gradually and became the major product in a 2-h reaction

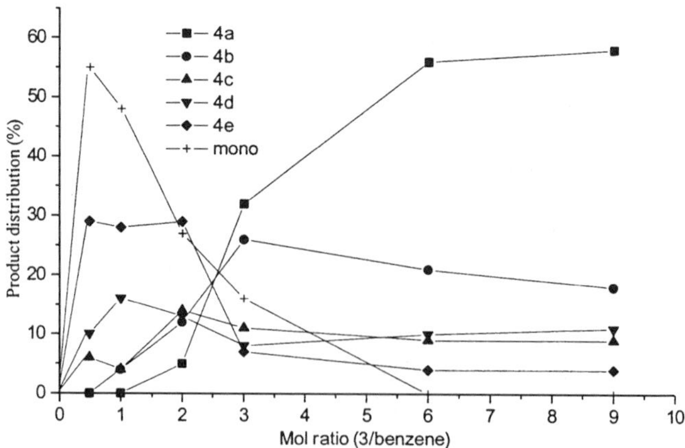

FIG. 1. Alkylation product distribution vs mole ratio of **3**/benzene.

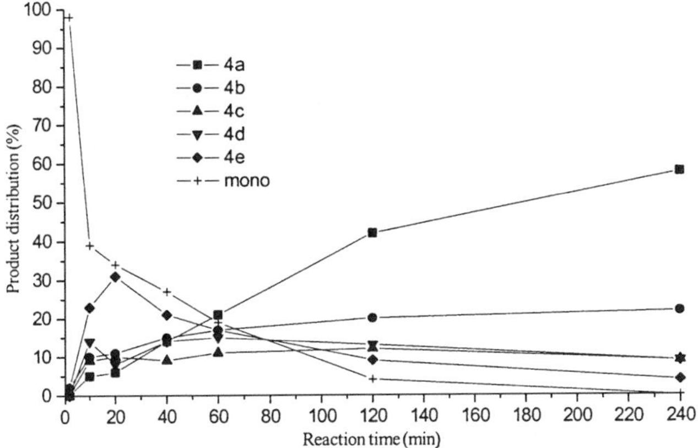

FIG. 2. Alkylation product distribution vs reaction time.

time. As similarly observed in Fig. 1, compounds **4c** and **4d** were obtained in approximately 10% yield throughout the rest of the reaction time. This indicates that they were being produced from monoalkylated and dialkylated products at almost the same rate at which they were consumed to give higher alkylation products. Therefore **4a** can be produced efficiently by using a 6-fold excess of **3** to benzene with a 2-h reaction time.

2. *Polyalkylation of Monoalkylbenzenes with Excess 3*

In the peralkylation of alkylbenzenes in the presence of aluminum chloride catalyst,[28] both polyalkylation and transalkylation reactions have been reported as competing reactions. The yields of peralkylation products decrease as the chain length of the alkyl substituents on the benzene ring increases, probably due to easier dealkylation under the reaction conditions. Toluene reacted with 5 mol of **3** in the presence of aluminum chloride at room temperature for 2 h to give pentakis[2-(methyldichlorosilyl)ethyl]toluene in 61% yield and less alkylated products: tetrakis[2-(methyldichlorosilyl)ethyl]toluene (12%), tris[2-(methyldichlorosilyl)ethyl]toluene (6%), bis[2-(methyldichlorosilyl)ethyl]toluene (2%), and [2-(methyldichlorosilyl)ethyl]toluene (2%) based on toluene used. The peralkylation reactions of alkylbenzenes having longer alkyl groups such as ethyl, *n*-propyl, and *n*-butyl with **3** gave the corresponding peralkylated products in low yields (about 25%) along with the transalkylation products (Eq. (7)). The results from the reaction of alkylbenzenes with **3** are summarized in Table VI.

The reaction of ethylbenzene with 5 mol of **3** under the same reaction conditions for the alkylation of toluene with **3,** gave pentakis[2-(methyldichlorosilyl)ethyl]-(25%), tetrakis[2-(methyldichlorosilyl)ethyl]-(9%), tris[2-(methyldichlorosilyl)-

RT, 2 h, $AlCl_3$; 3

m = 1 - 6, n = 0 - 3, m+n ≤ 6

m + n = 6; R = Et, *n*-Pr, *n*-Bu (8–22%)

R = Me (61%)

R = Et, *n*-Pr, *n*-Bu (25–27%)

(7)

ethyl]-(4%), and bis[2-(methyldichlorosilyl)ethyl]ethylbenzene (1%) as well as many transalkylated products as shown in Table VI. It is of interest that longer alkyl substituted benzenes exhibited different behavior in the peralkylation with **3** in comparison with toluene. Transalkylated products were expected because Anschutz first reported the disproportionation of alkylbenzenes in the presence

TABLE VI

POLYALKYLATION PRODUCTS OF ALKYLBENZENES WITH **3**

R	Products (%)					
	Peralkylated (n = 1, m = 5)	Less Alkylated (*n* = 1) m = 1	m = 2	m = 3	m = 4	Transalkylated[a] (n = 0 or 2)
Me	61	2	2	6	12	
Et	25	0	1	49	9	Tetrakis(sil)diethylbenzene (11)
						Tris(sil)diethylbenzene (11)
						Bis(sil)diethylbenzene (2)
						Hexakis(sil)benzene (trace)
						Pentakis(sil)benzene (8)
						Tetrakis(sil)benzene (7)
						Tris(sil)benzene (5)
						Bis(sil)benzene (trace)
n-Pr	26	3	8	10	12	Tetrakis(sil)di-*n*-propylbenzene (9)
						Bis(sil)benzene (2)
						(Sil)benzene (2)
n-Bu	27	7	10	11	13	Tetrakis(sil)di-*n*-butylbenzene (8)
						Bis(sil)benzene (2)
						(Sil)benzene (1)

[a] (sil) represents 2-(methyldichlorosilyl)ethyl group.

of Lewis acid as early as in 1886.[60] When toluene was heated with aluminum chloride at reflux, it was found that toluene was converted into a mixture of benzene and xylenes, said to be mainly the *m*- and *p*-isomers. Similarly, ethylbenzene gave benzene and a mixture of *m*- and *p*-diethylbenzene. In 1935, Baddeley and Kenner presented convincing evidence that the products obtained from the disproportionation of *p*-di(*n*-propyl)benzene in the presence of aluminum chloride were *n*-propylbenzene, *m*-bis(*n*-propyl)benzene, and 1,3,5-tris(*n*-propyl)benzene.[1,61] Thus, it may be concluded that the inter- and intra-transalkylation of ethylbenzene and alkylated products to the mixture of ethylbenzene, benzene, and diethylbenzene, followed by the polyalkylation with **3,** leads to the many byproducts. This also indicates that the transalkylation of ethylbenzene was responsible for the significantly lower yield (25%) of pentakis[2-(methyldichlorosilyl)ethyl]-benzene in comparison with those obtained from the alkylation of benzene[27] or toluene.

Peralkylation of *n*-propylbenzene and *n*-butylbenzene showed similar results to those of ethylbenzene.

3. *Peralkylation of Dimethyl- and Trimethylbenzenes with Excess* ***3***

o-, *m*-, or *p*-Xylene reacts with 4 mol of **3** to afford isomeric mixtures of peralkylated xylenes consisting of tetrakis[(methyldichlorosilyl)ethyl]-*o*-xylene or tetrakis[(methyldichlorosilyl)ethyl]-*m*-xylene and tetrakis[(methyldichlorosilyl)-ethyl]-*p*-xylene, and isomeric transalkylation products (Eq. (8)). These results indicate that reorientation occurred during the alkylation of xylenes. With longer reaction times and higher temperatures, tetrakis[(methyldichlorosilyl)ethyl]-*p*-xylene was favored due to less steric hindrance between the substituents on benzene.

o-, *m*-, or *p*-Xylene + 4 **3** (Cl–Si(Me)(Cl)–CH=CH$_2$) —RT, 2 h, AlCl$_3$→ xylene–(CH$_2$CH$_2$–SiCl$_2$Me)$_m$, m = 1 - 4 + tetrakis[2-(methyldichlorosilyl)ethyl]-*p*-xylene (25 - 64%) (8)

It has been demonstrated by several investigators that methylarenes undergo reorientation without any observable transalkylation. Baddeley *et al.* reported that

TABLE VII

POLYALKYLATION OF XYLENE WITH **3** AND PRODUCT DISTRIBUTIONS

Xylene	Product Yields (%)			
	Peralkylated[a] (m = 4)	Less Alkylated (Isomers)		
		m = 1	m = 2	m = 3
o-Xylene	75 (29)	1 (1)	5 (4)	9 (4)
m-Xylene	29 (25)	4 (1)	11 (4)	39 (4)
p-Xylene	76 (64)	1 (1)	4 (5)	8 (4)

[a] Tetrakis[2-(methyldichlorosilyl)ethyl]-*p*-xylene is presented in parentheses among the isomeric mixture of peralkylated products.

a mixture of *p*-xylene and aluminum bromide kept at room temperature for 1 h gave no change in the hydrocarbon, but in the presence of hydrogen bromide, reorientation occurred to produce a mixture of the three xylene isomers (in a ratio of ca. 26:7:27 = *o*/*m*/*p*) with no appreciable formation of transalkylation products.[61,62] These reports correspond well to the above results.

Because of its high structural symmetry, tetrakis[(methyldichlorosilyl)ethyl]-*p*-xylene was easily purified by recrystallization from THF. The less alkylated products, monoalkylated to trialkylated compounds, were also formed along with many isomers due to reorientation of the methyl group. The yields of peralkylated and less alkylated products including isomers are summarized in Table VII.

Peralkylation of mesitylene with **3** gave only reoriented products but no transalkylated products. It was found that the undesirable isomerization of 2,4,6-tris-(methyldichlorosilylethyl)mesitylene to the 3,5,6- or 4,5,6-isomers could be avoided by carrying out the reaction in hexane solution at room temperature for 12 h. In this reaction, peralkylated product, 2,4,6-tris[2-(methyldichlorosilyl)-ethyl]mesitylene, was obtained in 38% yield along with mono- and dialkylated mesitylene without any reoriented products in 7 and 37% yields, respectively.

The Friedel–Crafts type polyalkylation of alkyl-substituted benzenes with **3** becomes easier and faster as the number of electron-donating methyl groups on the phenyl group increases. This is consistent with the fact that the alkylation occurs in the fashion of electrophilic substitution. The tendency of starting methylbenzenes to form reoriented products also increases in the same order from toluene to mesitylene.

The alkylation of benzenes having electron withdrawing groups, such as chlorobenzene and anisole, with **3** gave only monoalkylated, dialkylated, and trialkylated compounds,[3,63] but no peralkylated products were obtained even upon heating of the reaction mixtures for long periods. This might be due to the deactivation effect of the electron-withdrawing substituents and the complexation between $AlCl_3$ catalyst and the lone pair electrons on the substituents.

IV
FRIEDEL–CRAFTS ALKYLATIONS OF ARENES WITH (CHLOROALKYL)SILANES

The Friedel–Crafts alkylation of aromatic compounds with alkyl halides in the presence of Lewis acid is well defined in organic chemistry.[1] However, alkylations with chlorosilanes[8,14–16,18–20] containing chlorinated alkyl groups have been relatively unexplored due to the lack of readily available starting materials. (Polychlorinated methyl)chlorosilanes have been produced as byproducts in the photochlorination process[64,65] of dimethyldichlorosilane for the preparation of (chloromethyl)methyldichlorosilane (**5**). Bis(4-fluorophenyl)methyl[(1H-1,2,4-triazol-1-yl)methyl]silane, known as a good fungicide, was prepared starting from **5** by Moberg in 1985,[66] and was commercialized by Du Pont. Jung *et al.* also studied the synthesis of new bioactive organosilicon compounds[67,68] starting from (chloromethyl)trichlorosilane which can be prepared by the chlorination[69] of methyltrichlorosilane. To find useful applications of (polychlorinated methyl)chlorosilanes [$(Cl_nH_{3\text{-}n}C)SiX_3$, n = 2, 3] formed as byproducts in the photochlorination of methylchlorosilanes, they studied the dechlorination of byproducts to re-form starting materials[70] and the Friedel–Crafts alkylation of benzene with the (polychlorinated methyl)chlorosilanes. In a series of these works, a variety of (chlorinated alkyl)chlorosilanes was synthesized[64,65,69,71–73] and their alkylation reactions with benzene derivatives were studied. This section will describe the Friedel–Crafts reactions of aromatic hydrocarbons with (chlorinated alkyl)chlorosilanes containing short chain alkyl groups such as methyl, ethyl, propyl groups. These reactions will be described in terms of substituent effect on the silicon and alkyl-chain length effect of the (chlorinated alkyl)silanes.

A. *Alkylation with (ω-Chloroalkyl)silanes*

(Chloroalkyl)silanes have been reported to react with aromatic compounds in the presence of aluminum chloride catalyst to give the corresponding alkylation products (Eq. (9)).[8,14–16,18–20] However, the details of the alkylation were not reported. Thus, Jung *et al.* undertook a systematic investigation of the alkylation of silanes containing a terminal chlorinated alkyl group [$Cl{-}(CH_2)_mSiX_3$, n = 1, 2, 3] to benzene in the presence of aluminum chloride catalyst. In this work, the

$$X^2\text{-}\underset{X^3}{\overset{X^1}{Si}}\text{-}(CH_2)_mCl + C_6H_6 \xrightarrow[\text{rt - 200°C}]{\text{20 mol \% AlCl}_3} \left(X^2\text{-}\underset{X^3}{\overset{X^1}{Si}}\text{-}(CH_2)_mCH_2\right)_n\text{-}C_6H_{6-n} \quad (9)$$

$m = 1\text{–}3$
$n = 1\text{–}6$

alkylation reactivity of (ω-chloroalkyl)silanes was found to depend on the substituents on the silicon atom and the spacer length between the silicon and C—Cl bond. The results obtained for the alkylation with (ω-chloroalkyl)silanes at various reaction temperatures are summarized in Table VIII.

As shown in Table VIII, the spacer length between C—Cl and the silicon and substituents on the silicon atom of (ω-chloroalkyl)silanes affects the reactivity of (ω-chloroalkyl)silanes in the alkylation of benzene. The alkylation reactions occurred at lower temperature as the spacer length between C—Cl and the silicon increased from (chloromethyl)silane to (β-chloroethyl)silane and to (γ-chloropropyl)silane. These results show that the alkylation rate of (ω-chloroalkyl)silanes increases drastically as the alkyl-chain length of (ω-chloroalkyl)silanes increases from methylene to ethyl and to propylene, reflecting that the activation of C—Cl bond of (ω-chloroalkyl)silanes depends on both the electron-withdrawing and steric effects of the neighboring trichlorosilyl-group. In the case of substituent effects on the silicon of (chloromethyl)silanes, the reactions occurred at higher reaction temperatures as the number of chlorine groups on the silicon increased from

TABLE VIII
REACTION CONDITIONS OF ω-CHLOROALKYLSILANES ($Cl(CH_2)_mSiX^1X^2X^3$) WITH BENZENE AND ALKYLATION PRODUCTS

Silane		Reaction Conditions[a]		Alkylation Products (%)		Unreacted Silane
m	$X^1X^2X^3$	Temp. (°C)	Time (h)	Mono-	Di-	
1	Cl_3	200	20	31	—	55
2	Cl_3	rt	2	82	—	—
3	Cl_3	rt	1	70	—	—
1	Cl_2,Me	80	16	71	13	10
1	Me_3	rt	4	Decomposition products[b]		

[a] The reaction was carried out in the presence of 20 mol% aluminum chloride catalyst.

[b] Toluene, xylene, and trimethylchlorosilane were observed as major products.

(chloromethyl)trimethylsilane to (chloromethyl)methyldichlorosilane and to (chloromethyl)trichlorosilane. In particular, (chloromethyl)trichlorosilane reacted with benzene under drastic conditions to afford the alkylated product: a reaction temperature of 200°C in a sealed stainless tube. After a 20-h reaction period, only 49% of the (chloromethyl)trichlorosilane was consumed to give (trichlorosilylmethyl)benzene in 31% yield. No alkylated products of ((trimethylsilyl)methyl)-benzene were observed in the alkylation with (chloromethyl)trimethylsilane, but lower boiling products such as toluene, xylene, and trimethylchlorosilane were obtained instead. The reaction with (chloromethyl)silanes containing one chlorine on silicon also gave alkylation products, along with lower boiling compounds as observed in the alkylation with trimethylchlorosilane. The alkylation rate of benzene was faster with (ω-chloroalkyl)trichlorosilanes having longer spacers between C—Cl and the silicon and fewer chlorine substituents on the silicon. This indicates that the C—Cl bond is deactivated by inductive effects since the electron-withdrawing ability of silyl group increases with increasing number of chloro-groups on the silicon.[14] Other alkylation products obtained from the reaction of (ω-chloroalkyl)silanes with aromatic compounds such as benzene derivatives and naphthalene in the presence of aluminum compounds as catalyst are summarized in Table IX.[8,14–16,18,20]

To prepare multifunctionalized symmetric organosilicon compounds by the polyalkylation of benzene, (2-chloroethyl)trichlorosilane and (3-chloropropyl)trichlorosilane were reacted with benzene. Polyalkylations of benzene with (2-chloroethyl)silane and (3-chloropropyl)silane were carried out in the presence of aluminum chloride catalyst at a reaction temperature of 80°C. The reaction of benzene with excess (2-chloroethyl)trichlorosilanes afforded peralkylated product, hexakis(2-(trichlorosilyl)ethyl)benzene in good yield (70%).[74]

RT, $AlCl_3$ → + others

80°C/$AlCl_3$ → 70% (10)

TABLE IX

ALKYLATION CONDITIONS OF BENZENE DERIVATIVES (R–Ph) WITH (CHLOROALKYL)SILANES ($X^3X^2X^1Si(CH_2)_mCHXCl$) AND PRODUCT YIELDS

Reactants				Reaction Conditions			Alkylation Products (%)		
Silane									
$X^1X^2X^3$	m	X	R[a]	Cat.	Temp. (°C)	Time (h)	Mono-	Di-	Ref.
Cl_3	0	H	H	$AlCl_3$	Reflux	60–100	42–85		8, 12, 20
Cl_3	0	H	Cl	$AlCl_3$	Reflux	80–100	65		8
Cl_3	0	H	Me	$AlCl_3$	Reflux	100	13		12
Cl_3	0	Me	H	$AlCl_3$	Reflux	80–100	65		8
Cl_3	0	Et	H	Al	Reflux	20	18		18
Cl_3	0	$SiCl_3$	H	$AlCl_3$			no		12
Cl_3	1	H		Al	200	20		—	
Cl_3	1	H	H	$AlCl_3$	Reflux	33–100	33–83	—	8, 12, 15
Cl_3	1	H	H	Al	70		72		15
Cl_3	1	H	Cl	$AlCl_3$	Reflux	50	77	—	12
Cl_3	1	H	Me	$AlCl_3$	Reflux	33	72		12
							55		8,
Cl_3	1	H	Ph	Al	90		34		15
Cl_3	1	H	naph	$AlCl_3$	130–5	3–4	21–28		15, 18
Cl_3	1	H	OPh	$AlCl_3$	130	25	23		15
Cl_3	1	Me	H	$AlCl_3$	Reflux	20	53		18
Cl_3	1	Me	Me	$AlCl_3$	Reflux	20	58		18
Cl_3	1	Me	Ph	Al	Reflux	20	23		18
Cl_3	1	Me	Ph	$AlCl_3$	Reflux	20	58		18
Cl_3	1	Me	OPh	$AlCl_3$	Reflux	20	42		18
Cl_3	2	H	H	$AlCl_3$			47(16)		18
Cl_3	2	H	Cl	$AlCl_3$			73		18
Cl_3	2	H	Me	Al			34		18
Cl_3	2	H	Ph	$AlCl_3$			45		18
Cl_3	2	H	OPh	$AlCl_3$			45		18
Cl_3	2	H	naph	$AlCl_3$			6		18
Me,Cl_2	0	H	H	$AlCl_3$	Reflux	60	26–65	8	8, 16, 20
Me,Cl_2	0	H	Cl	$AlCl_3$	Reflux	60	26	—	16
Me,Cl_2	0	H	Me	$AlCl_3$	Reflux	60	16	—	16
Me,Cl_2	1	Me	H	Al			44	—	18
Me,Cl_2	1	Me	Cl	$Al Cl_3$			50	—	18
Me,Cl_2	1	Me	Me				40	—	18
Me,Cl_2	1	Me	Ph				40	—	18
Me,Cl_2	1	Me	OPh				27	—	18
Me,Cl_2	1	Me	naph				26	—	18
Me,Cl_2	2	Me	H	$AlCl_3$			34	—	18
Me,Cl_2	2	Me	Cl	Al			44	—	18
Me,Cl_2	2	Me	Me	$AlCl_3$			61	—	18
Me,Cl_2	2	Me	naph	$AlCl_3$			39	—	18
Et,Cl_2	0	Me	H	$AlCl_3$	Reflux	60	50	—	16
Et,Cl_2	0	H	Cl	$AlCl_3$	Reflux	60	25	—	16
Et,Cl_2	0	H	Me	$AlCl_3$	Reflux	60	25	—	16
Et,Cl_2	1	H	H	$AlCl_3$	Reflux	8–60	25–34		12, 16
Et,Cl_2	1	H	Cl	$AlCl_3$	Reflux	60	42	—	16
Et,Cl_2	1	H	Me	$AlCl_3$	Reflux	60	41	—	16

[a] naph = naphthalene.

The reaction of benzene with excess (3-chloropropyl)trichlorosilane afforded peralkylated product, hexakis(3-(trichlorosilyl)propyl)benzene, in moderate yield (53%).[74]

$$C_6H_6 + Cl(CH_2)_3SiCl_3 \xrightarrow[AlCl_3]{RT} C_6H_5(CH_2)_3SiCl_3 + \text{others} \xrightarrow{80°C/AlCl_3} C_6[(CH_2)_3SiCl_3]_6 \; (53\%) \quad (11)$$

B. *Alkylation with (Dichloroalkyl)silanes*

The alkylation of benzene with (ω,ω-dichloroalkyl)silanes was also studied in the presence of aluminum chloride catalyst. The alkylation gave diphenylated products, (ω,ω-diphenylalkyl)chlorosilanes in fair to good yields (Eq. (12)).

$$X\text{–}SiCl_2\text{–}(CH_2)_m\text{–}CCl_2\text{–}H + 2\,C_6H_6 \xrightarrow{20\text{ mol \% }AlCl_3} X\text{–}SiCl_2\text{–}(CH_2)_m\text{–}CH(C_6H_5)\text{–}C_6H_4\text{–}\left(CH(C_6H_5)\text{–}(CH_2)_m\text{–}SiCl_2\text{–}X\right)_n \quad n = 0, 1 \quad (12)$$

The results obtained from the alkylation reaction of (ω,ω-dichloroalkyl)silanes with excess benzene are summarized in Table X.

As observed in the alkylation with (ω-chloroalkyl)silanes,[14] both the spacer length between the C–Cl and silicon and the substituents on the silicon atom of

TABLE X

ALKYLATION CONDITIONS OF BENZENE WITH (ω,ω-DICHLOROALKYL)SILANES ($Cl_2XSi(CH_2)_mCHCl_2$) AND PRODUCT YIELDS

Silane		Reaction Conditions[a]		Products (%)[b]		Unreacted Silane
m	X	Temp. (°C)	Time (h)	n = 0	n = 1	
0	Cl	80	1	3	—	95
1	Cl	rt[c]	2	82	—	—
0	Me	80	0.5	74	12	—

[a] Reaction was carried out in the presence of 20 mol% $AlCl_3$ based on silane used.

[b] Refer to Eq. (12) for the product structures ($n = 0, 1$).

[c] 10 mol% $AlCl_3$ was used based on silane.

(ω,ω-dichloroalkyl)chlorosilanes affect the reactivity of (ω,ω-dichloroalkyl)silanes for the alkylation of benzene in the presence of aluminum chloride catalyst. The alkylation of benzene with (ω,ω-dichloroalkyl)trichlorosilanes occurred at lower temperature as the spacer length between the C—Cl and silicon increased from (dichloromethyl)silane to (2,2-dichloroethyl)silane. The alkylation proceeded at room temperature with (2,2-dichloroethyl)silane, while the alkylation with (dichloromethyl)silane occurred at 80°C. As the number of chlorine substituents on the silicon decreased from (dichloromethyl)trichlorosilanes to (dichloromethyl)methyldichlorosilanes, the products were obtained in good yield after a shorter reaction period.

In the alkylation of benzene with (dichloroalkyl)chlorosilanes in the presence of aluminum chloride catalyst, the reactivity of (dichloroalkyl)silanes increases as the spacer length between the C—Cl and silicon and as the number of chloro-groups on the silicon of (dichloroalkyl)chlorosilanes decreases as similarly observed in the alkylation with (ω-chloroalkyl)silanes. The alkylation of benzene derivatives with other (dichloroalkyl)chlorosilanes in the presence of aluminum chloride gave the corresponding diphenylated products in moderate yields.[14,17] Those synthetic data are summarized in Table XI.

The alkylation of halogen-substituted benzenes such as fluorobenzene and dichlorobenzenes with other (dichloroalkyl)silanes in the presence of aluminum chloride catalyst afforded isomeric mixtures of the corresponding (dihalogen-substituted phenyl)alkylsilanes in moderate yields (Eq. (13)). These results are summarized in Table XII.

$$X^1C_6H_4X^2 + Cl_2H_{2n-1}C_n{-}SiCl_2{-}R \xrightarrow{AlCl_3} (X^1X^2C_6H_4)_2H_{2n-1}C_n{-}SiCl_2{-}R \qquad (13)$$

TABLE XI

ALKYLATION CONDITIONS OF BENZENE DERIVATIVES (R–Ph) WITH (DICHLOROALKYL)SILANES ($Cl_2X^1Si(CH_2)_mCX^2Cl_2$) AND PRODUCT YIELDS

Reactants: Silane				Reaction Conditions			Alkylation Products[a] (%)		
X^1	m	X	R	Cat.	Temp (°C)	Time (h)	Di-	Mono	Ref.
Cl	0	H	H	$AlCl_3$	reflux	60	10	—	12
Cl	0	H	Cl	$AlCl_3$	reflux	64	50	—	17
Cl	1	H	H	$AlCl_3$	reflux	—	15	—	17
Cl	1	H	Cl	$AlCl_3$	reflux	—	26	—	17
Cl	1	H	Me	$AlCl_3$	reflux	—	3	11	17
Me	0	Me	H	$AlCl_3$	reflux	—	38	—	17
Me	0	Me	Cl	$AlCl_3$	reflux	—	40	9	17
Me	0	Me	Me	$AlCl_3$	reflux	—	5	41	17
Me	0	Me	H	$AlCl_3$	reflux	—	6	—	17
Me	0	Me	Cl	$AlCl_3$	reflux	—	42	—	17
Me	0	Me	Me	$AlCl_3$	reflux	—	4	—	17

[a] Di and mono stand for (diphenylalkyl)- and (monophenylalkyl)chlorosilanes.

In extension of the alkylation reactions to polychlorobenzenes, polychlorinated benzenes such as 1,2,4-trichlorobenzene and 1,2,3,4-tetrachlorobenzene were alkylated with (1,2-dichloroethyl)trichlorosilanes in the presence of aluminum chloride catalyst. Although the electron-withdrawing chlorine substituents on the ring deactivated the electrophilic substitution reaction, the alkylation

TABLE XII

ALKYLATION CONDITIONS OF BENZENE DERIVATIVES (X^1X^2-C_6H_4) WITH (DICHLOROALKYL)SILANES ($Cl_2RSi(C_nH_{2n-1}Cl_2)$) AND YIELDS OF PRODUCTS ($(X^1X^2C_6H_4)_2H_{2n-1}C_nSiRCl_2$)

Reactants: Silane			Reactants: X^1X^2-C_6H_4		Reactants Conditions		
R	n	Position of Cl_2	X^1	X^2	Temp. (°C)	Time (h)	Alkylation Products (%)
Me	1	1,1-	H	F	Reflux	2.0	65
Me	1	1,1-	Cl	4-Cl	150	7.0	47
Cl	2	1,2-	Cl	3-Cl	120	0.5	39
Cl	2	1,2-	Cl	4-Cl	120	0.5	43
Cl	2	2,2-	Cl	4-Cl	80	1.0	59
Cl	3	2,3-	H	H	70	0.1	76
Cl	3	2,3-	H	F	70	0.2	41

proceeded for 0.3 h at 120°C to give unusual 5-membered ring products.[63] In the case of alkylation of 1,2,4-trichlorobenzene, dialkylation products, [2,2-bis(trichloro phenyl)ethyl]trichlorosilanes, and cyclic products, 1-silylmethyl-2-silyl-3-trichlorophenyl-4,5(6),7-trichloroindanes, were obtained in 57 and 6% yields (Eq. (14)), respectively.[63] The alkylation of 1,2,3,4-tetrachlorobenzene gave cyclic products, 1-silylmethyl-3-(2,3,4,5-tetrachlorophenyl)-4,5,6,7-tetrachloroindane and 1-silylmethyl-3-(2,3,4,5-tetrachlorophenyl)-4,5,6,7-tetrachloroindane, in 58 and 5% yields, respectively (Eq. (15)).[63]

AlCl3, 120°C; 6%; 57% (14)

58%; 5% (15)

(Dichloroalkyl)chlorosilanes undergo the Friedel–Crafts alkylation type reaction with biphenyl in the presence of aluminum chloride catalyst to afford 9-((chlorosilyl)alkyl)fluorenes through two step reactions (Eq. (16)). The results obtained from the alkylation of biphenyl and the cyclization reaction to 5-membered-ring product are summarized in Table XIII.

$$\text{biphenyl} + Cl_2H_{2n-1}C_n{-}SiCl_2{-}X \xrightarrow{AlCl_3} \text{9-R-9-}(CH_2)_{n-1}{-}SiCl_2{-}X\text{-fluorene} \tag{16}$$

R = H, Me

TABLE XIII

ALKYLATION CONDITIONS OF BIPHENYL WITH (DICHLOROALKYL)SILANES ($Cl_2XSi(C_mH_{2m-1}Cl_2)$) AND YIELDS OF PRODUCTS ($(C_{13}H_8R)$-$(CH_2)_nSiXCl_2$)

Silane			Reaction Conditions		Alkylation Products[a]		
X	m	Position of Cl_2	Temp. (°C)	Time (h)	R	n	(%)
Cl	2	1,2-	130	3	H	1	34
Me	1	1,1-	120	2.5	H	0	59
Me	2	1,2-	120	0.2	H	1	18
Me	3	2,3-	140	2.0	H	2	32
Me	4	2,3-	100	1.0	Me	2	34

[a] Alkylation products stands for (9-fluorenylalkyl)silanes.

C. *Alkylation with (Trichloromethyl)silanes*

When (trichloromethyl)silanes reacted with excess benzene in the presence of aluminum chloride at reflux temperature, (triphenylmethyl)silanes were obtained as the major products along with (diphenylmethyl)silanes as minor products (Eq. (17)). Excess benzene was used to avoid the production of polymeric materials due to polyalkylation of one phenyl group.

$$X\text{-}SiCl_2\text{-}CCl_3 + 3\,C_6H_6 \xrightarrow[80^{\circ}C]{20\text{ mol \% }AlCl_3} X\text{-}SiCl_2\text{-}CH_{3-n}(C_6H_5)_n \qquad n = 2, 3 \tag{17}$$

The results obtained from the alkylation of benzene with (trichloromethyl)silanes are summarized in Table XIV.

As shown in Table XIV, the reactivity of (trichloromethyl)silanes varied depending upon the substituent on silicon. The reactivity and yields of (trichloromethyl)-methyldichlorosilanes were slightly higher than those of (trichloromethyl)trichlorosilanes in the aluminum chloride-catalyzed alkylation as similarly observed in the alkylations with (ω-chloroalkyl)silanes and (dichloroalkyl)silanes. The electron-donating methyl group on the silicon facilitates the alkylation more than the electron-withdrawing chlorine. The minor products, (diphenylmethyl)chlorosilanes, were presumably derived from the decomposition of (triphenylmethyl)-chlorosilanes.

To examine the decomposition of (triphenylmethyl)chlorosilanes to (diphenylmethyl)chlorosilanes during the alkylations of benzene with (trichloromethyl)-

TABLE XIV

ALKYLATION CONDITIONS[a] FOR BENZENE WITH (TRICHLOROMETHYL)SILANES ($Cl_2XSiCCl_3$) AND YIELDS OF PRODUCTS ($Cl_2XSiCH_{3-n}(Ph)_n$)

Silane (X)	Reaction Time (h)	Products (%) n = 3	Products (%) n = 2	Unreacted Silane
Cl	1	60	2	7
Cl	2	64	3	—
Me	0.5	72	18	18
Me	1	68	25	—

[a] Reaction was carried out at 80°C in the presence of 20 mol% $AlCl_3$ based on silane used.

chlorosilanes in detail, methyl(triphenylmethyl)dichlorosilane was stirred under the same alkylation conditions but without starting material. As expected, methyl-(diphenylmethyl)dichlorosilane was obtained (Eq. (18)).

$$\text{MeSiCl}_2\text{–CPh}_3 \xrightarrow[\text{in benzene}]{AlCl_3} \text{MeSiCl}_2\text{–CHPh}_2 \quad (18)$$

The results obtained from the decomposition reaction of (triphenylmethyl)-methyldichlorosilane to (diphenylmethyl)methyldichlorosilane in benzene solvent in the presence of aluminum chloride are summarized in Table XV.

TABLE XV

DECOMPOSITION OF (TRIPHENYLMETHYL)METHYLDICHLOROSILANE (**I**) TO (DIPHENYLMETHYL)METHYLDICHLOROSILANE (**II**) IN THE PRESENCE OF $AlCl_3$

Reaction Conditions[a] Temp. (°C)	Time (h)	Compounds (%) **I**	**II**	**II/I**
rt	2	100	—	0
80	1	85	10	0.12
80	2	72	20	0.28

[a] The reaction was carried out in the presence of 20 mol% aluminum chloride catalyst.

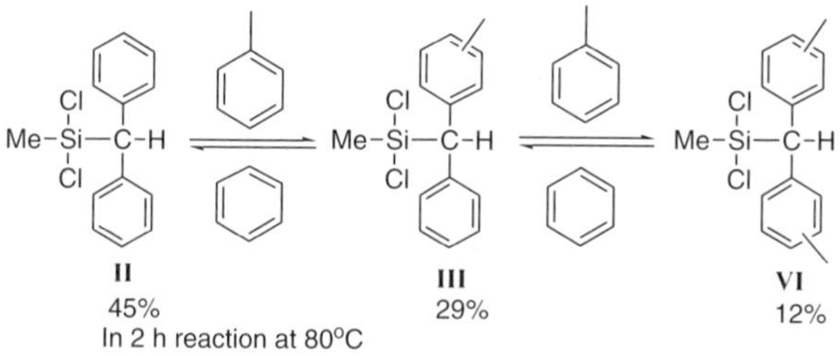

II 45%

III 29%

VI 12%

In 2 h reaction at 80°C

SCHEME 1. Exchange reaction between (phenylalkyl)silane and toluene.

As shown in Table XV, the decomposition of (triphenylmethyl)methyldichlorosilane did not occur at room temperature, but occurred at the reflux temperature of benzene to give (diphenylmethyl)methyldichlorosilane in 10 and 20% yields after 1 and 2 h reaction periods. The results indicate that the decomposition occurs in the alkylation reaction conditions of benzene with (trichloromethyl)chlorosilanes as observed in the decomposition of tetraphenylmethane to triphenylmethane.[75,76]

To confirm the production of benzene from the decomposition reaction of methyl(triphenylmethyl)dichlorosilane, the decomposition reaction of methyl(diphenylmethyl)dichlorosilane in the presence of aluminum chloride was carried out in toluene solvent at 80°C. In this reaction, the exchange reaction between phenyl groups on the methyl group of (diphenylmethyl)(methyl)dichlorosilane and toluene occurred to give [phenyl(tolyl)methyl](methyl)dichlorosilane and (ditolylmethyl)(methyl)dichlorosilane (Scheme 1).[75,76]

a. *Mechanism for alkylation and exchange reactions* The mechanism for the alkylation of benzene, as an example, with chloroalkylsilanes in the presence of aluminum chloride as catalyst is outlined in Scheme 2. The C—Cl bond of chloroalkylsilanes interacts with aluminum chloride catalyst to give a polar $^{\delta+}C{-}Cl^{\delta-}$

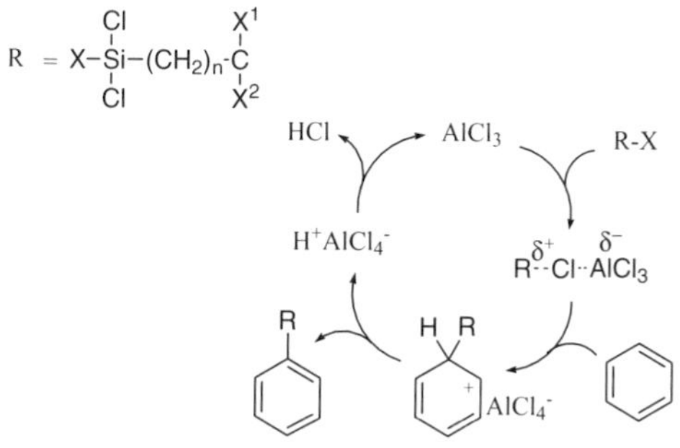

$R = X{-}SiCl_2{-}(CH_2)_n{-}CX^1X^2$

HCl, $AlCl_3$, R-X, $H^+AlCl_4^-$, $R^{\delta+}\cdots Cl\cdots AlCl_3^{\delta-}$, $AlCl_4^-$

SCHEME 2. Mechanism for the Friedel–Crafts alkylation of benzene with (chloroalkyl)silane.

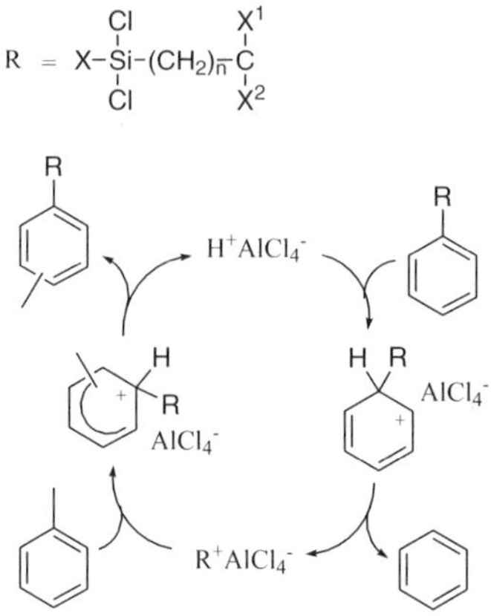

SCHEME 3. Mechanism for the transalkylation reaction between the phenyl group of phenylalkylsilane and toluene.

($^{\delta+}$C—Cl---Al$^{\delta-}$Cl$_3$) intermediate or a carbocation $C^+AlCl_4^-$.[77,78] This intermediate electrophilically attacks the benzene ring to generate a benzenonium ion intermediate which gives alkylated benzene through deprotonation by aluminum tetrachloride anion. Finally the hydrogen aluminum tetrachloride complex affords aluminum chloride and hydrogen chloride gas. This aluminum chloride is recycled in the catalytic cycle of alkylation.

The mechanism for the transalkylation of (diphenylmethyl)silane with toluene in the presence of aluminum chloride as catalyst is outlined in Scheme 3. In the aluminum chloride-catalyzed reactions, a small amount of hydrogen chloride, resulting from the reaction of anhydrous aluminum chloride with moisture inevitably present in the reactants, initiates the reaction.[32] The proton resulting from hydrogen chloride and aluminum chloride interacts with the benzene ring to generate a benzenonium ion intermediate through protonation to the *ipso*-carbon of benzene.[77,78] This intermediate is dealkylated to give a alkyl cation intermediate ($R^+AlCl_4^-$) and benzene. The alkyl cation intermediate interacts with toluene to give an alkylated toluenonium ion which is deprotonated to generate a proton ($H^+AlCl_4^-$) and give (tolylated methyl)silane. This proton is recycled in the catalytic cycle of alkylation.

The mechanism for the production of 9-((chlorosilyl)alkyl)fluorenes from the Friedel–Crafts alkylation reaction of biphenyl with (1,2-dichloroethyl)silane in the presence of aluminum chloride as catalyst is outlined in Scheme 4. At the beginning stage of the reaction, one of two C—Cl bonds of (1,2-dichloroethyl)silane ($ClCH_2$—ClCH—SiX_3) interacts with aluminum chloride catalyst to give intermediate **I** (a polar $^{\delta+}$C—Cl$^{\delta-}$ ($^{\delta+}$C—Cl—Al$^{\delta-}$Cl$_3$) or a carbocation $C^+AlCl_4^-$).[77,78]

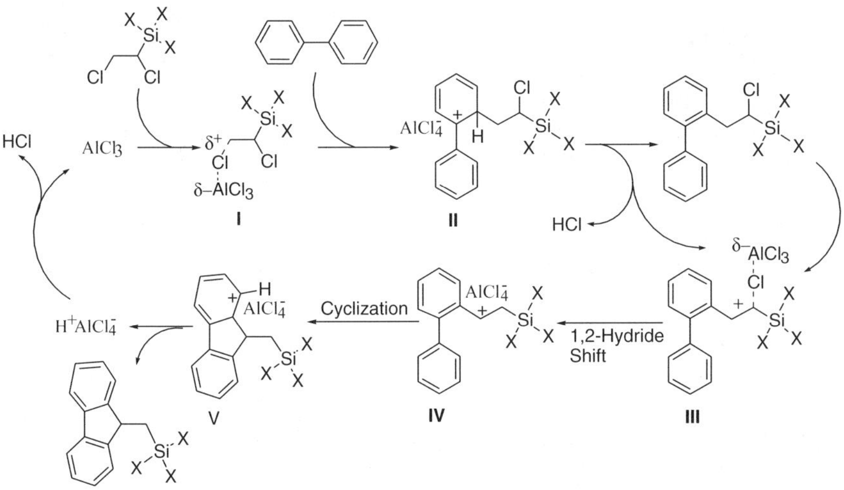

SCHEME 4. Mechanism for the production of 9-(silylmethyl)fluorenes from the reaction of biphenyl with (dichloroethyl)silane.

This intermediate **I** electrophilically attacks the 2-carbon of biphenyl to generate a 2-phenylbenzenonium ion intermediate **II.** This intermediate gives 2-(2-silyl-1-chloroethyl)biphenyl and hydrogen chloride through deprotonation by aluminum tetrachloride anion and dehydrochlorination of hydrogen aluminum tetrachloride complex to generate aluminum chloride catalyst. Then, the C—Cl bond of 2-(2-silyl-1-chloroethyl)biphenyl interacts again with aluminum chloride to generate intermediate **III** which rearranges to give the more stable benzylic carbocation intermediate **IV** through a 1,2-hydrogen-shift. Intermediate **IV** undergoes an intramolecular Friedel–Crafts cycloalkylation with the other phenyl ring to give intermediate **V** which deprotonates to afford 9-(silylmethyl)fluorene and a hydrogen aluminum tetrachloride complex. In the final step, the hydrogen aluminum tetrachloride complex decomposes to aluminum chloride and hydrogen chloride gas. This aluminum chloride is recycled in the catalytic cycle of alkylation.

Among the Friedel–Crafts alkylations of aromatic compounds with (chlorinated alkyl)silanes, the alkylation of benzene with (ω-chloroalkyl)silanes in the presence of aluminum chloride catalyst was generally affected by two factors: the spacer length between the C—Cl and silicon and the electronic nature of substituents on the silicon atom of (ω-chloroalkyl)silanes. As the spacer length between the C—Cl and silicon increases from (chloromethyl)silane to (β-chloroethyl)silane to (γ-chloropropyl)silane, the reactivity of the silanes increases. As the number of chloro-groups on the silicon decreases from (chloromethyl)trichlorosilanes to (chloromethyl)methyldichlorosilanes to (chloromethyl)trimethylsilanes, the reactivity also increases. But no ((trimethylsilyl)methyl)benzene as an alkylated product was observed in the alkylation with (chloromethyl)trimethylsilanes. The reactivities of (chlorinated alkyl)silanes increase as the number of chloro-groups on the carbon of (chlorinated alkyl)silanes increase from (chloromethyl)chlorosilane to (dichloromethyl)chlorosilane to (trichloromethyl)chlorosilane.

V

CONCLUSION AND PROSPECTS

Studies on the alkylation reaction of aromatic compounds with organosilicon compounds are summarized in this review. A variety of chlorosilanes containing alkenyl and chloroalkyl groups can be applied in the Friedel–Crafts alkylations of aromatic hydrocarbons to give the corresponding aromatic compounds containing Si—Cl bonds as functionality. Such organosilicon compounds with Si—Cl bonds are expected to be useful starting materials for the silicone industry.

ACKNOWLEDGMENTS

We thank the Ministry of Science and Technology of Korea and Dow Corning Corporation for support of our research reported herein. We also thank Professor D. Son of Southern Methodist University, Dallas, Texas for his help and discussions in the preparation of this review.

REFERENCES

(1) (a) Roberts, R. M.; Khalaf, A. A. In *Friedel–Crafts Alkylation Chemistry;* Marcel Dekker, Inc.: New York and Basel, 1984; (b) Olah, G. A. In *Friedel–Crafts and Related Reactions*; Interscience Publishers: John Wiley & Sons: New York–London–Sydney, 1963; Volumes I–IV.

(2) Friedel, C.; Crafts, J. M. *Compt. Rend.* **1877,** *84,* 1392.

(3) Friedel, C.; Crafts, J. M. *Bull. Soc. Chim. Fr.* **1877,** *27,* 530.

(4) Yoneda, N. *Sekiyu, Gakkaishi* **1972,** *15,* 894; *Chem. Abstr.* **1972,** *78,* 743,69j.

(5) Tiltscher, H.; Faustmann, J. *Chem.-Ing.-Tech.* **1981,** *53,* 393.

(6) Huellmann, M.; Mayr, H.; Becker, R. *Ger. Offen. DE* 4,008,694, 1991; *Chem. Abstr.* **1991,** *116,* 6262x.

(7) Wager, G. H.; Bailey, D. L.; Pines, A. N.; Dunham, Mcintire, M. L. *Ind. Eng. Chem.* **1953,** *45,* 367.

(8) Petrov, A. D.; Mironov, V. F.; Ponomarenko, V. A.; Chernyshev, E. A. *Dokl. Akad. Nauk SSSR* **1954,** *97,* 687; *Chem. Abstr.* **1955,** *51,* 10,166g.

(9) Rochow, E. G. *J. Am. Chem. Soc.* **1945,** *67,* 963.

(10) Andrianov, K. A.; Zhdanov, A. A.; Odinnets, V. A. *Zh. Obshch. Khim.* **1961,** *31,* 4033; *Chem. Abstr.* **1962,** *57,* 9874a.

(11) Andrianov, K. A.; Zhdanov, A. A.; Odinnets, V. A. *Zh. Obshch. Khim.* **1962,** *32,* 1126; *Chem. Abstr.* **1963,** *58,* 1484a.

(12) Nametkin, N. S.; Vdovin, V. M.; Findel-Shtein, E. S.; Oppengeim, V. D.; Chekalina, N. A. *Isv. Akad. Nauk. SSSR., Ser. Khim.* **1966,** *11,* 1998.

(13) Voronkov, M. G.; Kovrigin, V. M.; Lavrent'ev, V. I.; Moralev, V. M. *Dokl. Akad. Nauk. SSSR.* **1985,** *281(6),* 1374.
(14) Petrov, A. D.; Chernyshev, E. A.; Dolgaya, M. E. *Zh. Obshch. Khim.* **1955,** *25,* 2469; *Chem. Abstr.* **1956,** *50,* 9319a.
(15) Chernyshev, E. A.; Dolgaya, M. E. *Zh. Obshch. Khim.* **1957,** *27,* 48; *Chem. Abstr.* **1957,** *51,*12,045g.
(16) Chernyshev, E. A.; Dolgaya, M. E.; Egorov, Yu. P. *Zh. Obshch. Khim.* **1957,** *27,* 267; *Chem. Abstr.* **1958,** *52,* 7187b.
(17) Chernyshev, E. A.; Dolgaya, M. E.; Egorov, Yu. P. *Zh. Obshch. Khim.* **1958,** *28,* 613; *Chem. Abstr.* **1958,** *52,* 17,150b.
(18) Chernyshev, E. A.; Dolgaya, M. E.; Egorov, Yu. P. *Zh. Obshch. Khim.* **1958,** *28,* 2829; *Chem. Abstr.* **1959,** *53,* 9110c.
(19) Chernyshev, E. A.; Dolgaya, M. E. *Zh. Obshch. Khim.* **1959,** *29,* 1850; *Chem. Abstr.* **1960,** *54,* 8604d.
(20) Cer, L.; Vaisarova, V.; Chvaiovsky, V. *Collect. Czech. Chem. Commun.* **1967,** *32,* 3784; *Chem. Abstr.* **1968,** *68,* 21,992u.
(21) Hurd, D. T. *J. Am. Chem. Soc.* **1945,** *67,* 1813.
(22) Mironkov, V. F.; Zelinskii, D. N. *Izv. Akad. Nauk SSSR, Ser. Khim.* **1957,** 383.
(23) Krieble, R. H.; Elliott, J. R. *J. Am. Chem. Soc.* **1947,** *67,* 1810.
(24) Runge, F.; Zimmermann, W. *Chem. Ber.* **1954,** *87,* 282; *Chem. Abstr.* **1955,** *49,* 6088a.
(25) Yeon, S. H.; Lee, B. W.; Kim, S. -I.; Jung, I. N. *Organometallics* **1993,** *12,* 4887.
(26) Lee, B. W.; Yoo, B. R.; Kim, S. -I.; Jung, I. N. *Organometallics* **1994,** *13,* 1312.
(27) Cho, E. J.; Yoo, B. R.; Jung, I. N.; Sohn, H.; Powell, D. R.; West, R. *Organometallics* **1997,** *16,* 4200.
(28) Cho, E. J.; Lee, V.; Yoo, B. R.; Jung, I. N. *J. Organomet. Chem.* **1997,** *548,* 237.
(29) Jung, I. N.; Yoo, B. R.; Han, J. S.; Cho, Y. S. *US Patent* 5,847,182, 1998.
(30) Yeon, S. H.; Lee, B. W.; Yoo, B. R.; Suk, M. Y.; Jung, I. N. *Organometallics* **1995,** *14,* 2361.
(31) Thomas, C. A. In *Anhydrous Aluminum Chloride in Organic Chemistry;* Reinhold: New York, 1941.
(32) Wierschke, S. G.; Chandrasekhar, J.; Jorgensen, W. L. *J. Am. Chem. Soc.* **1985,** *107,* 1496.
(33) Kresge, A. J.; Tobin, J. B. *Angew. Chem., Int. Ed. Engl.* **1993,** *32,* 721.
(34) Eabon, C.; Bott, R. W. In *Organometallic Compounds of the Group IV Elements;* MacDiarmid, A. G., Ed.; Dekker: New York, 1968; Vol. 1, Part 1, pp. 359–437.
(35) Hassner, A. *J. Org. Chem.* **1968,** *33,* 2684.
(36) Jenkins, P. R.; Gut, R.; Wetter, H.; Eschenmoser, A. *Helv. Chim. Acta* **1979,** *62,* 1922.
(37) For a review, see: Jung, I. N.; Yoo, B. R. *Synlett* **1999,** 519.
(38) Cho, B. G.; Choi, G. M.; Jin, J. I.; Yoo, B. R.; Suk, M. Y.; Jung, I. N. *Organometallics* **1997,** *16,* 3576.
(39) Choi, G. M.; Yeon, S. H.; Jin, J. I.; Yoo, B. R.; Jung, I. N. *Organometallics* **1997,** *16,* 5158.
(40) Choi, G. M.; Yoo, B. R.; Lee, H. -J.; Lee, K. B.; Jung, I. N. *Organometallics* **1998,** *17,* 2407.
(41) Kim, K. M.; Yoo, B. R.; Jung, I. N., unpublished result.
(42) Jung, I. N. In *Chemistry of the Third Generation Silicones* (Korean); Freedom Academic: Seoul, 1997.
(43) Brown, H. C.; Grayson, M. *J. Am. Chem. Soc.* **1953,** *75,* 6285.
(44) Allen, R. H.; Yats, L. D. *J. Am. Chem. Soc.* **1961,** *83,* 2799.
(45) Olah, G. A.; Flood, S. H.; Moffatt, M. E. *J. Am. Chem. Soc.* **1964,** *86,* 1060.
(46) Swain, C. G.; Unger, S. H.; Rosenquist, N. R.; Swain, M. S. *J. Am. Chem. Soc.* **1983,** *105,* 492.
(47) Lee, B. W.; Yoo, B. R.; Jung, I. N., unpublished results.
(48) Cunane, N. M.; Leyshon, D. M. *J. Chem. Soc.* **1954,** 2942.

(49) Kuwestha, G. N.; Shanker, U.; Pathania, B. S.; Negi, J.; Bhattacharyya, K. K. *Indian J. Chem., Sect. B* **1981,** *20B(4),* 298.

(50) Tsuge, O.; Tashiro, M. *Bull. Chem. Soc. Jpn.* **1967,** *40,* 119.

(51) Hansch, C.; Leo, A.; Taft, R. W. *Chem. Rev.* **1991,** *91,* 165.

(52) Brown, H.; Okamoto, Y. *J. Am. Chem. Soc.* **1958,** *80,* 4979.

(53) Hammet, L. P. In *Physical Organic Chemistry;* McGraw-Hill: New York, 1940; p. 40.

(54) Deeming, A. J. In *Comprehensive Organometallic Chemistry;* Wilkinson, G.; Stone, F. G. A., Eds.; Pergamon Press: New York, 1982; Vol. 4, p. 475.

(55) Rosenblum, M.; Santer, J. O.; Howells, W. G. *J. Am. Chem. Soc.* **1963,** *85,* 1450.

(56) Olah, G. A., In *Friedel–Crafts and Related Reactions,* Interscience Publishers; 1965, Vol. IV, p. 128, and references therein.

(57) Tverdokhlebov, V. P.; Polyakov, B. V.; Tselinskii, I. V.; Golubena, L. I. *Zh. Obshch. Khim.* **1982,** *52,* 2032.

(58) Jung, I. N.; Yoo, B. R.; Ahn, S. Y.; Han, J. S. *US Patent* 5,760,263, 1998; *Chem. Abstr.* **1998,** *129,* 67878.

(59) Herberhold, M. In *Ferrocenes;* Togni, A.; Hayashi, T., Eds.; VCH: New York, 1995; p. 219.

(60) Anschutz, R.; *Justus Liebigs Ann. Chem.* **1886,** *235,* 177.

(61) Baddeley, G.; Kenner, J. *J. Chem. Soc.* **1935,** 303.

(62) Baddeley, G.; Holt, G.; Voss, D. *J. Chem. Soc.* **1952,** 100.

(63) Jung, I. N. *et al.,* unpublished results.

(64) Krieble, R. H.; Elliott, J. R. *J. Am. Chem. Soc.* **1947,** *67,* 1810.

(65) Runge, F.; Zimmermann, W. *Chem. Ber.* **1954,** *87,* 2882; *Chem. Abstr.* **1955,** *49,* 6088a.

(66) Moberg, W. K. *US Patent* 4,510,136, 1985; *Chem. Abstr.* **1986,** *104,* 207438k.

(67) Yoo, B. R.; Suk, M. Y.; Han, J. S.; Yu, Y.-M.; Hong, S-G.; Jung, I. N. *Pestic. Sci.* **1998,** *52,* 138.

(68) Yoo, B. R.; Suk, M. Y.; Yu, Y.-M.; Hong, S.-G.; Jung, I. N. *Bull. Korean Chem. Soc.* **1998,** *19,* 358.

(69) Nozakura, S. *J. Chem. Soc. Jpn., Pure Chem. Sec.* **1954,** *75,* 427; *Chem. Abstr.* **1955,** *49,* 10167.

(70) Cho, Y. S.; Han, J. S.; Yoo, B. R.; Kang, S. O.; Jung, I. N. *Organometallics* **1998,** *17,* 570.

(71) Sommer, L. H.; Bailey, D. C.; Goldberg, G. H.; Buck, C. E.; Bye, T. C.; Evans, F. J.; Whitmore, F. C. *J. Am. Chem. Soc.* **1954,** *76,* 1613.

(72) Sommer, L. H.; Bailey, D. C.; Whitmore, F. C. *J. Am. Chem. Soc.* **1948,** *70,* 2532.

(73) Sommer, L. H.; Dorfman, E.; Goldberg, G. H.; Whitmore, F. C. *J. Am. Chem. Soc.* **1946,** *68,* 488.

(74) Kim, K. M.; Yoo, B. R.; Jung, I. N. manuscript in preparation.

(75) Fonken, G. J. *J. Org. Chem.* **1963,** *28,* 1909.

(76) Hine, J. In *Physical Organic Chemistry;* McGraw-Hill: New York, 1962.

(77) Brown, H. C.; Grayson, M. *J. Am. Chem. Soc.* **1953,** *75,* 6285.

(78) Brown, H. C.; Junk, H. *J. Am. Chem. Soc.* **1955,** *77,* 5584.

ADVANCES IN ORGANOMETALLIC CHEMISTRY, VOL. 46

Transition-Metal Systems Bearing a Nucleophilic Carbene Ancillary Ligand: from Thermochemistry to Catalysis

LALEH JAFARPOUR and STEVEN P. NOLAN*

Department of Chemistry
University of New Orleans
New Orleans, Louisiana 70148

I
INTRODUCTION

Although tertiary phosphine ligands are useful in controlling reactivity and selectivity in organometallic chemistry and homogeneous catalysis,[1,2] they often are sensitive to air oxidation and therefore require air-free handling in order to minimize ligand oxidation. More importantly, significant P—C bond degradation occurs when these ligands are subjected to higher temperatures which in certain catalytic processes results in the deactivation of the catalyst, and as a consequence requires the use of higher phosphine concentration.[3] Furthermore, specific applications benefit from or require the use of sterically demanding phosphine ligation to stabilize reactive intermediates.[4–7] Therefore, there is a need for strongly nucleophilic (electron-rich), bulky ligands that are resistant to oxidizing agents and form stable bonds with metals.

*Email: snolan@uno.edu

0065-3055/01 $35.00

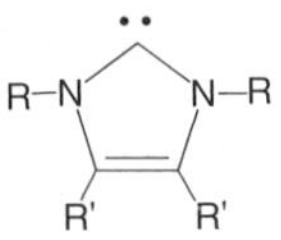

FIG. 1. Nucleophilic carbene.

Nucleophilic carbene ligands imidazol-2-ylidenes are two-coordinate carbon compounds that have two nonbonding electrons and no formal charge on the carbon.[8] In other words they are considered as neutral, two-electron donor ligands with negligible π-back-bonding tendency. As such they can be viewed as alternatives to phosphines.[9–11] Wanzlick was the first to recognize that the electron-rich imidazole framework should be able to stabilize a carbene center at the carbon center situated between the two nitrogens (Fig. 1).[9] Although much of his work was on the saturated imidazoline ring (no double bond between C4 and C5), there are some reports involving unsaturated analogues.[12–23] He succeeded in desulfurizing 1,3-disubstituted imidazole-2-thiones to produce 1,3-disubstituted imidazolium salts but no carbenes were ever isolated in any of these early systems.[24,25] Herrmann and co-workers have synthesized rhodium and palladium complexes bearing 1,3-disubstituted imidazol-2-ylidenes ligands as catalyst precursors in *Heck* coupling reactions.[26,27] These nucleophilic carbenes are generally generated *in situ* from their imidazolium salts (Eq. (1)); thus their general use in catalysis becomes difficult compared to that of an isolable ligand.

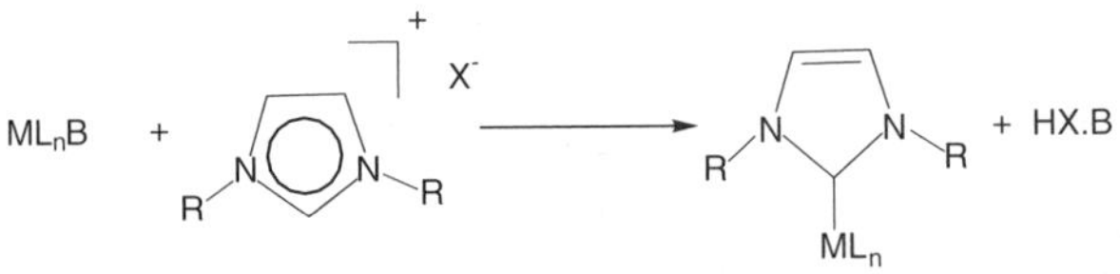

Arduengo and co-workers have solved the isolation problem by flanking the carbene functionality with sterically demanding groups which provide steric protection from carbene degradation pathways.[28] They developed a new one-step synthesis of imidazolium salts that allows the production of substituted imidazolium salts that were not previously accessible by conventional routes (Eq. (2)).[29] 1,3-Bis(1-adamantyl)imidazole-2-ylidene was the first stable carbene isolated by reacting the corresponding imidazolium chloride with sodium hydride in the presence of a catalytic amount of dimethyl sulfoxide (Eq. (3)).[28] Arduengo's carbenes have been used to isolate homoleptic 14-electron bis(carbene)nickel and -platinum complexes analogous to $M(PCy_3)_2$, where M = group 10 metal centers.[30]

$$2\,RNH_2 + HCOH + HX + R'C(O)C(O)R' \xrightarrow{-H_2O} [\text{1,3-R}_2\text{-4,5-R'}_2\text{-imidazolium}]^+ X^- \quad (2)$$

$$[\text{1,3-R}_2\text{-imidazolium}]^+ X^- + NaH \xrightarrow[\text{DMSO}]{\text{THF}} \text{1,3-R}_2\text{-imidazol-2-ylidene} + H_2\uparrow + NaCl\downarrow \quad (3)$$

R = 1-adamantyl

Adducts of a related carbene ligand with Group 2 and 12 metallocenes have also been reported.[31]

The stability of these bulky nucleophilic carbenes has been the subject of many theoretical studies and several factors such as the π-interaction in the imidazole ring, electronegativity effects from the nitrogens, steric effects, kinetic factors, and the large singlet-triplet gap in imidazole-2-ylidenes (80 kcal/mol) have been put forward to explain the stability of these ligands.[32–40] Further investigations show that the resonance stabilization is not an important factor in the stability of these carbenes because stable saturated imidazol-2-ylidenes can also be isolated.[41] These studies also indicated that the unusual stability of these carbenes is due to electron donation from the nitrogen pairs into the formally vacant p(π) orbital of the carbene carbon which makes the carbene more nucleophilic.[42] X-ray crystallographic data of the metal complexes bearing these carbene ligands show that M-C(carbene) bond lengths are in the range of single bonds and ab initio studies also indicate that π-back bonding is not significant in these ligands.[43] These ligands are similar to the tertiary phosphines but appear to be more strongly coordinating ligands.[44]

The present account follows a journey in this arena from solution calorimetric studies dealing with nucleophilic carbene ligands in an organometallic system to the use of these thermodynamic data in predicting the feasibility of exchange reactions to applications in homogeneous catalysis.

II

THERMOCHEMICAL AND STRUCTURAL STUDIES

Tilley and co-workers reported the isolation of coordinatively unsaturated complexes Cp*Ru(L)Cl (Cp* = η^5-C_5Me_5, L = PCy_3 and P^iPr_3) by a simple reaction with the tetrameric species $[Cp^*RuCl]_4$.[45] This synthetic pathway has allowed us to measure the binding affinity of these two bulky phosphine ligands.[46] With the goal of describing the stereoelectronic properties of a series of nucleophilic carbene ligands, we conducted a thermochemical and structural study dealing with the binding of these imidazole-based ligands to the Cp*RuCl moiety.[47]

The versatile starting material $[Cp^*RuCl]_4$ (**1**) reacts rapidly with sterically demanding phosphines (PCy_3 and P^iPr_3) as well as with the nucleophilic carbene ligands (L) to give deep blue, coordinatively unsaturated Cp*Ru(L)Cl complexes **2-8** (L = 1,3-bis(2,4,6-trimethylphenyl) (IMes, **2**); 1,3-R_2-imidazol-2-ylidene = cyclohexyl (ICy, **3**); 4-methylphenyl (ITol, **4**); 4-chlorophenyl (IPCl, **5**); adamantyl (IAd, **6**); 4,5-dichloro-1,3-bis(2,4,6-trimethylphenyl) (IMesCl, **7**); and 1,3-bis(2,6-diisopropylphenyl)imidazol-2-ylidene (IPr, **8**) in high yields according to Eq. (4).

$$1/4[Cp^*RuCl]_4 + \text{(1,3-R}_2\text{-4,5-X}_2\text{-imidazol-2-ylidene)} \xrightarrow{\text{THF}} Cp^*Ru(L)Cl \qquad (4)$$

R	X	L	
mesityl	H	IMes	**2**
cyclohexyl	H	ICy	**3**
4-methylphenyl	H	ITol	**4**
4-chlorophenyl	H	IPCl	**5**
adamantyl	H	IAd	**6**
2,4,6-trimethylphenyl	Cl	IMesCl	**7**
2,6-diisopropylphenyl	H	IPr	**8**

III

CALORIMETRIC STUDIES

The reactions depicted in Eq. (1) are suitable for calorimetric investigations since they proceed rapidly and quantitatively as monitored by NMR spectroscopy. The solution calorimetric protocol has been described elsewhere.[48] The enthalpy values were determined by anaerobic solution calorimetry in THF at 30°C by reacting 4 equivalents of each carbene with one equivalent of tetramer. The results of this study are presented in Table I.

The enthalpies of reaction can be converted to relative enthalpies of reaction on a mole of product basis by dividing the enthalpies by 4 which represents the number of bonds made in the course of reaction. The difference between two relative enthalpy values in Table I represents the enthalpic driving force for a substitution of one for another ligand listed. With the exception of IAd (**6**), all reactions involving carbene ligands show more exothermic reaction enthalpy values than do PCy_3 and P^iPr_3. From the relative enthalpy data it is apparent that $Cp^*Ru(PCy_3)Cl$ should undergo a substitution of phosphine ligand by nucleophilic carbenes with more

TABLE I

ENTHALPIES OF LIGAND SUBSTITUTION, RELATIVE REACTION ENTHALPY (kcal/mol), AND NMR DATA FOR THE REACTION:

$$\underset{\mathbf{1}}{[Cp^*RuCl]_4} + 4L \xrightarrow[30^\circ C]{THF} 4Cp^*Ru(L)Cl$$

Complex	L	$-\Delta H_{rxn}$ (kcal/mol)[a]	Relative BDE (kcal/mol)	$\delta\ ^1H$ (Cp*) (400 MHz, 25°C, d_8-THF)
2	IMes	62.6(0.2)	15.6	1.07
3	ICy	85.0(0.2)	21.2	1.67
4	ITol	75.3(0.4)	18.8	0.99
5	IpCl	74.3(0.3)	18.6	1.03
6	IAd	27.4(0.4)	6.8	1.49
7	IMesCl	48.5(0.4)	12.1	1.06
8	IPr	44.5(0.4)	11.1	1.20[b]
9	PCy_3	41.9(0.2)	10.5	1.48[b]
10	P^iPr_3	37.4(0.3)	9.4	1.43[b]

[a] Enthalpy values are reported with 95% confidence limits.
[b] In C_6D_6.

exothermic reaction enthalpy and it is indeed the case (Eq. (5)).

$$Cp^*Ru(PCy_3)Cl + \text{(R–N–C(:)–N–R imidazol-2-ylidene)} \xrightarrow{THF} Cp^*Ru(L)Cl + PCy_3 \quad (5)$$

R	L
mesityl	IMes
cyclohexyl	ICy
4-methylphenyl	ITol
4-chlorophenyl	IPCl
2,4,6-trimethylphenyl	IMesCl
2,6-diisopropylphenyl	IPr

IV

STRUCTURAL STUDIES

The enthalpies of reaction for nucleophilic carbenes depend on the stereoelectronic properties of the ligands affecting the availability of the carbene lone pair.[11] An example of electronic influence is the 3.5 kcal/mol enthalpy difference between the isosteric pair IMes and IMesCl that shows the electron-withdrawing nature of Cl compared to H. This trend again is in line with electron donor/withdrawing ability of arene substituents. The effect in this last case is a long range electronic

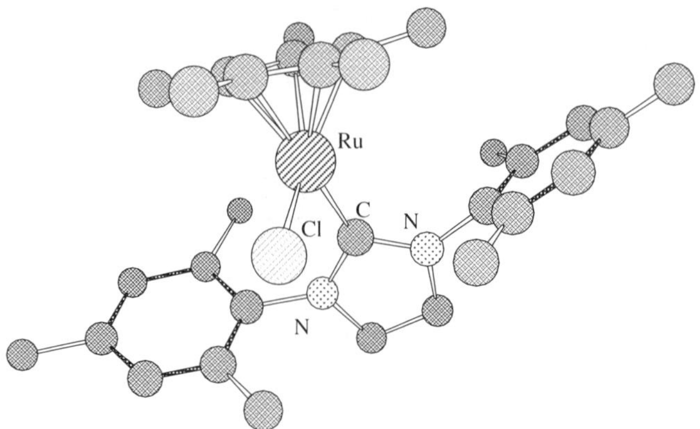

FIG. 2. Molecular structure of Cp*Ru(IMes)Cl (**2**). Hydrogen atoms are omitted for clarity.

effect and is relatively small in view of the distance separating the aryl and the carbene lone pair. Substituting an alkyl group in lieu of an aryl increases the donor ability of the carbene ligand. The case in point is the ICy which is some 5.6 kcal/mol more exothermic than IMes. The other alkyl-substituted carbene investigated is the adamantyl derivative IAd which is the least exothermic ligand examined. Steric effects are at the origin of this low enthalpy of ligand substitution. Increase in the steric congestion around the carbene carbon atom hinders a closer approach of the ligand, therefore affording smaller metal-lone pair overlap. The calorimetric results offer a clear picture of the electronic properties of the nucleophilic carbene moieties as ancillary ligands. To gauge the steric factors at play in the Cp*Ru(L)Cl system, structural studies were carried out on samples of complexes **2**[49], **3**[47], **4**[47], **6**[47], **7**[49], **8**[50], and **9**[49] (Figs. 2–8) and were compared with the structural data

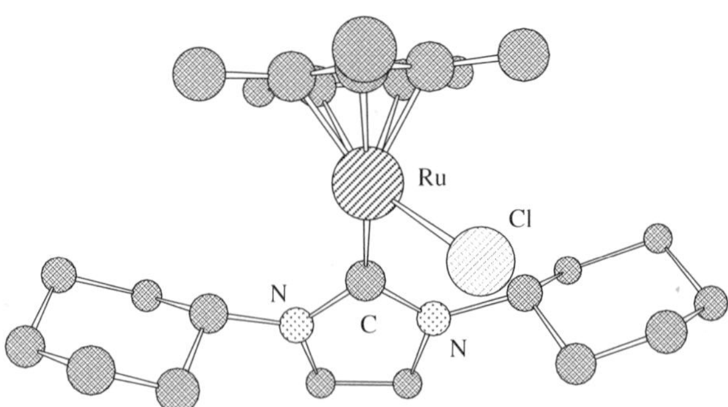

FIG. 3. Molecular structure of Cp*Ru(ICy)Cl (**3**). Hydrogen atoms are omitted for clarity.

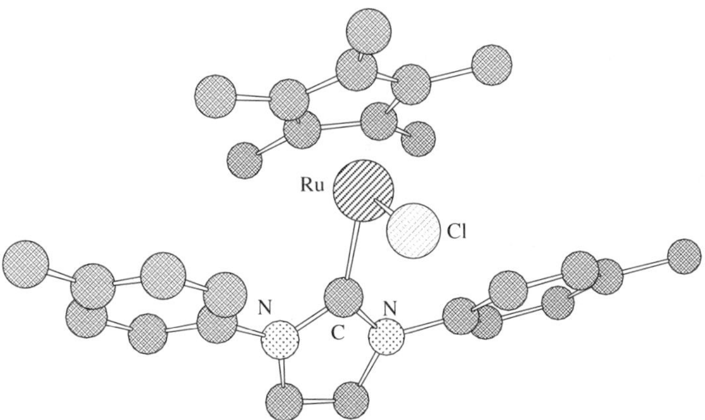

FIG. 4. Molecular structure of Cp*Ru(ITol)Cl (**4**). Hydrogen atoms are omitted for clarity.

already available for Cp*Ru(P^{i}Pr$_3$)Cl[51]. Selected bond angles and bond lengths are presented in Table II.

Ru—C(carbene) bond distances are shorter than Ru—P bond lengths, but this can simply be explained by the difference in covalent radii between P and C.[52] The variation of Ru—C(carbene) bond distances among ruthenium carbene complexes illustrates that nucleophilic carbene ligands are better donors when alkyl, instead of aryl, groups are present, with the exception of **6.** This anomaly can be explained on the basis of large steric demands of the adamantyl groups on the imidazole framework which hinder the carbene lone pair overlap with metal orbitals. Comparison of the Ru—C(carbene) bond distances among the aryl-substituted carbenes show

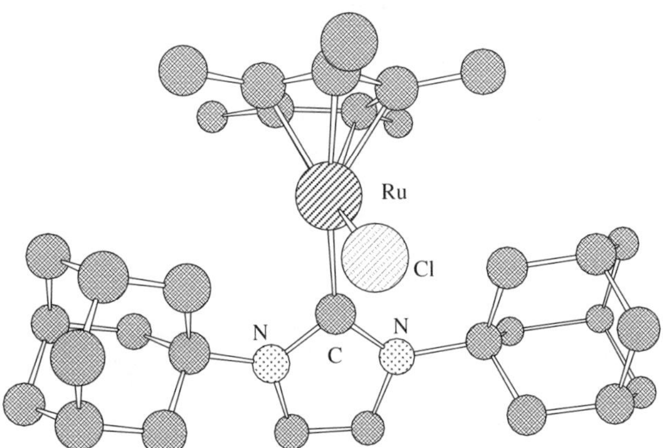

FIG. 5. Molecular structure of Cp*Ru(IAd)Cl (**6**). Hydrogen atoms are omitted for clarity.

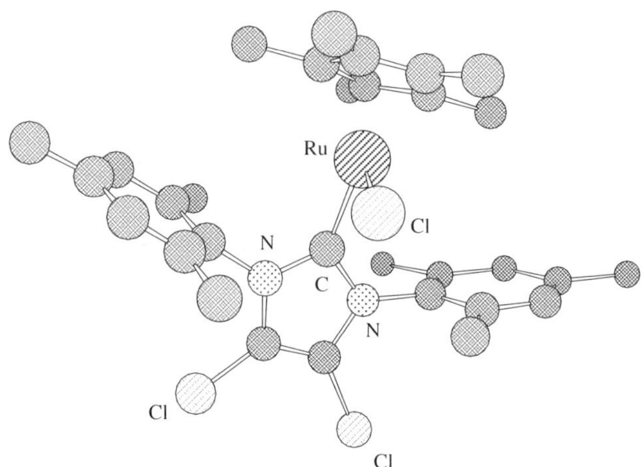

FIG. 6. Molecular structure of Cp*Ru(IMesCl)Cl (**7**). Hydrogen atoms are omitted for clarity.

that the least sterically encumbered carbene, ITol, has the shortest bond length. The difference in the aforementioned bond length between IMes and IMesCl results from the electronic effect of Cl imidazol-2-ylidene substituents. It should be kept in mind here that the enthalpy of reaction cannot be directly converted into an absolute gauge of the bond disruption energy of the Ru—Carbene in view of the existence of significant reorganization energies as depicted by variation in the

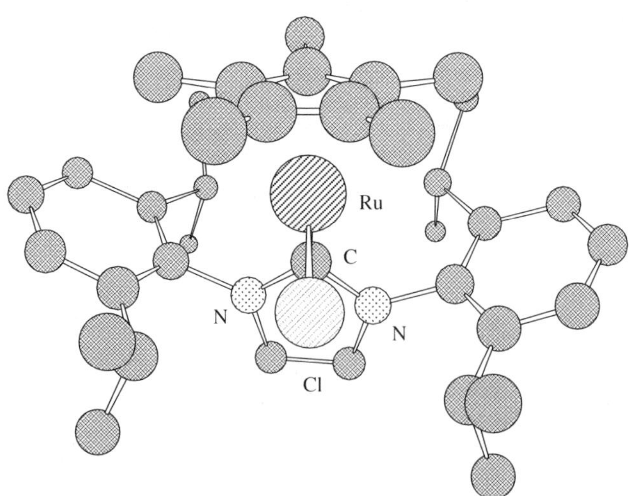

FIG. 7. Molecular structure of Cp*Ru(IPr)Cl (**8**). Hydrogen atoms are omitted for clarity.

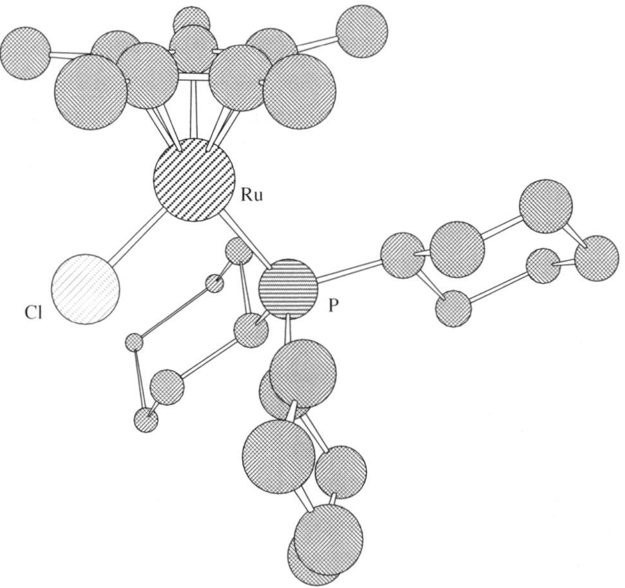

FIG. 8. Molecular structure of Cp*Ru(PCy_3)Cl (**9**). Hydrogen atoms are omitted for clarity.

Ru–Cp* and Ru–Cl bond distances. A very clear example of the presence of this reorganization energy is evident when comparing the ICy and the ITol containing complexes **3** and **4.** Here, complex **3** is more stable than **4** by some 10 kcal/mol, yet the Ru–C(carbene) distances are statistically identical. The largest difference between the two complexes resides in the differences in Ru–Cp* centroid bond

TABLE II

CARBENE STERIC PARAMETERS (deg) AND SELECTED BOND LENGTHS (Å) AND ANGLES (deg) FOR COMPLEXES **2–4** AND **6–10**

Complex	Carbene Sterics (A_L and A_H)	Ru–C	Ru–Cl	Ru–Cp*	C–Ru–Cl	C–Ru–Cp*	Cp*–Ru–Cl
2	150.7, 70.4	2.105	2.376	1.766	90.6	140.7	128.6
3	126.7, 31.8	2.070	2.524	1.658	93.7	129.3	15.5
4	155.2, 30.8	2.068	2.340	1.755	96.0	130.8	132.8
6	149.0, 41.4	2.153	2.438	1.778	87.9	138.7	130.4
7	152.0, 69.9	2.074	2.375	1.765	90.0	142.0	128.0
8	134.0, 137.6	2.086	2.371	1.754	89.32	141.5	129.2
9	115.8	2.383[a]	2.378	1.771	91.2[a]	138.9[a]	129.9
10	100.8	2.395[a]	2.365	1.810	91.4[a]	139.2[a]	129.3

[a]Replace C with P.

distance [a differences of 0.1 Å with **4** being longer than **3**] and the Ru—Cl bond distance (a difference of 0.18 Å with **3** being longer than **4**). It appears that the electron density has been pushed back into the Cp* ligand. When the aryl-substituted carbene ruthenium complexes are examined, the Ru—C(carbene) vs enthalpy trend is evident and makes sense in terms of the electronic explanations discussed above. Here, there appears to be a small variation in the Ru—Cl and Ru—Cp* bond distances from one complex to the other. Aryl- and alkyl substituted carbenes behave differently. We propose that this difference in bonding behavior can be attributed to the presence of a π system in the aryl case which localizes (or contributes as an acceptor) the effect on the carbene ligand, which in turn diminishes the large reorganization effects present in the alkyl cases where a π system on the carbene substituent is absent. An additional piece of evidence supporting this explanation is the position of the Cp* protons in the ^{1}H NMR spectra of these complexes which are reported in Table I. The aryl-substituted carbene complexes center around 1 ppm while the alkyl derivatives (including phosphines which are known as good donor ligands) are at ca. 1.5 to 1.7 ppm. The increased electron density on the Cp* ring leading to greater shielding affords shifts of the Cp* resonance of higher frequency.[2] No straightforward bond strength/bond length correlation can be made in this system in view of the presence of the reorganization energy. The steric factors have been qualitatively addressed. It would be of use to quantify the steric factors characterizing this class of ligands. They cannot be viewed in the same light as phosphine ligands since a cone angle (as defined by Tolman[53]) cannot be defined in the present system. In terms of steric effects, the nucleophilic carbenes can be considered as "fences," with "length" and "height." As a first model to describe the steric profile, we propose that two parameters be used to quantify the steric effects afforded by this ligand class. These two quantities can be taken directly from the crystallographic data. The two views presented in Fig. 9 depict the method used to extract the two parameters.

Numerical values defining length and height of the carbene "fences" are listed in Table II. Not surprisingly, all aryl-substituted carbenes possess nearly the same

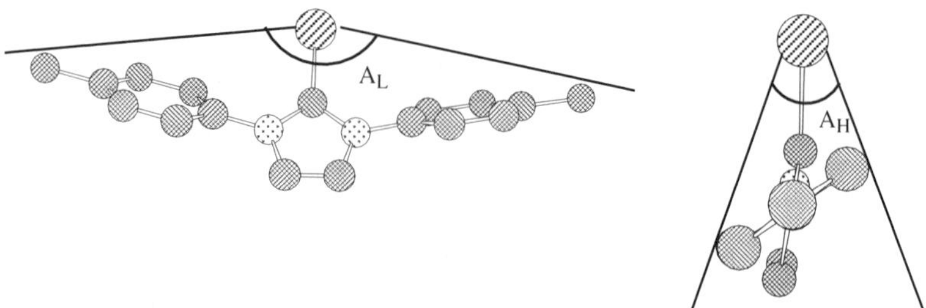

FIG. 9. Determination of two steric parameters (A_L and A_H) associated with carbene ligands in Cp*Ru(L)Cl complexes.

large A_L value since the "length" of the ligand is measured using the aryl para methyl group. The magnitude of the "height" parameter (A_H) depends on the presence or absence of ortho substituents. The most demanding aryl-substituted carbene ligand is the IMes ligand with 150.7° (A_L) and 70.4° (A_H) as steric parameters. The alkyl-substituted carbene ICy is the least sterically demanding ligand investigated in the present series. The IAd ligand, although bearing sterically demanding adamantyl groups, has an A_L parameter comparable to the aryl ligands examined and a smaller A_H parameter. The ICy ligand appears to be unique in the series investigated. In complex **3**, the AL angle is 126.3°. In the phosphine complexes **9** and **10** these angles were measured as 115.8° (8) and 100.8° (9). The ICy is sterically more closely related to these two phosphine complexes.

V

APPLICATIONS OF NUCLEOPHILIC CARBENES AS CATALYST PRECURSORS IN HOMOGENEOUS CATALYSIS

A. *Olefin Metathesis*

The last decade has witnessed the growing use of olefin metathesis in organic synthesis.[54–59] Ring closing metathesis [RCM, Eq. (6)] and ring opening metathesis [ROM, Eq. (7)] as well as a combination of these transformations have resulted in providing opportunities to build molecules of interest and importance.

M= (6)

+ R R' M= R R' (7)

Catalysts of the Grubbs type (benzylidene and vinylalkylidene) are of special interest, since they are only moderately sensitive to air and moisture and show significant tolerance of functional groups.[60–62] The ruthenium carbene complex, $RuCl_2(=CHPh)(PCy_3)_2$ (**11**), developed by Grubbs *et al.* is a highly efficient catalyst precursor and its use is widespread in organic and polymer chemistry.[6,60–69] We have shown that most of the substituted imidazol-2-ylidene ligands studied form stronger covalent bonds to the ruthenium center than PCy_3 (*vide supra*). Therefore, it is possible to replace one or both of the phosphine ligands in **11** with nucleophilic carbene ligands.[49,70–72] It has been noted that an increase in the ancillary phosphine electron donor ability leads to increased catalytic activity.[4] Since we have determined that imidazol-2-ylidene carbenes such as IMes and IPr are better donors than PCy_3[30], the catalytic behavior should reflect this increased electron donor ability. When both of the phosphines were replaced by imidazol-2-ylidene ligands they showed only little if any improvements in applications to ROM and RCM.[70] Upon using sufficiently bulky imidazol-2-ylidene carbenes only one PCy_3 is replaced (see Eq. (8)).[47,49,50]

$$\underset{\mathbf{11}}{\text{Cl}_2(\text{PCy}_3)_2\text{Ru}{=}\text{C(H)Ph}} + \text{L} \longrightarrow \text{Cl}_2(\text{L})(\text{PCy}_3)\text{Ru}{=}\text{C(H)Ph} + \text{PCy}_3 \qquad (8)$$

L = IMes, **12**
L = IPr, **13**

The identities of **12** and **13** were confirmed by single crystal X-ray diffraction studies. Structural models of **12** and **13** along with selected metrical parameters are presented in Figs. 10 and 11 and Table III, respectively.

The structural analyses reveal distorted square pyramidal coordination with a nearly linear Cl(1)–Ru–Cl(2) angle (168.62 and 170.42° for **12** and **13**,

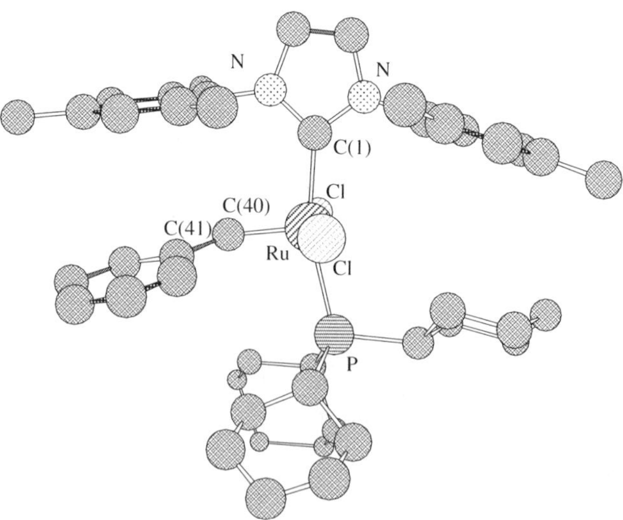

FIG. 10. Molecular structure of $RuCl_2(=C(H)Ph)(IMes)$ (**12**). Hydrogen atoms are omitted for clarity.

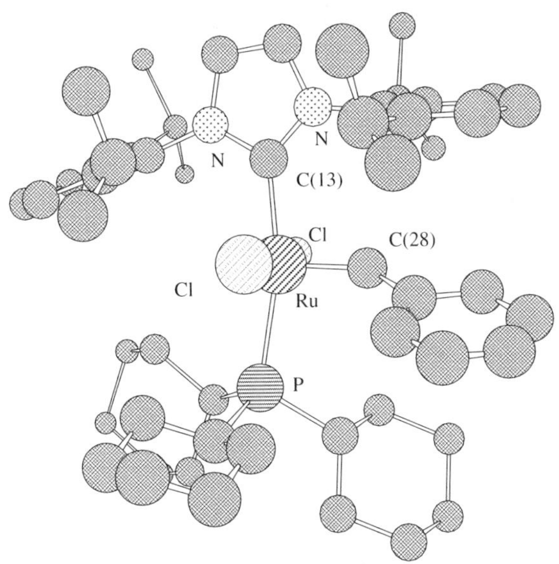

FIG. 11. Molecular structure of $(PCy_3)(IPr)Cl_2Ru(=C(H)Ph)$ (**13**). Hydrogen atoms are omitted for clarity.

respectively). The carbene unit in each structure is perpendicular to the C(1)—Ru—P plane, and the carbene aryl moiety is only slightly twisted out of the Cl(1)—Ru—Cl(2)—C(carbene) plane. The Ru—C(carbene) bond distances (1.841(11) Å and 1.817 Å for **12** and **13,** respectively) are the same as in $RuCl_2(=CH\text{-}p\text{-}C_6H_4Cl)(PCy_3)_2$ (1.838 (3) Å)[61] and marginally shorter than in $RuCl_2(=CHCH=CPh_2)$

TABLE III

SELECTED BOND LENGTHS (Å) AND ANGLES (deg) FOR $(PCy_3)\ RuCl_2(=C(H)Ph)(IMes)$ (**12**), $(PCy_3)RuCl_2(=C(H)Ph)(IPr)$ (**13**) AND $(PCyp_3)RuCl_2(=CHCH=CMe_2)(IMes)$ (**14**)

	12	**13**	**14**
Ru—C(L)	2.069(11)	2.088(2)	2.081(3)
Ru—Cl(1,2)	2.393(3), 2.383(3)	2.3822(7), 2.4008(7)	2.4012(10), 2.3950(8)
Ru—C(carbene)	1.841(11)	1.817(3)	1.764(4)
Ru—P	2.419(3)	2.4554(7)	2.4487(10)
C(carbene)—Ru—C(L)	99.2(5)	97.56(10)	102.46(15)
C(carbene)—Ru—P	97.1(4)	96.64(8)	95.55(13)
C(carbene)—Ru—Cl(1,2)	104.3(5), 87.1(5)	99.49(9), 88.87(9)	91.16(17), 97.37(17)
P—Ru—Cl(1,2)	89.51(10),89.86(9)	93.06(2), 90.59(2)	93.33(4), 87.28(4)
C(L)—Ru—Cl(1,2)	90.4(3), 86.9(3)	90.46(6), 83.77(6)	89.98(7), 86.75(7)

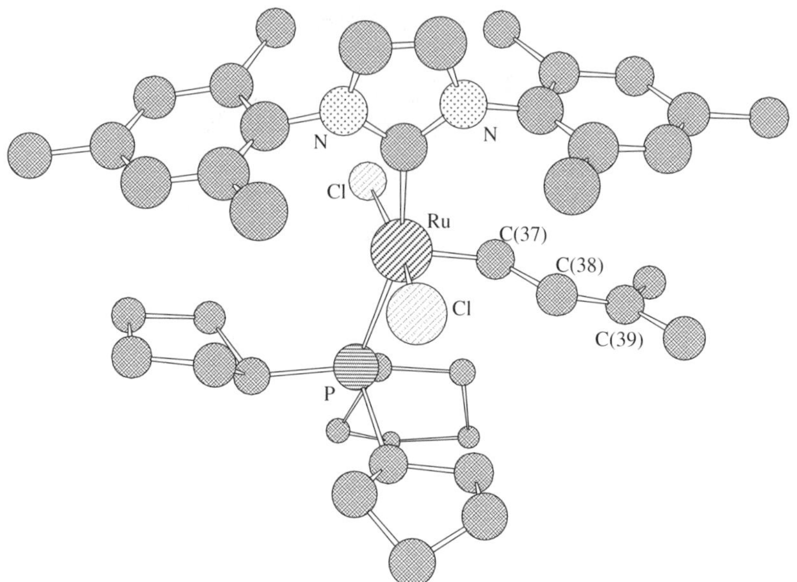

FIG. 12. Molecular structure of $(PCyp_3)\ Cl_2(IMes)Ru(=C(H)-C(H)=CMe_2)$ (**14**). Hydrogen atoms are omitted for clarity.

$(PCy_3)_2$ (1.851 (21) Å).[6] While two (formally) carbene fragments are present in **12** and **13**, they display different Ru–C distances (e.g., in **12** Ru–C(carbene), 1.841(11), and Ru–C(L), 2.069(11) Å). These important metrical parameters clearly distinguish two metal–carbene interactions: a covalently bound benzylidene and a datively bound imidazol-2-ylidene-carbene, the latter acting as a simple two-electron donor. From Figs. 10 and 11, it is also clear that the IMes and IPr ligands are sterically more demanding than PCy_3.

The exchange reaction of one phosphine ligand ($PCyp_3$, Cyp=cyclopentyl) in $(PCyp_3)_2RuCl_2(=CHCH=CMe_2)$ with IMes results in the formation of $(PCyp_3)$ $Ru(IMes)Cl_2(=CHCH=CMe_2)$ (**14**).[47] This compound was structurally characterized by X-ray crystallography. A structural model of **14** along with some selected bond lengths and bond angles are presented in Fig. 12 and Table III. The metrical data for compounds **12** and **14** show similar bond distances and angles (Table III). The coordination sphere around the metal center forms a distorted square pyramid with the benzylidene and vinylmethylene moiety at the apex. However, the distance of the apical carbene carbon to the metal center in **14** is shorter than that in **12.** Therefore, it can be inferred that the vinylmethylene moiety is more strongly bound to ruthenium than the benzylidene. All other bond distances around the metal center are slightly longer in **14** than in complex **12.** The bulk of the fence

TABLE IV
RING CLOSING METATHESIS OF DIETHYLDIALLYLMALONATE USING CATALYST PRECURSORS **11, 12, 13,** AND **14**

Catalyst Precursor[a]	Temp. (°C)	Time (min)	Yield (%)[b]
11	RT	15	85
12	RT	15	92
13	RT	15	100
14	RT	15	10

[a]Catalyst precursor loading = 5 mol%.
[b]Calculated from ^{1}H NMR spectra.

created by the IMes ligand sterically interferes with the carbene moiety and causes the bond angles to undergo large deviations (Table III).

The catalytic activity of these new complexes (**12, 13,** and **14**) was tested by using the standard RCM substrate, diethyldiallylmalonate (Eq. (9)).[4] Results are shown in Table IV.[47,49,50] Moreover, the thermal stability of the imidazol-2-ylidene bearing complexes **12** and **13** and **14** were compared to that of **11** and $(PCyp_3)_2RuCl_2$ (=CHCH=CMe_2). Results show that where the original phosphine bearing catalyst precursors **11** and $(PCyp_3)_2RuCl_2$ (=CHCH=CMe_2) decompose within 1 and 2 h, respectively, at 60°C, **12, 13,** and **14** are more robust. In fact their decomposition starts after 2 weeks at this temperature.[49,50] This preliminary study shows that the imidazol-2-ylidene analogues of the Grubbs system can act as olefin metathesis catalyst precursors, displaying significant activity and improved thermal stability compared to those of existing catalysts.

Grubbs *et al.* have prepared the saturated version of imidazol-2-ylidenes, 4,5-dihydroimiazol-2-ylidenes (Fig. 13).[73] They proposed that the higher basicity of the saturated imidazole ligand compared to its unsaturated analogues would translate into an increased reactivity of the desired catalysts. The ruthenium complexes formed ($RuCl_2$(=C(H)Ph)(PCy_3)(1,3-R_2-4,5-dihydroimidazol-2-ylidene)) showed increased ring closing metathesis activity compared to their bis(phosphine) analogues.[73]

Recently, the possibility that complexes of unsaturated "C_α" ligands other than alkylidenes might also serve as catalyst precursors in olefin metathesis has

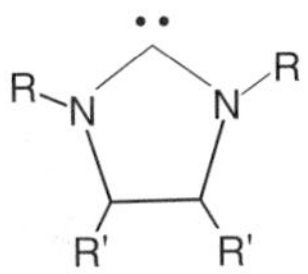

R–N N–R R' R'

FIG. 13. 1,3-R_2-4,5-dihydroimidazol-2-ylidene.

received more attention.[72,74–78] For example, it has been shown that (*p*-cymene)$RuCl_2$ (PCy_3) (**15**), and its cationic, 18-electron allenylidene derivative, [(*p*-cymene)RuCl(PCy_3) (=C=C=CPh_2)]PF_6 (**16**), are active catalyst precursors for various RCM reactions.[74,75,79] In every example mentioned above the use of sterically demanding and electron donating phosphines is required to stabilize reactive intermediates. Hence, it is of interest to synthesize the imidazol-2-ylidene analogues of these complexes and test their catalytic activity in olefin metathesis. Reaction of commercially available [(*p*-cymene)$RuCl_2$]$_2$ (**17**) with IMes or IPr in THF at room temperature results in the formation of (*p*-cymene)$RuCl_2$(IMes) (**18**) and (*p*-cymene)$RuCl_2$(IPr) (**19**) in good yields. Treating **18** with 1,1-diphenylprop-2-ynyl alcohol in the presence of $NaPF_6$ results in the formation of [(*p*-cymene)RuCl(IMes)(=C=C=CPh_2)]PF_6 (**20**) (Scheme 1).[78]

The X-ray crystal structure of **20** has been determined and a structural model of **20** is shown in Fig. 14. The coordination geometry around the Ru center can be

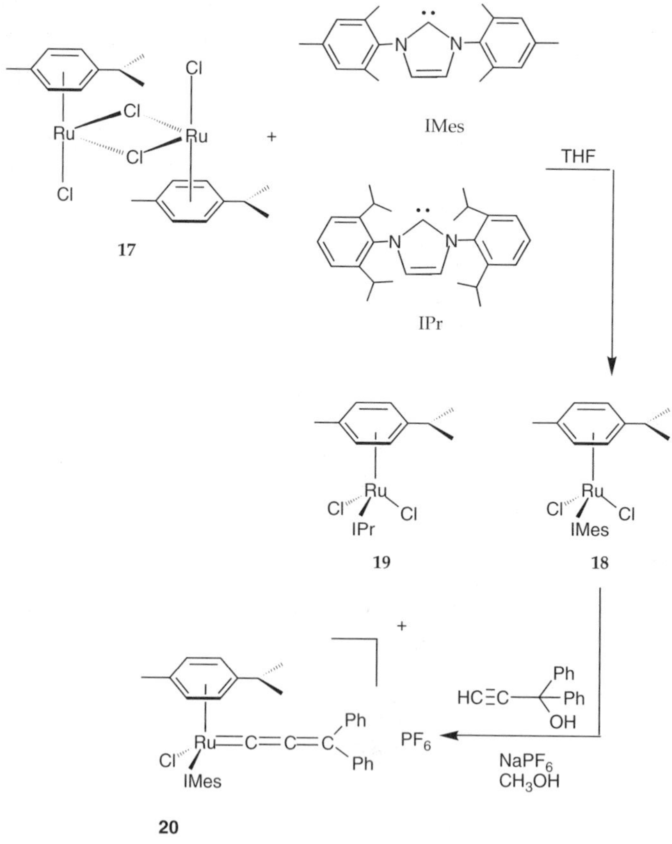

SCHEME 1.

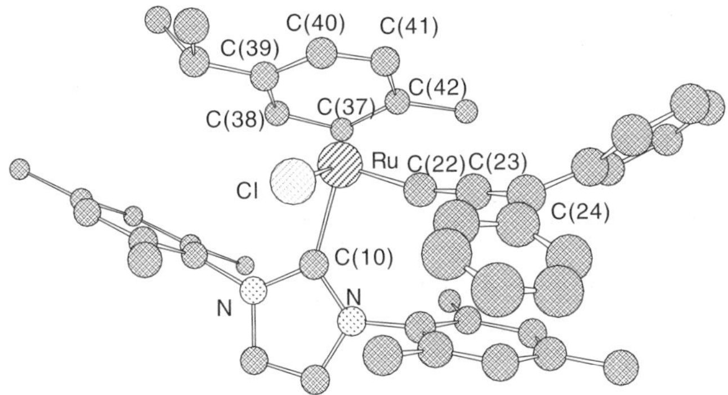

FIG. 14. Molecular structure of [(*p*-cymene)RuCl(IMes)(=C=C=CPh$_2$)]PF$_6$ (**20**). Hydrogen atoms and PF$_6$ anion have been omitted for clarity.

considered as a three-legged piano stool. *p*-Cymene is bound to ruthenium in a η^6 fashion, the isopropyl groups on the arene are distorted away from the metal center, presumably due to unfavorable steric factors. The two mesityl groups on the IMes ligand are bent toward the ruthenium center with the dihedral angles of 78.2°(2) and 89.9°(2) providing steric crowding which appears beneficial in RCM reactions. The allenylidene group is not linear but rather bent at the middle carbon (C22—C23—C24 = 171.8°). The Ru—C22 (1.890(4) Å) bond distance is considerably shorter than the Ru—C10 single bond (2.077(4) Å). Selected bond lengths and bond angles are reported in Table V.

The catalytic activities of **7**, **8,** and **9** have been tested by using the RCM substrate, diethyldiallylmalonate (Scheme III) and compared to those of (*p*-cymene) RuCl$_2$ (PR$_3$) (R=Cy (**21**) and *i*-Pr (**22**)). Results are presented in Table VI. When

TABLE V

SELECTED BOND DISTANCES (Å) AND BOND ANGLES (deg) FOR (*p*-cymene)RuCl(IMes)(=C=C=CPh$_2$)PF$_6$ (**20**)

Ru(1)-C(22)	1.890(4)	C(22)—Ru(1)—Cl(1)	84.95(12)
Ru(1)—C(10)	2.077(4)	C(22)—Ru(1)—C(10)	92.62(15)
Ru(1)—C(39)	2.224(5)	C(10)—Ru(1)—Cl(1)	85.38(11)
Ru(1)—C(40)	2.249(4)	C(22)—C(23)—C(24)	171.8(4)
Ru(1)—C(38)	2.356(4)	Ct*—Ru—C(34)	131.8
Ru(1)—C(37)	2.383(4)	Ct*—Ru—C(59)	122.9
Ru(1)—Ct*	1.804(4)	Ct*—Ru—Cl(1)	125.4
Ru(1)—Cl(1)	2.3903(11)		
Ru(1)—C(42)	2.246(4)		
Ru(1)—C(41)	2.258(4)		

*Ct is the center of the arene ring.

TABLE VI

RING CLOSING METATHESIS OF DIETHYLDIALLYLMALONATE USING CATALYST PRECURSORS **18, 19, 20, 21,** AND **22**

Entry	Catalyst Precursor	Solvent	Temp. (°C)	Time (h)	Conversion (%)[a]
1	**20**	CD_2Cl_2	40	27	85
2	**18**	CD_2Cl_2	40	27	78
3	**19**	CD_2Cl_2	40	27	40
4	**19**[b]	CD_2Cl_2	40	27	40
5	**21**	CD_2Cl_2	40	27	47
6	**22**	CD_2Cl_2	40	27	48
7	**18**	d_8-toluene	80	2	100
8	**18**[b]	d_8-toluene	80	2	100
9	**19**	d_8-toluene	80	2	100

[a]Calculated from NMR spectra.
[b]The experiment was performed in absence of light.

the reactions were carried out in CD_2Cl_2 and heated to 40°C, **20** catalyzed reaction 9 with a conversion of 85%, whereas **18** showed a conversion of 78%. The use of **21** and **22** as catalyst precursors led to the yields of 48 and 47%, respectively, after 27 hours (Table VI entries 1, 2, 5, and 6). The catalytic activity of **19** at this temperature (40% yield, Table VI entry 3) was in the range of those of the phosphine containing complexes **21** and **22.** To investigate the role of solvent, temperature, and light, RCM was performed with **18** and **19** as the catalyst precursors and d_8-toluene as the solvent. Upon heating the reaction mixtures to 80°C, a 100% conversion to product was observed after only 2 h in both cases (Table VI entries 7 and 9). Performing the reactions in the dark did not change the outcome and yields of the reactions (Table VI entries 4 and 8), which would indicate that the catalytic reactions are not photo-induced. This is in contrast with the complexes of the type M(*p*-cymene)$Cl_2(PR_3)$ (M = Ru, Os; R = Cy, *i*-Pr) which have been reported to become active ROM catalysts only when activated by UV irradiation.[80] It has also been reported that RCM in the presence of [(*p*-cymene)(PCy_3)ClRu=C=C=CPh_2]PF_6 and Ru(*p*-cymene)$Cl_2(PCy_3)$ is accelerated by exposure to UV or neon light.[75,79] No such effect is observed for our system. Examination of data gathered in Table VI shows that the IMes containing complexes **18** and **20** are the best catalyst precursors found in this study whereas the ruthenium complex incorporating the IPr ligand, **19**, showed reactivity similar to those of **21** and **22.** From the solution calorimetric data the IMes ligand proved to be a stronger binder than IPr ligand whose relative enthalpy is comparable to that of PCy_3 ligand.[49] The initial step in the ring closing metathesis mechanism using the (*p*-cymene)RuLCl_2 complexes must involve the formation of a ruthenium-carbene complex;[60,62] and in the case of ruthenium-arene complexes the carbene moiety can presumably be formed by the change in the hapticity of the arene ring, leading to vacant sites on Ru. The more electron donating ligand (IMes) can facilitate this process more easily than either IPr or the phosphines, and this is proposed to explain the origin of the higher catalytic activity of **18** compared to those of **19, 21,** and **22** at 40°C. When the temperature is raised to 80°C both **18** and **19** show the same activity. It could be argued that at higher temperatures the activation barrier for the change in arene hapticity has already been overcome and under these conditions the electronic differences between the ligands are not very important.

We were also interested in developing the synthesis of imidazol-2-ylidene analogues of the previously synthesized neutral Ru—allenylidene complexes, $RuCl_2(=C=C=CPh_2)(PR_3)_2$, R=Ph, Cy[76] via substitution reactions and in comparing their RCM activity to those of the cationic 18-electron ruthenium allenylidene complexes. Analysis of the product of simple substitution reactions showed that the "C_α" unsaturated moiety in this complex is not an allenylidene but rather a cyclized vinyl carbene "an indenylidene" (Scheme 2).[81,82] The X-ray crystal structures of the IPr bearing complexes **26** and **28** have been determined (Figs. 15 and 16) and clearly show the coordination of Ru to an indenylidene moiety. In each compound, the coordination geometry around the ruthenium center is distorted square pyramidal with the strongest π-acidic ligand (indenylidene) assuming the unique apical site. The square base is defined by the two chlorides and the donor atoms of the phosphine and the imidazol-2-ylidene ligands with the ruthenium center lying 0.293(8) Å in **26** and 0.3443(12) Å in **28** above this plane.

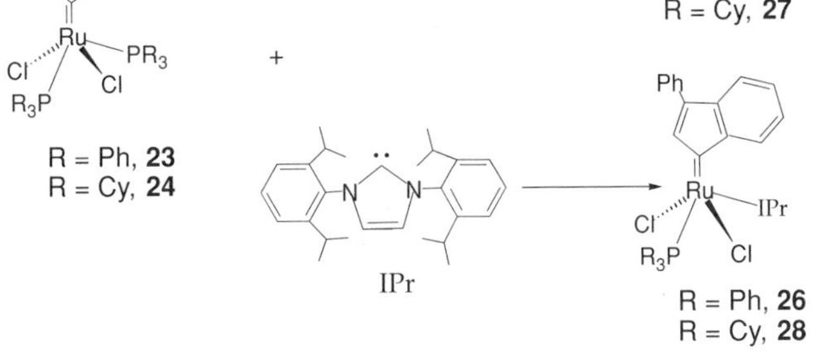

R = Ph, **23**
R = Cy, **24**

IMes

IPr

R = Ph, **25**
R = Cy, **27**

R = Ph, **26**
R = Cy, **28**

SCHEME 2.

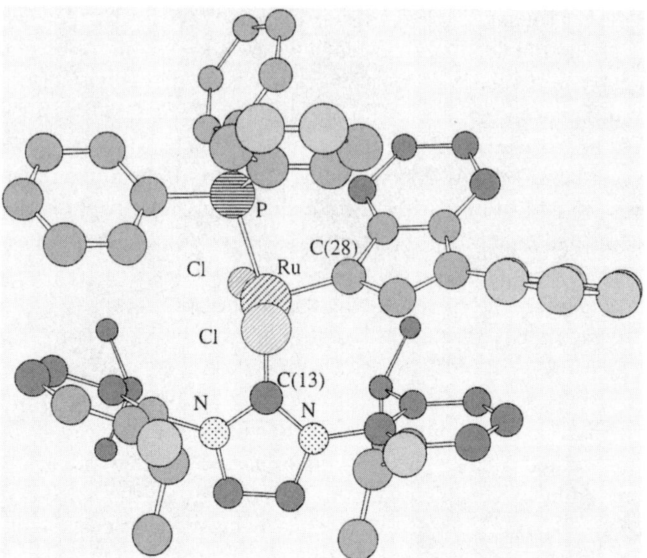

FIG. 15. Molecular structure of (IPr)(PPh_3)Cl_2Ru(3-phenylindenylid-1-ene) (**26**). Hydrogen atoms are omitted for clarity.

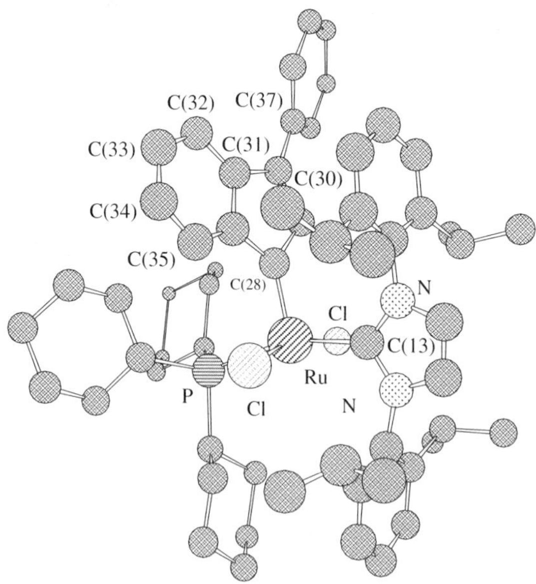

FIG. 16. Molecular structure of (IPr)(PCy_3)Cl_2Ru(3-phenylindenylid-1-ene) (**28**). Hydrogen atoms are omitted for clarity.

TABLE VII

SELECTED BOND DISTANCES (Å) AND BOND ANGLES (deg) FOR (IPr)(PPh_3)Cl_2Ru(3-phenylindenylid-1-ene) **26**

Ru(1)–C(28)	1.852(3)	C(34)–C(35)	1.397(5)
Ru(1)–C(13)	2.102(3)	C(35)–C(36)	1.385(5)
Ru(1)–Cl(1)	2.3895(9)	C(28)–C(36)	1.517(5)
Ru(1)–Cl(2)	2.3688(8)	C(30)–C(37)	1.478(5)
Ru(1)–P(1)	2.3977(10)	C(31)–C(36)	1.397(4)
C(28)–C(29)	1.456(4)	C(28)–Ru(1)–C(13)	104.47(13)
C(29)–C(30)	1.337(5)	C(28)–Ru(1)–Cl(1)	98.10(10)
C(30)–C(31)	1.483(5)	C(13)–Ru(1)–Cl(2)	84.48(8)
C(31)–C(32)	1.377(5)	C(28)–Ru(1)–P(1)	92.55(10)
C(32)–C(33)	1.406(7)	C(13)–Ru(1)–P(1)	162.98(8)
C(33)–C(34)	1.379(7)	Cl(1)–Ru(1)–Cl(2)	167.15(3)

The Ru–C_α(indenylidene) bond distances are significantly shorter than the bond length between the ruthenium and the weaker π-acid imidazol-2-ylidene (1.852(3) Å vs 2.102(3) Å in **26** and 1.861(4) Å vs 2.113(4) Å in **28**). Selected bond lengths and bond angles for **25** and **27** are presented in Tables VII and VIII, respectively.

Investigating thermal stabilities of compounds **23–28** revealed that when heated to 80°C the compounds containing PPh_3 i.e., **23, 25,** and **27,** were less stable than their PCy_3 bearing analogues (**24, 26,** and **28,** respectively). The least stable was (PPh_3)$_2$$Cl_2$Ru(3-phenylindenylid-1-ene) (**23**) which decomposed after 2 h at 80°C (Table IX, entry 1) and the most stable was (IPr)(PPh_3)Cl_2Ru(3-phenylindenylid-1-ene) (**27**) (Table IX, entry 4) which showed decomposition after 42 h at the same temperature. It can be concluded that the presence of the nucleophilic carbene

TABLE VIII

SELECTED BOND DISTANCES (Å) AND BOND ANGLES (deg) FOR (IPr)(PCy_3)Cl_2Ru(3-phenylindenylid-1-ene) **28**

Ru(1)–C(28)	1.861(4)	C(34)–C(35)	1.387(7)
Ru(1)–C(13)	2.113(4)	C(35)–C(36)	1.378(6)
Ru(1)–Cl(1)	2.3833(11)	C(28)–C(36)	1.494(6)
Ru(1)–Cl(2)	2.3903(11)	C(30)–C(37)	1.473(6)
Ru(1)–P(1)	2.4264(12)	C(31)–C(36)	1.423(6)
C(28)–C(29)	1.474(6)	C(28)–Ru(1)–C(13)	102.12(17)
C(29)–C(30)	1.350(6)	C(28)–Ru(1)–Cl(1)	100.13(14)
C(30)–C(31)	1.484(6)	C(13)–Ru(1)–Cl(2)	90.16(11)
C(31)–C(32)	1.381(6)	C(28)–Ru(1)–P(1)	96.59(13)
C(32)–C(33)	1.406(7)	C(13)–Ru(1)–P(1)	161.29(12)
C(33)–C(34)	1.379(7)	Cl(1)–Ru(1)–Cl(2)	164.50(4)

TABLE IX
THERMAL STABILITY OF CATALYST PRECURSORS **23–28** AT 80°C

Entry	Compound	Start of Decomposition (h)
1	**23**	4
2	**24**	—
3	**25**	2
4	**26**	42
5	**27**	—
6	**28**	—

ligand IPr in the coordination sphere of ruthenium stabilizes the complex significantly. The complexes incorporating PCy_3 were more robust and did not decompose at elevated temperatures even after about 10 days (256 h) (Table IX, entries 2, 5, and 6). This result is hardly surprising because if the decomposition pathway involves the dissociation of phosphines (other ligands are less likely to dissociate) the less electron-releasing and hence the less tightly bound PPh_3 ligand should undergo dissociation faster than the more tightly bound PCy_3 moiety.

The role of complexes **23–28** as catalyst precursors in the ring closing metathesis reactions was investigated. Three different diene substrates diethyldiallylmalonate (**29**), diallyltosylamine (**30**), and diethyldi(2-methylallyl)malonate (**31**) were added to the NMR tubes containing a solution of 5 mol% of catalyst precursor in an appropriate deuterated solvent. The NMR tubes were then kept at the temperatures reported in Table X. Product formation and diene disappearance were monitored by integrating the allylic methylene peaks in the ^1H NMR spectra and the results are presented in Table X and the catalytic transformations are depicted in Scheme 3.

$(PPh_3)_2Cl_2Ru$(3-phenylindenylid-1-ene) (**23**) did not show any catalytic activity with **28** (Table X, entry 1) and therefore was eliminated from the list of catalyst precursors in this study. The ruthenium-indenylidene-imidazol-2-ylidene complexes that contained PPh_3, (IMes)$(PPh_3)Cl_2Ru$(3-phenylindenylid-1-ene) (**25**) and (IPr)$(PPh_3)Cl_2Ru$(3-phenylindenylid-1-ene) (**26**) showed good catalytic activity with diethyldiallymalonate (**29**) and diallyltosylamine (**30**) as substrates (Table X, entries 3, 4, 8, and 9). It should be noted that **25** catalyzes this reaction only when heated to 40°C, whereas **26** does so at room temperature. Sterically hindered diethyldi(2-methylallyl)malonate (**31**) does not easily undergo ring closing metathesis reaction: catalyst precursor **25** can convert 60% of this substrate into the product after 2 h and this is only achieved when the reaction is heated to 80°C. The rate of conversion with **26** is only 17% at this temperature. Heating the reaction mixtures for longer periods of time does not increase the yields of the reactions, it

TABLE X

RING CLOSING METATHESIS RESULTS USING CATALYST PRECURSORS **23–28**

Entry	Substrate	Catalyst Precursor	Solvent	Temp. (°C)	Time (min)	Yield (%)[a]
1	**29**	**23**	CD_2Cl_2	40	25	0
2	**29**	**24**	CD_2Cl_2	RT	25	84
3	**29**	**25**	CD_2Cl_2	40	25	65
4	**29**	**26**	CD_2Cl_2	40	25	56
5	**29**	**27**	CD_2Cl_2	RT	25	88
6	**29**	**28**	CD_2Cl_2	RT	25	75
7	**30**	**24**	CD_2Cl_2	RT	25	96
8	**30**	**25**	CD_2Cl_2	40	25	94
9	**30**	**26**	CD_2Cl_2	RT	25	94
10	**30**	**27**	CD_2Cl_2	RT	25	30
11	**30**	**28**	CD_2Cl_2	RT	25	89
12	**31**	**24**	d_8-toluene	80	120	0
13	**31**	**25**	d_8-toluene	80	120	66
14	**31**	**26**	d_8-toluene	80	120	17
15	**31**	**27**	d_8-toluene	80	120	20
16	**31**	**28**	d_8-toluene	80	120	19

[a]Calculated from ^{1}H NMR spectra.

can be inferred, therefore, that the catalyst is disabled after a certain period of time at higher temperatures (Table X, entries 12–16). $(PCy_3)_2Cl_2Ru$(3-phenylindenylid-1-ene) (**24**) exhibited high reactivity when used with substrates **29** and **30** but was not effective with substrate **31** (Table X, entries 2, 7, and 12). The indenylidene derivatives of (PCy_3)Ru(imidazol-2-ylidene) compounds (IMes)$(PCy_3)Cl_2Ru$(3-phenylindenylid-1-ene) (**27**) and (IPr)$(PCy_3)Cl_2Ru$(3-phenylindenylid-1-ene) (**28**)

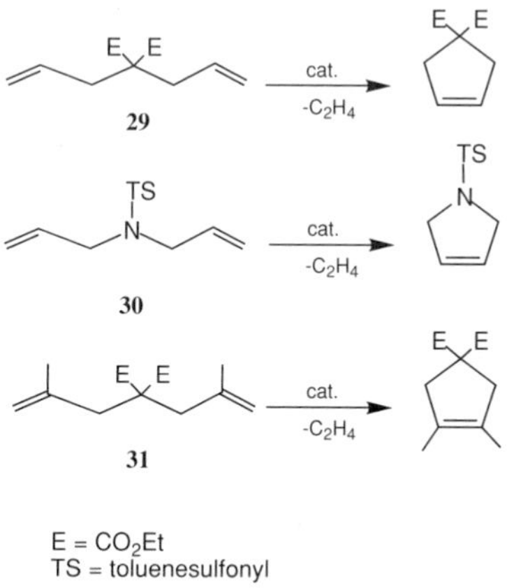

E E

29

cat.

$-C_2H_4$

TS

N

30

31

E = CO_2Et
TS = toluenesulfonyl

SCHEME 3.

were comparable to **24** in their reactivity toward **29** (Table X, entries 5 and 6). The IPr analogue (**28**) converted **30** with a very high yield, whereas the IMes compound (**27**) was much less reactive toward the same substrate (Table X, entries 10 and 11). The yields of the reactions with **31** were low for both compounds (Table X, entries 15 and 16). The indenylidene complexes **24–28** are shown to be good catalyst precursors in the RCM of sterically unhindered substrates (**29** and **30**), comparable to the alkylidene complexes developed by Grubbs[72] and Herrmann.[70]

To develop a method for the synthesis of true neutral ruthenium allenylidene complexes, $(PPh_3)_2RuCl_2$ (**32**) was allowed to react with 3,3-diphenylpropyn-3-ol in the presence of a better donating ligand such as PCy_3.[83] This reaction led exclusively to the formation of $(PCy_3)_2Ru({=}C{=}C{=}CPh_2)Cl_2$ (**33**). It is interesting to note that in the absence of PCy_3, the sole product is the indenylidene complex **23** which undergoes substitution with PCy_3 to yield **24.** Complex **32** is also accessible from the reaction of $[(p\text{-cymene})RuCl_2]_2$ **17** with 3,3-diphenylpropyn-3-ol and two equivalents of PCy_3 via loss of *p*-cymene. However, the product (85% **33** based on ^{31}P NMR data) contains two side products, one identified as the 3-phenyl-1-indenylidene complex **24** (8% by NMR data) and one unknown (7% by NMR data). When two equivalents of PPh_3 instead of PCy_3 were used no carbene moiety was formed. All attempts to convert the allenylidene into the indenylidene by addition of protic acids or by subjecting the allenylidene to elevated temperatures were unsuccessful. The exchange of one PCy_3 ligand for IMes affords the allenylidene complex **34** in high yields. The reactions are summarized in Scheme 4.[83] X-ray crystal structures of **33** and **34** were determined and their structural models are shown in Figs. 17 and 18, respectively. In both structures the five coordinated ruthenium center is located at the bottom of a square pyramid. The allenylidene moiety is located at the apex and the trans chlorides and PCy_3 ligands (**33**), PCy_3 and IMes ligands (**33**), form the base. In both complexes the Ru—C_α bond distances are nearly identical (1.794(11) Å). This is in the usual range (1.76 to 1.84 Å) for carbene moieties in this kind of 16-electron ruthenium complex.[47,49] However, these bond distances are much shorter than the bond lengths determined for cationic 18-electron ruthenium allenylidene complexes (1.87 to 1.92 Å).[84–87] This indicates a better overlap and a significantly higher bond strength of the carbene moiety to the metal center in complexes **33** and **34.** The comparable metal–ligand bond distances at the base also give very similar values indicating no significant change for the electronic environment of the metal center. These structural features may explain the similar catalytic properties of complexes **33** and **34** (*vide infra*). The bond angles in both complexes at the base do not deviate more than 4° from ideal 90°. However, steric interference with the allenylidene moiety causes widening of one of each C_α—Ru—Cl angles (96.2(5)° (**33**), 95.89(4)° (**34**)), one C_α—Ru—P angle (101.4(5)°) in complex **33** and the C_α—Ru—C(IMes) angle (98.89(5)°) in complex **34.** The allenylidene chain is only slightly bent in complex **33** (Ru—C_α—C_β = 175.36(11)°, C_α—C_β—C_γ = 175.29(13)°). This indicates a

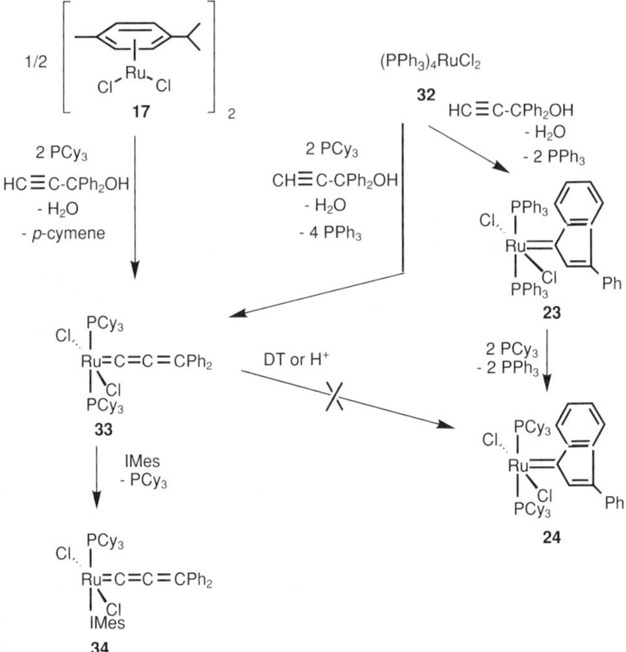

SCHEME 4.

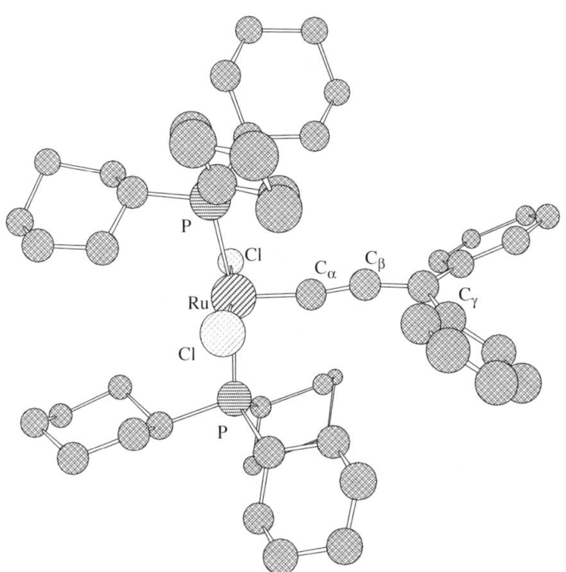

FIG. 17. Molecular structure of $(PCy_3)_2Ru(=C=C=CPh_2)Cl_2$ (**33**). Hydrogen atoms are omitted for clarity.

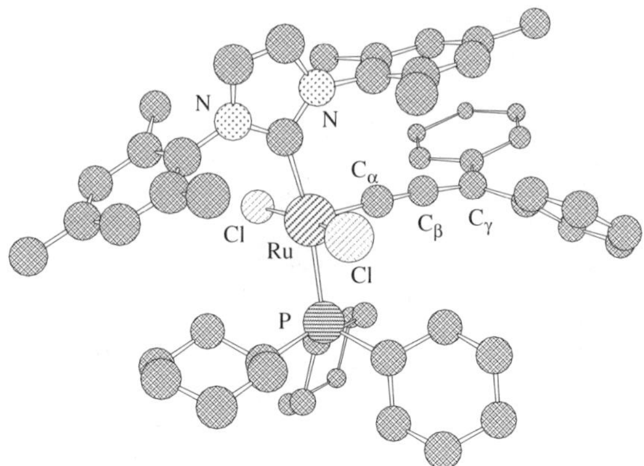

FIG. 18. Molecular structure of $(PCy_3)Ru(IMes)(=C=C=CPh_2)Cl_2$ (**34**). Hydrogen atoms have been omitted for clarity.

strong conjugation along the spine excluding C—H π-interaction to neighboring hydrogen atoms as observed in other complexes.[86] Complex **33**, however, shows significantly stronger bending along the spine (Ru—C_α—C_β = 169.20(11)°, C_α—C_β—C_γ = 167.20(18)°). C—H π-interaction to hydrogen atoms of the PCy_3 ligands may be present. This interaction might also have a weak influence on the bond distances along the spine causing a slight extension [C_α—C_β = 1.27 Å (**33**), 1.26 Å (**33**) and C_β—C_γ = 1.35 Å (**32**), 1.34 Å (**34**)], but these values are in the usual range for ruthenium allenylidene complexes.[84–87] Selected bond distances and angles for complexes **33** and **34** are reported in Table XI.

TABLE XI

SELECTED BOND DISTANCES [Å] AND ANGLES [°] FOR THE COMPLEXES **33** AND **34**

	33	**34**
Ru–C_α	1.79	1.79
Ru–C(IMes)	—	2.09
Ru–P	2.35, 2.41	2.41
Ru–Cl	2.37, 2.38	2.37, 2.39
C_α–C_β	1.27	1.26
C_β–C_γ	1.35	1.34
C_α–Ru–C(IMes)	—	98.9
C_α–Ru–P	91.4, 101.4	92.2
C_α–Ru–Cl	91.6, 96.2	93.1, 95.9
C(IMes)–Ru–Cl	—	88.9, 89.6
P–Ru–Cl	86.4, 87.6, 91.6, 92.6	88.1, 91.7
Ru–C_α–C_β	169.2	175.4
C_α–C_β–C_γ	167.2	175.3

TABLE XII

RING CLOSING METATHESIS MEDIATED BY **33** AND **34**

Entry	Substrate	Catalyst Precursor	Solvent	Temp. (°C)	Time	Yield (%)[a]
1	**29**	**33**	CD_2Cl_2	40	25 min	12
2	**29**	**34**	CD_2Cl_2	40	25 min	8
3	**30**	**33**	CD_2Cl_2	40	25 min	4
4	**30**	**34**	CD_2Cl_2	40	25 min	0
5	**31**	**33**	d_8-toluene	80	2 h	0
6	**31**	**34**	d_8-toluene	80	2 h	0

[a]Monitored by ^1H NMR spectroscopy.

Compounds **33** and **34** were subjected to elevated temperatures. Both compounds turned out to be relatively robust at 80°C.[83] Even after 32 h of constant heating no signs of decomposition products were found. Initial signs of decomposition were noticed for complex **33** after 64 h and for complex **34** after 128 h. A similar increased thermal stability has been observed for the $Cl_2(PCy_3)(IMes)Ru(=C(H)Ph$ complex compared to $Cl_2(PCy_3)_2Ru$ (=C(H)Ph.[47,49]

The role of complexes **33** and **34** as catalyst precursors in the RCM reactions was investigated.[83] Three different diene substrates, diethyl diallylmalonate (**29**), diallyltosylamine (**30**), and diethyl di(2-methylallyl)malonate (**31**), were added to the NMR tubes containing a solution of 5 mol% of catalyst precursor in an appropriate deuterated solvent. The catalytic reactions are depicted in Scheme 3 and results of the RCM reactions are presented in Table XII. Product formation and diene disappearance were monitored by integrating the allylic methylene peaks in the ^1H NMR. Both complexes perform very poorly in these reactions compared to cationic 18-electron arene-ruthenium allenylidene complexes.[74] The significantly higher bonding energy of the allenylidene moiety at the metal center as inferred from the single crystal X-ray data may be at the origin of the lower catalytic activity displayed by **33** and **34.** The sterically hindered substrate diethyl di(2-methylallyl)malonate shows no sign of ring closing, using either complexes even after 2 h at 80°C. In order to get detectable conversion of the other substrates, reaction mixtures were heated to 40°C in CD_2Cl_2. The turnover rates after 25 min indicated slightly lower catalytic activity for the IMes substituted complex **34** (diethyl diallylmalonate 8%, diallyltosylamine 0%) compared to complex **33** (diethyl diallylmalonate 12%, diallyltosylamine 4%). The disappointingly low catalytic activities for ring closing metathesis reactions obtained for **33** and **34** could be attributed to the very similar metal–ligand bond distances in the solid state indicating a similar electronic environment at the metal center for both complexes.

B. *Carbon—Carbon and Carbon—Nitrogen Bond Formation*

Palladium catalyzed cross-coupling reactions of arylhalides or halide equivalents with various nucleophiles have been shown to be highly effective and practical methods for the formation of C—C bonds.[88–91] In this review we focus our attention on two of these methods, namely, Suzuki and Kumada reactions. The Suzuki reaction developed in the early 1980s is very versatile and is extensively used in natural product synthesis. The substrates in this reaction are boronic acids ($ArB(OH)_2$) and arylhalides or pseudo-halides such as triflates (Ar′X).[92–96] In Kumada reactions C—C bond formation is accomplished by reacting Grignard reagents with alkenyl or arylhalides in the presence of catalytic amounts of either palladium or nickel complexes.[97–99] In a closely related area, palladium or nickel mediated coupling of arylhalides with amines has attracted significant interest because of the important use of this methodology in organic synthesis and materials science.[100–103] It has been shown that the supporting ligands on the metal center play a crucial role in dictating the efficiency of the catalytic system.[100–103] Bulky monodentate phosphines or bidentate PX (X = P, N, O for amination) have been the ligand of choice in all the above-mentioned methods.[93–95,100–115] Although arylchlorides undergo coupling reactions more slowly than do arylbromides or aryliodides (C—Cl > C—Br > C—I), their use as chemical feedstock would economically benefit a number of industrial processes.[116–118]

Nucleophilic *N*-heterocyclic carbenes as ligands have the primary advantage that they do not easily dissociate from the metal center, and as a result an excess of the ligand is not required in order to prevent aggregation of the catalyst usually affording the bulk metal.[11,27,42,47,49,119,120] Imidazol-2-ylidenes exhibit a considerable stabilizing effect in organometallic systems (*vide supra*).[27,31,119,120] Herrmann and co-workers have reported Suzuki cross-coupling activity of carbene ancillary ligands involving arylbromides and activated arylchlorides.[121] In view of the stereoelectronic factors required in the metal-mediated C—C and C—N bond formation[105] and the results from the thermochemical studies[47], the activity of a bulky nucleophilic carbene as supporting ligation in these catalytic processes was investigated.

VI

SUZUKI CROSS-COUPLING OF ARYLCHLORIDES WITH ARYLBORONIC ACIDS

The Suzuki reaction has proved extremely versatile and has found extensive use in natural product synthesis.[122] Arylboronic acids [$ArB(OH)_2$] are the usual substrates in this reaction together with arylhalides or triflates (Ar′X, X = halogen or triflate). In our initial experiments in the coupling of phenylboronic acid and 4-chlorotoluene in the presence of catalytic amount of $Pd_2(dba)_3$, the carbene IMes and Cs_2CO_3 as a base was observed in a modest yield (59% isolated yield). Since carbenes are considerably less stable to air and moisture than the corresponding imidazolium salts, to avoid preparation and isolation of the carbene we sought to develop a protocol in which the carbene ligand would be generated *in situ* from the salt. Under the same general conditions, using IMes.HCl resulted in the isolated yield of 96%, which represents a significant improvement over the procedure employing the nucleophilic IMes in terms of both isolated yields and ease of execution (see Eq. (9)).[96]

Me, Cl (4-chlorotoluene) + B(OH)2 (phenylboronic acid) → Me (4-methylbiphenyl)

Pd2(dba)3 (1.5 mol%)
IMes.HCl (3.0 mol%)
base (2 equivalents)
dioxane, 80˚C
1.5 h

(9)

An investigation of the base for the *in situ* generation of the carbene ligand IMes from the salt IMes.HCl revealed that Cs_2CO_3 was the reagent of choice (Table XIII). Other inorganic bases such as Na_2CO_3, KOAc, K_2CO_3, and CsF resulted in longer reaction times for complete consumption of the arylchloride and afforded moderate to low yields of coupling products. When the organic base triethylamine was employed, the reaction ceased within minutes and precipitation of palladium black was observed.[96] As illustrated in Table XIV, the palladium catalyzed Suzuki reaction with the ancillary ligand IMes.HCl was exceptionally tolerant of a variety of functional groups substituted on the arylchlorides and arylboronic acids. Electron-donating and electron-withdrawing substituents were both tolerated by the catalytic system and provided the corresponding coupling products in excellent yields.[96]

It is a significant challenge to avoid the use of a glovebox altogether in organic synthesis and especially in process chemistry.[100] Therefore, we set out to examine the reactivity of the air stable palladium(II) complex ($Pd(OAc)_2$) in the Suzuki cross-coupling reactions. As illustrated in Table XV, the palladium (II) catalyzed Suzuki reaction with the ancillary ligand IMes.HCl exhibited tolerance toward a variety of functional groups (electron-donating or -withdrawing) substituted on the arylchlorides and arylboronic acids and the corresponding coupling products were formed in very good yields.[96]

An investigation of the ligands for *in situ* generation of the carbene ligands from their salts revealed that IMes.HCl is our best choice (Table XVI, entry 1); all other imidazolium salts investigated resulted in longer reaction times for complete

TABLE XIII

EFFECT OF BASE ON THE RATE OF $Pd_2(dba)_3$/IMes.HCl CATALYZED BY SUZUKI CROSS-COUPLING REACTIONS OF 4-CHLOROTOLUENE WITH PHENYLBORONIC ACID

Entry	Base	Time (h)	Yield (%) [a,b]
1	Et_3N	24	<5[c]
2	Na_2Co_3	43	6[c]
3	KOAc	43	42[c]
4	K_2CO_3	24	53
5	**Cs_2CO_3**	**1.5**	**96**
6	CsF	2	65

[a] Isolated yields.

[b] All reactions were monitored by TLC.

[c] 4-Chlorotoluene was not completely consumed within the indicated reaction time.

consumption of the arylchloride and afforded moderate to low yields of coupling products. An interesting example is, however, when the commercially available 1-ethyl-3-methyl imidazolium hydrochloride was employed under longer reaction time (39 h), 75% of isolated yield was obtained (Table XVI, entry 4). This means the catalyst bearing carbene ligand is thermally stable as we have observed in other systems.[96]

Use of the imidazolium salt IMes.HCl allowing for the *in situ* generation of IMes is a significant improvement upon existing Suzuki cross-coupling reaction methods. Furthermore, $Pd(OAc)_2$ can similarly affect the Suzuki cross-coupling reaction of arylchlorides with arylboronic acids. However, reactions performed with other imidazolium salts required longer times and afforded lower yields, while $Pd(OAc)_2$ was found to be equally as effective as $Pd_2(dba)_3$ under the reaction conditions described above. So the current catalytic system represents the simplest Suzuki coupling so far by employing arylchlorides as feedstock and handling catalyst precursors without the use of a glovebox.

TABLE XIV

FUNCTIONAL GROUP TOLERANCE OF $Pd_2(dba)_3$/IMes.HCl CATALYZED SUZUKI CROSS-COUPLING REACTIONS OF ARYL CHLORIDES WITH PHENYLBORONIC ACID DERIVATIVES

Entry	X	Y	Time (h)	Isolated Yield (%)
1	4-Me	H	1.5	99
2	4-Me	H	1.5	97[a]
3	4-Me	4-OMe	1.5	99
4	4-Me	2-Me	1.5	88
5	4-Me	3-OMe	1.5	91
6	H	4-OMe	1.5	99
7	2,5-$(Me)_2$	H	1.5	89
8	4-OMe	H	1.5	93
9	4-CO_2Me	H	1.5	99

[a]6.0 mol% of **2**.HCl.

VII

CROSS-COUPLING OF ARYLCHLORIDES WITH ARYL GRIGNARD REAGENTS (KUMADA REACTION)

In the previous section the result of employing Pd(0) or Pd(II) compounds and imidazolium salts as the catalyst system in the Suzuki cross-coupling reactions of arylchlorides and arylboronic acids was discussed.[96] Considering that arylboronic acids and other organometallic reagents used in this type of C—C coupling are generally made from the corresponding Grignard or lithium reagents,[92] it would prove valuable to find a general method for Kumada coupling.

Based on the success with IMes.HCl in Suzuki coupling of arylchlorides with arylboronic acids,[96] we employed a similar protocol for Kumada coupling. In order to form free carbene IMes from its salt instead of using an external base (Cs_2CO_3 in Suzuki reactions), a slight excess of Grignard reagent was employed. Unfortunately, using IMes.HCl and $Pd_2(dba)_3$ as the catalyst system in the reaction of 4-chlorotoluene with phenylmagnesiumbromide afforded a moderate yield (Table XVII, entry 1).[123] Since the use of bulky ligands may improve catalytic performance in this C—C coupling procedure,[93,95,96,105,124,125] the carbene salt

TABLE XV

$Pd(OAc)_2$/IMes.HCl CATALYZED SUZUKI CROSS-COUPLING REACTIONS OF ARYLCHLORIDES WITH PHENYLBORONIC ACID DERIVATIVES

Entry	X	Y	Time (h)	Isolated Yield (%)
1	4-Me	H	2	99
2	4-Me	4-OMe	2	80
3	4-Me	2-Me	2	50
4	4-Me	3-OMe	19	93
5	4-CO_2Me	H	2	99
6	4-Me	H	8	85
7	3-NMe_2	H	2	98[a]
8	2,5-$(Me)_2$	H	2	94
9	2,5-$(Me)_2$	H	2	98[b]
10	2,6-$(Me)_2$	H	2	69[a]

[a] 5.0 mol% of **2**.HCl.
[b] 5.0 mol% of $Pd(OAc)_2$ and 5.0 mol% of **2**.HCl.

IPr.HCl was employed in this reaction and afforded a 99% yield of isolated product (Table XVII, entry 2). The result is consistent with the stereoelectronic properties of this family of ligands.[8,47,49] The ligands are excellent donors and therefore facilitate the oxidative addition of arylhalides. The steric bulk provided by the ortho position substituents on the imidazolyl aryl groups may serve to stabilize the active catalyst which we feel is a "$LPd(Ar)_2$ " intermediate. This intermediate may very well be stabilized by solvent or substrate binding to complete the square planar metal coordination. The role of the solvent was also investigated and the result showed that optimum yield is achieved by using dioxane as the co-solvent and heating the reaction mixture to 80°C. The reaction in THF/toluene and THF required a longer reaction time and lead to lower yields (Table XVII, entries 3 and 4).[123]

A survey of catalytic cross-coupling of arylhalides with arylmagnesiumbromides using IPr.HCl is presented in Table XVIII. Reactions were faster for arylbromides and aryliodides than for arylchlorides (Table XVIII, entries 3, 8, and 10). As in similar coupling reactions, the significant challenge in practical Kumada reactions is the functional group tolerance.[122,126] The halides bearing methoxy (Table XVIII, entries 4, 10–14) or hydroxy (Table XVIII, entries 8 and 9) groups react with Grignard reagents to form biaryl in excellent yields. When methyl 4-bromobenzoate was used, methyl 4-phenylbenzoate was isolated in 69%

TABLE XVI

LIGAND EFFECTS ON $Pd(OAc)_2$/IMes.HCl CATALYZED SUZUKI CROSS-COUPLING REACTIONS

Entry	L	Time (h)	Isolated Yield (%)
1	IMes	2	99
2	IAd	2	44
3	IEtMe[a]	2	16
4	IEtMe	39	75
5	ICy	2	14
6	IBenMes[b]	2	12
7	ITol	2	5

[a]IEtMe = 1,3-bis(2-ethyl-6-methylphenyl)imidazol-2-ylidene.

[b]IBenMes = 1,3-bis(2,6-dibenzylphenyl)imidazol-2-ylidene.

TABLE XVII

CROSS-COUPLING OF CHLOROTOLUENE WITH PHENYLMAGNESIUMBROMIDE UNDER VARIOUS CONDITIONS

Entry	L	Solvent	Temp. (°C)	Time (h)	Isolated Yield (%)
1	IMes	Et_2O/THF	45	20	35
2	IPr	Et_2O/THF	45	20	97
3	IPr	Toluene/THF	80	20	10
4	IPr	THF	80	5	86
5	None	Dioxane/THF	80	3	0
6	IMes	Dioxane/THF	80	3	41
7	IPr	Dioxane/THF	80	3	99

TABLE XVIII

PALLADIUM/IMIDAZOLIUM SALT-CATALYZED CROSS-COUPLING OF ARYLHALIDES WITH ARYL GRIGNARD REAGENTS.[a]

$$\text{Ar-X} + \text{Ar'MgBr} \xrightarrow[\text{dioxane/THF, 80°C}]{\text{Pd}_2(\text{dba})_3 \text{ (1 mol\%)}, \text{ IPr.HCl (4 mol\%)}} \text{Ar-Ar'} + \text{MgBrX}$$

Entry	Ar−X	Ar′	Time (h)	Yield (%)[b]
1	4-MeC_6H_4−Cl	C_6H_5	3	99
2	4-MeC_6H_4−Cl	C_6H_5	3	96[c]
3	4-MeC_6H_4-Br	C_6H_5	1	99
4	4-$MeOC_6H_4$−Cl	C_6H_5	3	97
5	2,5-$(Me)_2C_6H_3$−Cl	C_6H_5	3	85
6	2,6-$(Me)_2C_6H_3$−Cl	C_6H_5	5	87[d]
7	4-$MeO_2CC_6H_4$−Br	C_6H_5	5	69
8	4-HOC_6H_4-I	C_6H_5	3	96[e]
9	4-HOC_6H_4−Cl	C_6H_5	5	95[e]
10	6-MeO-Np-2−Br	C_6H_5	1	98
11	4-$MeOC_6H_4$−Cl	4-MeC_6H_4	3	99
12	4-MeC_6H_4−Cl	3-MeC_6H_4	3	83
13	4-MeC_6H_4−Cl	2-FC_6H_4	3	99
14	4-MeC_6H_4−Cl	2,4,6-$(Me)_3C_6H_2$	3	95
15	2,6-$(Me)_2C_6H_3$−Cl	2,4,6-$(Me)_3C_6H_2$	24	0
16	2,4,6-$(Me)_3C_6H_2$−Br	2,4,6-$(Me)_3C_6H_2$	24	0

[a]The reactions were carried out according to the conditions indicated by the above equation; 1.2 equivalent of PhMgBr (1.0 M solution in THF) unless otherwise stated.

[b]Isolated yields after flash chromatography.

[c]2.0 mol% of $Pd(OAc)_2$ used instead of 1.0mol% of $Pd_2(dba)_3$.

[d]1.8 equivalent of PhMgBr was used.

[e]2.5 equivalent of PhMgBr was used.

yield (Table XVIII, entry 7), even though more complicated byproducts were formed when the less reactive chloro analog was used. Although sterically hindered substrates are often problematic,[96,107,109] the ortho-substituted 2-fluoro- or 2,4,6-trimethylphenylmagnesiumbromides coupled with 4-chloroanisole without any difficulty. When ortho-substituents were present in arylchlorides, good yields were obtained by using a slight excess of Grignard reagent (1.8 equivalent instead of 1.2 equivalent). However, the coupling of 2-chloro-*m*-xylene or 2-bromomesitylene with mesitylmagnesiumbromide failed because of steric congestion around both reactive centers. Homocoupling products of Grignard reagent were observed in all other reactions as minor products with the exception of mesitylmagnesiumbromide where no homocoupling was detected.[123]

While the mechanism remains to be elucidated, it seems apparent that at least diarylpalladium intermediates are involved in this reaction.[127] At some point the steric bulk of the ligand becomes more important than its electron donor contribution, as IPr is a poorer donor[50] than IMes but results in more effective couplings. We have recently observed that using a Pd/L ratio of 1/1 affords the products in similar high yields in a shorter time. This observation makes sense in terms of an understanding of the "active catalytic species" bearing only one nucleophilic carbene ligand. The methodology presented above proves effective for unactivated arylchlorides, arylbromides, and aryliodides by simply employing Pd(0) or Pd(II) and an imidazolium chloride as the catalyst precursor.

VIII

AMINATION OF ARYLCHLORIDES

In the previous sections it was shown that employing bulky carbene ligands as part of a catalytic system has a major positive effect on the ease of C–C bond formation.[96,123] Therefore, we were interested in examining what effect the use of judiciously selected bulky imidazolium ligands might have on catalytic arylamination reactions. Based on the success with IMes.HCl and IPr.HCl as ancillary ligand precursors in Suzuki[96] and Kumada[123] coupling involving arylchlorides, a similar protocol was used to perform the amination of arylchlorides. In an effort to select the most effective imidazolium salt, a number of 1,3-diaryl-substituted imidazolium chlorides were used in a model reaction and the results are shown in Table XIX. The bulky IPr.HCl was found to be the most effective

TABLE XIX

AMINATION OF 4-CHLOROTOLUENE USING DIFFERENT IMIDAZOLIUM CHLORIDES

Me–C$_6$H$_4$–Cl + PhN(Me)H → PhN(Me)Ph

$Pd_2(dba)_3$ (1 mol%)
L.HCl (4 mol%)
KO^tBu, dioxane
100°C

Entry	Ligand **L**	Time (h)	Yield (%)[a]
1	None	3	0
2	ITol	3	<5
3	IXy[b]	3	11
4	IMes	3	22
5	IPr	3	98

[a]Isolated yield represents the average of two runs.
[b]IX_y = 1,3-bis(2,6-dimethylphenyl)imidazol-2-ylidene.

imidazolium salt examined, leading to the isolation of the coupled product in a 98% yield (Table XIX, entry 5).

A survey of catalytic cross-coupling of arylhalides with primary and secondary cyclic or acyclic amines using IPr.HCl as the supporting ligand is provided in Table XVIII. The role of the added base KO^tBu is two-fold: it initially deprotonates the imidazolium chloride to form the free carbene ligand *in situ* which then coordinates to Pd(0) and also serves as a strong base to neutralize the HX formed in the course of the coupling reaction. This catalytic system proves general and efficient as shown by results presented in Table XX.

The less reactive unactivated arylchlorides reacted with various amines including primary (Table XX, entries 6, 7, and 9) and secondary cyclic (Table XX, entries 4, 5, and 11) amines in high yields. Ortho-substituted arylchlorides reacted with amine without difficulty. The reaction of 4-chlorotoluene with highly hindered amines (Table XX, entry 7) leads to lower yields. The generality of the method is illustrated by the efficient coupling of unhindered arylchlorides with both acyclic primary and secondary alkylamines. To the best of our knowledge no reported catalyst allows this transformation.

Generally, aminations involving arylbromides and aryliodides proceed under milder conditions than those involving arylchlorides. In order to examine the halide substituent effect, the efficiency and selectivity of the present catalytic system for amination of arylbromides and aryliodides was examined. Both arylbromides and aryliodides (Table XXI) reacted with amines smoothly at room temperature. Most interesting in these studies involving aryls bearing both chloro and iodo (or bromo) substituents is the observation that bromo and iodo functionalities can be converted at room temperature (Table XXI, entries 3 and 4) and the remaining chloro functionality can subsequently be converted at more elevated temperatures. This could prove a significant advantage in process chemistry.

While the detailed mechanism remains to be elucidated, it seems apparent that at least arylamidopalladium intermediates are involved in this reaction as previously observed.[93,100,101,103,104,106,110–115,128] It appears to us that both steric and electronic effects combine to mediate the coupling process. Initially, the electron donor properties of the carbene facilitate the activation of arylchlorides and a secondary effect may be provided by the bulk located at the ortho positions of the carbene aryl group. Results of structural studies in related platinum complexes of general formulation L_2PtMe_2 (L = nucleophilic carbene) indicate that these ortho positions are oriented directly toward the groups involved in the reductive elimination step. Steric hindrance imparted by the ligands would therefore favor this reductive elimination step. At this time we feel the L_2PdCl_2 may very well represent the resting state of the catalyst. We believe a combination of effects are at play: once the electron donor carbene imparts enough electron density to the metal to enable it to perform the oxidative addition, the ligand sterics subsequently stabilize the reactive intermediates. The steric effects provided by ortho-substituents on

TABLE XX

AMINATION OF ARYLCHLORIDES WITH VARIOUS AMINES[a]

$$\mathrm{ArCl} + \mathrm{HNR'R''} \xrightarrow[\mathrm{KO^tBu,\ dioxane}]{\mathrm{Pd_2(dba)_3,\ IPr.HCl}} \mathrm{Ar\text{-}NR'R''}$$

Entry	Ar–Cl	HNR′R″	Ar-NR′R″	Yield (%)[b]
1	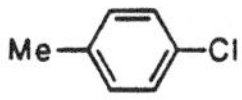 Me— —Cl	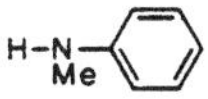 H–N(Me)—	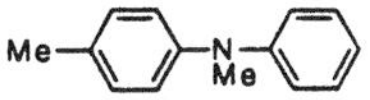 Me— —N(Me)—	99
2		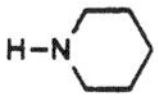 H–N	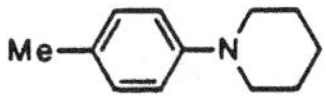 Me— —N	96
3		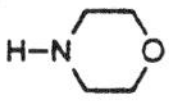 H–N O	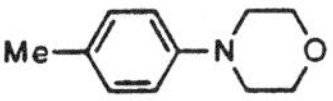 Me— —N O	82
4		 H–N( )$_2$	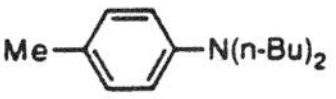 Me— —N(n-Bu)$_2$	95
5		 H_2N	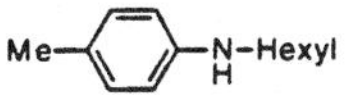 Me— —N(H)–Hexyl	86[c]
6		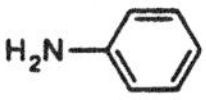 H_2N—	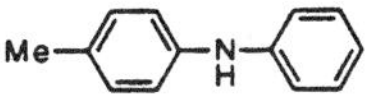 Me— —N(H)—	96
7		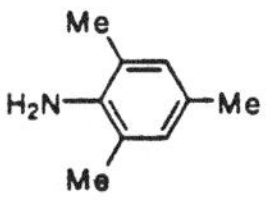 H_2N— (Me, Me, Me)	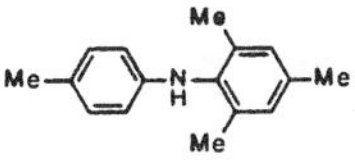 Me— —N(H)— (Me, Me, Me)	59
8	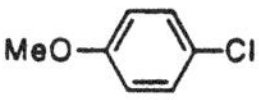 MeO— —Cl	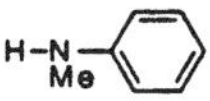 H–N(Me)—	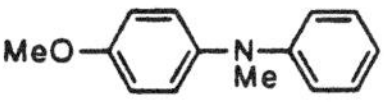 MeO— —N(Me)—	91
9		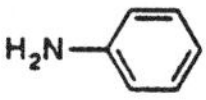 H_2N—	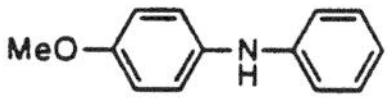 MeO— —N(H)—	91
10		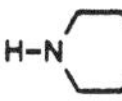 H–N O	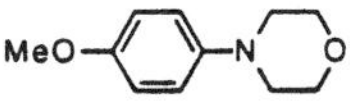 MeO— —N O	80
11		 H–N( )$_2$	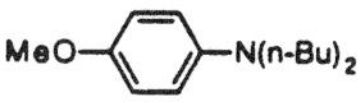 MeO— —N(n-Bu)$_2$	98
12	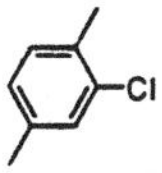 —Cl	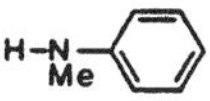 H–N(Me)—	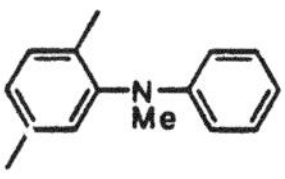 —N(Me)—	94

[a] Reaction conditions: 1.0 mmol of aryl chloride, 1.2 mmol of amine, 1.5 mmol of KO^{t}Bu. 1.0 mmol% $Pd_2(dba)_3$, 4.0%IPrHCl (2 L/Pd), 3 mL of dioxane, 100°C. Reactions were complete in 3–30 h and reaction times were not minimized.

[b] Average isolated yields of two runs.

[c] Dialkyl aniline was isolated as a byproduct in 5%.

the carbene aryl groups influence the efficiency of the catalytic transformation in a dramatic fashion. Our thermochemistry studies on ruthenium systems involving these carbenes[47,50] show that ITol is the best electron donor (ITol > IMes ≈ IXy > IPr) while IPr is the most bulky ligand (IPr > IMes ≈ IXy > ITol). The catalytic activity of systems involving these various imidazolium salts follows the steric trend. When a Pd/L ratio of 1/1 is used the products are formed in similar high

TABLE XXI

AMINATION OF ARYLBROMIDES AND IODIDES[a]

$$\mathrm{ArCl + HNR'R''} \xrightarrow[\mathrm{KO}^t\mathrm{Bu,\ dioxane}]{\mathrm{Pd_2(dba)_3,\ IPr.HCl}} \mathrm{Ar\text{-}NR'R''}$$

Entry	Ar-X	HNR′R″	Ar-NR′R″	Yield (%)[b]
1	$Me-C_6H_4-Br$	$H-N(Me)-C_6H_5$	$Me-C_6H_4-N(Me)-C_6H_5$	89
2	$Me-C_6H_4-Br$	H-N (piperidine)	$Me-C_6H_4-N$ (piperidine)	83
3	$Cl-C_6H_4-Br$	H-N (piperidine)	$Cl-C_6H_4-N$ (piperidine)	94
4	$Cl-C_6H_4-I$	H-N (piperidine)	$Cl-C_6H_4-N$ (piperidine)	97

[a] Reaction conditions: 1.0 mmol of aryl halides, 1.2 mmol of amine, 1.5 mmol of KO^{t}Bu. 1.0 mmol% $Pd_2(dba)_3$, 4.0% IPrHCl (2 L/Pd), 3 mL of dioxane, room temperature. Reactions were complete in 3–30 h and reaction times were not minimized.

[b] Average isolated yields of two runs.

yields but in a shorter time. As previously discussed, this ratio appears optimum for generating the active catalyst intermediates. Without excess ligand, the resting state composition "L_2 Pd(Ar)(X)" can not be reached.

In summary, a general and efficient methodology for the amination of arylchlorides (and bromides and iodides) has been developed. The simple methodology makes use of a combination of a palladium(0) complex and an imidazolium chloride forming the catalytic precursor which proves effective for unactivated arylchlorides as well as arylbromides and aryliodides in high isolated yields. This methodology provides the first report of arylamination involving arylchlorides with both acyclic primary and secondary alkylamines.

IX

CONCLUSION

This journey started where most of our studies are initiated: from our interest in fundamental thermodynamic properties quantified by solution calorimetry. From these data it became clear that nucleophilic carbenes (most of them anyway) are better donors than the best donor phosphines. This stability scale provided by the calorimetry allowed us to put on paper "feasible" reactions. Developments in both olefin metathesis and palladium-mediated coupling reactions can be attributed to the thermochemistry guiding us in the right direction. Both these catalytic processes benefit from the steric protection provided by the nucleophilic carbene fences, yet they differ in the electronic requirements: olefin metathesis benefit from as much electron donation as possible. We feel the electron donation provided by IMes or IPr are more than enough to allow C—Cl bond activation in coupling catalysis. The fact that IAd can effectively mediate these processes is a clue as to how much electron donation is really needed to mediate this conversion.

The nucleophilic carbenes are "phosphine-mimics" but they are much more. They reside at the upper end of the Tolman electronic and steric parameter scales. Much remains to be explored with these ligands. With a rudimentary understanding of ligand stereoelectronic properties, we feel confident much exciting chemistry remains to be explored.

Acknowledgment

S.P.N. acknowledges the National Science Foundation, Louisiana Board of Regents, and Dupont for support of this work.

References

(1) Parshall, G. W.; Ittel, S. D. *Homogeneous Catalysis;* Wiley-Interscience: New York, 1992.
(2) Pignolet, L. H. *Homogeneous Catalysis with Metal Phosphine Complexes;* Plenum: New York, 1983.
(3) Collman, J. P.; Hegedus, L. S.; Norton, J. R.; Finke, R. G. *Principles and Applications of Organotransition Metal Chemistry,* 2nd ed.; University Science: Mill Valley, CA, 1987.
(4) Kubas, G. J. *Acc. Chem. Res.* **1988,** *21,* 120.
(5) Gonzalez, A. A.; Mukerjee, S. L.; Chou, S.-J.; Zhang, K.; Hoff, C. D. J. *J. Am. Chem. Soc.* **1988,** *110,* 4419.
(6) Nguyen, S. T.; Grubbs, R. H.; Ziller, J. W. *J. Am. Chem. Soc.* **1993,** *115,* 9858.
(7) Fu, G. C.; Nguyen, S. T.; Grubbs, R. H. *J. Am. Chem. Soc.* **1993,** *115,* 9856.
(8) Arduengo, A. J., III; Rasika Dias, H. V.; Harlow, R. L.; Kline, M. *J. Am. Chem. Soc.* **1992,** *114,* 5530.
(9) Wanzlick, H.-W. *Angew. Chem., Int. Ed. Engl.* **1962,** *1,* 75.
(10) Lappert, M. F. *J. Organomet. Chem.* **1988,** *358,* 185.
(11) Arduengo, A. J., III; Krafczyk, R. *Chem. Z.* **1998,** *32,* 6.
(12) Wanzlick, H.-W.; Schikora, E. *Angew. Chem.* **1960,** *72,* 494.
(13) Wanzlick, H.-W.; Schikora, E. *Chem. Ber.* **1961,** *94,* 2389.
(14) Wanzlick, H.-W.; Kleiner, H.-J. *Angew. Chem.* **1961,** *73,* 493.
(15) Wanzlick, H.-W.; Esser, F.; Kleiner, H.-J. *Chem. Ber.* **1963,** *96,* 1208.
(16) Wanzlick, H.-W.; Kleiner, H.-J. *Chem. Ber.* **1963,** *96,* 3024.
(17) Wanzlick, H.-W.; Ahrens, H. *Chem. Ber.* **1964,** *97,* 2447.
(18) Wanzlick, H.-W.; Konig, B. *Chem. Ber.* **1964,** *97,* 3513.
(19) Wanzlick, H.-W.; Lachmann, B.; Schikora, E. *Chem. Ber.* **1965,** *98,* 3170.

(20) Wanzlick, H.-W.; Ahrens, H. *Chem. Ber.* **1966,** *99,* 1580.
(21) Wanzlick, H.-W.; Lachmann, B. *Z. Naturforsch* **1969,** *24b,* 574.
(22) Lachmann, B.; Wanzlick, H.-W. *Liebigs Ann. Chem.* **1969,** *729,* 27.
(23) Lachmann, B.; Steinmaus, H.; Wanzlick, H.-W. *Tetrahedron* **1971,** *27,* 4085.
(24) Wanzlick, H.-J.; Schonherr, H. *Angew. Chem., Int. Ed. Engl.* **1968,** *7,* 141.
(25) Wanzlick, H.-J.; Schonherr, H. *Chem. Ber.* **1970,** *103,* 1037.
(26) Herrmann, W. A.; Goossen, L. J.; Spiegler, M. *J. Organomet. Chem.* **1997,** *547,* 357.
(27) Herrmann, W. A.; Kocher, C. *Angew. Chem., Int. Ed. Engl.* **1997,** *36,* 2163.
(28) Arduengo, A. J., III; Harlow, R. L.; Kline, M. J. *J. Am. Chem. Soc.* **1991,** *113,* 361.
(29) Arduengo, A. J., III; *U.S. Patent No. 5182405,* **1993.**
(30) Arduengo, A. J., III; Gamper, S. F.; Calabrese, J. C.; Davidson, F. *J. Am. Chem. Soc.* **1994,** *116,* 4391.
(31) Arduengo, A. J., III; Davidson, F.; Krafczyk, R.; Marshall, W. J.; Tamm, M. *Organometallics* **1998,** *17,* 3375.
(32) Dixon, D. A.; Arduengo, A. J., III *J. Phys. Chem.* **1991,** *95,* 4180.
(33) Cioslowski, J. *J. Quantum Chem., Quantum Chem. Symp.* **1993.** *27,* 309.
(34) Arduengo, A. J., III; Dias, H. V. R.; Dixon, D. A.; Harlow, R. L.; Klooster, W. T.; Koetzle, T. F. *J. Am. Chem. Soc.* **1994,** *116,* 6812.
(35) Arduengo, A. J., III; Dixon, D. A.; Kumashiro, K. K.; Lee, C.; Power, W. P.; Zilm, K. W. *J. Am. Chem. Soc.* **1994,** *116,* 6361.
(36) Arduengo, A. J., III; Bock, H.; Chen, H.; Dixon, D. A.; Green, J. C.; Herrmann, W. A.; Jones, N. L.; Wagner, M.; West, R. *J. Am. Chem. Soc.* **1994,** *116,* 6641.
(37) Heineman, C.; Thiel, W. *Chem. Phys. Lett.* **1994,** *217,* 11.
(38) Sauers, R. R. *Tetrahedron Lett.* **1996,** *37,* 149.
(39) Heinemann, A. E.; Muller, T.; Apeloig, Y.; Schwarz, H. *J. Am. Chem. Soc.* **1996,** *118,* 2023.
(40) Boehme, C.; Frenking, G. *J. Am. Chem. Soc.* **1996,** *118,* 2039.
(41) Arduengo, A. J., III; Goerlich, J. R.; Marshall, W. J. *J. Am. Chem. Soc.* **1995,** *117,* 11,027.
(42) Herrmann, W. A.; Elison, M.; Fischer, J.; Kocher, C.; Artus, G. R. J. *Angew. Chem., Int. Ed. Engl.* **1995,** *34,* 2371.
(43) Frohlich, N.; Pidun, U.; Stahl, M.; Frenking, G. *Organometallics* **1997,** *16,* 442.
(44) Lappert, M. J. *J. Orgaonmet. Chem.* **1975,** *100,* 139.
(45) Fagan, P. J.; Ward, M. D.; Caspar, J. V.; Calabrese, J. C.; Krusic, P. J. *J. Am. Chem. Soc.* **1988,** *110,* 2981.
(46) Luo, L.; Nolan, S. P. *Organometallics* **1994,** *13,* 4781.
(47) Huang, J.; Schanz, H.-J.; Stevens, E. D.; Nolan, S. P. *Organometallics* **1999,** *18,* 5375.
(48) Serron, S. A.; Huang, J.; Nolan, S. P. *Organometallics* **1998,** *17,* 534.
(49) Huang, J.; Stevens, E. D.; Nolan, S. P.; Petersen, J. L. *J. Am. Chem. Soc.* **1999,** *121,* 2674.
(50) Jafarpour, L.; Stevens, E. D.; Nolan, S. P. *J. Organomet. Chem.,* in press.
(51) Campion, B. K.; Heyn, R. H.; Tilley, T. D. *J. Chem. Soc., Chem. Commun.* **1998,** 278.
(52) Shannon, R. D. *Acta Crystallogr.* **1976,** *A32,* 751.
(53) Tolman, C. A. *Chem. Rev.* **1977,** *77,* 313.
(54) Grubbs, R. H. *Comprehensive Organometallic Chemistry;* Pergamon: New York, 1982; Vol. 8.
(55) Leconte, M.; Basset, J. M.; Quignard, F.; Larroche, C. *Reactions of Coordinated Ligands;* Plenum: New York, 1986; Vol. 1.
(56) Feldman, J.; Schrock, R. R. *Progress in Inorganic Chemistry;* John Wiley & Sons: New York, 1991; Vol. 39.
(57) Ivin, K. J.; Mol, J. C. *Olefin Metathesis and Metathesis Polymerization;* Academic Press: San Diego, 1997.
(58) Mol, J. C.; Moulijin, J. A. *Catalysis: Science and Technology;* Springer-Verlag: Berlin, 1987; Vol. 8.

(59) Grubbs, R. H.; Miller, S. J.; Fu, G. C. *Acc. Chem. Res.* **1995,** *28,* 446.
(60) Schwab, P.; France, M. B.; Ziller, J. W.; Grubbs, R. H. *Angew. Chem., Int. Ed. Engl.* **1995,** *34,* 2039.
(61) Schwab, P.; Grubbs, R. H.; Ziller, J. W. *J. Am. Chem. Soc.* **1996,** *118,* 100.
(62) Diaz, E. L.; Nguyen, S. T.; Grubbs, R. H. *J. Am. Chem. Soc.* **1997,** *119,* 3887 (and the references cited therein).
(63) Nguyen, S. T.; Johnson, L. K.; Grubbs, R. H. *J. Am. Chem. Soc.* **1992,** *114,* 3974.
(64) Stumpf, A. W.; Saive, E.; Demonceau, A.; Noles, A. F. *J. Chem. Soc., Chem. Commun.* **1995,** 1127.
(65) Wu, Z.; Nguyen, S. T.; Grubbs, R. H.; Ziller, J. W. *J. Am. Chem. Soc.* **1995,** *117,* 5503.
(66) Herrmann, W. A.; Schattenmann, W. C.; Nuyken, O.; Glander, S. C. *Angew. Chem., Int. Ed. Engl.* **1996,** *35,* 1087.
(67) Mohr, B.; Lynn, D. M.; Grubbs, R. H. *Orgaonometallics* **1996,** *15,* 4317.
(68) Demonceau, A.; Stumpf, A. W.; Saive, E.; Noles, A. F. *Macromolecules* **1997,** *30,* 3127.
(69) Kingsbury, J. S.; Harrity, J. P. A.; Bonitatebus, P. J., Jr.; Hoveyda, A. H. *J. Am. Chem. Soc.* **1999,** *121,* 791.
(70) Weskamp, T.; Schattenmann, W. C.; Spiegler, M.; Herrmann, W. A. *Angew. Chem., Int. Ed. Engl.* **1998,** *37,* 2490.
(71) Weskamp, T.; Kohl, K. J.; Hieringer, W.; Gleich, D.; Herrmann, W. A. *Angew. Chem., Int. Ed. Engl.* **1999,** *38,* 2416.
(72) Scholl, M.; Trnka, T. M.; Morgan, J. P.; Grubbs, R. H. *Tetrahedron Lett.* **1999,** *40,* 2247.
(73) Scholl, M.; Ding, S.; Lee, C. W.; Grubbs, R. H. *Organic Letters* **1999,** *1,* 953.
(74) Fürstner, A.; Picquet, M.; Bruneau, C.; Dixneuf, P. H. *Chem. Commun.* **1998,** 1315.
(75) Fürstner, A.; Ackermann, L. *Chem. Commun.* **1999,** 95.
(76) Harlow, K. J.; Hill, A. F.; Wilton-Ely, J. D. E. T. *J. Chem. Soc., Dalton Trans.* **1999,** 285.
(77) Fürstner, A.; Hill, A. F.; Lieble, M.; Wilton-Ely, J. D. E. T. *Chem. Commun.* **1999,** 601.
(78) Jafarpour, L.; Huang, J.; Stevens, E. D.; Nolan, S. P. *Organometallics* **1999,** *18,* 3760.
(79) Picquet, M.; Bruneau, C.; Dixneuf, P. H. *Chem. Commun.* **1998,** 2249.
(80) Hafner, A.; Muhlebach, A.; van der Schaaf, P. A. *Angew. Chem., Int. Ed. Engl.* **1997,** *36,* 2121.
(81) Jafarpour, L.; Schanz, H.-J.; Stevens, E. D.; Nolan, S. P. *Organometallics* **1999,** *18,* 5416.
(82) Fürstner, A.; Hill, A. F.; Jafarpour, L.; Stevens, E. D.; Liebl, M.; Nolan, S. P.; Wilton-Ely, J. D. E. T. **1999,** manuscript in preparation.
(83) Schanz, H.-J.; Jafarpour, L.: D., S. E.; Nolan, S. P. *Orgaonmetallics* **1999,** *18,* 5187.
(84) Selegue, J. *Organometallics* **1982,** *1,* 217.
(85) Gamasa, M. P.; Gimeno, J.; Gonzalez-Bernardo, C.; Borge, J.; Garcia-Granada, S. *Organometallics* **1997,** *16,* 2483.
(86) Buriez, B.; Burns, I. D.; Hill, A. F.; White, A. J. P.; Williams, D. J.; Wilton-Ely, J. D. E. T. *Organometallics* **1999,** *18,* 1504.
(87) Cadierno, V.; Gamasa, M. P.; Gimeno, J.; Iglesias, L.; Garcia-Granada, S. *Inorg. Chem.* **1999,** *38,* 2874.
(88) Trost, B. M.; Verhoeven, T. R. *Comprehensive Organometallic Chemistry;* Pergamon: Oxford, 1982; Vol. 8.
(89) Heck, R. F. *Palladium Reagents in Organic Synthesis;* Academic Press: New York, 1985.
(90) Tsuji, J. *Palladium Reagents and Catalysis;* Wiley: Chichester, 1995.
(91) Tsuji, J. *Synthesis* **1990,** 739.
(92) Miaura, N.; Suzuki, A. *Chem. Rev.* **1995,** *95,* 2457.
(93) Hamann, B. C.; Hartwig, J. F. *J. Am. Chem. Soc.* **1998,** *120,* 7369.
(94) Reetz, M. T.; Lohmer, G.; Schwickardi, R. *Angew. Chem., Int. Ed. Engl.* **1998,** *37,* 481.
(95) Littke, A. F.; Fu, G. C. *J. Org. Chem.* **1999,** *64,* 10.
(96) Zhang, C.; Huang, J.; Trudell, M. L.; Nolan, S. P. *J. Org. Chem.* **1999,** *64,* 3804.

(97) Tamao, K.; Sumitani, K.; Kumada, M. *J. Am. Chem. Soc.* **1972,** *94,* 4374.
(98) Corriu, R. J.; Masse, J. P. *J. Chem. Soc., Chem. Commun.* **1972,** 144.
(99) Yamamura, M.; Moritani, I.; Murahashi, S. *J. Organomet. Chem.* **1975,** *91,* C39.
(100) Wolfe, J. P.; Wagaw, S.; Marcoux, J.-F.; Buchwald, S. L. *Acc. Chem. Res.* **1998,** *31,* 805.
(101) Hartwig, J. F. *Acc. Chem. Res.* **1998,** *31,* 852.
(102) Hartwig, J. F. *Synlett.* **1997,** 329.
(103) Yang, B. H.; Buchwald, S. L. *J. Organomet. Chem.* **1999,** *576,* 125.
(104) Old, D. W.; Wolfe, J. P.; Buchwald, S. L. *J. Am. Chem. Soc.* **1998,** *120,* 9722.
(105) Littke, A. F.; Fu, G. C. *Angew. Chem., Int. Ed. Engl.* **1998,** *37,* 3387.
(106) Bei, X.; Guram, A. S.; Turner, H. W.; Weinburg, W. H. *Tetrahedron Lett.* **1999,** *40,* 1237.
(107) Saito, S.; Ohtani, S.; Miyaura, N. *J. Org. Chem.* **1997,** *62,* 8024.
(108) Saito, S.; Sakai, M.; Miyaura, N. *Tetrahedron Lett.* **1996,** *37,* 2993.
(109) Indolese, A. F. *Tetrahedron Lett.* **1997,** *38,* 3513.
(110) Bei, X.; Uno, T.; Norris, J.; Turner, H. W.; Weinburg, W. H.; Guram, A. S.; Petersen, J. L. *Organometallics* **1999,** *18,* 1840.
(111) Brenner, E.; Fort, Y. *Tetrahedron Lett.* **1998,** *39,* 5359.
(112) Yamamoto, T.; Nishiyama, M.; Koie, Y. *Tetrahedron Lett.* **1997,** *39,* 2367.
(113) Wolfe, J. P.; Buchwald, S. L. *J. Am. Chem. Soc.* **1997,** *119,* 6054.
(114) Riermeier, T. H.; Zapf, A.; Beller, M. *Top. Catal.* **1997,** *4,* 301.
(115) Reddy, N. P.; Tanaka, M. *Tetrahedron Lett.* **1997,** *38,* 4807.
(116) Cornils, B.; Herrmann, W. A. *Applied Homogeneous Catalysis with Organometallic Compounds;* VCH: Weinheim, 1996.
(117) 1, *J. Chem. Eng. News* **1998,** 24.
(118) 13, *J. Chem. Eng. News* **1998,** 71.
(119) Regitz, M. *Angew. Chem., Int. Ed. Engl.* **1996,** *35,* 725.
(120) Voges, M. H.; Romming, C.; Tilset, M. *Organometallics* **1999,** *18,* 529.
(121) Herrmann, W. A.; Reisinger, C.-P.; Spiegler, M. J. *J. Organomet. Chem.* **1998,** *557,* 93.
(122) Stanforth, S. P. *Tetrahedron* **1998,** *54,* 263.
(123) Huang, J.; Nolan, S. P. *J. Am. Chem. Soc.* **1999,** *121,* 9889.
(124) Tamao, K.; Sumitani, K.; Kisa, Y.; Zambayashi, M.; Fujioka, A.; Kodama, A.; Nakajima, I.; Minato, A.; Kumada, M. *Bull. Chem. Soc. Jpn.* **1976,** *49,* 1958.
(125) Hayashi, T.; Konishi, M.; Kobori, Y.; Kumada, M.; Higuchi, T.; Hirotsu, K. *J. Am. Chem. Soc.* **1984,** *106,* 158.
(126) Miller, J. A.; Farrell, R. P. *Tetrahedron Lett.* **1998,** *39,* 7275.
(127) Busacca, C. A.; Eriksson, M. C.; Fiaschi, R. *Tetrahedron Lett.* **1999,** *40,* 3101.
(128) Wolfe, J. P.; Buckwald, S. L. *Angew. Chem., Int. Ed. Engl.* **1999,** *38,* 2413.

ADVANCES IN ORGANOMETALLIC CHEMISTRY, VOL. 46

Organometallic Chemistry of Transition Metal Porphyrin Complexes

PENELOPE J. BROTHERS*

Department of Chemistry, The University of Auckland, Private Bag 92019, Auckland, New Zealand

*E-mail: p.brothers@auckland.ac.nz.

0065-3055/01 $35.00

I

INTRODUCTION

A. *Background*

The systematic chemistry of metalloporphyrin complexes, encompassing structure, reactivity, and spectroscopy, grew at an astonishing rate during the 1960s and 1970s.[1,2] This growth was spurred in large part by the central role played by porphyrin species in biology, and facilitated by the simultaneous increase in sophistication and availability of structural and spectroscopic techniques. During the same period, the importance and scope of transition metal organometallic chemistry became firmly established, although the two areas remained rather unconnected, despite the occasional report of metalloporphyrin compounds containing metal-carbon σ-bonds. However, with the report in 1977 of an iron porphyrin dichlorocarbene complex,[3,4] followed during the early 1980s by further examples of metalloporphyrin carbene, alkene, alkyne, and hydride complexes,[5] it became apparent that there was much to be learned from the use of the porphyrin macrocycle as a supporting ligand in organometallic chemistry. For example, the redox-induced migrations of organic fragments between the coordinated metal and a pyrrole nitrogen atom was shown to be relevant to the function of iron porphyrins in cytochrome P450.[6] Classical organometallic reactions were shown to proceed in porphyrin complexes by unanticipated radical reactions, and one example of this, the insertion of CO into a rhodium hydride bond to form a formyl complex, was a process hitherto not observed in organometallic chemistry. The publication, in 1986, 1987, and 1988, of three reviews focusing on the organometallic chemistry of transition metal porphyrin complexes was an indication that this area had become firmly established.[5,7,8]

Over a decade on, this area is now beginning to reach maturity, with some of the early new developments subjected to intensive research and now consolidated. New areas have been opened up, for example, organometallic complexes of group 3 and 4 metals which were unknown at the time of the earlier reviews. Aspects of organometallic chemistry which have become important in the last decade such as C—H bond activation, the relationship between metal-hydride and dihydrogen complexes, and supramolecular interactions have resulted in parallel developments in porphyrin complexes. The use of porphyrin complexes in catalytic organometallic processes has begun to be developed, as in the cobalt porphyrin-catalyzed alkene polymerization and the shape selective catalysis of alkene cyclopropanation by rhodium porphyrins.

This review essentially comprises a survey of the developments in the organometallic chemistry of porphyrin complexes over the last decade, continuing on from the three reviews published during 1986–1988. Literature since the mid-1980s has been surveyed, and work reported prior to this will be touched on primarily to put the more recent developments into context and will not be described in depth. A new multivolume set encompassing the entire range of porphyrin chemistry has been recently published, and this contains a chapter on organometallic porphyrin chemistry.[9]

B. *Scope*

The broadest description of an organometallic porphyrin complex is one which contains both a porphyrin macrocycle and a metal carbon bond. A useful classification of the types of organometallic porphyrin complexes has been described by Richter-Addo in a recent paper,[10] and his system is summarized here (Fig. 1). Type A complexes contain one or two metal-carbon single or multiple bonds. This class comprises the majority of the organometallic porphyrin complexes which contain σ-alkyl, σ-aryl, or carbene ligands. Both five-coordination and six-coordination is possible, with typical formulae as follows: M(Por)(R), M(Por)(R)(L), M(Por)$(R)_2$, M(Por)(CR_2), M(Por)(CR_2)(L), M(Por)$(CR_2)_2$. Notably absent from this category is an example of a metalloporphyrin carbyne complex. Type B complexes contain π-ligands, including η^2-alkene, η^2-alkyne, η^5-cyclopentadienyl,

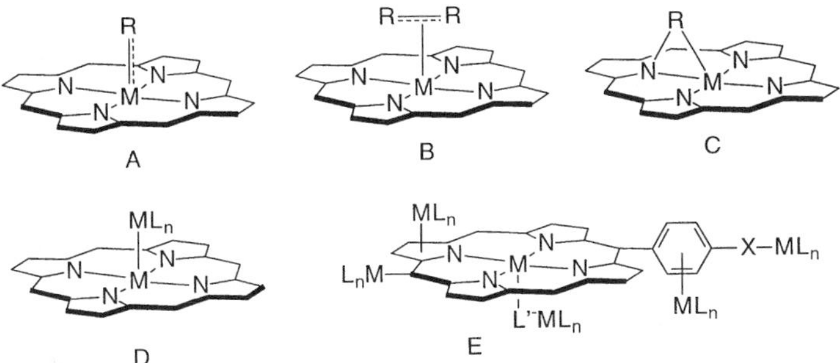

FIG. 1. Classification of the types of organometallic porphyrin complexes.

and related ligands. The newest example in this category, reported in 1999, is the first metalloporphyrin π-allyl complex. Type C complexes involve bridging μ-M,N groups in which the alkyl group bridges between the coordinated metal and one porphyrin nitrogen atom. The M-N bond may or may not be retained in these complexes. Type D complexes contain a metal–metal bond, with the metal–carbon bonds associated with the metal not coordinated to the porphyrin. Type E is a composite structure showing possible sites for coordination of a second metal fragment. It can be π-coordinated to one of the porphyrin pyrrole rings[11] or to an aryl substituent on the periphery of the porphyrin. A number of examples contain a metal coordinated to the periphery of the porphyrin through a metal–carbon σ bond. Types A, B, and C comprise the most important types or organometallic porphyrin complexes, and this review attempts to cover all the important examples from these categories. Types D and E are represented by a much smaller number of examples, most reported in the late 1990s. However, the growing interest in the use of metalloporphyrin moieties as building blocks in supramolecular chemistry should lead to an increase in the number of examples of these types. However, complexes of types D and E are largely beyond the scope of this review. The exceptions are some examples of type E complexes which comprise a subset of the area sometimes described as "inorganometallic chemistry"[12] in which complexes contain metal–heteroatom bonds in which the heteroatom moiety is isolobal with an alkyl or carbene ligand. Metalloporphyrin stannyl and stannylene complexes are examples from this category, and some examples of these are discussed in the review.

Defining the scope of the review has not been easy, and the central criterion has been to focus on compounds containing a bond between the coordinated metal and carbon. However, the central role played by organometallic compounds and reactions in modern chemistry means that many reactions may involve intermediates or even transition states with a metal–carbon interaction. A good example of this is the widely studied oxidation chemistry catalyzed by high-valent metalloporphyin oxo species, spurred by the remarkable biological transformations mediated by cytochrome P450.[13] This chemistry has links to organometallic chemistry on several fronts. Metal oxo (M=O), metal imido (M=NR), and metal carbene ($M{=}CR_2$) fragments are formally isolobal. As a result, alkene epoxidation and cyclopropanation reactions mediated by metalloporphyrin oxo and carbene complexes, respectively, are closely related reactions, yet while the latter is an organometallic process, the former is not. On a more subtle level, the reactions involving metal oxo species themselves may or may not meet the criterion for a metal–carbon interaction. The alkene epoxidation process, for example, is proposed by some to involve interaction of the alkene substrate with the metal center prior to, or concerted with, oxygen atom transfer, in which case there is a direct metal–carbon interaction.[13–15] Other researchers invoke a direct transfer of oxygen to the substrate without interaction of the alkene with the metal. Overall, metalloporphyrin oxo chemistry and related developments in imido chemistry by themselves carry sufficient weight that they will not be covered here, even where it has been suggested on the basis of mechanistic studies that metal–carbon interactions may be part of the process.

For the purposes of this review the criterion has been refined to include only those porphyrin complexes where there is direct structural or spectroscopic evidence for a metal–carbon interaction. This interaction will not, however, be limited to covalent bonds. The last decade has seen the rise in importance of supramolecular chemistry and non-covalent interactions, and a small set of examples involving porphyrin complexes will be included as the last section in the review.

Another area where defining the scope of this review is difficult arises in the boundary between organometallic and coordination chemistry. Transition metal carbonyl complexes form one of the cornerstones of organometallic chemistry. Metalloporphyrin complexes containing an axial carbonyl ligand are also very significant, particularly for iron porphyrins where coordination of CO in the site usually occupied by O_2 in hemoglobin and myoglobin is of enormous importance in biology. Although these complexes clearly meet the metal–carbon bond criterion, much of the work on iron porphyrin carbonyl complexes has focused on structure, spectroscopic properties, kinetic studies, and thermodynamic data and has little direct relevance to organometallic chemistry. A further family of ligands isoelectronic with CO which form M—C bonds are CN^- and isocyanides (CNR). Porphyrin complexes containing the NO ligand have received a recent burst of attention, arising from new understanding about the role of NO in biological systems. This ligand also enjoys a special role in classical organometallic chemistry because NO^+ is isoelectronic with CO. As with CO itself, complexes containing the CN^-, CNR, and NO ligands will only be considered when there is something to be learned in the context of organometallic chemistry.

One family of porphyrin complexes that will be treated in the review, even though they do not contain metal–carbon bonds, are metalloporphyrin hydride and dihydrogen complexes. As in classical organometallic chemistry, hydride complexes play key roles in some reactions involving porphyrins, and the discovery of dihydrogen complexes and their relationship to metal hydrides has been an important advance in the last decade.

The development of both the coordination chemistry and organometallic chemistry of main group porphyrin complexes has been slower than that of the transition metals. As a generalization, the chemistry of the middle elements from groups 13 and 14 is well established (Al, Ga, In, Ge, Sn) with that of the lightest and heaviest elements (B, Tl, Si, Pb) less well known.[16] Advances in the coordination of the group 15 elements (excluding nitrogen) are more recent, and porphyrin complexes containing group 16 or 17 elements are as yet unknown. A feature of main group elements is the coordination of non-metals (boron, silicon, phosphorus) to porphyrins, with some of these (for example, boron) reported only within the last few years.[17,18] Organometallic main group porphyrin complexes will be discussed in a companion review article.[19]

C. *The Porphyrin as a Supporting Ligand in Organometallic Chemistry*

The porphyrin macrocycle provides four nitrogen donor ligands in a square planar arrangement, with a hole size close to 2.0 Å. Coordination numbers for a metal coordinated to a porphyrin are usually 4, 5, or 6, although higher coordination numbers are occasionally observed. For five-coordination (one additional ligand), square pyramidal geometry in which the metal is displaced from the N_4 plane toward the axial ligand is common. For six-coordinate complexes the *trans* arrangement of the 5th and 6th ligands above and below the plane of the porphyrin is by far the most common. Although the *cis* geometry is not precluded, it is relatively rare and observed largely for complexes involving the early transition metals.[20] The availability of *cis* coordination sites is important in many of the classic organometallic processes like oxidative addition, reductive elimination, migratory insertion, and agostic interactions. One of the great advantages of using the porphyrin macrocycle as the supporting ligand in organometallic chemistry is the high degree of control over coordination geometry, and the expectation that some these processes might be strongly inhibited by the enforced *trans* geometry.

A further advantage is that the 24-atom core and hence the immediate coordination environment of the central element can be maintained intact, while substitution on the periphery of the porphyrin macrocycle allows a high degree of control in tuning the steric and electronic properties of the ligand. This property has for a long time been used to great advantage in the coordination chemistry of porphyrin complexes, but has only more recently begun to be used as a tool for manipulating and understanding the chemistry of organometallic porphyrin complexes. Structural variations within the 24-atom porphyrin core are possible, with fully planar, saddle-shaped, or ruffled distortions among the common arrangements.[21] Given the importance of X-ray crystallography in organometallic chemistry in general, relatively few organometallic porphyrin complexes have been structurally characterized, and this was especially true last time this area was reviewed. Exhaustive compilations of physical data for organometallic porphyrin complexes have not been included in this review, with the exception of X-ray data, for which there are now sufficient examples to warrant the inclusion of tables collecting selected data for structurally characterized organometallic porphyrin complexes. Simple sketches, using an oval shape to represent the porphyrin ligand, are used in the schemes and equations. However, within each section a representative group of complete structures are shown, using examples taken from the Cambridge Structural Database. These are also useful for illustrating the effect of substitution on the porphyrin periphery and distortions of the porphyrin ligands.

A particular feature of transition metal porphyrin chemistry is that the energies of the metal *d* orbitals and the frontier orbitals of the porphyrin ligand are often quite close, with the result that the redox chemistry of the porphyrin ligand and the

TABLE I

ABBREVIATIONS FOR PORPHYRIN DIANIONS AND TETRAPYRROLE ANIONS

Abbreviations for porphyrin dianions	
Por	Porphyrin dianion, unspecified
OEP	2,3,7,8,12,13,17,18-Octaethylporphyrin
TPP	5,10,15,20-Tetraphenylporphyrin
TTP	5,10,15,20-Tetra-*p*-tolylporphyrin
OETPP	2,3,7,8,12,13,17,18-Octaethyl-5,10,15,20-tetraphenylporphyrin
TAP	5,10,15,20-Tetra-*p*-methoxyphenylporphyrin
T*p*CF_3PP	5,10,15,20-Tetra-*p*-trifluoromethylphenylporphyrin
T*p*ClPP	5,10,15,20-Tetra-*p*-chlorophenylporphyrin
TMP	5,10,15,20-Tetramesitylporphyrin
TTEP	5,10,15,20-Tetrakis(1,3,5-triethylphenyl)porphyrin
TTiPP	5,10,15,20-Tetrakis(1,3,5-triisopropylphenyl)porphyrin
TXP	5,10,15,20-Tetraxylylporphyrin
Abbreviations for anions of tetrapyrrole macrocycles	
Pc	Phthalocyanine (dianion)
EtioPc	2,7,12,17-Tetraethyl-3,6,13,16-tetramethylporphycene (dianion)
OEC	2,3,7,8,12,13,17,18-Octaethylcorrole (trianion)
OETAP	Octaethyltetraazaporphyrin (dianion)

coordinated metal become entwined. Fine tuning by altering peripheral substitution on the porphyrin ligand can have a dramatic effect on both the site of initial electron transfer (kinetic control) and the ultimate product of redox processes (thermodynamic control). As a very simple example, the two most commonly employed porphyrin ligands are the dianions of tetraphenylporphyrin (TPP) and octaethylporphyrin (OEP). The complementary substitution patterns of the two ligands result in different redox potentials, with OEP complexes typically reduced at potentials *ca.* 200 mV more negative than their TPP counterparts. The commonly used porphyrin ligands and their abbreviations are given in Table I. The numbering system for the periphery of the porphyrin macrocycle, and sketches of the two most common porphyrin ligands, H_2OEP and H_2TPP, are shown in Fig. 2. Less common porphyrins will be defined as they are encountered in the text.

Porphyrin complexes, besides being amenable to the usual structural and spectroscopic tools favored by organometallic chemists, have characteristics that are particularly suited to certain physical methods. The porphyrin ligand itself absorbs strongly in the visible and near UV regions, and perturbations of the bands in the electronic absorption spectrum can give useful information about the local electronic environment. The aromatic ring current associated with the macrocycle has an effect on the NMR chemical shift of axial ligands, which usually exhibit marked upfield shifts in diamagnetic complexes. In paramagnetic complexes, analysis of the chemical shifts of both the porphryin and the axial ligands can give useful information about the location of unpaired spin density. Six-coordinate

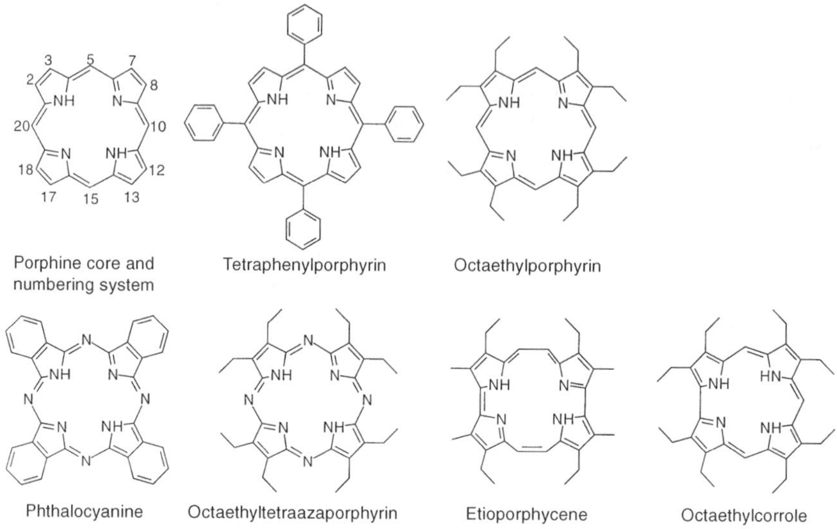

FIG. 2. Porphyrin and tetrapyrrole macrocycles.

porphyrin complexes with *trans* geometry and two equivalent axial ligands display idealized D_{4h} symmetry. This reduces to C_{4v} in five-coordinate complexes and in six-coordinate complexes containing two different axial ligands. This can often be identified clearly in the ^{1}H NMR spectra of OEP complexes, where the CH_2 protons of the peripheral ethyl groups become diastereotopic and display a characteristic AB pattern. Similarly, the *ortho* and *meta* protons on the peripheral phenyl groups in TPP complexes are in chemically different environments (assuming rotation of the phenyl rings is slow on the NMR time scale) and two sets of resonances are observed in the ^{1}H NMR spectrum. EPR spectroscopy is useful for paramagnetic complexes, and g values can be used to determine whether unpaired spin is located on the metal or the porphyrin ring.

General approaches to the synthesis of organometallic porphyrin complexes have been described well in the original three review articles. By far the bulk of the chemistry has been reported for the groups 8 (Fe, Ru, Os) and 9 (Co, Rh, Ir) elements. The organometallic chemistry of iron and cobalt has been established for the longest time, and σ-alkyl complexes are generally formed using one of two methods: a metal(III) halide M(Por)X with an alkyl or aryl Grignard or lithium reagent, or a metal(I) anion $M(Por)^{-}$ with an alkyl or aryl halide. Although these methods are also useful for rhodium, the major developments in the organometallic chemistry of ruthenium, osmium, rhodium, and iridium complexes have paralleled the preparation and exploration of the chemistry of the M(II) dimers $[M(Por)]_2$. The corresponding Mo and W dimers are also useful in this context. The structural and spectroscopic features of these dimers is an interesting story in its own right, and has been reviewed recently.[22]

Organometallic complexes are known which contain a wide variety of macrocycles closely related to porphyrins: corroles, porphycenes, tetraazaporphyrins, phthalocyanines, tetraazaannulenes, and porphyrinogens are the best known examples. Examples containing the first four of these macrocycles are included where there is a useful comparison or contrast with the relevant porphyrin chemistry, but the discussion will not be comprehensive. The four macrocycles are shown in Fig. 2.

A group-by-group descriptive approach suffers from being somewhat pedestrian, but given the huge scope of the chemistry covered, and the big differences in the chemistry even of elements within the same group, this approach to organization of the review is the most user-friendly. The last section comprises a short discussion on non-covalent interactions between metalloporphyrins and organic molecules, an area which is destined to become more significant with the current growth of supramolecular chemistry.

II

THE EARLY TRANSITION METALS (GROUPS 3 AND 4) AND LANTHANIDES

A. *Overview*

One class of compounds which simply did not feature the last time organometallic porphyrin complexes were reviewed were organometallic porphyrin complexes containing the early transition metals.[5,7,8] An important breakthrough which greatly facilitated progress in this area was the development of reliable synthetic routes to the alkali metal complexes $M_2(Por)L_n$ (M = Li, Na, K; L = coordinating solvent).[23,24] These could then be used as precursors to the early transition metal porphyrin halides through simple salt elimination routes. This eliminated the need for the strongly polar or acidic solvents, high temperatures, and chromatographic purification commonly required in conventional methods for insertion of metals into the porphyrin macrocycle, and which would not suit the oxophilic early transition elements. These advances in synthetic methods have allowed the organometallic chemistry of the groups 3, 4, and 5 elements scandium, titanium, zirconium, hafnium, and tantalum to flourish, resulting in a variety of complexes with unprecedented ligand types, stoichiometries, and structures. Recent developments in the chemistry of early transition metal porphyrin compounds were reviewed in 1995.[25]

Some of the key differences between the early transition metals and their later metal counterparts are the tendency of the elements to form compounds in their highest oxidation states, the relatively large size of the elements, and the wider range of coordination numbers observed. Coordination numbers ranging from 5 to 8 are encountered for the early metal porphyrin complexes, compared to 4, 5, and 6 commonly observed for the later metals. In addition, large out-of-plane displacements are often observed for the central metal, reflecting the larger size of the elements. Finally, preferences for different geometries, for example *cis* coordination of the fifth and sixth donor groups in six-coordinate complexes, are quite commonly observed. Coordination number 7 is found in a variety of tris-μ-oxo or -hydroxo complex, for example $\{[Zr(OEP)]_2(\mu\text{-}OH)_3\}^+$,[26] and a number of structural types are found with coordination number 8 including $Zr(TPP)(OAc)_2$[27] which contains two η^2-acetate ligands on the same face of the porphyrin, the bisporphyrin zirconium, and hafnium complexes $M(Por)_2$,[28,29] and the very unusual bridging dichalcogenide complexes $[M(TPP)(E_2)]_2$ (M = Zr, Hf; E = S, Se)[30] which contain two bridging E_2 units between each pair of metalloporphyrin units.

Bis(cyclopentadienyl) complexes are central to the organometallic chemistry of the early transition metals and feature in applications such as alkene polymerization chemistry. Parallels can be drawn between a porphyrin ligand and two cyclopentadienyl ligands, in that they both contribute a 2− formal charge and exert a considerable steric influence on other ligands in the same molecule. Several of the metalloporphyrin complexes discussed below have bis(cyclopentadienyl) counterparts, and authors in some cases have drawn quite detailed comparisons, although these discussions will not be repeated here.

To date, the only organometallic lanthanide porphyrin complexes to be reported contain yttrium and lutetium, and they will be considered in the section on scandium. Representative structural types of porphyrin complexes containing groups 3 and 4 metals are shown in Fig. 3 and selected data for all the structurally characterized complexes are given in Table II.

B. *Scandium, Yttrium, and Lutetium*

The first organometallic scandium porphyrin complex to be prepared was of particular significance as it represented the first example of a porphyrin complex containing a π-cyclopentadienyl ligand. The first report was the result of the reaction of Sc(TTP) with LiCp*, giving a complex for which NMR data alone was collected, but which showed a resonance at −0.42 ppm attributed to the Cp* protons, shifted markedly upfield by the porphyrin ring current.[31] This was confirmed by the preparation and structural characterization of $Sc(OEP)(\eta^5\text{-}Cp)$, which showed the π-bonded Cp ring coordinated to scandium (Fig. 3).[31] The Sc-C(av) distance is 2.494(4) Å, with the scandium atom displaced from the N_4 plane toward the Cp ring by 0.80 Å. This complex, along with the Cp* and C_5H_4Me analogues which were also prepared, is air stable and sublimable and surprisingly inert to hydrolysis.[32,33] A further example, the indenyl complex $Sc(OEP)(\eta^5\text{-}C_9H_7)$

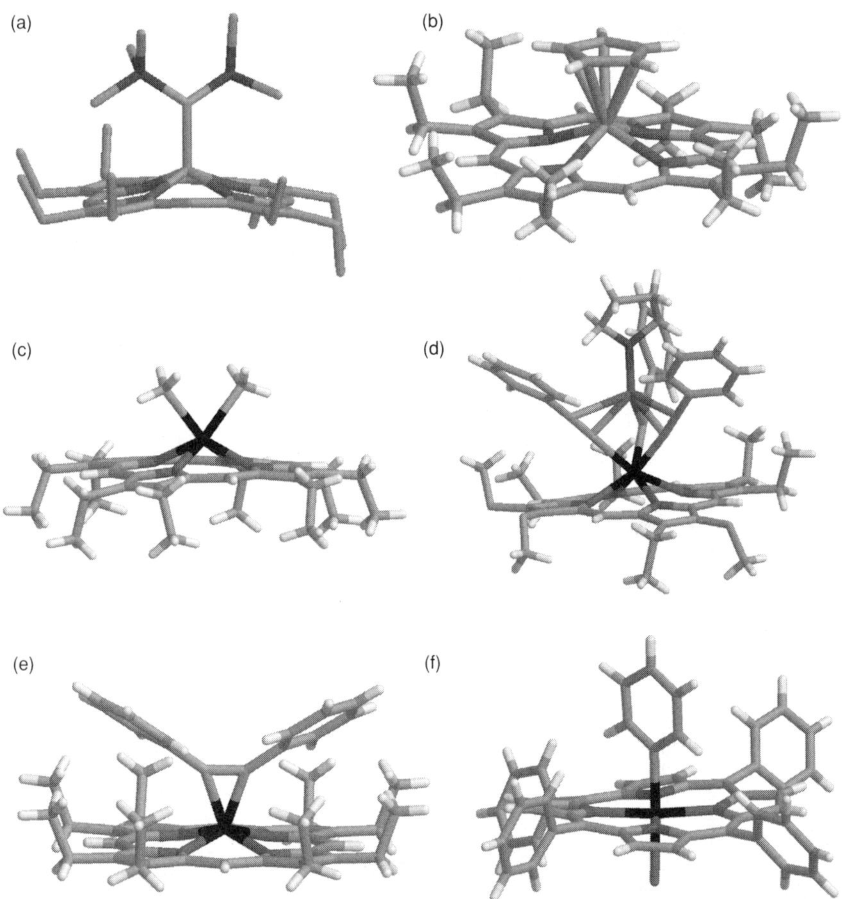

FIG. 3. Molecular structures of selected organometallic groups 3, 4, and 6 porphyrin complexes: (a) $Sc(OEP)CH(SiMe_3)_2$,[33] (b) $Sc(OEP)(\eta^5\text{-}Cp)$,[32] (c) $Zr(OEP)Me_2$,[48] (d) $Zr(OEP)(C{\equiv}CPh)_3Li(THF)$,[52] (e) $Ti(OEP)(\eta^2\text{-}PhC{\equiv}CPh)$,[36] (f) Mo(TPP)(Ph)Cl.[59]

(prepared from Sc(OEP)Cl with NaC_9H_7) contains a tilted, slipped η^5-indenyl ring with slightly longer Sc-C distances, ranging from 2.504(8)–2.580(8) Å, and a similar Sc-N_4 plane displacement of 0.78 Å.[33]

The scandium alkyl complexes Sc(OEP)R, prepared from Sc(OEP)Cl with $MgMe_2$, $Mg(CH_2CMe_3)_2$, or $LiCH(SiMe_3)_2$, are much more air and moisture sensitive than their cyclopentadienyl counterparts, arising perhaps from their electron poor nature, being formally only 10-electron complexes. The molecular structures of Sc(OEP)Me and $Sc(OEP)CH(SiMe_3)_2$ (Fig. 3) showed similar Sc—C bond lengths (2.246(3) and 2.243(8) Å) and Sc-N_4 out-of-plane distances (0.66 and

TABLE II

SELECTED DATA FOR STRUCTURALLY CHARACTERIZED COMPLEXES FROM GROUPS 3, 4 AND THE LANTHANIDES

	M–C Bond Length/Å	M–N(av) Bond Length/Å	$M-N_4$ Plane/Å	Other	Reference
Sc(OEP)(η^5-Cp)	2.494(4)	2.190(3)	0.80		32
Sc(OEP)(η^5-C_9H_7)	2.499(8)–2.580(8)	2.182(6)	0.78		33
Sc(OEP)Me	2.246(3)	2.155(2)	0.66		33
Sc(OEP)CH($SiMe_3$)$_2$	2.243(8)	2.163(6)	0.71		33
Lu(OEP)CH($SiMe_3$)$_2$	2.374(8)	2.260(7)	0.918		34
Ti(OEP)(η^2-Ph≡C(CPh)[a]	2.017(7)(av)	2.095(8)	0.54	C≡C, 1.30(1)Å C–C≡C, 142.4(6)°	36
Zr(OEP)Me_2	2.343(4)(av)	2.231(3)	0.89	C–Zr–C, 78.21(12)°	48
Zr(OEP)(CH_2SiMe_3)$_2$	2.287(4)(av)	2.242(3)	0.93	C–Zr–C, 82.7(1)° Zr–C–Si, 125.5(2)°	46
Zr(TPP)(η^5-Cp)	2.49(2)(av)	2.177(10)	0.773		55
Zr(OEP)(η^5-1,2-$C_2B_9H_{11}$)	2.586(9) 2.549(7)	2.214(4)	0.904	Zr–B, 2.503 (10)(av)Å	45, 51
Zr(OEP)(C≡CPh)$_3$Li(THF)	2.323(10)(av)	2.259(7)	1.008	C≡C, 1.218(12)	52
Zr(OEP)(CH_2SiMe_3)	2.216(8)	2.150(6)	0.63	Zr–C–Si, 124.1(4)°	54
Zr(OEP)(η^2-PhC≡CPh)	2.160(4)	2.204(4)	0.90	C≡C, 1.333(8) C–C≡C, 133.4(3)	55
Mo(TPP)(Ph)Cl	2.241(1)	2.070(3)	0.089[b]	Mo–Cl, 2.382Å	59
Mo(TPP)(η^2-PhC≡CPh)	1.974(4)		0.63		60

[a] Data for two independent molecules averaged.
[b] Mo displaced toward the Cl ligand.

0.71 Å).[33] The alkyl complexes react rapidly with CO and isocyanides, but pure products could not be isolated. Fast reactions of CO_2 and acetone with Sc(OEP)Me gave the well-characterized acetate and *t*-butoxide products, Sc(OEP)OAc and Sc(OEP)O-*t*-Bu, respectively, formed by insertion into the Sc—C bond.[33]

Organometallic yttrium and lutetium complexes are prepared from free base H_2OEP with $M(CH(SiMe_3)_2)_3$ or $M(OAr)_3$ (M = Y, Lu; R = 2,6-C_6H_3-t-Bu_2) to form the alkyl or aryloxide complexes M(OEP)$CH(SiMe_3)_2$ or M(OEP)OAr.[34] A crystal structure of Lu(OEP)$CH(SiMe_3)_2$ shows a highly dished porphyrin with a Lu—C distance of 2.374(8) Å and the lutetium atom displaced from the N_4 plane by 0.918 Å. The aryloxide and alkyl complexes can be interconverted by treatment with $LiCHSiMe_3$ or HOAr, as appropriate, and the alkyl complexes are readily hydrolyzed to the bis-μ-hydroxo dimers $[M(OEP)]_2(\mu\text{-}OH)_2$. Protonolysis of the alkyl complexes using *t*-BuC≡CH gives dimeric alkynyl complexes, $[M(OEP)(C{\equiv}C\text{-}t\text{-}Bu)]_2$. The alkyl complexes do not react with hydrogen to give the hydride complex, unlike the Cp_2MR counterparts which do react with H_2 to produce Cp_2MH.[34]

Y(OEP)(OAr) reacts with two equivalents of MeLi to give the Y,Li-bridged complex Y(OEP)$(\mu\text{-}Me)_2Li(OEt_2)$, essentially an adduct of Y(OEP)Me with LiMe·OEt_2. If just one equivalent of MeLi is used in the reaction, a reduced yield of the same product is formed, illustrating the strong tendency in this system to form "ate" complexes. The coordinated MeLi can be removed by treatment with $AlMe_3$, this time producing the Y,Al-bridged complex Y(OEP)$(\mu\text{-}Me)_2AlMe_2$. Even at −60°C, only one peak is observed in the ^1H NMR spectrum for all four Al—CH_3 groups in this molecule, indicating that the $AlMe_4$ moiety is highly fluxional on the NMR time scale. The bridging (but not the terminal) methyl groups insert oxygen when the compound is exposed to O_2, forming Y(OEP)$(\mu\text{-}OMe)_2AlMe_2$ which shows distinct signals for the O—CH_3 and Al—CH_3 groups in the ^1H NMR spectrum.[34,35] The chemistry of the yttrium complexes is summarized in Scheme 1.

C. *Titanium*

Organometallic titanium porphyrin complexes center around a single class of compound containing a Ti(II) porphyrin coordinated to an η^2-alkyne ligand. For the early transition metals, low valent organometallic porphyrin species are unusual, with complexes in the highest available oxidation state predominating for the other groups 3, 4, 5 and lanthanide porphyrin complexes surveyed here. For example, Zr(IV) porphyrin dialkyl complexes feature strongly, but the Ti(IV) analogues are as yet unknown. The titanium alkyne complexes complement the only other porphyrin alkyne complexes known, Mo(Por)(PhC≡CPh).

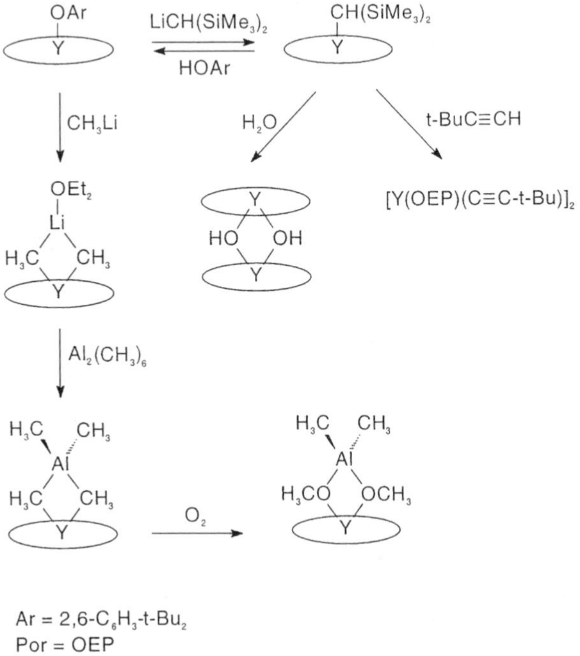

Ar = 2,6-C_6H_3-t-Bu_2
Por = OEP

SCHEME 1.

Reduction of $Ti(Por)Cl_2$ with excess $LiAlH_4$ in the presence of an alkyne produces the series of alkyne complexes Ti(Por)(RC≡CR), where Por is OEP or TTP, and the alkyne is diphenylacetylene, 2-butyne, 2-pentyne, or 3-hexyne. An alternative synthesis begins with reduced titanium porphyrin complexes, treating Ti(TTP)Cl with $NaBEt_3H$ and alkyne, or $Ti(TTP)(THF)_2$ with alkyne alone. The complexes are diamagnetic and the porphyrin ligands retain their four-fold symmetry even at low temperature indicating that there is rapid rotation of the coordinated alkyne ligand. The extremely oxophilic titanium center reacts rapidly with O_2 to give the titanium oxo complex, Ti(Por)=O.[36–38] A crystal structure of Ti(OEP)(PhC≡CPh) (Fig. 3) reveals Ti—C distances averaging 2.017 Å (two independent molecules) and displacement of the titanium atom from the N_4 plane of 0.54 Å. The C≡C distance and C—C≡C angle in the alkyne are 1.30(1) and 142.1(6)°, respectively. These data, together with IR and ^{13}C NMR results are consistent with the alkyne ligand donating four electrons, as is the case for the alkyne ligand in Mo(TTP)(PhC≡CPh), and contrasts with $Cp_2Ti(CO)(PhC{\equiv}CPh)$ for which the 18-electron rule requires that the alkyne is a two-electron donor.[36]

The alkyne ligand is not displaced by CO or ethene but is substituted by stronger σ-donor ligands like pyridine and 4-picoline, which give the paramagnetic complexes *trans*-Ti(TTP)$(L)_2$. Diphenylacetylene and terminal alkynes will displace 2-butyne from Ti(TTP)(MeC≡CMe), perhaps because the stronger π-acids will bind more readily to the electron rich Ti(II) center. This method has been used to prepare the phenylacetylene and ethyne complexes Ti(TTP)(PhC≡CH) and Ti(TTP)(HC≡CH).[37] The 3-hexyne complex Ti(TTP)(EtC≡CEt) in C_6D_6 with excess THF exists in equilibrium with Ti(TTP)$(THF)_2$ ($K_{eq} = 3.8$) but is quantitatively converted to the bis(isocyanide) complex with two equivalents of CN-*t*-Bu. These observations have allowed the relative preference for binding of a neutral ligand to the Ti(TTP) moiety to be established as follows: pyridine ~ picoline > CNR > PhC≡CPh > EtC≡CEt > THF.[38]

Ti(Por)(RC≡CR) acts as a source of the Ti(Por) fragment in atom transfer reactions, accepting a chalcogenide atom from E=PPh_3 (E = S, Se) to give Ti(Por)=E, and reacting with elemental sulfur or selenium to give Ti(Por)(η^2- E_2), which in turn react with Ti(Por)(RC≡CR) to give two equivalents of Ti(Por)=E. Ti(Por)(η^2-O_2) behaves similarly, and the oxo complex Ti(Por)=O forms the new, paramagnetic μ-oxo binuclear species $[Ti(Por)]_2(\mu\text{-}O)$ upon reaction with Ti(Por)(RC≡CR).[37,39,40] Imine and organic azide reagents transfer an imido fragment to Ti(TTP)(EtC≡CEt), giving Ti(TTP)=NPh or Ti(TTP)=$NSiMe_3$ from PhN=NPh or N_3SiMe_3, respectively.[41,42] Diazoalkanes act as carbene sources with group 8 metal porphyrins, giving M(Por)=CR_2 (M = Fe, Ru, Os), an interesting contrast with the reaction of Ti(TTP)(PhC≡CPh) with N_2CAr_2 which gives not the carbene but rather the hydrazido complex (Ti(TTP)=N–N=CAr_2 (Ar = 4-C_6H_4Me), reflecting perhaps the preference of an early transition metal for the harder nitrogen donor atom.[37]

The Ti(Por) fragment is formally acting as a reductant in the atom transfer reactions, and this can occur in a one-electron or two-electron sense. The former is illustrated by the reductive dechlorination of vicinal dihalo-substituted organic substrates, in which Ti(TTP)(EtC≡CEt) reacts with 1,2-dichloroethane, 1,2-dichlorocyclohexane, or 1,2-dichloroethene to give ethene, cyclohexene, or ethyne, respectively, together with Ti(TTP)Cl as the inorganic product. Two-electron reductions occur in the reactions of Ti(TTP)(EtC≡CEt) with epoxides or alkyl- or arylsulfoxides (RS(O)R), producing Ti(TTP)=O and alkenes or sulfides (RSR), respectively.[43] The reactions of the titanium alkyne complexes are summarized in Scheme 2.

D. *Zirconium and Hafnium*

The first organometallic zirconium complexes to be prepared all contained Zr(IV) and were reported almost contemporaneously.[44–46] The reaction of Zr(TPP)-$(OAc)_2$ with RLi or RMgBr produced the dialkyl complexes Zr(TPP)R_2 (R = Me, Et, *n*-Bu or Ph), characterized by spectroscopy.[44,47] The development of the chlorozirconium complexes opened up the chemistry further, with

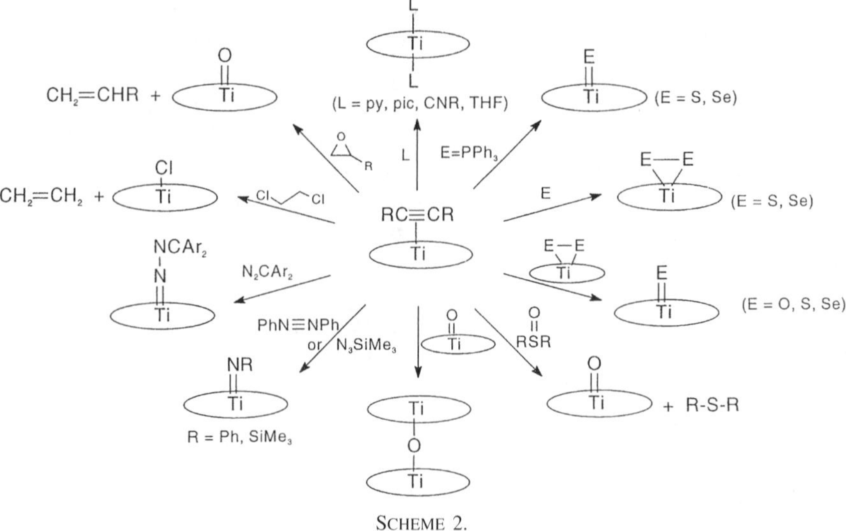

SCHEME 2.

$Zr(TPP)Cl_2(THF)$ and $Zr(OEP)Cl_2$ (prepared from $Li_2(OEP)(THF)_4$ with $ZrCl_4L_2$ where L = THF or tetrahydrothiophene) reacting with $Li(p\text{-}C_6H_4\text{-}t\text{-}Bu)$ to give $Zr(OEP)(p\text{-}C_6H_4\text{-}t\text{-}Bu)_2$, and with LiR or MgR_2 to give the dialkyl complexes $Zr(Por)Me_2$ (Por = OEP, TPP), $Zr(OEP)Et_2$ and $Zr(OEP)(CH_2SiMe_3)_2$.[45,46,48] $Hf(TPP)Me_2$ was prepared similarly from $Hf(TPP)Cl_2$ with MeLi.[49,50] All of the complexes were assigned *cis* geometry on the basis of the lack of symmetry in the plane containing the porphyrin as evidenced by the diastereotopic CH_2 protons observed for the OEP ethyl groups. This was confirmed by X-ray crystal structures for $Zr(OEP)Me_2\cdot CH_2Cl_2$ (Fig. 3) and $Zr(OEP)(CH_2SiMe_3)_2$ which showed Zr—C(av) bond lengths of 2.343(4) and 2.287(4) Å, Zr—N_4 plane displacements of 0.89 and 0.93 Å, and C—Zr—C angles of 82.7(1) and 78.21(12)°, respectively.[46,48]

Zirconium(IV) π-complexes containing the η^8-cyclooctatetraenyl (COT) or η^5-dicarbolide anions, $Zr(OEP)(\eta^8\text{-COT})$ or $Zr(OEP)(\eta^5\text{-}1,2\text{-}C_2B_9H_{11})$, were prepared from $Zr(OEP)Cl_2$ with the salts K_2COT or $Na_2C_2B_9H_{11}$, respectively. An X-ray crystal structure of the dicarbolide complex shows the η^5-coordination through the planar C_2B_3 face (Zr—C(av) 2.57 Å, Zr—B(av) 2.50 Å, Zr—N_4 plane 0.904 Å) and can be considered an analogue of the scandium Cp complex Sc(OEP)-(η^5- Cp).[45,46,48,51]

The tendency to form "ate" complexes observed for yttrium is also seen in the preparation of a novel zirconium alkynyl complex. Reaction of $Zr(Por)Cl_2$ (Por = OEP or TPP) with either two or three equivalents of LiC≡CR (R = Ph or $SiMe_3$) gives the trisalkynyl complexes $Zr(Por)(C\equiv CR)_3Li(THF)$. An X-ray crystal structure of $Zr(OEP)(C\equiv CPh)_3Li(THF)$ shows that all three alkynyl ligands are on the same face of the molecule, with the Li atom interacting with the six alkynyl carbon atoms, and the THF oxygen in turn coordinated to the Li along the Li···Zr axis (Fig. 3). The average Zr—C and C≡C distances are 2.323(10) and 1.218(12) Å and the Zr—N_4 plane displacement is very large, 1.008 Å.[52]

All of the complexes containing Zr—C and Hf—C bonds are extremely sensitive to oxygen, water, and light, and the aryl complex Zr(OEP)(p-C_6H_4-t-Bu$)_2$ is so sensitive it cannot even tolerate normal room light. The complexes are more stable in solution in non-coordinating solvents, as coordinating solvents like THF, pyridine, or ether promote decomposition. The products of hydrolysis are the oxo- or hydroxo-bridged complexes [M(Por)$]_2$ (μ-O)(μ-OH$)_2$ (M = Zr, Hf), {[Zr(Por)$]_2$ (μ-OH$)_3\}^+$ or [Zr(Por)$]_2$(μ-OH$)_4$.[45,47,48,50] Reaction of the dialkyl complexes with acetic acid produces Zr(OEP)(OAc$)_2$, which can also be produced by reaction of Zr(OEP)Me_2 with CO_2. Zr(OEP)Me_2 reacts with acetone to give the alkoxide complex Zr(OEP)(OC$Me_3)_2$. The diethyl complex Zr(OEP)Et_2 slowly decomposes in solution to generate ethane and ethene.[48] Protonolysis of the OEP (but not the TPP) complexes M(OEP)($CH_2SiMe_3)_2$ using the ammonium salt [HNMe_2Ph][BPh_4] produces the cationic Zr and Hf complexes [M(OEP)(CH_2SiMe_3)$]^+$ as the BPh_4^- salts.[53]

The trisalkynyl complexes Zr(OEP)(C≡CPh$)_3$Li(THF) react with anhydrous HCl in THF to produce quantitatively HC≡CPh and the C—C coupled organic product CH_2=C(Ph)C≡CPh along with Zr(OEP)Cl_2. In the presence of an added alkyne HC≡CR′ no cross-coupling was observed, indicating that the coupling reaction is likely to be intramolecular. This kind of reactivity has also been observed for the related zirconocene complex Li[Cp_2Zr(C≡CPh$)_3$] with HCl.[52]

The silylmethyl complex Zr(OEP)($CH_2SiMe_3)_2$ does not react directly with ethene, but does react with ethene when hydrogen is added, and catalyzes the production of ethane. Formation of the ethyl complexes Zr(OEP)(Et)(CH_2SiMe_3) and Zr(OEP)Et_2 was observed by NMR during the course of the reaction, but no direct spectroscopic evidence for intermediate zirconium hydride complexes could be obtained. Similar reactivity occurred for propene, with only n-propyl and no isopropyl zirconium products formed.[46] In the absence of the alkene, the zirconium complex degrades to produce a green paramagnetic compound which has been identified as the Zr(III) complex Zr(OEP)CH_2SiMe_3.[54] This is unusual for two reasons. First, there are very few alkylzirconium(III) complexes known, and second, reduction of Zr(IV) to Zr(III) by H_2 was unprecedented. The compound can be prepared directly from treatment of Zr(OEP)($CH_2SiMe_3)_2$ with H_2 in toluene, and was shown by X-ray crystallography to contain five-coordinate zirconium with a Zr—C bond length of 2.216(8) Å and the Zr atom displaced from the N_4 plane by 0.63 Å. Decrease of both these distances relative to the six-coordinate precursor is consistent with relief of steric strain. To describe Zr(OEP)CH_2SiMe_3 as a Zr(III) complex is an oversimplification of its electronic structure. ^{1}H NMR, EPR, and UV-visible spectra, together with small changes in the N—C and C—C bond lengths in the porphyrin core, provide evidence that there is spin density associated with both the Zr atom and the *meso* carbon atoms of the porphyrins. This indicated that the complex may have some porphyrin radical anion character and that its electronic structure can be described by two resonance forms, $Zr^{III}(OEP)CH_2SiMe_3$ and $Zr^{IV}(OEP^{\cdot-})CH_2SiMe_3$.[54]

Two more very important low valent organozirconium complexes have been reported recently. Reaction of the trisalkynyl complex $Zr(TPP)(C{\equiv}CPh)_3Li(THF)$ with $Cp_2Ti(Me_3SiC{\equiv}CSiMe_3)$ produced, along with other products, the paramagnetic cyclopentadienyl zirconium complex $Zr(TPP)(\eta^5\text{-}Cp)$. This can also be directly prepared from $Zr(TPP)Cl_2$ with TlCp using Na/Hg amalgam as the reductant. A crystal structure of $Zr(TPP)(\eta^5\text{-}Cp)$ shows the Zr—C(av) bond length to be 2.49(2) Å, and the Zr—N_4 plane distance to be 0.773(5) Å. The EPR and UV-visible spectra of this compound are very similar to those reported for $Zr(OEP)CH_2SiMe_3$, suggesting that this compound too might have Zr(III) and Zr(IV) porphyrin radical anion character.[55]

The only zirconium(II) compound to be reported is analogous to the titanium porphyrin alkyne complexes, and was prepared from $Zr(OEP)Cl_2$ with diphenylacetylene and magnesium metal as the reductant. The diamagnetic product, $Zr(OEP)(PhC{\equiv}CPh)$, like the titanium complex, contains a formally four-electron donor alkyne ligand. The Zr—C bond length and Zr—N_4 plane distances, 2.160(4) and 0.90 Å, respectively, are larger than those in the titanium analogue as would be expected for the larger zirconium atom. The C≡C distance of 1.333(8) Å is very similar, but the Ph—C≡C angle of 133.4(3) Å indicates that the alkyne ligand in the zirconium complex is slightly more bent than in the titanium complex.[55]

Zirconium porphyrin dicarboxylates have been shown to catalyze the stereo- and regioselective ethylalumination of alkynes. Although little is known about likely reaction intermediates, NMR evidence for ethylzirconium complexes was observed for the products of the reaction of $Zr(TPP)(O_2C\text{-}t\text{-}Bu)_2$ with $AlEt_3$, attributed to $Zr(TPP)(O_2C\text{-}t\text{-}Bu)Et$ and $[Zr(TPP)(\mu\text{-}O_2C\text{-}t\text{-}Bu)(\mu\text{-}Et)]AlMe_2$.[56]

III

THE MIDDLE TRANSITION ELEMENTS (GROUPS 5, 6, AND 7)

A. *Overview*

The last decade has seen the development of a rich and varied chemistry for organometallic porphyrin complexes of the early transition metals (groups 3 and 4). However, there have been many fewer developments in the organometallic chemistry of the middle transition elements. Despite the paucity of its organometallic porphyrin compounds, molybdenum has played a very important role in organometallic porphyrin chemistry. The diphenylacetylene complex Mo(TPP)-(η^2-PhC≡CPh), first reported in 1981, was the first example of a porphyrin η^2-alkyne complex, and this class is still limited to a handful of compounds containing molybdenum, tungsten, titanium, or zirconium.

At the time this area was last reviewed the molybdenum complex was also the only example of an organometallic porphyrin complex from groups 5, 6, or 7.[5,7,8] Since then, only a small number of new organometallic porphyrin complexes from these groups have been reported. These include tantalum(V) complexes, which are most closely related to the high valent zirconium and hafnium complex, a tungsten diphenylacetylene complex, three examples of chromium(III) σ-aryl or σ-alkyl complexes, a molybdenum(IV) phenyl complex, and a molybdenum dichlorocarbene complex. The molybdenum and tungsten alkyne complexes are closely related to the titanium and zirconium examples, all with the formula M(Por)(η^2-PhC≡CPh), and all formally in the +2 oxidation state. The chromium and molybdenum σ-bonded complexes and the manganese carbene have more parallels to the later transition metal complexes containing these ligand types. Organometallic porphyrin complexes containing vanadium, niobium, technetium, or rhenium are as yet unknown. X-ray data for the two structurally characterized molybdenum complexes are included in Table II along with the early transition metal complexes.

B. *Tantalum*

Tantalum porphyrins are rare and there has been only one report of their organometallic derivatives.[57] Treatment of Li_2 (OEP)$(THF)_4$ with $TaMe_3Cl_2$ in CH_2Cl_2 at room temperature gave a good yield of Ta(OEP)Me_3. The series of ^{1}H NMR chemical shifts for the methyl groups in Sc(OEP)Me, Zr(OEP)Me_2, and Ta(OEP)Me_3 (−4.0, −4.5, and −6.0 ppm) illustrates nicely the decreasing effect of the ring current as the ligands move farther away from the center of the porphyrin. Although the compound was not characterized structurally, the lack of mirror symmetry in the porphyrin plane (as evidenced by NMR) indicates that, like the zirconium complexes, the methyl groups are likely to lie on one face of the compound. Ta(OEP)Me_3 reacts only slowly with water to form the tris-μ-oxo complex $[Ta(OEP)]_2(\mu\text{-}O)_3$, in contrast to the extremely water-sensitive zirconium alkyl complexes. Unlike the scandium and zirconium complexes, the tantalum complex does not insert CO_2 or acetone into the Ta—Me bond. However, the compound is moderately light sensitive.

A more reactive cationic dimethyl tantalum derivative is produced from Ta(OEP)Me_3 using one equivalent of the weak acid $[HNMe_2Ph][BPh_4]$. $[Ta(OEP)Me_2]^+$ reacts cleanly with CO to produce a cationic enediolate compound, containing the MeC(O^-)=C(O^-)Me ligand which results from both insertion and

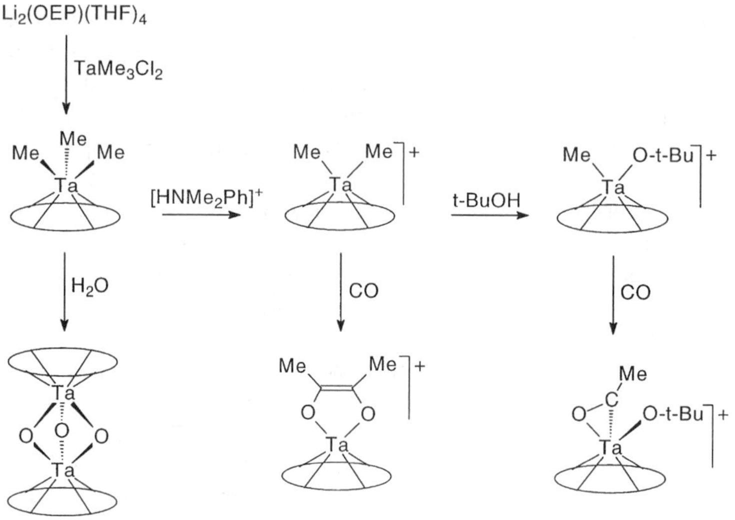

SCHEME 3.

coupling reactions, in a process well known in cyclopentadienyl zirconocene and tantalocene systems. Treatment of $[Ta(OEP)Me_2]^+$ with *t*-butanol gives $[Ta(OEP)Me(O\text{-}t\text{-}Bu)]^+$, which then reacts with CO to form $[Ta(OEP)(\eta^2\text{-}C(O)Me)(O\text{-}t\text{-}Bu)]^+$. The enediolate and η^2-acyl products are proposed on the basis of their NMR spectra, as neither could be isolated in a pure form.[57] Reactions of tantalum porphyrins are shown in Scheme 3.

C. *Chromium, Molybdenum, and Tungsten*

The reaction of Cr(TTP)Cl with PhLi, $MgBr(p\text{-}t\text{-}BuC_6H_4)$, or $MgCl(CH_2SiMe_3)$ produced Cr(TTP)Ph, $Cr(TTP)(p\text{-}t\text{-}BuC_6H_4)$ or $Cr(TTP)(CH_2SiMe_3)$, respectively.[58] The Cr(III)complexes are paramagnetic with $S = 3/2$, and the broadened NMR spectra show temperature dependent shifts which follow Curie behavior. Unlike the earlier transition metals, the compounds can be purified by chromatography on a short plug of alumina. The complexes are moderately air stable in the solid state but very air sensitive in solution. For example, Cr(TTP)Ph reacted with O_2 in dry $CDCl_3$ in the dark to give PhD together with the diamagnetic oxo complex Cr(TTP)=O. A radical mechanism involving Cr—C bond homolysis was proposed for this reaction. The aryl complexes react with acetic acid, $HgCl_2$, or I_2 to give Cr(TTP)X (X = O_2CCH_3, Cl or I).[58]

A single example of a σ-alkyl molybdenum complex was serendipitously prepared from a solution of the very light-sensitive dicarbonyl complex $Mo(TPP)(CO)_2$.

During the course of an attempted recrystallization of this complex from benzene containing chlorinated impurities the solution was exposed to light. The crystalline compound that formed from this solution was identified by X-ray crystallography as Mo(TPP)(Ph)Cl (Fig. 3). The complex contains *trans* phenyl and chloro ligands, and the Mo—C and Mo—N_4 plane distances are 2.241(1) and 0.125 Å, respectively. A systematic synthesis of the complex could not subsequently be developed and consequently other spectroscopic and magnetic data were not collected.[59]

The diamagnetic Mo(II) diphenylacetylene complex Mo(TPP)(η^2-PhC≡CPh) was prepared by reduction of Mo(TPP)Cl_2 by $LiAlH_4$ in toluene in the presence of excess diphenylacetylene.[60] The OEP and TTP analogues containing both molybdenum and, more recently, tungsten, have been prepared similarly.[61,62] The short Mo—C and long C≡C distances in Mo(TPP)(η^2-PhC≡CPh) were consistent with the alkyne ligand acting as a four-electron donor. This has been supported by a theoretical study, and contrasts with non-porphyrin complexes such as Cp_2Mo(HC≡CH) which are constrained by the 18-electron rule to contain a two-electron donor alkyne ligand.[63] The tungsten porphyrin complexes W(Por)(η^2-PhC≡CPh) are also expected to contain four-electron donor alkyne ligands.[62]

The molybdenum and tunsten diphenylacetylene compounds have been chemically useful primarily as precursors to the quadruple metal–metal bonded dimers $[M(Por)]_2$, formed by solid-state vacuum pyrolysis reactions.[61,62] However, Mo(TTP)(η^2-PhC≡CPh) is also a useful substrate in atom-transfer reactions, reacting with S_8 or Cp_2TiS_5 to form Mo(TTP)=S. The reaction can be reversed by treatment of Mo(TTP)=S with PPh_3 (which removes sulfur as Ph_3P=S) and PhC≡CPh. The order of preference for ligand binding to molybdenum(II) has been established to be PPh_3 < PhC≡CPh < 4-picoline.[64]

D. *Manganese*

Manganese porphyrin complexes have been widely used as catalysts for oxidation reactions, and the atom transfer properties of the high valent oxo, nitrido, and imido species have received a lot of attention. However, organometallic manganese porphyrin complexes are limited to one example, a dichlorocarbene complex prepared by the same method as that used for the iron porphyrin dichlorocarbene first reported by Mansuy in 1978. Mn(TPP)Cl was treated with CCl_4 in CH_2Cl_2 under a nitrogen atmosphere, using either iron powder or $NaBH_4$ as a reductant. The carbene carbon atom in the resulting complex, Mn(TPP)(=CCl_2), was observed at 264 ppm by ^{13}C NMR spectroscopy. Broad signals were observed for the porphyrin ring in the ^{1}H NMR spectrum, and this was attributed to the presence of low spin Mn(II). When only one equivalent of a reductant was used in the preparation, a compound assigned as Mn(TPP)CCl_3 was observed by NMR.[65]

IV

IRON

A. *Overview*

Iron porphyrins continue to be the most intensely scrutinized group of metalloporphyrin complexes, a fact which is no surprise given the importance of iron porphyrins in electron transfer systems, as oxygen carriers and as biological oxidation catalysts. Interest in organometallic iron porphyrins was sparked by the observation that iron σ-bonded alkyl or aryl complexes, and iron carbene complexes, are formed during cytochrome P_{450} metabolism of certain substrates. Reversible migration of a σ-bonded ligand to a pyrrole nitrogen also features in the deactivation of cytochrome P450, and has led to close scrutiny of the migration process.[6]

The synthesis, reactivity, spectroscopy, and electrochemistry of organometallic iron porphyrins was described in some detail in the three reviews published in the period from 1986 to 1988.[5,7,8] Although a brief synopsis of the early chemistry will be given here, this review will focus on more recent developments.

Iron(III) σ-bonded complexes containing alkyl or aryl ligands can be prepared by the two classical methods also widely used for the synthesis of Co(III) and Rh(III) σ-bonded porphyrins. The reaction of Fe(Por)X (X = halide) with RMgBr or RLi, or the reaction of electrochemically generated Fe(I) porphyrins, $[Fe(Por)^-]$ with organohalides, RX gives Fe(Por)R (R = alkyl, aryl, vinyl, etc.). Capture of alkyl radicals by iron(II) porphyrins is a further method of synthesis.

Complexes with the formula Fe(Por)R constitute a relatively small class of paramagnetic iron alkyl complexes containing 15 electrons at the iron center. The spin state of the iron atom in the d^5 Fe(III) complexes Fe(Por)R is affected by the properties of both the axial ligand R and the porphyrin. In general, Fe(Por)R where R is a simple alkyl or aryl ligand and Por = OEP or TPP are low spin, $S = 1/2$ complexes. Where $R = C_6F_5$ or C_6F_4H, high spin $S = 5/2$ complexes are observed. Low spin/high spin conversion can occur as a function of temperature or solvent. Useful tables of EPR data for σ-bonded organoiron porphyrins are given in two of the reviews.[7,8] NMR spectroscopy has also been used as a probe, with chemical shift differences useful for assigning both the spin state and the location of spin density. Mössbauer data for σ-bonded organoiron porphyrins have been collected and compared to non-porphyrin iron complexes.[7] The low-spin organoiron(III) complexes are generally very chemically reactive, being light and thermally sensitive, and susceptible to facile Fe—C bond homolysis.

One-electron oxidation of the Fe(III) complexes Fe(Por)R can either give an observable Fe(IV) complex, $[Fe(Por)R]^+$, or can induce migration of the R group to a pyrrole nitrogen, essentially an intramolecular reductive elimination process, giving the Fe(II) species $[Fe(Por\text{-}N\text{-}R)]^+$. The migration is reversible, and one-electron reduction of $[Fe(Por\text{-}N\text{-}R)]^+$ initiates the reverse reaction to re-form Fe(Por)R. Whether oxidatively induced migration or reversible oxidation occurs depends on a range of factors, including the nature of the R group. Migration was observed for phenyl and tolyl groups, but not for fluorophenyl (C_6F_5 or C_6F_4H) or methyl groups. The reasons for the different behavior were not well understood when this area was last reviewed.[8] For example, it was not believed to be a simple spin state phenomenon, since phenyl- and methyl-iron porphyrin complexes are both low spin, yet phenyl migration occurred whereas methyl migration was not observed. Metal-to-pyrrole nitrogen migration is also induced by one-electron oxidation of organocobalt porphyrins, although a significant difference is that whereas the initial one-electron oxidation in Fe(Por)R occurs at iron, in Co(Por)R it occurs at the porphyrin ligand. The electrochemistry of σ-bonded organoiron porphyrin complexes and also of iron N-alkyl- or N-aryl-porphyrin complexes has been examined in some detail, and early work has been discussed in the reviews.[7,8] More recent investigations into oxidatively induced Fe-to-N migration have served to clarify the situation and will be discussed below.

The dichlorocarbene complex, $Fe(TPP){=}CCl_2$, was doubly significant as it was both the first fully characterized transition metal dihalocarbene complex and the first metalloporphyrin carbene complex to be reported.[66] It was prepared from Fe(TPP) with CCl_4 using an excess of iron powder,[3] a method which has been used to synthesize a range of halocarbene complexes, Fe(Por)=CXY, by the reaction of Fe(TPP) with CX_3Y as the carbene source and iron powder or sodium dithionite as the reductant (X = halide, Y = halide or CN, CO_2Et, CF_3). These carbene complexes containing electron-withdrawing ligands are susceptible to nucleophilic attack, and the carbene ligand is transformed to a CO or CNR (isocyanide) ligand on reaction with water or an amine, respectively. Where Y = SR or SeR, reaction of the carbene with Lewis acids leads to thio- or selenocarbonyl complexes, Fe(Por)(CS) or Fe(Por)(CSe). When CI_4 was used as the carbene source the product was the μ-carbido complex (TPP)Fe=C=Fe(TPP), and similarly $Cl_3CCH(p\text{-}C_6H_4Cl)_2$ (the insecticide DDT) gave the vinylidene complex $Fe(TPP){=}C{=}C(p\text{-}C_6H_4Cl)_2$.

One-electron oxidation of the vinylidene complex transforms it from an Fe=C axially symmetric Fe(II) carbene to an Fe(III) complex where the vinylidene carbon bridges between iron and a pyrrole nitrogen. Cobalt and nickel porphyrin carbene complexes adopt this latter structure, with the carbene fragment formally inserted into the metal-nitrogen bond. The difference between the two types of metalloporphyrin carbene, and the conversion of one type to the other by oxidation in the case of iron, has been considered in a theoretical study. The comparison is especially interesting for the iron(II) and cobalt(III) carbene complexes $Fe(Por)CR_2$ and $[Co(Por)(CR_2)]^+$ which both contain d^6 metal centers yet adopt different structures.[5,67]

B. *Iron σ-Bonded Alkyl and Aryl Complexes*

1. *Synthesis and Spectroscopic and Electrochemical Properties*

There has been relatively little activity in the development of new synthetic methods for σ-bonded organoiron porphyrin complexes over the last 10–15 years, and most studies have concentrated on detailed spectroscopic and electrochemical investigations. Some new examples that have been reported are the bridged alkyl or aryl complexes, (TPP)Fe(CH_2)$_4$Fe(TPP) or (TPP)Fe(μ-1,4-C_6H_4)Fe(TPP), prepared from Fe(TPP)Cl with Li(CH_2)$_4$Li or 1,4-$C_6H_4Li_2$ in toluene-d_8 and characterized by ^{1}H NMR spectroscopy. An interesting contrast is between Fe(TPP)C_6H_5, which is EPR active, and (TPP)Fe(μ-1,4-C_6H_4)Fe(TPP), which is not; presumably the two paramagnetic Fe(III) centers in the latter interact.[68] Secondary and tertiary alkyl complexes are difficult to prepare by the classical methods as they rapidly decompose through Fe—C bond homolysis. However, they can be prepared at low temperature from the reaction of Fe(TTP)Cl with RMgBr at −80°C (R = *t*-Bu, 1-adamantyl).[69]

Complexes with unsaturated ligands (σ-vinyl, σ-allyl, and alkynyl) have been reported, each prepared from Fe(TPP)Cl with the appropriate Grignard (vinyl, 2-methylvinyl, 2,2-dimethylvinyl, allyl, or 2-methylallyl) or lithium reagent (LiC≡C-n-Pr or LiC≡CPh) and observed by NMR spectroscopy (Scheme 4).[70–73] The vinyl and alkynyl complexes are stable in solution at 25°C, whereas the allyl species decompose quickly if allowed to warm to room temperature. All were too reactive to be purified by chromatography. The vinyl and allyl complexes show characteristic low spin behavior, although the temperature dependence of the vinyl

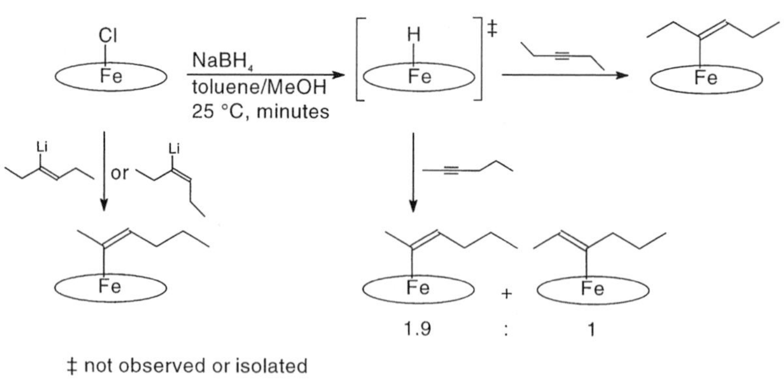

Cl
Fe
NaBH$_4$
toluene/MeOH
25 °C, minutes
H
Fe
‡
Fe
Li
or
Li
Fe
Fe + Fe
1.9 : 1
‡ not observed or isolated

SCHEME 4.

complexes show Curie behavior while that of the allyl complexes shows non-Curie behavior, possibly linked to a spin state change or thermal population of the high spin state.[70] Both *cis* and *trans* isomers of the 2-methylvinyl derivatives could be identified by NMR.[72] In toluene or CH_2Cl_2 the alkynyl derivatives are high spin, $S = 5/2$ complexes, but are converted to six-coordinate, low spin ($S = 1/2$) complexes on addition of pyridine or THF.[71]

Iron porphyrins containing vinyl ligands have also been prepared by hydrometallation of alkynes with Fe(TPP)Cl and $NaBH_4$ in toluene/methanol. Reactions with hex-2-yne and hex-3-yne are shown in Scheme 4, with the former giving two isomers. Insertion of an alkyne into an Fe(III) hydride intermediate, Fe(TPP)H, formed from Fe(TPP)Cl with $NaBH_4$, has been proposed for these reactions.[72] In superficially similar chemistry, Fe(TPP)Cl (present in 10 mol%) catalyzes the reduction of alkenes and alkynes with 200 mol% $NaBH_4$ in anaerobic benzene/ethanol. For example, styrene is reduced to 2,3-diphenylbutane and ethylbenzene. Addition of a radical trap decreases the yield of the coupled product, 2,3-diphenylbutane. Both Fe(III) and Fe(II) alkyls, Fe(TPP)CH(Me)Ph and $[Fe(TPP)CH(Me)Ph]^-$, were proposed as intermediates, but were not observed directly.[74]

Six-coordinate organoiron porphyrin nitrosyl complexes, Fe(Por)(R)(NO), were prepared from Fe(Por)R (Por = OEP or TPP; R = Me, n-Bu, aryl) with NO gas. The NMR chemical shifts were typical of diamagnetic complexes, and the oxidation state of iron was assigned as iron(II).[75]

The electroreduction of iron porphyrins in the presence of alkyl halides is one method for the preparation of σ-alkyl iron porphyrin complexes, and the details of these electrochemical processes have received considerable attention. An electrochemical study using four porphyrins, OEP, TTP and two amide-linked "baskethandle" porphyrins showed that, beginning with Fe(Por), both the singly reduced $[Fe(Por)]^-$ and doubly reduced $[Fe(Por)]^{2-}$ species are alkylated at the iron atom. No porphyrin ring alkylation is observed, even though this is a possibility if the reduced species have some porphyrin radical anion character. At low alkyl halide concentrations, a reduced iron alkyl species, formally containing Fe(I), could be observed as a transient. At higher concentrations, catalytic reduction of the alkyl halide by the $[Fe^{II}(Por)R]^-/[Fe^{I}(Por)R]^{2-}$ couple occurs. The porphyrins containing the amide-linked chains greatly stabilize the reduced species.[76] The use of sterically hindered porphyrins or alkyl halides was investigated as a means of determining the stability of the four accessible oxidation states for σ-alkyl iron porphyrins, which encompass the neutral Fe(III) complexes Fe(Por)R, the one-electron oxidized and one- and two-electron reduced species. The doubly-reduced complex $[Fe(Por)]^{2-}$ acts as an outer-sphere electron transfer agent, thus the electroreduction method allows the synthesis of secondary and tertiary alkyl iron complexes normally inaccessible by other means (Fe(I) + R^+ source or Fe(III) + R^- source).[77]

The reduction of *n*-, *sec*-, and *t*-butyl bromide, of *trans*-1,2-dibromocyclohexane and other vicinal dibromides by low oxidation state iron porphyrins has been used as a mechanistic probe for investigating specific details of electron transfer *vs.* S_N2 mechanisms, redox catalysis *vs* chemical catalysis and inner sphere *vs* outer sphere electron transfer processes.[78–80] The reaction of reduced iron porphyrins with alkyl-containing supporting electrolytes used in electrochemistry has also been observed, in which the electrolyte (tetraalkyl ammonium ions) can act as the source of the R group in electrogenerated Fe(Por)R.[81]

^{1}H NMR spectroscopy studies of iron(III) σ-alkyl and σ-aryl porphyrins have been very important in elucidating spin states. Alkyl and most aryl complexes with simple porphyrin ligands (OEP, TPP, or TTP) are low spin, $S = 1/2$ species. NMR spectra for the tetraarylporphyrin derivatives show upfield resonances for the porphyrin pyrrole protons (*ca.* −18 to −35 ppm), and alternating upfield and downfield hyperfine shifts for the axial alkyl or aryl resonances. For *n*-alkyl complexes, the α-protons show dramatic downfield shifts (to *ca.* 600 ppm), upfield shifts for the β-protons (−25 to −160 ppm) and downfield shifts for the γ-protons (12 ppm).[82] The α-protons of alkyliron porphyrins are not usually detected as a result of their large downfield shift and broad resonance. These protons were first detected by deuterium NMR in the deuterated complexes Fe(TPP)CD_3 (532 ppm) and Fe(TPP)CD_2CD_3 (562, −117 ppm).[83]

For low spin aryl iron porphyrins at 25°C, the *ortho* and *para* protons are shifted upfield (*ca.* −80 and −20 ppm) and the *meta* protons downfield (*ca.* 13 ppm). The contact shifts for the iron-bonded aryl ligands indicated the presence of π-spin density on these groups. This is consistent with the electronic structure of low spin iron(III), with orbital occupancy $(d_{xy})^2(d_{xz,yz})^3(d_{z^2})^0(d_{x^2-y^2})^0$ which places the unpaired spin in an orbital of π symmetry with respect to the iron-ligand bonding. The five-coordinate σ-alkyl and σ-aryl complexes will bind a neutral donor ligand (for example, pyridine or imidazole) to become six-coordinate, but remain as low spin complexes.[84] The fluorophenyl derivatives Fe(Por)(C_6F_5) and Fe(Por)(C_6F_4H) show spectral characteristics of high spin, $S = 5/2$ iron(III) centers. For example, the porphyrin pyrrole CH protons in high spin Fe(TPP)(C_6F_5) appear at 66 ppm, compared to *ca.* −30 ppm in Fe(TPP)(C_6H_5).[85]

The electronic absorption spectra of the products of one-electron electrochemical reduction of the iron(III) phenyl porphyrin complexes have characteristics of both iron(II) porphyrin and iron(III) porphyrin radical anion species, and an electronic structure involving both resonance forms $[Fe^{II}(Por)Ph]^-$ and $[Fe^{III}(Por^{\cdot -})Ph]^-$ has been proposed.[86] Chemical reduction of Fe(TPP)R to the iron(II) anion $[Fe(TPP)R]^-$ (R = Et or *n*-Pr) was achieved using Li[$BHEt_3$] or K[BH(i-Bu)$_3$] as the reductant in benzene/THF solution at room temperature in the dark. The resonances of the *n*-propyl group in the ^{1}H NMR spectrum of $[Fe(TPP)(n\text{-}Pr)]^-$ appear in the upfield positions (−0.5 to −6.0 ppm) expected for a diamagnetic porphyrin complex. This contrasts with the paramagnetic, $S = 2$ spin state observed for other anionic iron(II) porphyrins containing halide or alkoxide axial ligands. The reduced alkyl complexes are reoxidized by O_2 to the iron(III) alkyls.[87] The corresponding diamagnetic phthalocyanine iron(II) alkyl complexes, $[Fe(Pc)R]^-$, were prepared by two-electron reduction of Fe(Pc) by $LiAlH_4$ to give $[Fe(Pc)]^{2-}$ (actually the Fe(I) phthalocyanine radical anion) followed by reaction with MeI, EtI or i-PrBr. The methyl compound, $[Fe(Pc)CH_3]^-$ was characterized by X-ray crystallography.[88]

Electrochemical and spectroelectrochemical studies on high spin and low spin five-coordinate aryl and perfluoroaryl (Ar_F) iron porphyrins containing a variety of porphyrin ligands demonstrated a direct correlation between the spin state of the iron(III) center and the stability of the oxidized or reduced species. The high spin complexes are unstable upon undergoing a one-electron reduction, but moderately stable if oxidized by one electron. The low spin aryl complexes can be singly reduced to produce a very stable species, but the singly oxidized species undergo a rapid migration of the aryl group.[85] These studies have been complemented by Mössbauer and magnetic studies. The fluoroaryl complexes are pure high spin over a wide temperature range. However, for some of the alkyl and aryl complexes which are low spin at 25°C, some high spin sites are observed by EPR spectroscopy in frozen solution or in the solid state. The amount is dependent on the nature of the axial ligand and the porphyrin, and in some cases even on the method of sample preparation.[89] For example, Fe(T*p*CF_3PP)(C_6H_5) contains a mixture of high- and low-spin iron(III). The spin state of five-coordinate, low spin Fe(Por)Ph (OEP or TPP) complexes remains unchanged on coordination of pyridine. However, five-coordinate fluoroaryl complexes Fe(Por)(Ar_F) ($Ar_F = C_6F_5$, C_6F_4H) are high spin in non-coordinating solvents, but form six-coordinate low spin complexes in pyridine solution.[90] The mechanistic details for the electroreduction of the high spin fluorophenyl porphyrins has been elucidated. The first reduction of Fe(Por)(Ar_F) is followed by loss of Ar_F^-, which then reacts with trace water to produce Ar_FH and OH^-, which in turn displaces Ar_F^- from the substrate. This chain reaction continues until all the substrate is consumed and converted to Fe(Por)OH which itself reacts to form the μ-oxo iron(III) species. This is a mechanism which is unique to the fluoroaryl σ-bonded organoiron porphyrin complexes.[91]

Even more control over the spin state has been reported by varying the number and substitution pattern of the fluorine atoms in fluorophenyl complexes. Phenyl complexes containing no fluorine substituents are low spin, whereas the compounds containing four ($Ar_{F4} = 2,3,5,6\text{-}C_6F_4H$) or five ($Ar_{F5} = C_6F_5$) fluorine substituents are high spin. The crossover point occurs for the trifluorophenyl ligand, where the iron complex containing 3,4,5-$C_6F_3H_2$ is low spin, whereas its isomer containing 2,4,6-$C_6F_3H_2$ is high spin.[92] A similar situation is observed for iron complexes of the porphyrin isomers porphycenes, in which the pyrrole rings are linked by alternating two- or zero-carbon bridges (Fig. 2). A series of σ-aryl iron porphycene (EtioPc) complexes containing none (C_6H_5), two (Ar_{F2} = 3,5-$C_6F_2H_3$), three (Ar_{F3} = 3,4,5-$C_6F_3H_2$), or four (Ar_{F4} = 2,3,5,6-C_6F_4H) fluorine substituents was prepared. A detailed spectroscopic (NMR, EPR, UV-visible) and electrochemical study showed that Fe(EtioPc)Ph, Fe(EtioPc)Ar_{F2}, and Fe(EtioPc)Ar_{F3} are low spin whereas Fe(EtioPc)Ar_{F4} and Fe(EtioPc)Cl are high spin. Fe(EtioPc)Ar_{F2} was characterized crystallographically.[93,94]

Unusual properties have been observed for σ-bonded organoiron porphyrin complexes containing a highly substituted porphyrin ligand, OETPP (octaethyltetraphenylporphyrin), which contains eight ethyl groups at the β- pyrrolic positions in addition to four phenyl groups at the *meso* positions of the porphyrin periphery. The resulting steric congestion induces a saddle shaped distortion in the porphyrin, which dramatically changes the electronic effects of the macrocycle, rendering it more electron rich than OEP or TPP. For example, the first oxidation wave for Fe(OETPP)Ph is at a very low potential (0.28 V *vs* SCE) and is the most facile oxidation ever observed for an iron(III) porphyrin in non-aqueous media. This oxidation is metal-centered and the product, which can be prepared independently by chemical oxidation, has spectral features consistent with an iron(IV), $S = 1$ center. $[Fe(OETPP)Ph]^+[SbCl_6]^-$ is the first σ-bonded organoiron(IV) porphyrin stable in non-aqueous media at room temperature. Altogether, three reversible oxidations are observed for Fe(OETPP)Ph, the first at iron and the second two at the ring. At longer scan times the Fe-to-N phenyl group migration occurs after the second oxidation. The first reduction, as for the OEP and TPP analogues, is metal-centered.[95,96]

Corroles are tetrapyrrole macrocycles similar to porphyrins but contain one *meso*-carbon replaced by a direct bond between two α-pyrrolic carbon atoms. A corrole macrocycle bears a 3-charge in its deprotonated form and is more effective than a porphyrin ligand at stabilizing high oxidation states. The corrole macrocycle is easier to oxidize and harder to reduce than a porphyrin. The octaethylcorrole (OEC) complex, Fe(OEC)Ph, was prepared from the reaction of Fe(OEC)Cl with four equivalents of PhMgBr. The paramagnetic compound exhibits an $S = 1$ spin state, consistent with the presence of an Fe(IV) center, and represents the first example of an air-stable Fe(IV) compound containing a tetrapyrrole ligand.[97] Fe(OEC)Ph can be oxidized, either electrochemically or chemically, using $Fe(ClO_4)_3$ to give the cationic complex [Fe(OEC)Ph]ClO_4. EPR, UV-visible, and Mössbauer spectroscopy are all consistent with an Fe(IV) corrole cation radical formulation for this complex. $[Fe(OEC)Ph]^+$ has an $S = 1/2$ spin state arising from strong antiferromagnetic coupling of an $S = 1$, d^4 Fe(IV) center with an $S = 1/2$ corrole radical cation.[98] Both [Fe(OEC)Ph] and [Fe(OEC)Ph]ClO_4 have been characterized by X-ray crystallography (Table III). The Fe—N(av), Fe—N_4 plane and Fe—C distance in the cationic complexes are all slightly shorter than those in the neutral complex. Both complexes exist as π-π dimers with the same mean interplanar separations of 3.53 Å.[97,98]

Only a handful of σ-bonded iron porphyrin complexes have been structurally characterized, listed in Table III, and four of these contain porphycene, corrole, or phthalocyanine ligands rather than porphyrins.[99–102] Selected data are given in Table III, and X-ray crystal structures of methyl- and phenyliron porphyrin complexes are shown in Fig. 4. All of the iron(III) porphyrin complexes exhibit

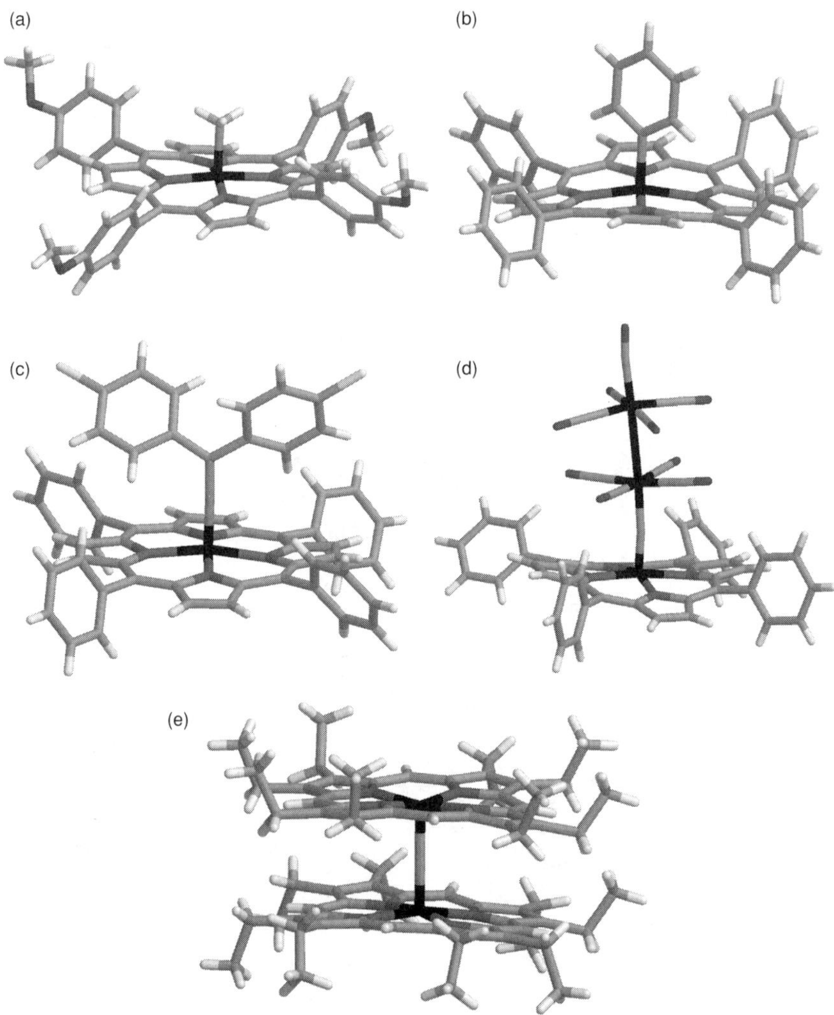

FIG. 4. Molecular structures of selected organometallic iron porphyrin complexes: (a) Fe(TAP)CH_3,[100] (b) Fe(TPP)Ph,[99] (c) Fe(TPP)(=C=CAr_2) (Ar = p-C_6H_4Cl),[131] (d) Fe(TPP)=C=Re-$(CO)_4Re(CO)_5$,[120] (e) $[Fe(OEP)]_2(\mu\text{-}C)$.[123]

TABLE III

SELECTED DATA FOR STRUCTURALLY CHARACTERISED IRON PORPHYRIN COMPLEXES

	M−C Bond Length/Å	M−N(av) Bond Length/Å	M−N_4 Plane/Å	Other	Reference
σ-Bonded complexes					
Fe(EtioPc)(3,5-$C_6F_2H_3$)	1.950(4)	1.930(5)	0.12		94
Fe(TPP)Ph	1.955(3)	1.962(2)	0.17		99
Fe(TAP)C(O)-*n*-Bu	1.965(12)	1.974(12)	0.19		100
Fe(TAP)CH_3	1.979(9)	1.967(5)	0.146		100
Fe(OEC)Ph	1.984(3)	1.871(3)	0.272	Mean OEC···OEC plane separation, 3.53 Å	97
[Fe(OEC)Ph]ClO_4	1.965(5)	1.864(3)	0.242	Mean OEC···OEC plane separation, 3.53 Å	98
[Fe(Pc)CH_3][Na(18-crown-6)]	2.025(3)	1.905(3)	0.10		88
Fe(TPP){μ-(Fe,N)−C=CAr_2}Cl (Ar = *p*-C_6H_4Cl)	1.921(5)	1.993(4)[a]		Fe···NC=CAr_2, 2.529(4) Fe−Cl, 2.299(1)	101,102
Carbene and μ-carbido complexes					
Fe(TPP)(=CCl_2)(H_2O)	1.83(3)	1.980	~0	Fe−O, 2.13(3) Å Fe−C−O, 177.10(15)°	4
Fe(TPP)(=C=CAr_2) (Ar = p-C_6H_4Cl)	1.699(3)	1.984(1)	0.23	C=C, 1.336(4) Fe−C−C, 176.7(3) Å	131
Fe(TPP)=C=Re$(CO)_4$Re$(CO)_5$	1.605(13)	1.983(11)		Re−C, 1.957(12) Re−Re, 3.043(1) Fe−C−Re, 172.8(4)°	120

$[Fe(TPP)]_2(\mu\text{-}C)$	1.675	1.980	0.26	Fe–C–Fe, 180° mean Por···Por separation, 3.87 Å	122
$[Fe(TPP)]_2(\mu\text{-}C)$	1.683(1)	1.978(3)	0.27	Fe–C–Fe, 180° mean Por···Por separation, 3.91 Å	125
$[Fe(OEP)]_2(\mu\text{-}C)$	1.664	1.986	0.192	mean Por···Por separation, 3.71 Å	123
$[Fe(TPP)(THF)](\mu\text{-}C)[Fe(Pc)(THF)]$	1.65(1) (TPP) 1.71(1) (Pc)	2.00(1) (TPP) 1.94(1) (Pc)	0.12 (TPP) 0.09 (Pc)	Fe–O, 2.27(1) (TPP), 2.23(1) (Pc) Fe–C–Fe, 179(1)° mean Por···Pc separation, 3.57 Å	125
$[Fe(Pc)(THF)]_2(\mu\text{-}C)$	1.64(2) 1.71(2)	1.95(1) 1.94(1)	0.08	Fe–O, 2.21(1), 2.23(1) Fe-C-Fe, 180(1)° mean Pc···Pc separation, 3.50 Å	125
$[Fe(Pc)(py)]_2(\mu\text{-}C)$	1.69(2)	1.94(1)	0.039	Fe–C–Fe, 177.5(8)° mean Pc···Pc separation, 3.44 Å	123
$\{[Fe(Pc)F]_2(\mu\text{-}C)\}[NBu_4]_2$	1.687(4)	1.939(4)	0.053	Fe–F, 2.033(2)° Fe–C–Fe, 179.5(3)° mean Pc···Pc separation, 3.48 Å	124

[a] Three Fe–N bonds only. Pyrrole nitrogen involved in carbene bridge is not bonded to iron.

Fe—N_4 plane displacements of over 0.15 Å, consistent with low spin Fe(III) centers.[21] The Fe—C bond lengths in the iron(III) complexes are shorter (less than 2 Å) than that in the iron(II) complex (greater than 2 Å). The shorter bond length for the phenyl complex Fe(TPP)Ph relative to the methyl complex Fe(TAP)Me is consistent with a stronger aryl Fe—C bond in the former.

2. *Iron-to-Nitrogen Migration*

The oxidatively induced, reversible Fe-to-N migration of a phenyl group is summarized in Scheme 5.[86,103] The process is initiated by a one-electron oxidation of Fe(Por)Ph to $[Fe(Por)Ph]^{+}$, which is formulated as an iron(IV) complex with an $S = 1$, $(d_{xy})^2(d_{xz})^1(d_{yz})^1$ electron configuration. This then undergoes migration of the phenyl group from iron to a pyrrole nitrogen, effectively reductive elimination from iron, forming a new C—N bond. Both the iron(II) N-phenyl iron porphyrin $[Fe(Por\text{-}N\text{-}Ph)]^{+}$ and its one-electron oxidation product, the iron(III) dication $[Fe(Por\text{-}N\text{-}Ph)]^{2+}$, are stable complexes which have themselves been subject to extensive investigation.[104–106] However one-electron reduction of $[Fe(Por\text{-}N\text{-}Ph)]^{+}$ produces the formally iron(I) transient Fe(Por-N-Ph) which undergoes reverse migration (effectively oxidative addition of the C—N bond to iron) to regenerate Fe(Por)Ph.

While this overall process has been well established for quite some time,[5,7,8] recent attention has focused on the factors controlling the migration process. The

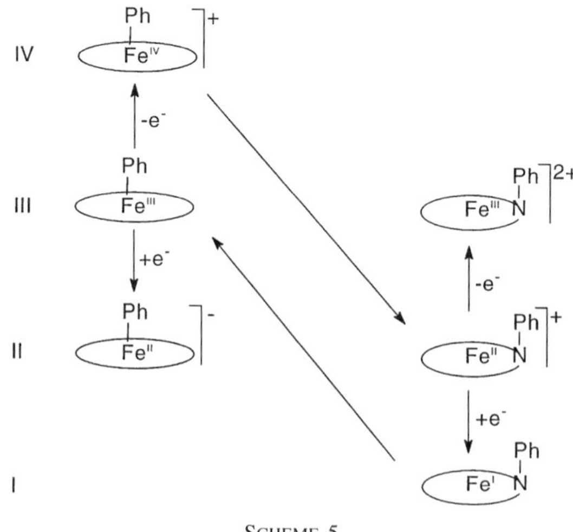

SCHEME 5.

observation that the low spin alkyl and aryl iron(III) complexes did undergo migration upon oxidation, whereas the high spin tetra- or pentafluoroaryl complexes apparently did not, led to speculation that either the spin state at iron or the site of the initial oxidation (iron-centered *vs* porphyrin ring-centered) controlled the migration process.

An unusual feature of fluoroaryl iron OETPP complexes, Fe(OETPP)Ar_F, is that they are low spin in contrast to the corresponding OEP and TPP complexes which are high spin. One-electron oxidation of the low spin porphyrin OEP and TPP complexes containing alkyl or aryl ligands leads to Fe-to-N migration. This reaction is normally not observed when the high spin Ar_F complexes are oxidized, leading to speculation that spin state may govern the migration process. However, despite the low spin configuration in the Fe(OETPP)Ar_F complexes, oxidatively induced migration is still not observed, indicating that spin state is not the only factor controlling the migration step.[96] Furthermore, the low spin phenyliron complex containing the OETPP ligand also does not undergo migration upon one-electron oxidation, but the phenyl group does migrate after a second oxidation, the only example where migration occurs from an intermediate two oxidation levels above iron(III).

A very detailed recent electrochemical and spectroscopic study focused on a series of iron aryl porphyrins Fe(OEP)Ar comprising the low spin complexes where Ar = Ph or 3,4,5-$C_6F_3H_2$, and the high spin complexes where Ar = 2,4,6-$C_6F_3H_2$, 2,3,5,6-C_6F_4H or C_6F_5.[92] The first oxidation leads to an iron(IV) porphyrin, and this undergoes Fe-to-N migration for all of the aryl groups, and for both the low and high spin complexes. The kinetics of the migration were investigated in both the absence and presence of pyridine as the sixth ligand. The important observation was that the rate of migration varied along the series, with the faster rates for the low spin complexes (with fewer fluorine substituents) and slower rates for the high spin complexes. The slow rate for the migration of the four- and five-fluorine substituted groups was the reason these processes had not previously been observed. In other words, the rate constant for migration correlates with the electron donor ability of the aryl group, with faster rates for more electron rich ligands. Slower migration rates are observed when pyridine is coordinated, thus σ-donation by the aryl group accelerates migration, whereas coordination of a donor in the sixth position decelerates migration. Overall, this study concluded that neither the iron(III) spin state nor the field strength of the aryl (or alkyl) group is the key to controlling the migration, but that the donor ability of the σ-aryl ligand influences the rate.[92]

3. *Iron–Carbon Bond Homolysis*

Alkyliron porphyrin complexes are not particularly stable in solution, and over a period of several hours at room temperature the iron(II) porphyrin appears, presumably as a result of Fe—C bond homolysis. The reaction shown in Eq. (1) using the deuterium-labeled alkyl iodides demonstrated alkyl group exchange, observed by ^1H NMR to have a half-life of 100 min for the solution conditions used. The mechanism proposed involved Fe—C bond homolysis, exchange of $\cdot CH_3$ radicals with CD_2CD_3I to give $\cdot CD_2CD_3$ radicals and CH_3I, followed by recombination of $\cdot CD_2CD_3$ with Fe(TPP).[83] The rate of the homolysis process is influenced by the nature of the porphyrin ring, with the half-life for decomposition of $Fe(Por)CH_2CH_3$ varying from 4 days for the electron rich TAP to 45 min for F_{20}-TPP.[107]

$$Fe(TPP)CH_3 + CD_2CD_3I \rightarrow Fe(TPP)CD_2CD_3 + CH_3I \qquad (1)$$

The kinetics of Fe—C bond homolysis were investigated using Ph_3SnH as a radical trap, which captures R· released from Fe(Por)R in competition with recombination. The overall reaction is shown in Eq. (2). The bond dissociation activation enthalpies were measured to be 33 kcal mol^{-1} for C_6H_5, 23 kcal mol^{-1} for CH_3, 19 kcal mol^{-1} for C_2H_5, and 17 kcal mol^{-1} for CH_2CMe_3, with the Fe—C(aryl) bond stronger than the Fe—C(alkyl) bonds. Assuming diffusion controlled recombination, the Fe—C bond dissociation energies are expected to be ca. 2 kcal mol^{-1} lower than these values. The trend in Fe—C(alkyl) bond dissociation energies can be rationalized on steric grounds. The effect of an added base (pyridine or PEt_3) was also investigated. Overall, the Fe—C bonds are 10–15 kcal mol^{-1} weaker than the corresponding organocobalt porphyrin metal-carbon bonds.[108]

$$Fe(Por)R + Ph_3SnH \rightarrow Fe(Por) + RH + \tfrac{1}{2}Sn_2Ph_6 \qquad (2)$$

4. *Insertion of Small Molecules into the Fe–C Bond*

Nitric oxide (NO) reacts with organoiron(III) porphyrins to form six-coordinate adducts, Fe(Por)(R)(NO).[75] Other small molecules (O_2, SO_2, CO) react by insertion into the Fe—C bond, although the nature of the reaction and the stability of the products varies greatly with both the molecule itself and the organoiron group.

Alkyliron(III) porphyrin complexes are air sensitive, and when exposed to oxygen under ambient conditions the products are the very stable iron(III) μ-oxo dimers, $[Fe(Por)]_2O$. A more careful investigation revealed that the reaction of the alkyl complexes with oxygen proceeds via insertion of O_2 into the Fe—C bond.[109] When a solution of Fe(Por)R (R = Me, Et, i-Pr) is exposed to O_2 at $-70°C$, the characteristic ^1H NMR spectrum of the low spin iron alkyl complex disappears and is replaced by a new, high spin species. The same species can be generated from the reaction of an alkyl hydroperoxide with Fe(Por)OH, and is formulated as

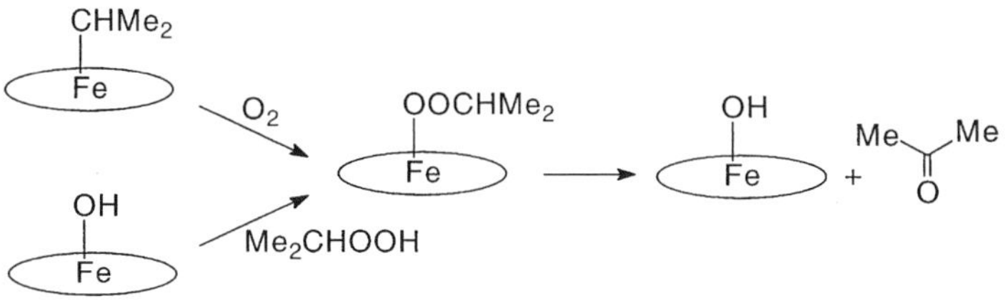

SCHEME 6.

an iron(III) alkylperoxide complex, Fe(Por)OOR. These complexes are very reactive and decompose upon standing or upon warming the solution, giving Fe(Por)OH (and eventually the μ-oxo complex) together with an aldehyde or ketone as the organic product (shown in Scheme 6 for R = i-Pr).[69,82,110,111] The oxygen insertion reaction proceeds similarly for primary, secondary, and tertiary alkyl groups, showing that it is not retarded by bulk in the axial ligand. However, it is sensitive to bulk in the porphyrin ligand, with the reaction proceeding more rapidly for OEP or TPP than for TMP, the latter requiring warming to $-50°C$ to proceed.[69] The products of the reaction of O_2 with iron(II) porphyrins are not observed, indicating that the mechanism does not proceed via Fe—C bond homolysis as the initial step. Insertion of O_2 into allyl and vinyl iron porphyrins occurs, and the allylperoxy compound was observed by NMR at low temperature, decomposing to acrolein ($O{=}CH{-}CH{=}CH_2$) and Fe(Por)OH upon warming.[70]

The low spin aryliron complexes Fe(Por)Ar behave differently than their alkyl counterparts toward oxygen. In chloroform the products of the reaction of Fe(Por)Ar with O_2 are the iron(IV) alkyl $[Fe(Por)Ar]^+$ and Fe(Por)Cl, while in toluene the aryloxide complexes Fe(Por)OAr are formed with no direct evidence for the formation of arylperoxo intermediates.[112] The high spin iron(III) alkynyl complexes which have more ionic character in the Fe—C bond do not react with oxygen, and the order of reactivity of organoiron(III) complexes toward O_2 is given by Fe(Por)(alkyl) > Fe(Por)(aryl) > Fe(Por)(C≡CR).[71]

One-electron reduction of the iron(III) alkyl complexes forms the diamagnetic iron(II) alkyl anions $[Fe(Por)R]^-$. The iron(II) anions do not react with oxygen directly, but are first oxidized by O_2 to the corresponding alkyliron(III) complexes, Fe(Por)R, which then insert O_2 as described above.[87]

Insertion of SO_2 into the Fe—C bond in $Fe(Por)CH_3$ was first reported in 1982, giving the sulfinato complexes $Fe(Por)SO_2CH_3$, which are moderately air stable but can be further oxidized by O_2 to give the sulfonato complexes $Fe(Por)SO_3CH_3$.[113] Alkyliron(III) porphyrins insert CO to give the acyl complexes Fe(Por)C(O)R. For example, Fe(TPP)C(O)-*n*-Bu was formed either by this method or by the reaction of $[Fe(TPP)]^-$ with ClC(O)-*n*-Bu, and was characterized by an X-ray crystal structure

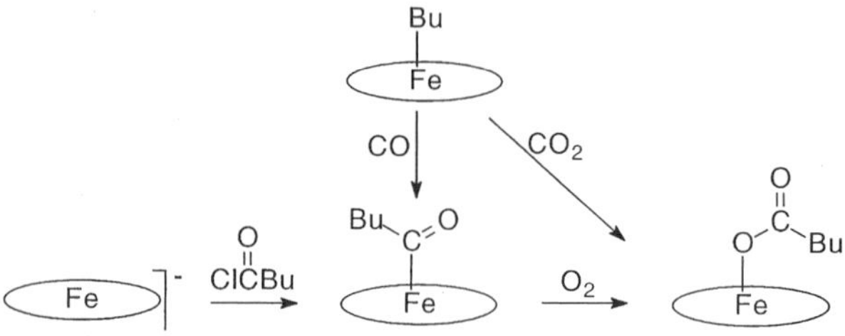

SCHEME 7.

determination ($\nu_{CO} = 1817\ cm^{-1}$).[100,114] The low spin iron(III) acyl products will themselves react with oxygen to give the high spin iron(III) carboxylato complexes Fe(Por)OC(O)R. The reaction of the acyl species with oxygen has more in common with the reaction of aryliron than alkyliron complexes with O_2, as no evidence for peroxoacyl intermediate species could be observed.[100] An iron(III) carboxylato complex, Fe(TPP)OC(O)-*n*-Bu, was also observed from the insertion of CO_2 into the Fe—C bond in the butyliron complex (Scheme 7).[114]

Iron(II) alkyl anions $[Fe(Por)R]^-$ (R = Me, *t*-Bu) do not insert CO directly, but do upon one-electron oxidation to Fe(Por)R to give the acyl species Fe(Por)C(O)R, which can in turn be reduced to the iron(II) acyl $[Fe(Por)C(O)R]^-$. This process competes with homolysis of Fe(Por)R, and the resulting iron(II) porphyrin is stabilized by formation of the carbonyl complex Fe(Por)(CO). Benzyl and phenyl iron(III) complexes do not insert CO, with the former undergoing decomposition and the latter forming a six-coordinate adduct, $[Fe(Por)(Ph)(CO)]^-$ upon reduction to iron(II). The failure of Fe(Por)Ph to insert CO was attributed to the stronger Fe—C bond in the aryl complexes. The electrochemistry of the iron(III) acyl complexes Fe(Por)C(O)R was investigated as part of this study, and showed two reversible reductions (to Fe(II) and Fe(I) acyl complexes, formally) and one irreversible oxidation process.[115]

Iron porphyrins (containing TPP, picket fence porphyrin, or a basket handle porphyrin) catalyzed the electrochemical reduction of CO_2 to CO at the Fe(I)/Fe(0) wave in DMF, although the catalyst was destroyed after a few cycles. Addition of a Lewis acid, for example Mg^{2+}, dramatically improved the rate, the production of CO, and the stability of the catalyst. The mechanism was proposed to proceed by reaction of the reduced iron porphyrin $[Fe(Por)]^{2-}$ with CO_2 to form a carbene-type intermediate $[Fe(Por){=}C(O^-)_2]$, in which the presence of the Lewis acid facilitates C—O bond breaking.[116] The addition of a Brönsted acid (CF_3CH_2OH, *n*-PrOH or 2-pyrrolidone) also results in improved catalyst efficiency and lifetime, with turnover numbers up to 350 per hour observed.[117]

5. *Iron Porphyrin Silyl and Stannyl Complexes*

The reaction of Fe(TPP)Cl with $LiSiMe_3$ at 0°C in HMPA produced the silyliron complex $Fe(TPP)SiMe_3$, along with a number of other products. The complex contains low spin iron(III), and the silyl methyl protons were observed at −1.2 ppm in the ^1H NMR spectrum—a surprisingly small paramagnetic shift compared to the β-protons in *n*-butyl iron(III) porphyrin, indicating that unpaired spin transfer through silicon is much less efficient compared to carbon. The complex has a half-life in solution of ca. 30 min at 25°C. It undergoes homolytic cleavage of the Fe—Si bond to form the iron(II) porphyrin in solution, which is then further reduced by $LiSiMe_3$ to form the diamagnetic iron(II) silyl complex $[Fe(TPP)SiMe_3]^-$. This can be reoxidized to $Fe(TPP)SiMe_3$ using I_2. $Fe(TPP)SiMe_3$ reacts with propylene oxide to form a complex assigned on the basis of its NMR spectrum to be $Fe(TPP)(CH_2)_3OSiMe_3$.[118]

Facile thermal homolysis of Fe—C bonds in iron porphyrin alkyls arises because of the relatively low Fe—C bond strength and produces a steady-state source of alkyl radicals, R·. Addition of excess Bu_3SnH to solutions of Fe(TAP)R lead to hydride abstraction from tin by the alkyl radical, and recombination of Fe(TAP) with the resulting tin-centered radical $Bu_3Sn\cdot$ generating a paramagnetic, low spin (tributyltin)iron(III) porphyrin complex. The same complex could be prepared from Fe(TAP)Cl with $LiSnBu_3$. The stannyl complex was not particularly stable in solution and decomposed over a period of hours to produce iron(II) porphyrin. However, ^1H NMR evidence was consistent with the existence of low spin $Fe(TAP)SnBu_3$, with signals assigned to the α- and β-protons of the butyl groups observed at 8.86 and 5.25 ppm.[119]

C. *Iron Porphyrin Carbene Complexes*

The first iron porphyrin carbene, $Fe(TPP)(=CCl_2)$, was reported in 1977[3] and its structure the following year.[4] Subsequent developments through the 1980s which resulted in a range of iron porphyrin carbene complexes, mostly with electron-withdrawing substituents on the carbene carbon, have been detailed in the overview of the iron porphyrins.[5,7,8] There have been relatively few new developments in this area since then. The reaction of Fe(TPP)Cl with $NaBH_4$ and internal alkynes (hex-2-yne or hex-3-yne) to give σ-vinyl complexes was described earlier in this section.[72] When the same reaction is carried out using the terminal alkynes hex-1-yne or pent-1-yne in toluene/methanol the products were proposed on the basis of NMR spectroscopy and mass spectrometry to be the dialkylcarbene complexes, $Fe(TPP)(=C(CH_3)R)$ (R = *n*-Bu or *n*-Pr, respectively). Characterization of the products as carbene complexes would have been more convincing had the distinctive chemical shift for the carbene carbon in the ^{13}C NMR been reported. Under different solvent conditions (using smaller amounts of methanol) the reaction with hex-1-yne gave a considerable amount of the σ-vinyl complex Fe(TPP)(C(=CH_2)-*n*-Bu) in addition to the carbene. The mechanism proposed for the carbene formation was insertion of the alkyne into the Fe—H bond of an intermediate hydride, Fe(TPP)H, to give the σ-vinyl complex, which is subsequently reduced (by $NaBH_4$) and protonated (by MeOH) to give the carbene complex (Eq. 3). This hypothesis was tested using deuteration studies.[72]

Cl–Fe $\xrightarrow[\text{toluene/MeOH}]{NaBH_4}$ [H–Fe]‡ $\xrightarrow{RC\equiv CH}$ R(CH₂=)C–Fe $\xrightarrow[(NaBH_4)]{+e^-}$ [R(CH₂=)C–Fe]⁻ $\xrightarrow[(MeOH)]{H^+}$ R(CH₃)C=Fe (3)

‡ not observed or isolated

Dihalocarbene complexes are useful precursors to new carbenes by nucleophilic displacement of the chlorine substituents.[66] This has been nicely illustrated for Fe(TPP)(=CCl_2) by its reaction with two equivalents of $[Re(CO)_5]^-$ to give the unusual μ-carbido complex Fe(TPP)=C=$Re(CO)_4Re(CO)_5$ which also contains a rhenium-rhenium bond.[120] The carbido carbon resonance was observed at 211.7 ppm in the ^{13}C NMR spectrum. An X-ray crystal structure showed a very short Fe=C bond (1.605(13) Å, shorter than comparable carbyne complexes) and a relatively long Re=C bond (1.957(12) Å) (Fig. 4, Table III).[120]

The Fe=C bond in this complex can be compared to that of 1.67 Å in the bis(porphyrinato)-μ-carbido iron complex [Fe(TPP)](μ-C), which was first prepared in 1981 from Fe(TPP) and CI_4 together with a reducing agent.[121,122] A number of other examples of iron μ-carbido complexes containing porphyrin or phthalocyanine macrocycles have been prepared and structurally characterized since then, and the structural data are summarized in Table III (Fig. 4).[122–125] The iron phthalocyanine complexes more readily coordinate axial donors (THF, nitrogen donors, or fluoride ion) than their porphyrin counterparts[123–127] One interesting example, [Fe(TPP)(THF)](μ-C)[Fe(Pc)(THF)], contains both porphyrin and phthalocyanine ligands, and was prepared from Fe(TPP)(=CCl_2) with $[Fe(Pc)]^{2-}$.[125] The structures of porphyrin and phthalocyanine iron μ-carbido complexes show similar Fe=C distances, but slightly shorter Fe—N(av) distances for the phthalocyanine complexes, reflecting the smaller hole size in the macrocycle. Separations between the mean macrocycle planes are also slightly less for the phthalocyanine complexes. The family of porphyrin and phthalocyanine iron μ-carbido complexes illustrate the importance of the steric protection afforded by the bulky macrocycle as only a small number of other transition metal μ-carbido complexes have been reported. A single ruthenium analog $[Ru(Pc)(py)]_2C$ has also been reported.[123]

The photochemistry of several of the iron porphyrin halocarbene complexes Fe(TPP)(=CXY) (CXY = CCl_2, CBr_2, CClF, CCl(CN) and the vinylidene complex Fe(TPP)=C=CAr_2 (Ar = p-C_6H_4Cl) has been studied in degassed benzene at 20°C using $\lambda > 360$ nm.[128,129] Irradiation of the dihalocarbene complexes produced Fe(TPP) and the free carbenes :CX_2 which were trapped as the cyclopropane derivatives using a variety of alkenes. The same product ratios were observed when CHX_3 and base was used as the carbene source, indicting that free carbenes are indeed produced from irradiation of the iron porphyrin carbene complexes. The complexes containing CCl(CN) or vinylidene ligands did not form the free carbene, the latter producing Fe(TPP) and the alkyne, ArC≡CAr. The photochemical generation of free carbenes from transition metal carbene complexes had not previously been observed, and is believed to result from mixing of the porphyrin π^* and Fe—C orbitals. This phenomenon is also responsible for the blue-shifted (hypso) spectrum observed for the iron porphyrin carbene complexes. Similar orbital mixing occurs in iron porphyrin carbonyl complexes, which also undergo photodissociation.[128,129]

The diamagnetic iron porphyrin vinylidene complex Fe(TPP)=C=CAr_2(Ar = p-C_6H_4Cl) is produced from Fe(TPP)Cl, the insecticide DDT (Cl_3CCH(p-$C_6H_4Cl)_2$), and iron powder[130] and the structure confirmed by X-ray crystallography (Fig. 4).[131] Oxidatively induced migration of the vinylidene ligand in this complex to give the iron(III) Fe—N-bridged vinylidene complex has been reviewed previously.[5,67] The intermediate spin ($S = 3/2$) iron(III) Fe—N-bridged compound Fe(TPP)(μ(Fe,N)—C=CAr_2)Cl and the closely related high spin ($S = 2$) iron(II) N-vinylporphyrin complex Fe(TPP—N—CH=CAr_2)Cl (see Scheme 8) were compared in ^{1}H NMR spectroscopy and crystallographic studies.[130,132] The

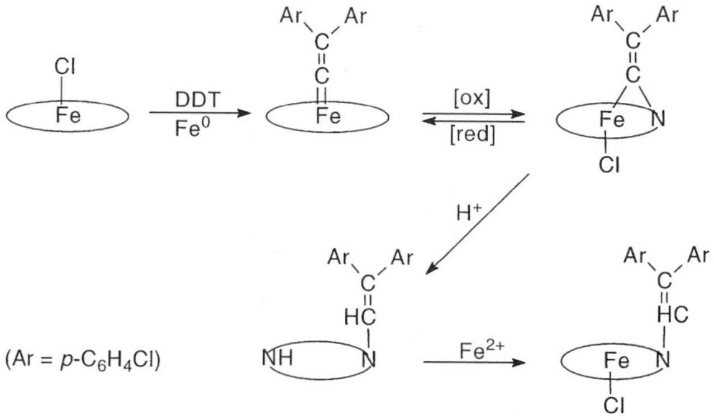

SCHEME 8.

two complexes were clearly distinct by ^{1}H NMR despite the fact that they both exhibit C_s symmetry and differ only by one proton and one electron.[132,133] NMR data were used to assign the ground state in the intermediate spin ($S = 32$) Fe—N-bridged compound as $(d_{xy})^2(d_{xz})^1(d_{yz})^1(d_{z^2})^1$. The Fe—N bonding is disrupted in this complex, with no formal Fe—N bond to the porphyrin nitrogen involved in the vinylidene bridge.[101,102,130] The Fe—C bond in the bridged structure is shorter than in the simple iron(III) porphyrin σ-bonded complexes (Table III).

An important reaction used to model cytochrome P450 chemistry is oxygen atom transfer from iodosylbenzene (PhI=O) to iron porphyrins to give high valent iron oxo intermediates. Since metal-carbene and metal-oxo complexes are formally isolobal, a parallel reaction utilizing iodonium ylides, $PhI{=}CR_2$, could potentially be used for the preparation of iron porphyrin carbenes. This was tested using the iodonium ylide shown in Eq. (4) in a reaction with Fe(TPP) which gave a diamagnetic complex which exhibited UV-visible and ^{1}H NMR data consistent with a carbene complex, but which was stable only below −30°C. Reaction of this with Br_2 led to the metallacyclic complex shown in Eq. (4), proposed to form via oxidatively induced Fe-to-N carbene migration followed by rearrangement of the Fe—N-bridged carbene to give a product containing a Fe—O bond.[134]

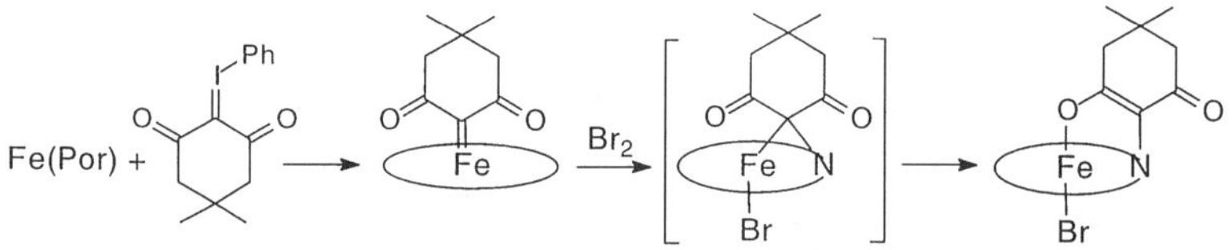

(4)

Diazoalkanes are useful as precursors to ruthenium and osmium alkylidene porphyrin complexes, and have also been investigated in iron porphyrin chemistry. In an attempt to prepare iron porphyrin carbene complexes containing an oxygen atom on the β-carbon atom of the carbene, the reaction of the diazoketone $PhC(O)C(N_2)CH_3$ with Fe(TpClPP) was undertaken. A low spin, diamagnetic carbene complex formulated as Fe(TpClPP)(=C(CH_3)C(O)Ph) was identified by UV-visible and ^{1}H NMR spectroscopy and elemental analysis. Addition of CF_3CO_2H to this rapidly produced the protonated N-alkyl porphyrin, and Br_2 oxidation in the presence of sodium dithionite gave the iron(II) N-alkyl porphyrin, both reactions evidence for Fe-to-N migration processes.[134]

The reaction of Fe(T*p*ClPP)Cl with the diazoalkane N_2CHCH_2Ph at −30°C produced the paramagnetic iron(III) Fe,N-bridged carbene complex Fe(T*p*ClPP){μ(Fe,N)-$CHCH_2Ph$}Cl. The spectroscopic data recorded for this complex were similar to those observed for the bridged vinylidene analog. Chemical or electrochemical reduction of the bridged complex produced the diamagnetic carbene complex Fe(T*p*ClPP)(=$CHCH_2Ph$), which in turn can be chemically or electrochemically oxidized to the iron(II) N-vinylporphyrin complex Fe(T*p*ClPP-N-CH=CHPh)(Cl). These reactions are significant in understanding the process in which N-vinyl-heme derivatives are produced during the cytochrome P450 dependent oxidative metabolism of a sydnone derivative which involves N_2CHCH_2Ph as a reactive metabolite.[135]

An Fe, N-bridged carbene complex was proposed in the reaction of Fe(OEP)-ClO_4 with $N_2CHSiMe_3$. The UV-visible spectrum was similar to that of the Fe, N bridged vinylidene complex, and was attributed to an iron(III) complex containing the $CHSiMe_3$ carbene fragment bridged between iron and a pyrrole nitrogen. Longer reaction times followed by workup with aqueous $HClO_4$ solution gave the N,N-ethano-bridged porphyrin as the final product.[136]

Cyclopropanation of alkenes catalyzed by rhodium or osmium porphyrins using ethyldiazoacetate (N_2CHCO_2Et) is known to involve metal-carbene intermediates, and will be discussed in detail in the sections on those elements. The reaction can also be catalyzed by iron porphyrins. Using Fe(TPP) as the catalyst, 1300 turnovers per hour were observed and gave *trans* and *cis* cyclopropane products in the ratio 8.8:1.[137] A mechanism involving iron porphyrin carbene intermediates was proposed, although no direct evidence for these was observed, and the stereochemistry of the products (predominantly *trans*) was explained in terms of stereochemical effects of the alkene substrate in the transition state. The Mansuy-type iron porphyrin carbenes bearing electron withdrawing substituents were not active for cyclopropanation.[137]

The isocyanide ligand, :CNR, is formally isolobal with a carbene ligand. Several studies have investigated iron(III) porphyrin isocyanide complexes which have the general formula $[Fe(Por)(CNR)_2]^+$.[138–140] However, these studies have mostly been concerned with spin state and spectroscopic properties rather than chemical transformations and will not be discussed in detail here. Crystallographic details are given for two of the complexes.[140]

D. *Oxidation Processes Catalyzed by Iron Porphyrins*

The cytochrome P450 family of enzymes activates molecular oxygen at a heme site using NADPH as an electron source. One oxygen atom of O_2 is reduced to water and the other is involved in mono-oxygenation reactions including hydroxylation of unactivated alkanes and arenes, and alkene epoxidation. The reductive dehalogenation of halogenated compounds is also mediated by cytochrome P450. The active intermediate in the oxygen atom transfer chemistry is a high valent iron(IV) oxo porphyrin radical cation species. Studies directed at elucidating the mechanism of these processes, ranging from investigations of the intact enzyme to model compounds prepared in the laboratory, now comprise a body of work. There are many links between this oxidation chemistry and organometallic chemistry, beginning with the formally isolobal relationship between iron oxo and iron carbene moieties. Furthermore, iron porphyrin carbene and alkyl complexes may play key roles in the so-called "suicide" inactivation of cytochrome P450 in which green pigments comprising N-alkyl porphyrins are formed through Fe-to-N alkyl and/or carbene migration processes. The intimate details of the transfer of the oxygen atom from the high valent iron intermediate to the organic substrate are still under discussion. Proposals for this step range from covalent Fe—C bonds, agostic interactions, some form of coordination of the alkane or alkene substrate to iron, or a direct oxygen atom transfer to the substrate with no direct iron···substrate interactions at all. Much of the discussion of carbene, bridging carbene, and Fe-to-N migration processes given in this review has relevance to cytochrome P450 chemistry. However, a detailed discussion of the organometallic aspects of cytochrome P450 chemistry is beyond the scope of this review, and in fact was the subject of a review in 1987.[6] Overall, the cytochrome P450 chemistry and oxidation chemistry mediated by high valent porphyrins (including, but not limited to, iron) has been thoroughly discussed in the literature.[13–15]

V

RUTHENIUM AND OSMIUM

A. *Overview*

The early period of ruthenium and osmium porphyrin chemistry was dominated by the coordination chemistry of the carbonyl and bis(pyridine) complexes, M(Por)(CO)L (where L is a neutral donor ligand) and $M(Por)(py)_2$. The development of reliable synthetic routes to the ruthenium(II) and osmium(II) porphyrin dimers, $[M(Por)]_2$, has been enormously important in opening up the organometallic chemistry of the porphyrin complexes of these elements. The dimers are paramagnetic with a $(\sigma)^2(\pi)^4(\delta)^4(\pi^*)^2$ configuration, consistent with two unpaired electrons and a double bond between the two metal atoms.[22] This bonding can be disrupted by strong donor ligands, so the reaction by which the dimers are formed, solid state pyrolysis of $M(Por)(py)_2$, can be reversed by treating benzene solutions of the dimers with pyridine. The 14-electron Ru(II) porphyrin complex Ru(TMP) is prevented by the bulky porphyrin ligand from dimerizing and exists as a monomer.[141] Four-electron reduction of the dimers fills the π^* and σ^* orbitals, giving a net M-M bond order of zero, and resulting in cleavage of the metal—metal bond to form the monomeric, formally zerovalent dianions, $[M(Por)]^{2-}$.[142,143] Alternatively, the M(II) dimers can be oxidized chemically or electrochemically to give the cationic complexes $\{[M(Por)]_2\}^+$ or $\{[M(Por)]_2\}^{2+}$ which have metal–metal bond orders of 2.5 and 3, respectively.[144] Finally, the ruthenium dimers react with anhydrous HX in CH_2Cl_2 to form paramagnetic Ru(IV) halide complexes, $Ru(Por)X_2$ (X = F, Cl, Br).[147] The neutral and oxidized dimers, the monomeric

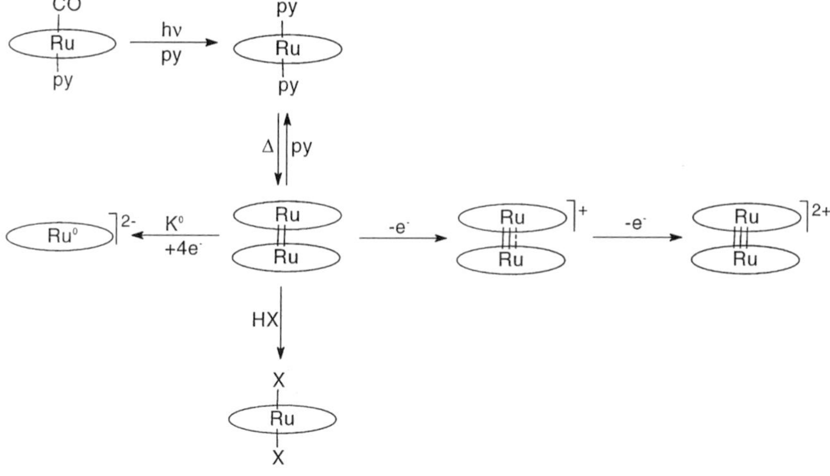

SCHEME 9.

dianions, and the Ru(IV) porphyrin halides have all been significant as precursors to new organometallic complexes. The formation and interconversion of these important precursors is shown for ruthenium in Scheme 9.

The first ruthenium porphyrin organometallic complexes were reported in 1985 and 1986,[142,143,145] and although these early examples have been reviewed,[5,7,8] more recent chemistry has helped to tie the whole area together. For this reason, the organometallic chemistry of ruthenium and osmium porphyrins will be discussed here in its entirety. In addition, a wider range of organometallic ligand types is found for ruthenium and osmium porphyrin complexes than for any other group of metalloporphyrin complexes.

The bulky ruthenium TMP complex Ru(TMP) is very electron deficient in the absence of any coordinating ligand, and a π-complex with benzene has been proposed. In fact, it readily coordinates dinitrogen, forming the mono- and bis-N_2 adducts Ru(TMP)(N_2)(THF) and Ru(TMP)$(N_2)_2$.[141,146] As a result, the use of the TMP ligand for careful stereochemical control of the chemistry at the metal center, which has been very successful for the isolation of elusive rhodium porphyrin complexes, is less useful for ruthenium (and osmium) because of the requirement to exclude all potential ligands, including even N_2.

Over the last decade a number of high oxidation state ruthenium porphyrin complexes containing oxo or imido ligands have been reported and have been thoroughly studied for their role in oxidation and atom-transfer chemistry. Although comparisons can be drawn with organometallic species (carbene, imido, and oxo ligands are formally isolobal) the chemistry of the oxo and imido complexes is beyond the scope of the review and will not be covered here.

B. *Ruthenium and Osmium σ-Bonded Alkyl and Aryl Complexes*

The first ruthenium porphyrin alkyls to be reported were prepared from the zerovalent dianion, $[Ru(Por)]^{2-}$ with iodomethane or iodoethane, giving the ruthenium(IV) dialkyl complexes $Ru(Por)Me_2$ or $Ru(Por)Et_2$ (Por = OEP, TTP).[143] Alternatively, the Ru(IV) precursors $Ru(Por)X_2$ react with MeLi or ArLi to produce $Ru(Por)Me_2$ or $Ru(Por)Ar_2$ (Ar = *p*-C_6H_4X where X = H, Me, OMe, F or Cl).[147–149] The osmium analogues can be prepared by both methods, and $Os(Por)R_2$ where R = Me, Ph and CH_2SiMe_3 have been reported.[143,150] Some representative structures are shown in Fig. 5, and the preparation and interconversion of ruthenium porphyrin alkyl and aryl complexes are shown in Scheme 10.

Attempts to exploit the reaction of the dianion with alkyl halides to produce a *cis*-dialkyl complex by using 1,2- or 1,3-dihaloalkanes did not indeed give this result. The reaction of $[Ru(Por)]^{2-}$ with 1,2-dibromoethane was sucessful, but the resulting metallacyclopropane product is better formulated as a π-complex of ethene, and will be discussed below in the section on alkene and alkyne complexes. The corresponding reaction of the dianion with 1,3-dichloropropane gave no evidence for a metallacyclobutane, but instead free cyclopropane was detected by GC analysis and the porphyrin product was $Ru(TTP)(THF)_2$.[143]

Thermolysis of the Ru(IV) dialkyl or diaryl porphyrins $M(Por)R_2$ at 100°C in benzene leads to loss of one of the axial ligands and formal reduction of the

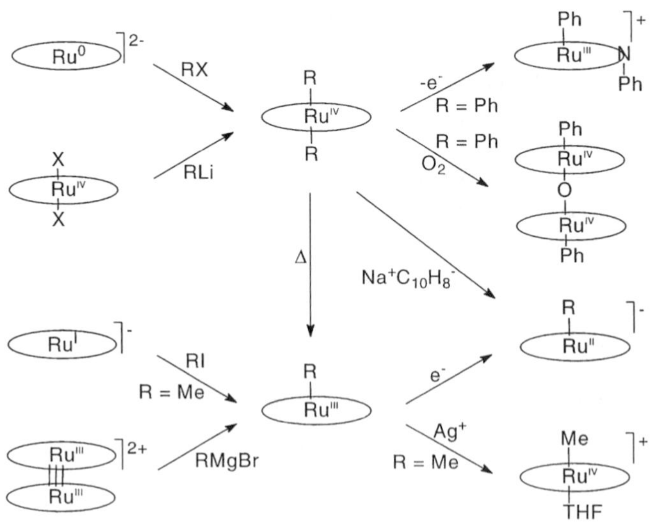

SCHEME 10.

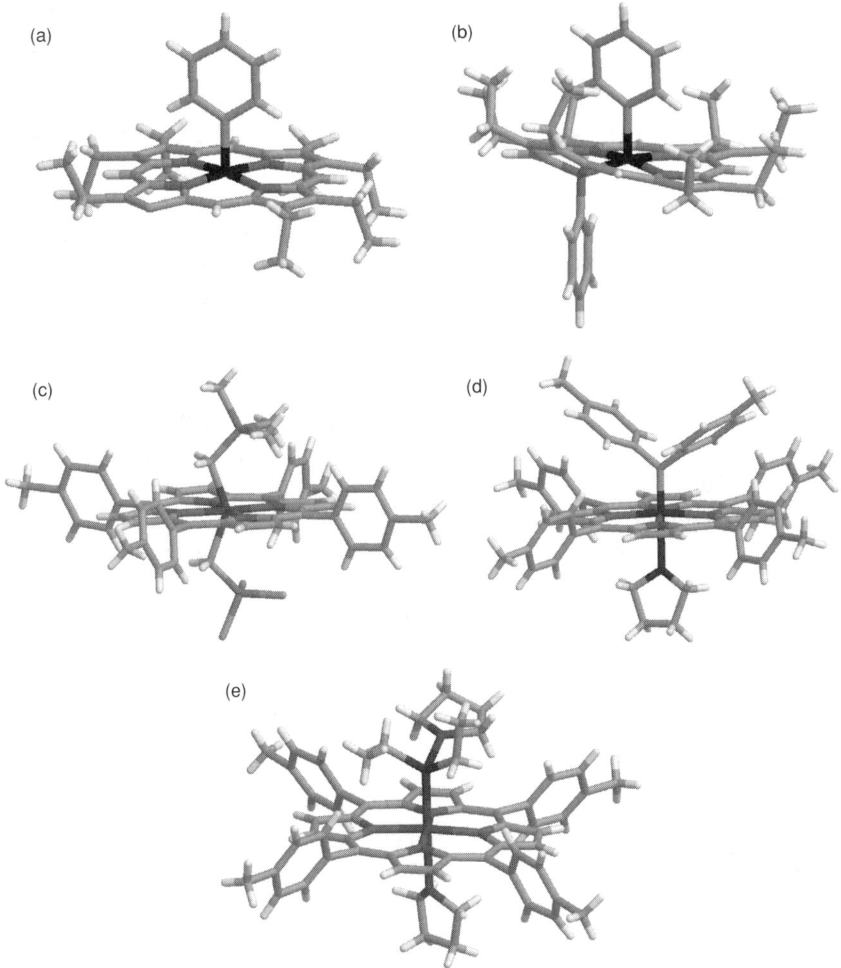

FIG. 5. Molecular structures of selected organometallic ruthenium and osmium porphyrin complexes: (a) Ru(OEP)Ph,[151] (b) $[Ru(OEP\text{-}N\text{-}Ph)Ph]^+$,[157] (c) $Os(TTP)(CH_2SiMe_3)_2$,[150] (d) $Os(TTP)(=C(p\text{-}C_6H_4Me)_2)(THF)$,[168] (e) $Os(TTP)(=SiEt_2\cdot THF)(THF)$.[167]

metal center to give the corresponding Ru(III) alkyl or aryl complexes Ru(Por)R (R = Me, Ph).[148,151] On closer inspection, the reaction of $Ru(OEP)Cl_2$ with PhLi actually gave both the Ru(IV) and Ru(III) phenyl products $Ru(OEP)Ph_2$ and Ru(OEP)Ph in 30 and 50% yield, respectively. The two complexes could be separated by chromatography.[152–154] The reaction of $Ru(OEP)Cl_2$ with neopentyl lithium gave only the ruthenium(III) product, $Ru(OEP)CH_2CMe_3$.[152] Reduction of the Ru(II) dimer $[Ru(OEP)]_2$ with two equivalents of potassium naphthalide gave the Ru(I) anion, $[Ru(OEP)]^-$ characterized on the basis of its reaction with MeI which gave Ru(OEP)Me. This method was also used to prepare the osmium analogue, Os(OEP)Me.[155] Alternatively, oxidation of the dimer by Ag^+ to form the Ru(III) dimer dication $[Ru(OEP)]_2^{2+}$ and reaction of this with Grignard or diorganozinc compounds gave Ru(OEP)R (R = Me, Et, Ph, p-C_6H_4Me).[155] This method was used to prepare cofacial bisorganometallic diporphyrins. The cofacial diporphyrin H_4(DPB) contains two octaethylporphyrin units linked through their *meso* positions by a biphenylene group. In Ru_2(DPB) the two Ru(II) centers are the correct distance apart to form a Ru=Ru double bond, which can be oxidized to form $[Ru_2(DPB)]^{2+}$ which reacts with Grignard reagents to form M_2R_2(DPB) where R = Me, p-C_6H_4Me or 3,5-$C_6H_3(CF_3)_2$.[156]

New complexes can be formed by redox chemistry. One-electron oxidation (by Ag^+) of $Ru(OEP)Ph_2$ led to migration of one phenyl group from ruthenium to a porphyrin nitrogen atom, giving the Ru(III) complex $[Ru(OEP\text{-}N\text{-}Ph)Ph]^+$,[157] while one-electron oxidation (by Ag^+) of Ru(OEP)(Me)(THF) gave the Ru(IV) complex $[Ru(OEP)(Me)(THF)]^+$.[158] $Ru(TPP)Ph_2$ is air stable in solution in the dark, but in ambient light slowly reacted with air in toluene to give the diamagnetic μ-oxo Ru(IV) complex $[Ru(TPP)Ph]_2O$. This is the first organometallic example of the well-known family of μ-oxo dinuclear complexes in which a halide, OH, or OR ligand takes the place of the phenyl group.[148]

Two organometallic complexes formally containing Ru(II) have been reported. $[Ru(OEP)R]^-$ was formed either by sodium naphthalide reduction of $Ru(OEP)R_2$ with loss of R^- (R = Me, Ph)[153,154,158] or by one-electron reduction of Ru(OEP)Ph.[155] In a second example, further reaction of $Ru(OEP)CH_2CMe_3$ with neopentyl lithium gave the unusual dinuclear complex $[Ru(OEP)CH_2CMe_3]_2(\mu\text{-}Li)_2$.[152] $[Ru(OEP)Ph]^-$ reacts with MeI to give the unsymmetrical Ru(IV) complex Ru(OEP)(Ph)(Me).[155] The paramagnetic diporphyrin complexes M_2R_2(DPB), containing two Ru(III) centers, undergo successive reductions to form $[M_2R_2(DPB)]^-$ and diamagnetic $[M_2R_2(DPB)]^{2-}$, containing Ru(III)/Ru(II) and two Ru(II) centers, respectively.[156]

One further category of σ-bonded ruthenium and osmium complexes contains nitrosyl ligands. Reaction of Ru(TTP)(NO)Cl with aryl or alkyl Grignard reagents led to Ru(TTP)(NO)(p-C_6H_4F) or Ru(TTP)(NO)Me.[159] The analogous osmium complexes were prepared beginning from Os(OEP)(NO) which was oxidized by $NO^+PF_6^-$ to the cationic intermediate $[Os(OEP)(NO)]^+$ (not isolated) which was then treated with Grignard reagents to give Os(OEP)(NO)R (R = Me, Et, i-Pr, t-Bu, p-C_6H_4F). The thionitrosyl osmium complex Os(OEP)(NS)Me was prepared from Os(OEP)(NS)Cl with MeMgBr.[10]

Most of the various complexes reported above have been isolated and characterized by a range of spectroscopic techniques, although some, like the redox products $[Ru(OEP)R]^-$ and $[Ru(OEP)(Me)(THF)]^+$, were only observed in solution by

^{1}H NMR spectroscopy. The complexes which have been structurally characterized are Ru(OEP)Ph, Ru(OEP)CH_2CMe_3, [Ru(OEP)CH_2CMe_3]$_2$(μ-Li)$_2$, [Ru(OEP-N-Ph)Ph]BF_4, Os(TTP)(CH_2SiMe_3)$_2$, Ru(TTP)(NO)(p-C_6H_4F), and Os(OEP)(NS)Me, and some structural data for these are give in Table IV. They are a sufficiently diverse set of compounds such that it is difficult to draw direct comparisons between them. However, the Ru(III) complexes appear to have shorter Ru—C bonds than the Ru(II) compounds. The OEP-N-phenyl complex [Ru(OEP-N-Ph)Ph]BF_4 exhibits an agostic Ru···H interaction (2.52 Å) involving the *ortho*-hydrogen of the OEP N-phenyl group.[157] The complexes containing bulky neopentyl or CH_2SiMe_3 groups show considerable distortion arising from steric factors, including N-M-C angles bent significantly away from the 90° expected in idealized geometry, and angles at the neopentyl CH_2 groups much larger than the sp^3 angle of 109°.[150,152] [Ru(OEP)CH_2CMe_3]$_2$(μ-Li)$_2$ is composed of two Ru(OEP)CH_2CMe_3 fragments doubly bridged in a face-to-face fashion by two Li^+ cations, which are each π-complexed to a portion of the Ru(OEP) group.[152] In Ru(TTP)(NO)(p-C_6H_4F) and Os(OEP)(NS)Me the nitrosyl and thionitrosyl groups are bent.[10,159]

The Ru(II) complexes [Ru(OEP)CH_2CMe_3]$_2$(μ-Li)$_2$ and [Ru(OEP)R]$^-$, and the nitrosyl and thionitrosyl complexes Ru(TTP)(NO)(p-C_6H_4F) and Os(OEP)(NS)Me (whose oxidation states are not trivial to assign) are diamagnetic.[10,152–155,158,159] The Ru(IV) and Os(IV) porphyrin complexes M(Por)R_2 are also diamagnetic, in contrast to the dihalide species Ru(Por)X_2 and the single example of a monoalkyl or -aryl Ru(IV) complexes, [Ru(OEP)(R)(THF)]$^+$ (R = Ph, Me, observed spectroscopically), which are paramagnetic.[143,147,148,150,154,158] A low spin electron configuration $(d_{xz})^2(d_{yz})^2$ is proposed for the diamagnetic dialkyl or diaryl complexes. In the dihalo species, π-donation from the halide ligands raises the energies of the d_{xz} and d_{yz} orbitals and results in a paramagnetic $(d_{xy})^2(d_{xz})^1(d_{yz})^1$ configuration.[148] As expected, the Ru(III) and Os(III) complexes M(Por)R are low spin, paramagnetic complexes with $S = 1/2$, although they exhibit well-resolved, paramagnetically shifted ^{1}H NMR spectra.[148,151,152,155] A linear dependence of the chemical shifts on temperature indicates Curie behavior and a single spin state over the range 200–350 K.[151] The N-phenyl Ru(III) complex [Ru(OEP-N-Ph)Ph]$^+$ is also paramagnetic.[153,154]

The chemical reactivity of the organoruthenium and -osmium porphyrin complexes varies considerably, with some complexes (M(Por)R_2, M(Por)R and Os(OEP)(NO)R) at least moderately air stable, while most are light sensitive and stability is improved by handling them in the dark. Chemical transformations directly involving the methyl group have been observed for Ru(TTP)(NO)Me, which inserts SO_2 to form Ru(TTP)(NO){OS(O)Me}[159] and Ru(OEP)Me which undergoes H· atom abstraction reactions with the radical trap TEMPO in benzene solution to yield Ru(OEP)(CO)(TEMPO). Isotope labeling studies indicate that the carbonyl carbon atom is derived from the methyl carbon atom.[160] Reaction of

TABLE IV

SELECTED DATA FOR STRUCTURALLY CHARACTERIZED RUTHENIUM AND OSMIUM PORPHYRIN COMPLEXES

	M–C Bond Length/Å	M–N (av) Bond Length/Å	M–N_4 Plane/Å	Other	Reference
σ-Bonded complexes					
[Ru(OEP–N–Ph)Ph]BF_4	1.999(4)	2.026(3)	0.147	Ru···H (N–Ph), 2.52 Å	157
Ru(OEP)Ph	2.005(7)	2.030(6)	0.122		151
Ru(OEP)CH_2CMe_3[a]	2.069(7), 2.12(1)	2.027(3)	0.11	Ru–C–C, 128.2(4), 126.2(6) Å N–Ru–C, 78.1(3)–107.9(3)°	152
Ru(TTP)(NO)(p-C_6H_4F)	2.095(6)	2.054		Ru–N–O, 152°	159
[Ru(OEP)CH_2CMe_3]$_2$(μ-Li)$_2$	2.100(3)	2.030(2)	0.0135	Ru–C–C, 124.2 Å N–Ru–C, 79.75(10)–107.06(10)°	152
Os(OEP)(NS)Me	1.999(8)[b]	2.056(7)		Os–N–S, 163.0(8)°	10
Os(TTP)(CH_2SiMe_3)$_2$	2.07(3), 2.17(3)	2.03(2)		C–Os–C, 140(1)° N–Os–C, 69.2(9)–111.6(9)°	150
Carbene, μ-carbido and silylene complexes					
Ru(TPP)(=C(CO_2Et)$_2$)(MeOH)	1.829(9)	2.046(6)	0.12	Ru–C–C, C–$C_{carbene}$–C, 112.2(7)°	164
[Ru(Pc)(py)]$_2$C	1.77(1)	2.010(8)	0.054	Ru–N(py), Ru–C–Ru, 174.5(8)° mean Pc···Pc separation, 3.65Å	123
Os(TTP)(=C(p-C_6H_4Me)$_2$)(THF)	1.865(5)	2.038(4)	0.14	Os–C–C, 123.5(3)° C–$C_{carbene}$–C, 113.0(4)°	168
Os(TTP)(=$CHSiMe_3$)(THF)	1.79(2)	2.052(7)	0.22	Os–C–Si, 142(2)°	168
Os(TTP)(=$SiEt_2$·THF)(THF)	Os–Si, 2.325(8)	2.03(2)	0.116	Si–O, 1.82(2) Os–Si–C, 117(1)°, 121(1)° C–Si–C, 110(1)°	167

[a] The neopentyl group is disordered, with two different orientations.

[b] Os–CH_3 and Os–NS ligands disordered, Os–C and Os–N distances could not be distinguished.

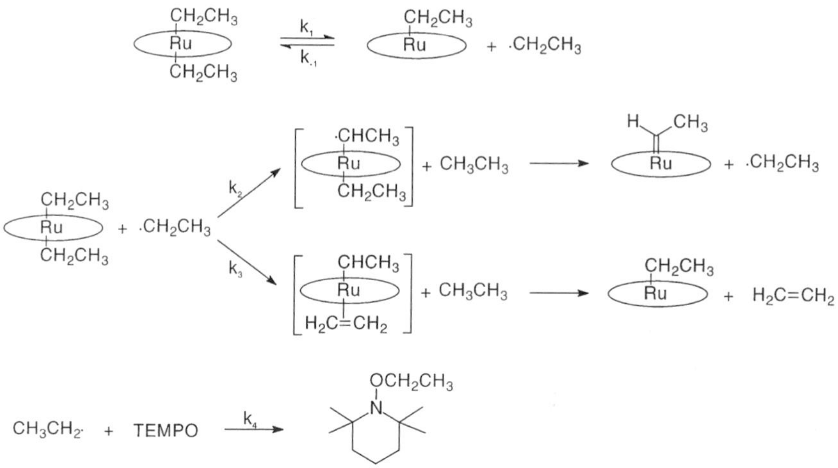

SCHEME 11.

the diorganoruthenium complexes $Ru(Por)R_2$ with HBr leads to cleavage of the Ru—C bond and formation of $Ru(Por)Br_2$.[147] The five-coordinate Ru(III) complexes Ru(Por)R will coordinate a sixth, neutral ligand to form Ru(Por)(R)L, where L = pyridine, PPh_3, etc.[151]

The diethyl ruthenium porphyrin compound $Ru(OEP)Et_2$, upon standing in benzene at room temperature for a few hours, converts to a 1:2 mixture of the ethylidene complex Ru(OEP)=CHMe and the Ru(III) ethyl complex Ru(OEP)Et, along with ethane and ethene. Kinetic studies using TEMPO as a radical trap are consistent with Ru—C bond homolysis in $Ru(OEP)Et_2$ as the first step in the mechanism shown in Scheme 11. The Ru—C bond dissociation energy ($\Delta H^{\ddagger}$) was determined to be 23.7 ± 0.5 kcal mol^{-1}, which represents the upper limit for the Ru—C bond energy in this complex.[145] Kinetic studies on the thermal decomposition of $Ru(Por)Ph_2$ to Ru(Por)Ph and biphenyl allowed determination of upper limits for the Ru—C bond energies in these complexes to be 31.6 ± 0.5 kcal mol^{-1} for the OEP complex and 34.2 ± 0.6 kcal mol^{-1} for the TPP complex. Comparing all these values, the phenyl Ru—C bond is ca. 8 kcal mol^{-1} stronger than the ethyl Ru—C bond, and the phenyl Ru—C bond is stronger by 2.6 kcal mol^{-1} in the TPP relative to the OEP complex, consistent with a stronger Ru—C bond in the complex containing the less basic porphyrin ligand.[148,151] The stronger Ru-aryl bond strength is reflected chemically in the decomposition of Ru(OEP)(Ph)(Me) which slowly forms the five-coordinate phenyl complex Ru(OEP)Ph on standing in the solid state.[155]

The redox chemistry of the ruthenium aryl and alkyl porphyrin complexes has been very thoroughly investigated, including both electrochemical and chemical oxidation and reduction. This has led not only to the syntheses of new complexes, as summarized above, but also to a very thorough understanding of the redox processes themselves. The redox chemistry of Ru(OEP)Ph_2, Ru(OEP)Ph, and $[Ru(OEP\text{-}N\text{-}Ph)Ph]^+$, as elucidated by electrochemistry in THF, is summarized in Scheme 12.[153,154] The scheme is arranged so that complexes at the same formal oxidation level appear on the same horizontal line. (The one-electron reduced complex $[Ru(OEP)Ph_2]^-$, although formally at the Ru(III) oxidation level, has spectroscopic features consistent with a Ru(IV) porphyrin radical anion.) The formulation of most of the species in the scheme has been corroborated by observation of their ^{1}H NMR spectra, and some of the electrochemical transformations have been repeated using chemical oxidants or reductants (shown by heavy arrows in the scheme). Of particular interest is the fact that one-electron oxidation of Ru(OEP)Ph_2 results in rapid migration of one phenyl group to a pyrrolic nitrogen to give the well-charcaterized (X-ray crystal structure) complex $[Ru(OEP\text{-}N\text{-}Ph)Ph]^+$

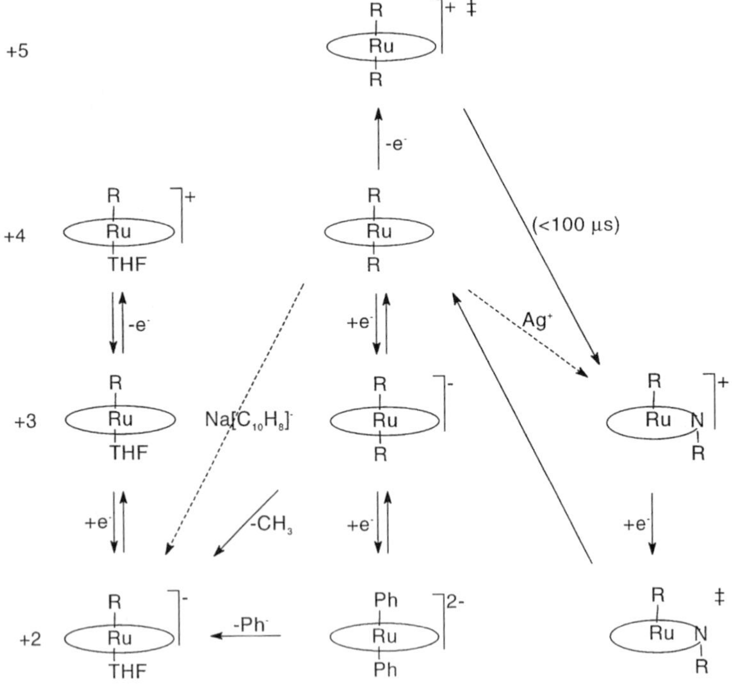

SCHEME 12.

(Fig. 5).[157] One-electron reduction of this complex results in the reverse migration and reforms $Ru(OEP)Ph_2$. In each case, the initial one-electron oxidized or reduced species has a very short lifetime and is not observed directly. This chemistry contrasts with the one-electron oxidation of Ru(OEP)Ph, which does not result in phenyl migration, and gives the Ru(IV) product $[Ru(OEP)Ph]^+$. Scheme 12 also applies for the redox chemistry of the dimethyl complex $Ru(OEP)Me_2$.[158] The only significant difference is that $[Ru(OEP)Me]^-$ is believed to form by loss of $\cdot CH_3$ from $[Ru(OEP)Me_2]^-$, in contrast to $[Ru(OEP)Ph]^-$ which is produced by loss of Ph^- from $[Ru(OEP)Ph_2]^{2-}$. The properties of the highly oxidized and reduced transient species $[Ru(OEP)Ar_2]^+$ and $[Ru(OEP)Ar_2]^-$, respectively, were investigated by studying the electrochemistry of a series of complexes where Ar = p-C_6H_4X, and X = H, Me, OMe, Cl or F.[149]

C. *Complexes with π-Bonded Ligands*

1. *Alkene and Alkyne Complexes*

The Ru(IV) dialkyl complexes $Ru(Por)R_2$ (R = Me, Et) were first prepared by the reaction of the dianion $K_2[Ru(Por)]$ with the corresponding alkyl halide in THF.[143] When 1,2-dibromoethane was used, the resulting product can be formulated either as the Ru(IV) metallacyclopropane, or as the Ru(II) ethene complex, $Ru(Por)(H_2C{=}CH_2)$, and indeed the reaction of the dimer $[Ru(Por)]_2$ with ethene gas gave the identical product (Scheme 13). Reaction of $[Os(TTP)]^{2-}$ with 1,2-dibromoethane gave the osmium analogue, $Os(TTP)(H_2C{=}CH_2)$. The ethene protons appear as a singlet at −4.06 ppm in the ^1H NMR spectrum of $Ru(TTP)(H_2C{=}CH_2)$, an upfield shift of 9 ppm relative to free ethene.[142,143] The dinitrogen ligands in $Ru(TMP)(N_2)_2$ can be displaced by alkene or alkyne ligands in C_6D_6. Ethene coordinated to Ru(TMP) has been observed by NMR (δ −3.27 ppm),

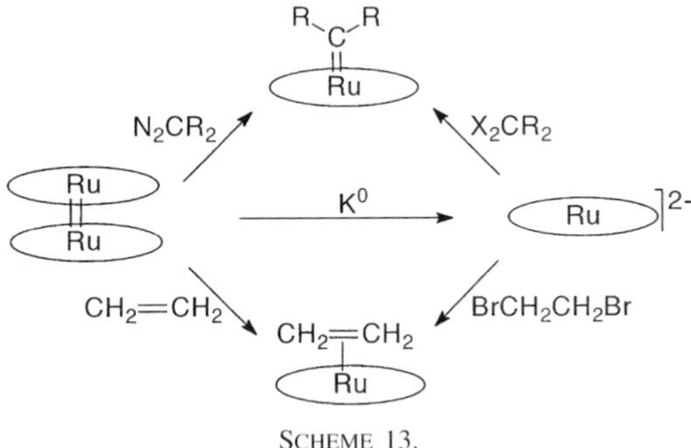

SCHEME 13.

and a broad peak appears for the alkene protons of cyclohexene in Ru(TMP)(C_6H_{10}).[161] Phenylacetylene and diphenylacetylene formed 1:1 π-complexes with Ru(TMP), again with clear ^{1}H NMR evidence for upfield shifted resonances for the phenyl groups in both complexes, and the alkyne hydrogen in PhC≡CH appears at −7.40 ppm in Ru(TMP)(PhC≡CH). The alkene complexes can be six-coordinate, with one equivalent of THF observed for Ru(TTP)(CH_2=CH_2)(THF) in C_6D_6 solution;[142] and the isopropanol adduct Ru(TMP)(CH_2=CH_2)(i-PrOH) has also been reported.[161]

In contrast, the reaction of Ru(TMP)$(N_2)_2$ with ethyne in C_6D_6 requires ambient light in order to proceed, and gave a downfield shifted peak at 10.44 ppm for the C_2H_2 group in the ^{1}H NMR spectrum ($\delta\,^{13}C = 264$ ppm) of the product, indicating a carbenoid rather than a π-alkene coordination mode. In addition, the C_2H_2 resonance integrated for 1 proton per Ru(TMP) unit, representing 2:1 stoichiometry, confirmed by a molecular weight determination. The product, which could be isolated, was deduced to be the bridging bis(carbene), (TMP)-Ru=CH−CH=Ru(TMP), and contrasts with the reaction of the rhodium dimer $[Rh(OEP)]_2$ with ethyne which gave the bridging alkenyl product (OEP)-Rh−CH=CH−Rh(OEP), with the CH protons at −9.92 ppm. Ethyne can be displaced from the ruthenium complex by CO or phosphites (L), giving the free alkyne and Ru(TMP)L_2.[161]

The most recent report of π-coordination to a ruthenium porphyrin fragment details the reaction of $[Ru(OEP)]_2$ with C_{60} in benzene/THF (100:1) solution. The UV-visible spectrum of the complex showed a new band at 780 nm, not observed in the spectrum of either $[Ru(OEP)]_2$ or C_{60}, and ^{1}H and ^{13}C NMR data also indicated the presence of a new complex. This has been formulated on the basis of the spectroscopic data as the fullerene complex Ru(OEP)(η^2-C_{60}).[162]

2. *Carbene Complexes*

Ruthenium porphyrin carbene complexes were reported at the same time as the first alkyl and alkene complexes, and can be prepared by the two complementary routes either using the dianion $[Ru(Por)]^{2-}$ with a geminal alkyl dihalide or by reaction of the dimer $[Ru(Por)]_2$ with a diazoalkane (Scheme 13).[142,143] The former route was used to prepare Ru(TTP)=$CHSiMe_3$ and the vinylidene complex Ru(TTP)=C=C(p-$C_6H_4Cl)_2$, using $Cl_2CHSiMe_3$ and Cl_2C=C(p-$C_6H_4Cl)_2$, respectively, and the latter method for Ru(Por)=$CHCO_2Et$, using N_2CHCO_2Et (ethyldiazoacetate) (Por = OEP, TTP, TMP).[142,143,163] The use of diazoalkane reagents to form ruthenium porphyrin carbene complexes has very recently been demonstrated to be effective using the carbonyl complex Ru(TPP)(CO)(EtOH) as the ruthenium precursor, avoiding the laborious conversion of the carbonyl complex to the ruthenium dimer. The preparation of Ru(TPP)(=CRR′) by this method was demonstrated using N_2CHCO_2Et, $N_2C(CO_2Et)_2$ and $N_2C(Me)CO_2Et$.[164]

The vinylidene complex Ru(TTP)=C=C(p-$C_6H_4Cl)_2$ has been prepared by an alternative route, using $Ru_3(CO)_{12}$ to insert ruthenium into a free-base porphyrin already containing the vinylidene fragment bridging between two adjacent pyrrole nitrogen atoms. Two further novel compounds which contain a disrupted porphyrin ring are also produced in this reaction. The first compound still essentially contains the vinylidene moiety bridged between two adjacent nitrogen atoms. However, a pyrrole N—C bond of one of the vinylidene-linked pyrrole groups has been ruptured by oxidative addition to ruthenium, forming Ru—N and Ru—C bonds. The ruthenium coordination sphere is completed by the other two pyrrole nitrogens and two CO ligands. The second compound is closely related, but contains a saturated N,N′-$CHCHAr_2$ bridging group. The chemistry is reversible, and heating the disrupted ruthenium porphyrin vinylidene complex at 130°C in chlorobenze repaired the porphyrin and reformed the axially symmetric vinylidene complex Ru(TTP)=C=C(p-$C_6H_4Cl)_2$.[165]

Neither the reaction of $[Ru(Por)]^{2-}$ with CH_2Cl_2 in THF nor that of $[Ru(Por)]_2$ with diazomethane in THF yielded the anticipated methylidene complex Ru(Por)=CH_2. Instead, each reaction resulted in equimolar mixtures of the ethylene complex Ru(Por)(CH_2=CH_2) and the bis(THF) complex $Ru(Por)(THF)_2$, possibly through bimolecular coupling of putative ruthenium methylidene fragments.[142,143] An ethylidene complex, Ru(TTP)=CHMe, can be synthesized from $[Ru(TTP)]^{2-}$ with Cl_2CHMe, or from $[Ru(TTP)]_2$ with N_2CHMe, and observed by ^1H NMR with the distinctive carbenoid proton resonating at 13.03 ppm.[142,143] However, this complex undergoes further reaction in solution, eventually rearranging to form the isomeric ethene complex, Ru(TTP)(CH_2=CH_2). The ethylidene and ethene complexes are also observed as the decomposition products formed from the diethyl complex $Ru(TTP)Et_2$. As illustrated in Scheme 11 these three species are linked by hydrogen abstraction processes initiated by Ru—C bond homolysis in the diethyl complex.[145]

Osmium porphyrin carbene complexes have been prepared by reaction of $[Os(TTP)]_2$ with diazoalkanes, forming Os(TTP)=C(p-$C_6H_4Me)_2$, Os(TTP)-(=$CHSiMe_3$), and Os(TTP)=$CHCO_2Et$.[166] In an important extension of this work, osmium silylene complexes were prepared either from the reaction of $[Os(TTP)]_2$ with hexamethylsiiacyclopropane (eliminating Me_2C=CMe_2 and forming Os(TTP)=$SiMe_2$·THF) or, more conveniently, from $[Os(TTP)]^{2-}$ with Cl_2SiR_2. This latter method was effective for R = Me, Et or i-Pr, but not for t-Bu or Ph, and in each case gave the silylene-THF adduct Os(TTP)(=SiR_2·THF).[167]

The porphyrin ligands in the diamagnetic ruthenium and osmium carbene complexes generally exhibit four-fold symmetry by NMR, indicating that the barrier to rotation about the M=C bond is low. The carbenoid protons appear shifted downfield in the ^1H NMR spectra, for example appearing for Ru(TTP)=$CHCO_2Et$ and Ru(TTP)=$CHSiMe_3$ at 13.43 and 19.44 ppm, respectively, and for the osmium analogues at 21.60 and 28.95 ppm, respectively.[142,143,166,168] Downfield shifts for the carbenoid proton in non-porphyrin complexes is common, and it is surprising that despite the ring current effect of the porphyrin the carbenoid protons are still significantly deshielded.

Several examples of carbene complexes have been structurally characterized (Fig. 5), and selected data for Ru(TPP)(=C(CO_2Et)$_2$)(MeOH), Os(TTP)-(=C(*p*-C_6H_4Me)$_2$)(THF), Os(TTP)(=CHSiMe_3)(THF), Os(TTP)(=SiEt_2·THF)-(THF) and a μ-carbido phthalocyanine complex, [Ru(Pc)(py)]$_2$C, are given in Table IV.[123,164,167,168] The ruthenium carbene complex has a Ru=C bond significantly shorter than those observed in the alkyl and aryl ruthenium porphyrin complexes. Ru(TPP)(=C(CO_2Et)$_2$)(MeOH) and Os(TTP)(=C(*p*-C_6H_4Me)$_2$)(THF) show, as expected, planar geometry at the carbene carbon atom, and the bulky SiMe_3 group in Os(TTP)(=CHSiMe_3)(THF) results in an enlarged Os—C—Si angle of 142(2)°. The osmium silylene complex is an example of a base-stabilized silylene, and shows a very short Os—Si bond and a long Si—O(THF) bond. The sum of the angles about silicon (involving osmium and the two ethyl groups) is 348°, indicating the silylene fragment is not planar but is significantly flattened relative to idealized sp^3 geometry (Fig. 5).[167]

The ruthenium and osmium carbene complexes are air sensitive, with the osmium silylene complex extremely so. The THF coordinated to silicon in the silylene complex can be displaced using pyridine, giving, for example, Os(TTP)=SiMe_2·py.[167] The silylene ligand itself is displaced by a carbene fragment when the silylene complex is treated with N_2CHCO_2Et.[166] The structurally characterized carbene complexes are six-coordinate, and exchange of the neutral THF ligand in the ruthenium and osmium carbenes by, for example, PPh_3 or pyridine has been reported. The THF ligand can be removed in vacuo and the five-coordinate complexes observed in solution by NMR.[164,168] Os(TTP)(=C(*p*-C_6H_4Me)$_2$)(THF) was unreactive toward a variety of regents (MeI, acetone, ethanol, silacyclopropane) but did react with pyridine N-oxide to give the diaryl ketone and Os(TTP)(=O)$_2$.[166]

The osmium carbene complex Os(TTP)=CHCO_2Et reacted with 4-picoline (4-pic) and other substituted pyridines to give ylide complexes, for example Os(TTP)(CHCO_2Et· 4-pic)(4-pic) which contains two picoline groups, one bonded to the carbene carbon and the other directly to the osmium. There is a dramatic shift in the chemical shift of the Os—CH proton, from 21.6 ppm in the carbene to −3.09 ppm in the ylide. Os(TTP)(=CHSiMe_3)(4-pic) exists in equilibrium with the ylide Os(TTP)(CHSiMe_3·4-pic)(4-pic) in C_6D_6 containing 4-picoline again with a large chemical shift difference for the Os—CH proton (28.95 vs −4.82 ppm). The zwitterionic formulation for the ylide complex involves a formal positive charge on the 4-picoline nitrogen and a formal negative charge on the osmium center.[169]

The most significant and widely studied reactivity of the ruthenium and osmium porphyrin carbene complexes is their role in catalyzing both the decomposition of diazoesters to produce alkenes and the cyclopropanation of alkenes by diazoesters. Ethyl diazoacetate is used to prepare the carbene complex Os(TTP)(=CHCO_2Et) from $[Os(TTP)]_2$, and it was noted that if excess N_2CO_2Et was used, the *cis* and *trans* alkenes $EtO_2CCH{=}CHCO_2Et$ (maleate and fumarate esters, respectively) were produced in a 2:16 ratio.[166] This reaction is catalyzed by the osmium porphyrin dimer, the carbene complexes, and even the carbonyl and bis(pyridine) complexes Os(Por)(CO)(py) and $Os(Por)(py)_2$, and was also reported for the ruthenium porphyrin dimers (OEP and TTP) and the monomeric complex Ru(TMP).[163] All give *cis:trans* ratios ranging from 15:1 to 26:1. The coupling reactions were very fast (complete in seconds) for N_2CHCO_2Et, but slower for $N_2C(p\text{-}C_6H_4Me)_2$ and were not observed for $N_2CHSiMe_3$. When the preformed carbene complexes $Os(TTP){=}CHSiMe_3$ or $Os(TTP){=}CAr_2$ ($Ar = p\text{-}C_6H_4Me$) were used as the catalysts with excess N_2CHCO_2Et, only the ethyldiazoacetate coupling products (maleate and fumarate) were observed, with no cross-coupling reactions, and the final osmium-containing products were the original carbene complexes. An osmium bis(carbene) complex has been proposed as a possible intermediate in these reactions, and one example, $Os(TTP)({=}CAr_2)_2$ ($Ar = p\text{-}C_6H_4Me$), has been observed by NMR, with the carbene carbon observed at 305.5 ppm in the ^{13}C NMR spectrum.[166]

The osmium carbene complex $Os(TTP){=}CHCO_2Et$ is capable of both stoichiometrically and catalytically cyclopropanating styrene using ethyl diazoacetate. This is important because carbene intermediates are proposed but have not been directly observed for the many transition metal complexes which catalyze this reaction, while carbene complexes which achieve the stoichiometric reaction do not show catalytic activity. In the stoichiometric reaction, $Os(TTP){=}CHCO_2Et$ with excess styrene gave the cyclopropane product in 73% yield with an *anti:syn* ratio of 11.5:1. Sample conditions for the catalytic process are $[Os(TTP)]_2$ (1.7 μmol), N_2CHCO_2Et (950 μmol) and styrene (960 μmol), giving an overall yield of cyclopropane with an *anti:syn* ratio of ca. 10:1.[170] The osmium carbonyl porphyrin complex also acted as a catalyst for the reaction. The similarity of the product ratios for the stoichiometric and catalyzed reactions were interpreted as evidence that the catalytic cycle involves a carbene complex. NMR evidence for the coordination of the styrene substrate to osmium was also reported. Other alkenes (α- or β-methylstyrene, 1-decene) were cyclopropanated, although with lower yields and stereoselectivities. Diphenylacetylene was doubly cyclopropanated (very rare) to give the bicyclobutane product.[170] The osmium ylide complex $Os(TTP)(CHCO_2Et{\cdot}4\text{-pic})(4\text{-pic})$ is also active for the stoichiometric cycloproapanation of styrene, with a high *anti:syn* ratio of 23:1, and is a potential model for metal diazoalkyl complexes proposed as intermediates in cycloproapanation reactions.[169]

Ruthenium porphyrin complexes are also active in cyclopropanation reactions, with both stoichiometric and catalytic carbene transfer reactions observed for $Ru(TPP)({=}C(CO_2Et)_2)$ with styrene.[164] Ru(Por)(CO) or $Ru(TMP)({=}O)_2$ catalyzed the cyclopropanation of styrene with ethyldiazoacetate, with *anti:syn* ratios of 13:1

for TPP but ca. 7:1 for the more crowded TMP ligand. Like the osmium system, aryl alkenes and terminal alkenes are more reactive than alkyl or internal alkenes. Diazoalkane coupling reactions which give maleate and fumarate are observed in place of cyclopropanation when a less reactive alkene is used as the substrate.[171] Moderate enantiomeric excesses are observed when a chiral ruthenium porphyrin is used as the catalyst.[171–174] A ruthenium carbene complex is presumed to be an intermediate, and when Ru(TMP)(CO) is used as the catalyst the carbene CH proton can be observed by ^{1}H NMR spectroscopy at 13.23 ppm.[171] Carbene intermediates have also been proposed in the ruthenium porphyrin-catalyzed insertion of the $CHCO_2Et$ fragment into the N—H and S—H bonds of amines and thiols.[175]

D. *Hydride and Dihydrogen Complexes*

A ruthenium porphyrin hydride complex was first prepared by protonation of the dianion, $[Ru(TTP)]^{2-}$ in THF using benzoic acid or water as the proton source. The diamagnetic complex, formulated as the anionic Ru(II) hydride $[Ru(TTP)(H)(THF)]^-$, showed by ^{1}H NMR spectroscopy that the two faces of the porphyrin were not equivalent, and the hydride resonance appeared dramatically shifted upfield to −57.04 ppm. The hydride ligand in the osmium analogue resonates at −66.06 ppm.[143] Reaction of $[Ru(TTP)(H)(THF)]^-$ with excess benzoic acid led to loss of the hydride ligand and formation of $Ru(TTP)(THF)_2$.

More details were elucidated by a study of the osmium OEP analogue. The anionic hydride $[Os(OEP)(H)(THF)]^-$ (δ −65.6) was protonated by benzoic acid to give a new complex containing two hydrogen atoms observed at (δ −30.00 in the ^{1}H NMR spectrum. The deuterium substituted complex showed $^1J_{HD} = 12$ Hz and $T_1 = 110 \pm 8$ ms at 20°C, consistent with an η^2-H_2 complex, $Os(OEP)(H_2)(THF)$. The dihydrogen ligand can be deprotonated using lithium diisopropylamide. The corresponding dihydrogen complex was more difficult to observe for ruthenium, although preparation of $[Ru(OEP)(H)(THF)]^-$ from $Ru(OEP)(THF)_2$ with H_2 and KOH was evidence for hydride formation through heterolytic activation of an intermediate dihydrogen complex.[176,177] A ruthenium porphyrin dihydrogen complex with a 2:1 Ru:H_2 ratio was observed using the cofacial diporphyrin $Ru_2(DPB)$, which coordinates the H_2 ligand in the pocket between the two ruthenium centers.[178] Osmium binds H_2 more tightly than ruthenium, and as a result the osmium η^2-H_2 complexes are more stable and more hydridic than their ruthenium counterparts. Variation of the σ-donor ligand in the sixth-coordination site showed that stronger σ-donors (imidazole derivatives) destabilize the H_2 complexes relative to weaker σ-donors such as THF. The osmium H_2 complex is more difficult to deprotonate than its ruthenium counterpart, requiring a stronger base to effect this (lithium diisopropylamide vs OH^-). Considerable attention was paid to the ability of this system to achieve H/D exchange between H_2 and D_2O as this is an important criterion for hydrogenase actvity.[179] Hydrogen can be released from the anionic hydride $[Ru(Por)(H)(THF)]^-$ either by protonation or by oxidation. The protonation reaction to release H_2 is not bimolecular, and the behavior of $[Ru(Por)(H)]^-$ contrasts with the isolectronic rhodium porphyrin hydride complex, Rh(Por)H, which readily undergoes bimolecular H_2 elimination, and in fact exists in equilibrium with H_2 and the dimer $[Rh(Por)]_2$.[180]

The sterically hindered ruthenium porphyrin complex Ru(TMP) and a hindered cofacial bisporphyrin complex Ru(DPAHM) (two trimesitylporphyrin ligands linked through the fourth *meso* position on each porphyrin by an anthracene bridge) are extremely reactive complexes containing 14-electron Ru(II) centers. They are sufficiently electron deficient that they appear to bind aromatic solvents such as benzene or toluene, and catalyze H_2/D_2 exchange in benzene solution or as solids. In addition, the complexes slowly catalyze exchange of H_2 into deuterated aromatic solvents, and in the absence of solvent, exchange of D_2 into CH_4.[181]

VI

COBALT

A. *Overview*

There are two predominant structural types for organometallic cobalt porphyrins. The first comprises 5- or 6-coordinate complexes (Por)Co(R) or (Por)Co(R)(L), where R is a σ-bonded alkyl, aryl, or other organic fragment, and L is a neutral donor. Most of the important synthetic developments and spectroscopic studies on cobalt σ-alkyl and -arylporphyrins were established prior to the last decade, and have been reviewed.[5,7,8] A source of considerable interest in σ-alkyl cobalt porphyrin species has been their close relationship with coenzyme B_{12}. Cobalt-adenosyl bond homolysis initiates the important hydrogen-abstraction functions of B_{12}, and as a result Co—C bond energies in cobalt porphyrin complexes are interesting for comparison. This has been a focus of continuing interest, and has been directly addressed in a paper by Halpern, intriguingly entitled "Why does nature not use the porphyrin ligand in Vitamin B_{12}?"[182] An exciting recent development for σ-alkyl cobalt porphyrins has been their use as catalysts for the free radical living polymerization of alkenes.

The second structural type found for organometallic cobalt porphyrins contains an organic fragment bridged between the cobalt and one pyrrolic nitrogen. Cobalt complexes of N-alkyl- or N-arylporphyrins are well established (but will not be specifically addressed here). The bridged complexes are derivatives of these where the N-alkyl group also forms a σ-bond to cobalt. They are also related to the axially symmetric carbene complexes observed for iron, ruthenium, and osmium in the sense that a carbene fragment, CR_2, can interact with a metalloporphyrin fragment M(Por) either by forming a double ($\sigma + \pi$) bond to the metal or by forming two σ-bonds, one to the metal and one to a pyrrole nitrogen. As discussed in the section on iron porphyrins, the two can be interconverted, where one-electron oxidation of an iron vinylidene complex induces formation of the Fe,N-bridged species. In cobalt (and nickel) porphyrin chemistry, only the bridging form of carbene is observed, although migration of a σ-bonded ligand from Co to N is also triggered by one-electron oxidation. Cobalt porphyrin complexes containing a Co, N-bridged carbene ligand were first reported over 25 years ago, and early developments are covered in previous review articles, including an interesting theoretical analysis that rationalizes the occurrence of the axially symmetric iron(II) porphyrin carbenes vs the Co,N-bridged cobalt(III) porphyrin carbenes, despite the fact that both types of complexes contain d^6, first-row transition metals.[5,7,8] The last decade has seen continuing developments in the chemistry of Co,N-bridged organometallic porphyrin species.

The chemistry of organorhodium and -iridium porphyrin derivatives will be addressed in a separate section. Much of the exciting chemistry of rhodium (and iridium) porphyrins centers around the reactivity of the M(II) dimers, $[M(Por)]_2$, and the M(III) hydrides, M(Por)H. Neither of these species has a counterpart in cobalt porphyrin chemistry, where the Co(II) porphyrin complex Co(Por) exists as a monomer, and the hydride Co(Por)H has been implicated but never directly observed. This is still the case, although recent developments are providing firmer evidence for the existence of Co(Por)H as a likely intermediate in a variety of reactions.

B. *Cobalt σ-Bonded Alkyl and Aryl Complexes*

1. *Synthesis and Reactivity*

Cobalt σ-alkyl or σ-aryl porphyrin complexes are generally prepared either by reaction of the Co(III) halides Co(Por)X with organolithium or Grignard reagents, or from the Co(I) anions $[Co(Por)]^-$ with alkyl or aryl halides. σ-Aryl complexes are best prepared by the former method, utilizing a Co(III) precursor.[5,7,8] A practical drawback to the use of organolithium and Grignard reagents is their air and moisture sensitivity. Alkylation of Co(TPP)Cl to produce Co(TPP)R using SnR_4 (R = Me, Et, *i*-Pr, *n*-Bu) or one of a selection of organocobalt complexes containing bis(dimethylglyoximato) or bipyridyl ligands avoids this problem.[183] Several other new methods for preparing σ-bonded complexes involve intermediate Co(Por)H species and will be discussed in that section.

An attempt to lithiate the Co-bonded aryl ring in Co(TPP)(p-C_6H_4Br) by reaction with *n*-BuLi led instead to exchange of the aryl group for the butyl group, giving Co(TPP)Bu as the product.[184] This reaction is more general, and a range of Co(TPP)R complexes containing aryl, alkyl, styryl, and alkynyl groups, when treated with organolithium compounds at −78°C, led either to rapid substitution of R or to an equilibrium mixture of products. Similar, but slower, exchange is observed with Grignard reagents. The process is postulated to proceed by attack of the carbanion R'^- on the substrate to give an intermediate "ate" complex of low stability, $[Co(TPP)RR']^-$, which then loses whichever of R or R′ is the better leaving group. The relative leaving group ability of R and R′ determines whether complete exchange or an equilibrium mixture is observed. Evidence for the "ate" complex comes from ^1H NMR observation of $[Co(TPP)(C{\equiv}CPh)_2]^-$ produced when R and R′ are both alkynyl. The much reduced *trans* effect of alkynyl groups (relative to simple aryl or alkyl groups) accounts for the reduced lability of this complex, which is stable in solution at 20°C.[184]

Very recently, a series of trihalomethyl cobalt porphyrin complexes $Co(OEP)CX_3$ was prepared either by the reaction of Co(OEP) with $CBrCl_3$, CBr_4 or CI_4, or by the reaction of $[Co(OEP)]^-$ with CCl_4 or CBr_4. The dihalomethyl complexes, formed in small amounts in these reactions, were prepared in larger amounts from $[Co(OEP)]^-$ with CHX_3.[185] In a similar fashion, R_3SnH reacted with $Co(OEP)CH_3$ or Co(OEP) to give $Co(OEP)SnR_3$ and CH_4 or H_2, respectively. Stannyl complexes were also prepared from R_3SnCl with $[Co(OEP)]^-$ (R = *n*-Bu, Ph). An X-ray crystal structure of $Co(OEP)SnPh_3$ confirmed its five-coordinate nature and exhibited a Co—Sn bond length of 2.510(2) Å (Fig. 6). This complex is very thermally stable and moderately air stable, and reacts with I_2 to give Co(OEP)I and $ISnPh_3$.[185] An

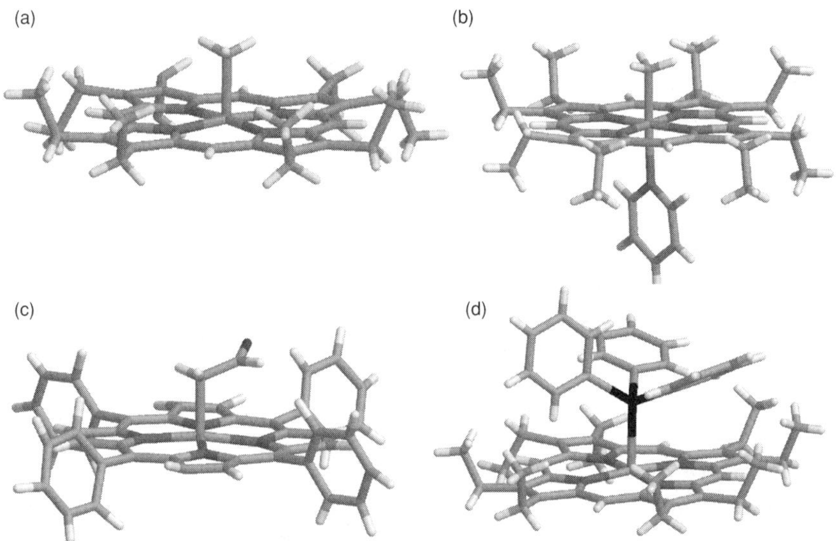

FIG. 6. Molecular structures of selected organometallic cobalt porphyrin complexes: (a) $Co(OEP)CH_3$,[190] (b) $Co(OEP)(CH_3)(py)$,[190] (c) $Co(TPP)CH_2CHO$,[188] (d) $Co(OEP)SnPh_3$.[185]

attempt to prepare a cobalt porphyrin silyl complex by the reaction of $[Co(TPP)]^-$ with Me_3SiCl in HMPA solvent led instead to the unexpected formation of a cobalt phosphoryl complex, $Co(TPP)P(O)(NMe_2)_2$.[186]

Two recent examples of Co,N-bridged carbene complexes have been prepared from the reactions of $[Co(OEP)(OH_2)]ClO_4$ with the diazoalkanes $N_2CHSiMe_3$ and $N_2C(CO_2Me)_2$ (Eq. (5)). Different reactivity is seen for the Co(III) halide complex Co(Por)Br (Por = OEP or TPP), which reacts, for example, with $N_2C(CO_2Me)_2$ to give a derivative containing a Co—O—C(OMe)=C(CO_2Me)—N bridged complex (Eq. (6)).[136,187]

(5)

(6)

2. *Structural and Spectroscopic Features*

Only a small number of structurally characterized organometallic cobalt porphyrin complexes have been reported, and selected data for these are collected in Table V.[188,189] In a recent study, structural features of the Co—C bond have been probed by a comparison of the molecular structures of $Co(OEP)CH_3$, $Co(OEP)(CH_3)(py)$, and Co(OEP)(py). The Co—C bond lengths are 1.973(6) and 2.018(12) Å, respectively, and both the Co—C and Co—N(py) bonds are longer in the six-coordinate complex. This has been attributed to the *trans* influence, with the Co-N(py) distance in this complex (2.214(9) Å) longer than that in other complexes of the type Co(Por)(X)(py), and in fact is one of the longest Co—N distances recorded. The five- and six-coordinate methyl complexes have been discussed as base-off and base-on models of coenzyme B_{12}.[190] Representative structures of cobalt alkyl complexes are shown in Fig. 6.

In diamagnetic porphyrin complexes the ^1H and ^{13}C NMR chemical shifts for axial ligands are usually observed upfield of their normal diamagnetic position due to the porphyrin ring current effect. The extent of the upfield shift usually reflects the distance of the nucleus in question from the macrocycle, so for *n*-alkyl axial ligands, the largest upfield shift is observed in the order ($\alpha > \beta > \gamma$, with the α-protons exhibiting the largest upfield shift. This is observed for simple *n*-alkyl complexes of Rh(III) porphyrins, for example, but cobalt *n*-alkyl porphyrins are anomalous with the α-proton resonance broadened and appearing downfield of the

TABLE V

SELECTED DATA FOR STRUCTURALLY CHARACTERIZED COBALT COMPLEXES

	M—C Bond Length/Å	M—N(av) Bond Length/Å	M—N_4 Plane/Å	Other	Ref.
-Bonded complexes					
'o(TPP)CH_2CHO	1.976(7)	1.969(9)	0.14		188
'o(OEP)CH_3[a]	1.976(6)	1.966(4)	0.094(3)		190
	1.970(6)		0.105(3)		
o(OEP)CH_2CH_3	1.988(2)	1.970(8)	0.102	Co—C—C, 114.25(14)°	192
o(OEP)(CH_3)(py)	2.018(12)	1.983(19)	0.008 (toward py)		190
o(TPP)$CH_2C(O)CH_3$	2.028(3)	1.948(4)	0.11		189
o(OEC)Ph	1.937(3)	1.856(3)	0.185		201
Co(OEC)Ph]ClO_4	1.970(7)	1.849(5)	0.165		201
i(TPP){μ-(Ni,N)—$CHCO_2Et$}	1.905(4)	1.916(3)[b]	0.19	Ni···N, 2.610(3) Å	288
'annyl complex					
o(OEP)$SnPh_3$	Co—Sn, 2.510(2)	1.966(5)	0.077		185

[a]Two independent molecules.
[b]Three Fe—N bonds only. Pyrrole nitrogen involved in carbene bridge is not bonded to iron.

β-protons. Various reasons have been proposed for this phenomenon, including a suggestion that an agostic interaction between cobalt and the β-hydrogens brings them into closer proximity to the porphyrin, and hence upfield of the α-protons.[191] A more detailed analysis finds that this is not the case,[192] based in part on an X-ray structure of Co(OEP)CH_2CH_3 which exhibits a Co—C bond of 1.988(2) (very similar to that observed for Co(OEP)CH_3[190]) and a Co—C—C angle of 114.25(14)°, inconsistent with a Co-β-H agostic interaction.[192] The broadening of the α-proton resonance in the ^1H NMR spectrum and reversal of its position relative to the β-protons is attributed to coupling to the quadrupolar ^{59}Co nucleus ($I = 7/2$), and a paramagnetic contact shift arising from a low-lying triplet state. In Co(OEP)CH_3 the methyl resonance exhibits anti-Curie behavior, consistent with a contact shift mechanism. Coordination of pyridine to Co(Por)(n-alkyl) complexes causes the α-protons to shift upfield and the β-protons to shift downfield. The pyridine ligand increases the energy difference between the ground state and the low-lying triplet state and decreases the size of the contact shifts.[192]

3. *Cobalt-Carbon Bond Energies*

Investigations of the cobalt-carbon bond energies in organometallic cobalt porphyrins continue to attract interest, originally because of their similarity to coenzyme B_{12}, and more recently because of their role in the catalysis of free radical living polymerization of alkenes. Bond energies have been measured from both kinetic and thermodynamic (equilibrium) studies. In the kinetic method, rates of Co—C bond homolysis are measured using TEMPO as a radical trap. The kinetic data can be used to calculate the Co—C bond energy after allowing for the diffusion-controlled Co—C recombination reaction (Eqs. (7) and (8) and the rate expression in Eq. (9)).[193] One particular study used this method to investigate the effect on the Co—C bond energy in $Co(OEP)(CH_2Ph)(PR_3)$ by varying the pK_a and size

$$Co^{III}(Por)R \underset{k_{-1}}{\overset{k_1}{\rightleftharpoons}} Co^{II}(Por) + R\cdot \tag{7}$$

$$R\cdot + TEMPO \xrightarrow{k_2} R\text{-}TEMPO \tag{8}$$

$$k_{obs} = \frac{k_1 k_2[TEMPO]}{k_{-1}[Co(Por)R] + k_2[TEMPO]} \tag{9}$$

(cone angle) of the phosphine ligand, PR_3.[182] The reason for this is that the vitamin B_{12} coenzyme contains a cobalt corrin complex, in which the puckered corrin ring, although closely related to a porphyrin, can adopt variable conformations. These changes in conformation affect the Co—C bond energy for the Co-5′-adenosyl group, and can trigger Co—C bond homolysis. Variation of the steric and electronic properties of the phosphine ligand in the cobalt porphyrin study modeled this process. The Co—C bond energies (in the range 24–30 kcal mol^{-1}) were found to be dependent on the phosphine pK_a values but independent of the size of the phosphine ligands. It was proposed that the porphyrin ligand is not sufficiently flexible, relative to a corrin, to transmit the steric effect of the bulky axial ligand to the Co—C bond. In other words, the porphyrin acts as a rigid barrier shielding the $Co—CH_2Ph$ group from steric perturbations, and this has been postulated as a reason nature does not use the porphyrin ring in vitamin B_{12}.[182] The importance of steric effects in vitamin B_{12} was further supported by a vibrational study on a series of complexes Co(Por)R, where Por = OEP, TPP or TMP, and R = CH_3 or CD_3. FT Raman data for $\nu_{Co\text{—}C}$ showed an isotopic shift for each pair of CH_3/CD_3 complexes of ca. 30 cm^{-1}. For $Co(OEP)CH_3$ and $Co(TPP)CH_3$, $\nu_{Co\text{—}C}$ was very similar (ca. 500 cm^{-1}) but for the much more crowded $Co(TMP)CH_3$ a big shift to lower energy, with $\nu_{Co\text{—}C} = 459\ cm^{-1}$, illustrating the steric effects on the Co—C bond.[194]

NMR line broadening is a suitable kinetic method for determining activation parameters for Co—C bond homolysis, and gave $\Delta H^{\ddagger}$ values in the range 18–22.5 kcal mol^{-1} for a selection of Co(Por)R complexes containing secondary or tertiary alkyl groups.[195] Bond dissociation enthalpies and entropies for several complexes Co(Por)R and their singly oxidized products $[Co(Por)R]^+$ (formed by chemical oxidation) were studied in a stopped flow kinetic study. The Co—C bonds in the oxidized products are weaker than in the neutral complexes.[196]

Equilibrium studies on Co—C bond dissociation energetics were investigated through a study on the reactions of the cobalt(II) radical complex Co(TAP)· with organic radicals, $\cdot C(CH_3)(R)CN$, in the presence of alkenes. Fast abstraction of H· from the organic radicals by Co(TAP)· formed an intermediate hydride Co(TAP)H (not observed) which added rapidly to alkenes to form organocobalt complexes. Equilibrium concentrations of the solution species present once the organic radical species had achieved steady-state concentrations allowed evaluation of the equilibrium constants for Co—C bond homolysis. For example, ΔH° values for Co—C bond homolysis in $(TAP)Co{-}C(CH_3)_2CN$ and $(TAP)Co{-}CH(CH_3)C_6H_5$ were determined directly by this method to be 17.8 ± 0.5 and 19.6 ± 0.6 kcal mol^{-1}, respectively. The value for $(TAP)Co{-}C_5H_9$ was measured indirectly by competition studies and found to be 30.9 kcal mol^{-1}.[197]

In terms of relative Co—C bond energies, those in the trihalomethyl complexes $Co(OEP)CX_3$ are observed to be qualitatively weaker than in Co-alkyl porphyrin complexes. The high thermal stability of the cobalt porphyrin stannyl complexes was interpreted as an indication that, surprisingly, the Co—Sn bond is stronger than the Co—C bond in Co(Por)(alkyl) complexes.[185]

4. *Redox Chemistry and Electrochemistry*

Organocobalt porphyrin complexes have been investigated by electrochemistry, and earlier studies are summarized in review articles.[7,8] Briefly, oxidation of Co(Por)R led to migration of R from cobalt to a porphyrin nitrogen atom, giving a Co(II) complex of an N-alkyl porphyrin, $[Co(Por\text{-}N\text{-}R)]^+$.[198] Reduction led to $[Co(Por)]^-$ or $[Co(Por)R]^-$, with the exact nature of the product dependent on the solvent, the scan rate, and the nature of R. In more recent developments, the initial site of electron transfer was elucidated in an electrochemical study on Co(TPP)R ($R = CH_3$, C_2H_5 or $CHCl_2$) and $Co(TPP)(C_2H_5)(py)$. Each complex undergoes up to two reductions and two oxidations, each of which occurs at the porphyrin π-system rather than at the cobalt center, giving Co(III) porphyrin anions and cations, respectively.[198] The oxidation of Co(TPP)R and Co(TPP)(R)(L) ($R = CH_3$, C_2H_5, C_4H_9 or C_6H_5; L = MeCN or a substituted pyridine) was investigated in more detail.[196] One- or two-electron oxidation of the cobalt complexes was achieved using $[Fe(phen)_3]^{3+}$ as the oxidant, and the products of the initial oxidation were characterized by EPR and electronic spectroscopy and stopped-flow kinetics. The initial singly oxidized products have some d^5 Co(IV) character for the six-coordinate complexes $[Co(TPP)(R)(L)]^+$, while the five-coordinate complexes $[Co(TPP)R]^+$ are best described as Co(III) porphyrin π-cation radicals. After two-electron initial oxidation, migration of the R group from cobalt to a porphyrin nitrogen is observed, and the kinetics of this process measured. The final product was the Co(III) N-substituted porphyrin complex $[Co(TPP\text{-}N\text{-}R)]^{2+}$.[196]

Chemical oxidation of vinyl- or aryl-cobalt porphyrins, for example by $[Ar_3N\cdot]^+[SbCl_6]^-$, also induces the migration of the σ-bonded group from cobalt to nitrogen.[199] The stereochemistry of the reversible metal-to-nitrogen transfer of alkyl and aryl groups was investigated using chiral cobalt(III) porphyrins, indicating that both the Co-to-N and the reverse migrations take place by intramolecular routes.[200]

A transient Co(IV) species is believed to be important in the Co-to-N alkyl or aryl migration process, although stable Co(IV) porphyrin complexes have not been reported (and indeed few stable non-porphyrin Co(IV) complexes are known). The corrole macrocycle, similar to a porphyrin but with one *meso*-carbon replaced by a direct bond between two α-pyrrolic carbon atoms and bearing a 3− charge in its deprotonated form, is more effective than a porphyrin at stabilizing high oxidation states. Recently prepared organocobalt corrole complexes show a number of unusual features.[201] The cobalt octaethylcorrole (OEC) complex Co(OEC)Ph, which formally contains Co(IV), was prepared by the reaction of the Co(III) complex Co(OEC) with PhMgBr. Co(OEC)Ph was further oxidized in CH_2Cl_2 solution using aqueous iron perchlorate as the oxidant, giving $[Co(OEC)Ph]ClO_4$ which was more stable than its neutral precursor. The molecular structures of both complexes were determined, and show Co—C bond lengths of 1.937(3) and 1.970(7) Å for Co(OEC)Ph and $[Co(OEC)Ph]^+$, respectively. The cobalt atom fits more neatly into the corrole plane in the oxidized complex, with shorter Co—N bonds. Both complexes exist in the solid state as π-π dimers with the two corrole planes ca. 3.5 Å apart. EPR data and magnetic measurements indicate that Co(OEC)Ph can be described as a d^5 Co(IV) complex, with one unpaired electron in a d orbital oriented toward the corrole nitrogens, giving the complex some Co(III) corrole π-cation radical character. $[Co(OEC)Ph]ClO_4$ was formulated as a Co(III) complex in which the corrole macrocycle is doubly oxidized, and as such exhibits unusual NMR behavior, including a paramagnetic ring current.[201]

Co(TPP) has been demonstrated to act as a catalyst for the electrocarboxylation of benzyl chloride and butyl bromide with CO_2, to give $PhCH_2C(O)OCH_2Ph$ and BuOC(O)C(O)OBu, respectively. The proposed mechanism involved Co(TPP)R and $[Co(TPP\text{-}N\text{-}R)]^+$ as intermediates (the latter detected by spectroscopy) in the catalytic production of free R^- or R·, which then reacted directly with CO_2.[202,203] Co(TPP) precipitated on graphite foil has been successfully used for the determination of organic halides, including DDT and 1,2,3,4,5,6-hexachlorocyclohexane (lindane), to sub-ppm level in aqueous solution. Deoxygenation of the solutions is not required, and the technique is moderately insensitive to the ionic composition of the solution.[204]

C. *The Search for Cobalt Porphyrin Hydride Complexes*

In contrast to the rhodium porphyrin hydride complexes, Rh(Por)H, which play a central role in many of the important developments in rhodium porphyrin chemistry, the corresponding cobalt porphyrin hydride complexes have been implicated as reaction intermediates in a variety of processes, but a stable, isolable example has yet to be achieved.

Cobalt(II) complexes of three water-soluble porphyrins are catalysts for the controlled potential electrolytic reduction of H_2O to H_2 in aqueous acid solution. The porphyrin complexes were either directly adsorbed on glassy carbon, or were deposited as films using a variety of methods. Reduction to $[Co(Por)]^-$ was followed by a nucleophilic reaction with water to give the hydride intermediate. Hydrogen production then occurs either by attack of H^+ on Co(Por)H, or by a disproportionation reaction requiring two Co(Por)H units. Although the overall feasibility of this process was demonstrated, practical problems including the rate of electron transfer still need to be overcome.[205,206]

Co(OEP) reacted with $NaBH_4$ and phenylacetylene or 1-hexyne in oxygenated benzene/methanol to give the vinyl complexes Co(OEP)C(R)=CH_2 (R = Ph or Bu). In a similar process using an alkene in place of the alkyne, alkylcobalt products were generated. For example, 1-hexene, 2-pentene, and 2-heptene each gave a secondary alkyl complex with the Co—C bond at the 2-position, and cyclopentene and cyclohexene resulted in the cobalt-cyclopentyl and -cyclohexyl products, respectively. More highly substituted alkenes did not react. The reactions were accelerated by the addition of an oxidant such as *t*-butyl hydroperoxide. The reaction was proposed to proceed through an intermediate hydride, Co(OEP)H, and the regiochemistry could be explained in terms of the stability of the organic radicals generated by addition of H· to the alkenes or alkynes. The reaction is summarized in Scheme 14, where the role of the oxidant is to oxidize Co(II) porphyrins formed from competing elimination of H_2.[207] The transient cobalt hydride can formally insert an unsaturated molecule (as above) or, alternatively, the hydride can react

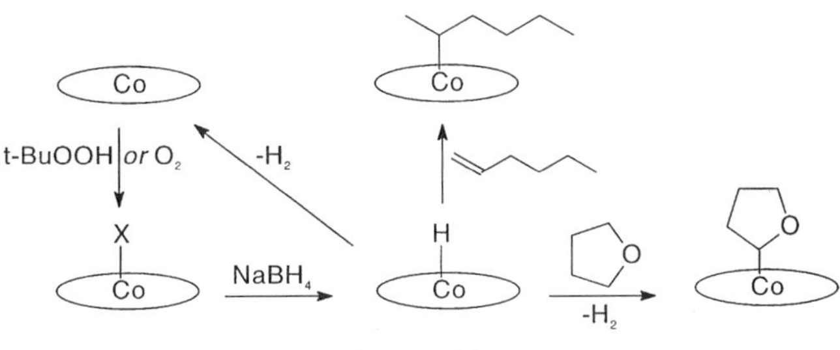

SCHEME 14.

with an organic substrate by hydrogen elimination to form a new organocobalt product. Several five- and six-membered ring cyclic ethers were metallated under the same conditions (using Co(OEP), $NaBH_4$, O_2, or an added oxidant) as shown in Scheme 14 for the case where THF is the substrate.[208]

With Co(OEP), $NaBH_4$, O_2, and propanal as the substrate, three products are observed, Co(OEP)C(O)Et (18%), Co(OEP)CH(Me)CHO (33%), and Co(OEP) OCH(Me)CHO (19%) (Eq. (10)), arising from nonselective hydrogen abstraction from the aldehyde to generate acyl and α-(formyl)alkyl radicals. If *t*-butyl hydroperoxide is added, then the acyl product Co(OEP)C(O)R is produced cleanly from a variety of aldehydes. Although these reactions have an apparent overall similarity to the reactions with alkene, alkyne, and cyclic ether substrates, it has been concluded that in the case of the aldehydes that a hydride intermediate, Co(OEP)H, is not involved. The acyl (and other products) are produced by Co(OEP) trapping of acyl radicals generated from reaction of *t*-BuO· with the aldehydes.[209,210] These reaction conditions have also been exploited to achieve the intramolecular cyclization of an aldehyde with an alkene.[210]

$$Co^{II} \xrightarrow[NaBH_4,\ O_2]{\text{propanal}} \text{Co–CH(Me)CHO (33\%)} + \text{Co–C(O)Et (18\%)} + \text{Co–OCH(Me)CHO (19\%)} \quad (10)$$

In the above examples the transient Co(OEP)H was generated by $NaBH_4$ reduction of a Co(III) porphyrin precursor, which then reacted with organic substrates in a radical mechanism through attack by H·. An alternative method for generation of Co(Por)H is to use organic radicals as the source of the hydride, by the reaction of Co(TAP) with tertiary organic radicals, ·C(Me)(R)CN, themselves produced from commercially available diazo radical initiators such as AIBN or VAZO-52.[211,212] In chloroform, Co(TAP) reacts with the organic radical and an alkene, $CH_2{=}CHX$, or alkyne, $HC{\equiv}CX$ to give alkyl or vinyl products, $Co(TAP)CH(CH_3)X$ or $Co(TAP)C(X){=}CH_2$, respectively. Reactions of Co(TAP) and $\cdot C(Me)_2CN$ with either alkyl halide or epoxide substrates occurred in DMF, giving alkyl or 2-hydroxyalkyl products. The reactions were conveniently followed by ^1H NMR spectroscopy, observing the loss of Co(TAP) and formation of the organocobalt(III) products. This process, which was exploited for the preparation of over 33 organocobalt derivatives, and which also works for Co(OEP), Co(TMP), and cobalt phthalocyanine, has been dubbed "tertiary radical synthesis."[211,212]

The products of the above reactions are consistent with Markovnikov addition of transient Co(Por)H to the unsaturated alkene or alkyne substrate. The regiochemistry is determined by formation of the most stable organic radical, which tends to produce secondary rather than primary alkyls, and thus the products are kinetically rather than thermodynamically favored. The rate of the reaction with alkenes matched the rate of radical production from the initiator. Yields were quantitative for very stable organocobalt products like Co(Por)CH(Me)CN or Co(Por)CH(Me)CO_2Et formed from acrylonitrile or ethyl acrylate, respectively. Yields dropped as the Co—C bond strength decreased as a result of competition with Co—C bond homolysis. For alkyne substrates, the kinetic product forms at short reaction times or low temperatures, with the thermodynamic product appearing after longer times (Eq. (11)). For example, HC≡CCH_2OH initially gives Co(TAP)C(CH_2OH)=CH_2, with CoCH(Me)CHO appearing at longer times as the result of a 1,3-hydrogen shift and keto-enol tautomerization.

Co^{II} —•C(CH_3)(R)CN→ [H–Co]‡ —HC≡CCH_2OH→ Co–C(=CH_2)CH_2OH → [Co–C(CH_3)=CHOH] → Co–CH(CH_3)CHO (11)

The corresponding reactions of transient Co(OEP)H with alkyl halides and epoxides in DMF has been proposed to proceed by an ionic rather than a radical mechanism, with loss of H^- from Co(OEP)H to give $[Co(TAP)]^-$, and products arising from nucleophilic attack on the substrates.[211,212] Overall, a general kinetic model for the reaction of cobalt porphyrins with alkenes under free radical conditions has been developed.[213] Cobalt porphyrin hydride complexes are also important as intermediates in the cobalt porphyrin-catalyzed chain transfer polymerization of alkenes (see below).

D. *Cobalt Porphyrins as Catalysts for the Polymerization of Alkenes*

The definition of a living polymerization process is one where each polymer unit contains an active site where chain growth occurs indefinitely without termination or chain transfer reactions. Living radical polymerization can be achieved using a combination of one radical that initiates polymerization, and a second that binds reversibly to the growing polymer radical. The use of Co(TMP)R to initiate and control the polymerization of acrylates to form hompolymers and block copolymers was first reported in 1994.[214,215] Indications of a living radical polymerization process were a linear increase in the number average molecular weight with monomer conversion, and relatively small polydispersities. Co(TMP)CH_2CMe_3 or Co(TMP)CH(Me)CO_2Me were used to initiate the formation of polymethylacrylate at 60°C in benzene under argon, with a ratio of methyl acrylate monomer

$$\mathrm{Co(TMP)R \longrightarrow Co(TMP)\cdot + R\cdot}$$

$$\mathrm{R\cdot + CH_2{=}CHX \longrightarrow RCH_2CHX\cdot}$$

$$\mathrm{RCH_2CHX\cdot + Co(TMP)\cdot \rightleftharpoons RCH_2CHXCo(TMP)}$$

$$\mathrm{RCH_2CHX\cdot + (n+1)CH_2{=}CHX \rightarrow RCH_2CHX(CH_2CHX)_nCH_2CHX\cdot}$$

$$\mathrm{RCH_2CHX(CH_2CHX)_nCH_2CHX\cdot + Co(TMP) \rightleftharpoons CH_2CHX(CH_2CHX)_nCH_2CHXCo(TMP)}$$

SCHEME 15.

to cobalt porphyrin of 2500:1. When this ratio was limited to 50:1, sequential formation of $\mathrm{Co(TMP)\{CH(CO_2Me)CH_2\}_nCH(CO_2Me)CH_3}$ units with $n = 0, 1, 2$, etc., was observed by ^{1}H NMR spectroscopy. The reaction sequence for this process is shown in Scheme 15. The initiating radical is R· and the radical which reversibly binds to the growing polymer chain is Co(TMP). Although the process is not fully living, because some bimolecular radical termination events do occur, the system is self-regulating through the persistent radical effect.[214,215]

Cobalt porphyrin complexes are involved in the chain transfer catalysis of the free-radical polymerization of acrylates. Chain transfer catalysis occurs by abstraction of a hydrogen atom from a growing polymer radical, in this case by Co(Por) to form Co(Por)H. The hydrogen atom is then transferred to a new monomer, which then initiates a new propagating polymer chain. The reaction steps are shown in Eqs. 12 (where R is the polymer chain, X is CN), (13), and (14).[213]

$$\mathrm{(Por)Co + \cdot C(R)(Me)X \rightarrow (Por)CoH + CH_2{=}C(R)X} \tag{12}$$

$$\mathrm{(Por)CoH + CH_2{=}C(Me)X \rightarrow (Por)Co + \cdot C(Me)_2X} \tag{13}$$

$$\mathrm{\cdot C(Me)_2X + (n+1)\{CH_2{=}C(Me)X\} \rightarrow X(Me)_2C\{CH_2C(Me)X\}_nCH_2C(Me)X\cdot} \tag{14}$$

This is closely related to the "tertiary radical synthesis" scheme for the preparation of organocobalt porphyrins, in which alkenes insert into the Co—H bond of Co(Por)H instead of creating a new radical as in Eq. (13). If the alkene would form a tertiary cobalt alkyl then polymerization rather than cobalt-alkyl formation is observed.[211,212] The kinetics for this process have been investigated in detail, in part by competition studies involving two different alkenes. This mimics the chain transfer catalysis process, where two alkenes (monomer and oligomers or the alkene formed from the radical initiator) are present.[213] Isotope studies using deuterated methyl methacrylate-d_8 showed a kinetic isotope effect greater than 3, indicating that hydrogen atom transfer occurs in the rate-limiting step of the catalytic cycle.[216] The dependence of free-radical propagation rate constants on the degree of polymerization in the cobalt porphyrin-catalyzed chain transfer polymerization of methyl methacrylate and methacrylonitrile has also been investigated.[217]

E. *Reactions of Cobalt(III) Porphyrins with Alkenes and Alkynes*

The interaction of alkynes with cobalt(III) porphyrins was first reported in 1986, with the reaction of $[Co(OEP)(H_2O)_2]ClO_4$ with a series of alkynes, including ethyne itself, in the presence of $Fe(ClO_4)_3$ as oxidant. The ultimate products of these reactions were the N^{21},N^{22}-etheno-bridged porphyrins, containing an N—C(R)=C(R)—N bridge between two adjacent nitrogen atoms in the porphyrin macrocycle.[218] Since that time, the details of the interactions of alkynes with cationic cobalt(III) porphyrins have been gradually unravelled.

The reaction of $[Co(OEP)(H_2O)_2]ClO_4$ with RC≡CR (R = CO_2Me) in CH_2Cl_2 rapidly gave a product containing an etheno bridge, Co—C(R)=C(R)—N, between cobalt and a pyrrolic nitrogen. Oxidation of this complex with $FeCl_3$ followed by an acidic workup gave the N^{21},N^{22}-etheno-bridged OEP complex as its $HClO_4$ salt.[219,220] Addition of ethyne to $[Co(OEP)(H_2O)_2]ClO_4$ in CH_2Cl_2, followed by 2,6-lutidine, gave the cationic vinyl complex $[Co(OEP)—CH=CH—N(C_7H_9)]^+$ bearing a pyridinium substituent on the β-carbon.[221] Reaction of ethyne with an equimolar mixture of $H_2(OEP)$ and $[Co(OEP)(H_2O)_2]ClO_4$ in CH_2Cl_2 gave a vinyl complex Co(OEP)—CH=CH—N—(OEP)H (as the $HClO_4$ salt), in other words a complex containing N-vinyl-OEP as the axial ligand on cobalt.[222]

These apparently disparate reactions can be tied together by a closer investigation of the nature of the interaction between $[Co(Por)(H_2O)_2]ClO_4$ and ethyne (Por = OEP or TPP). This reaction was accompanied by a color change from red-brown to green and is reversible, showing isosbestic behavior when followed by UV-visible spectroscopy. These observations suggested the formation of a cobalt porphyrin ethyne adduct. ^1H NMR spectroscopy of the adduct in the presence of excess ethyne showed the complex to be paramagnetic, with C_s symmetry.[223] The formally Co(III) precursor $[Co(Por)(H_2O)_2]^+$ had some Co(II) porphyrin radical cation character. Two possible structures can be proposed for the ethyne adduct: a Co(II) porphyrin radical cation π-ethyne complex and a Co(III) complex containing a cationic σ-vinyl ligand. The β-carbon atom in the latter form is subject to nucleophilic attack. Intramolecular attack by a pyrrolic nitrogen gave the Co, N-etheno-bridged product, while intermolecular attack by a nitrogen nucleophile,

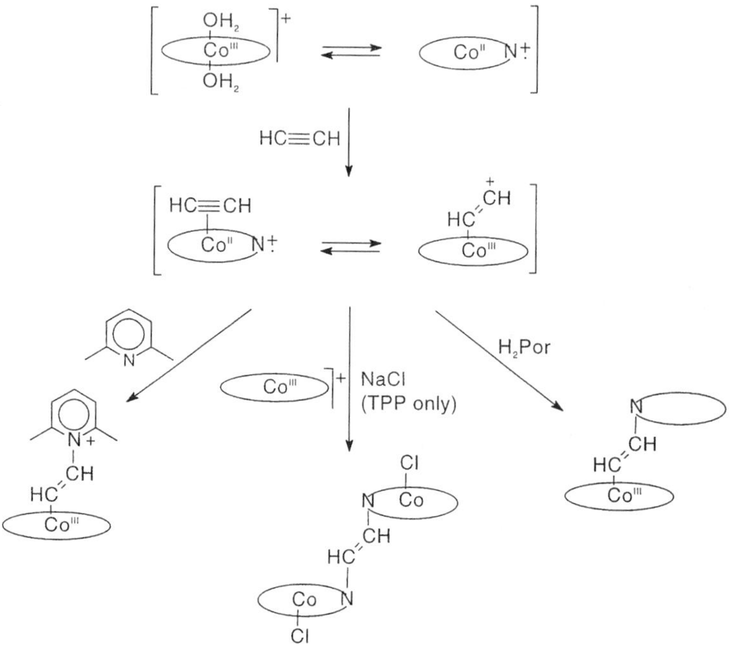

SCHEME 16.

either 2,6-lutidine or a free-base porphyrin, gave the cobalt-vinyl-pyridinium and Co,N-etheno-bridged bisporphyrin, respectively. These reactions are summarized in Scheme 16.[223]

The OEP and TPP complexes show some differences in reactivity. Reaction of ethyne with $[Co(TPP)(H_2O)_2]ClO_4$ in CH_2Cl_2, followed by addition of aqueous NaX (X = Cl, SCN) gave the biscobalt(II) bisporphyrin XCo(TPP—N—CH=CH—N—TPP)CoX, which could be demetallated to give free-base N,N′-etheno-linked bisporphyrin (Scheme 16).[222,223] Further reaction of the free-base bisporphyrin with $[Co(Por)(H_2O)_2]^+$ and ethyne gave tri- and tetraporphyrins, (Por)Co-R-(NPorN)-R-(NPor) and (Por)Co-R-(NPorN)-R-(NPorN)-R-Co(Por), where each –R- group is a —CH=CH— etheno bridge, and (NPorN) is a porphyrin linked through two adjacent nitrogens to an etheno bridge. The different reactivity of OEP and TPP could be exploited to create different combinations of homo- and heteroporphyrin oligomers. Oxidatively induced Co-to-N migration of the vinyl group can produce further variations in the bridge linkages.[224] The hetero-bisporphyrin complexes containing one cobalt and one free-base porphyrin display switching of the cobalt between the two porphyrin ligands during the course of demetallation/metallation sequences, and the transformation of an asymmetric, unsaturated N—CH=C—$N^{21'}$,$N^{22'}$ linkage to a symmetric, saturated N^{21},N^{22}—CH—CH—$N^{21'}$,$N^{22'}$ linkage was determined from an X-ray crystal structure of the free-base bisporphyrin containing the latter bridge.[225]

The electronic structure of cobalt(III) porphyrins depends on the nature of the axial ligand and the solvent. The simple halide complexes Co(Por)X are diamagnetic d^6 Co(III) complexes in both methanol and CH_2Cl_2. However, the aqua complex is also a diamagnetic d^6 Co(III) species in methanol, but in CH_2Cl_2 the axial ligands are lost and the species is best formulated as a d^7 Co(II) porphyrin π-cation radical.[219] As described above, $[Co(Por)(H_2O)_2]^+$ in CH_2Cl_2 reacts with substituted ethynes, RC≡CR, via an intermediate π-complex to give an intramolecular Co—C(R)=C(R)—N bridge if no other nucleophile is present. Different reactivity is observed for Co(Por)Cl (Por = OEP or TPP) which reacts with alkynes (HC≡CR, R = H, CO_2Me or CH_2OH) to give the β-chloro-vinyl product containing the alkyne inserted into the Co—Cl bond. For example, ethyne itself reacts with Co(TPP)Cl to generate Co(TPP)CH=CHCl. In the special case of $MeO_2CC{\equiv}CCO_2Me$, the product of its reaction with Co(OEP)Cl is an unusual Co(III) porphyrin complex where the intramolecular etheno bridge is between cobalt and a *meso*-carbon atom of the macrocycle.[226]

Silyl enol ethers react with Co(Por)X (Por = OEP or TPP, X = halide) to yield stable σ-alkyl cobalt porphyrins. For example, $H_2C{=}C(R)OSiMe_3$ reacted with Co(TPP)Cl with net loss of Me_3SiCl to give $Co(TPP)CH_2C(O)R$ (R = Ph, OEt or vinyl). The reactions were faster in CH_2Cl_2 than in MeOH, and faster for the weaker Co—X bonds (F $\geq$ Cl $\gg$ Br) reflecting the role of the $SiMe_3$ group in abstracting the halide. The butadiene derivative $CH_2{=}CHCH{=}CHOSiMe_3$ reacted with Co(Por)Cl to give only the primary alkyl product, $Co(Por)CH_2CH{=}CHCHO$, which could be identified by spectroscopy but was too labile to isolate.[227]

VII

RHODIUM AND IRIDIUM

A. *Overview*

Structural types for organometallic rhodium and iridium porphyrins mostly comprise five- or six-coordinate complexes (Por)M(R) or (Por)M(R)(L), where R is a σ-bonded alkyl, aryl, or other organic fragment, and L is a neutral donor. Most examples contain rhodium, and the chemistry of the corresponding iridium porphyrins is much more scarce. The classical methods of preparation of these complexes involves either reaction of Rh(III) halides Rh(Por)X with organolithium or Grignard reagents, or reaction of Rh(I) anions $[Rh(Por)]^-$ with alkyl or aryl halides. In this sense the chemistry parallels that of iron and cobalt porphyrins.

However, the significant key difference for rhodium arises from the chemistry of the Rh(II) dimer, $[Rh(Por)]_2$, which exhibits a relatively low Rh—Rh bond strength. It undergoes homolytic dissociation and exists in equilibrium with the monomer, Rh(Por)· (Eq. (15)). The rhodium dimer can also exist in equilibrium with the hydride Rh(Por)H (Eq. (16)), and thus the hydride complex can exhibit the chemistry of the dimer, driven by formation of the Rh(Por)· monomer formed as in Eqs. (15) and (16).

$$[Rh(Por)]_2 \rightleftharpoons 2Rh(Por)\cdot \quad (15)$$

$$2Rh(Por)H \rightleftharpoons [Rh(Por)]_2 + H_2 \quad (16)$$

The rhodium(II) porphyrin monomer Rh(Por) is an odd-electron species and can initiate radical reactions with a wide variety of organic substrates to give σ-bonded organometallic products. The seminal reaction of this type was the report in 1981 by Wayland *et al.* of the insertion of CO into the Rh—H bond in Rh(OEP)H to produce the formyl complex Rh(OEP)CHO.[228] This report generated considerable excitement among organometallic chemists as the insertion of CO into a metal-hydride bond had not been observed in classical organometallic chemistry. Furthermore, the hydride Rh(OEP)H apparently lacked the *cis* coordination site believed to be essential for migratory insertion reactions. The reaction was subsequently shown by Halpern *et al.* in 1985[229] to proceed by a radical chain mechanism, with the chain carrying Rh(OEP) fragment formed as in Eqs. (15) and (16). The key to this reaction is the balance between the Rh—Rh, Rh—C and Rh—H bond strengths, and in the intervening years Wayland's research group has further developed and refined the scope of reactions of this type, and assembled a useful array of thermochemical data. Another important development in this field has been the use of the bulky porphyrin TMP, which does not allow formation of the rhodium dimer, and thus monomeric Rh(TMP) can be prepared and used directly, instead of relying upon the equilibrium in Eq. (15) as the source of the odd-electron rhodium(II) monomer.

A key step proposed in the radical chain mechanism for the formation of the formyl complex is the coordination of CO to the Rh(OEP)· monomer, to give an intermediate carbonyl complex, Rh(OEP)(CO)· which then abstracts hydride from Rh(OEP)H to give the formyl product.[229] This mechanism was proposed without direct evidence for the CO complex, and since then, again from the research group of Wayland, various Rh(II) porphyrin CO complexes, Rh(Por)(CO)·, have been observed spectroscopically along with further reaction products which include bridging carbonyl and diketonate complexes.

While metalloporphyrin carbene complexes are well established for ruthenium and osmium, they are less well known for rhodium. Cationic rhodium porphyrin carbene intermediates were implicated in a report by Callot *et al.* in which rhodium(III) porphyrins catalyzed the decomposition of ethyl diazoacetate and transfer of the ethoxycarbonyl fragment to substituted alkenes to produce cyclopropane products.[230] The ratio of *syn:anti* products was higher than usually observed in reactions of this type. Further development of this reaction, spurred by its potential utility in organic synthesis, was undertaken by Kodadek *et al.*[273]

There have been very few developments in the chemistry of iridium porphyrins over the last decade. Synthesis and electrochemistry are covered in previous review articles.[7,8] The only recent report concerns activation of aldehydes and ketones by both $[Rh(OEP)]_2$ and $[Ir(OEP)]_2$. In general, iridium porphyrins show reactivity similar to rhodium porphyrins, although a key difference is that the insertion of CO into the Rh—H bond to give the formyl species, perhaps the seminal reaction in organometallic porphyrin chemistry, has not been observed to occur for iridium.

B. *Rhodium and Iridium σ-Bonded Alkyl and Aryl Complexes*

The syntheses and spectroscopic and electrochemical characterization of the rhodium and iridium porphyrin complexes (Por)M(R) and (Por)M(R)(L) have been summarized in three review articles.[5,7,8] The classical syntheses involve Rh(Por)X with RLi or RMgBr, and $[Rh(Por)]^-$ with RX. In addition, reactions of the rhodium and iridium dimers have led to a wide variety of rhodium σ-bonded complexes. For example, $[Rh(OEP)]_2$ reacts with benzyl bromide to give benzyl rhodium complexes, and with monosubstituted alkenes and alkynes to give σ-alkyl and σ-vinyl products, respectively. More recent synthetic methods are summarized below. Although the development of iridium porphyrin chemistry has lagged behind that of rhodium, there have been few surprises and reactions of $[Ir(Por)]_2$ and Ir(Por)H parallel those of the rhodium congeners quite closely.[231–233] Selected structural data for σ-bonded rhodium and iridium porphyrin complexes are collected in Table VI, and several examples are shown in Fig. 7.[234–236]

Electroreduction of the cationic Rh(III) complex $[Rh(Por)(MeNH_2)_2]^+$ in CH_2Cl_2 followed by reaction with alkyl halides has been utilized to form σ-alkyl products. The reaction scheme proposed for this reaction was one-electron reduction of Rh(III) to form Rh(Por)·. This can either dimerize or attack the carbon atom of the alkyl halide RCH_2X, the latter step involving elimination of either X· or X^-.[237,238] This parallels the reactions of Co(II) and Fe(II) porphyrins M(Por) with alkyl halides which also occur by radical reactions. However, the results of a recent electrochemical study in DMSO suggest that the Rh(II) porphyrin is a special case because in the polar solvents required for electrochemistry, the Rh(Por)· monomer disproportionates to Rh(III) and Rh(I) species, and the resulting $[Rh(Por)]^-$ anion then attacks the alkyl halide in a classical S_N2 reaction to give Rh(Por)R and X^-.[239,240] A similar process is implicated in the formation of Rh(Por)H from the electrochemical reduction of Rh(III) porphyrins in the presence of Brønsted acids.[241]

TABLE VI

SELECTED DATA FOR STRUCTURALLY CHARACTERIZED RHODIUM AND IRIDIUM COMPLEXES

	M–C Bond Length/Å	M–N(av) Bond Length/Å	M–N_4 Plane/Å	Other	Reference
σ-Bonded complexes					
Rh(OEP)CH_3	2.031(6)	2.032(6)	0.051	Rh···Rh, 7.22 Å	234
Rh(OEP)CH_3	1.970(4)	2.027(3)	0.024	Rh···Rh, 4.67	280
Rh(OEP)C(O)NH(2,6-$C_6H_3Me_2$)	1.988(5)	2.023(4)	0.072		263
Rh(TPP)(Ph)Cl	2.05	2.04			235
Rh(OEP)(3-C_6H_4CN)	2.0005(12)	2.035(8)	0.053		265
Rh(OEP)(μ-C,N-3-C_6H_4CN)	1.999(6)	2.022(5)		Rh-N(CAr), 2.2606(5)	266
Rh(OETAPP)(CH_3)	2.034(7)	1.962(4)	0.019		247
Ir(OEP)(C_8H_{13})	1.893(8)	2.05(1)	0.08		232
Ir(OEP)(C_3H_8)(PPh_3)	2.063(6)	2.039(4)	0.06 (toward P)		233
Ir(OEP)(C_3H_8)(DMSO)	2.08(1)	2.029(9)	0.01 (toward S)		233
Ir(Pc)(N_3)(CH_2C(O)CH_3)[a]	2.14(1) 2.12(1)	1.988 1.987			236
Carbene complex					
[Rh(TPP)(=C(NHR)$_2$)(CNR)]PF_6 (R = CH_2Ph)	2.030(11)	2.037(10)	~0	Rh–CNR, 2.064(3)	279
Heterobimetallic complexes					
Rh(TPP)$SiMe_3$	Rh–Si, 2.305(2)	2.016(5)			282
Rh(OEP)$SiEt_3$	Rh–Si, 2.32(1)	2.03(3)	0.090		269
Rh(OEP)$SnCl_3$	Rh–Sn, 2.450(1)	2.017(6)	0.02		283
Rh(OEP)In(OEP)	Rh–In, 2.584(2)	2.036(2)	0.01 (Rh) 0.83 (In)	OEP···OEP, 3.41 Å	284

[a] Two independent molecules.

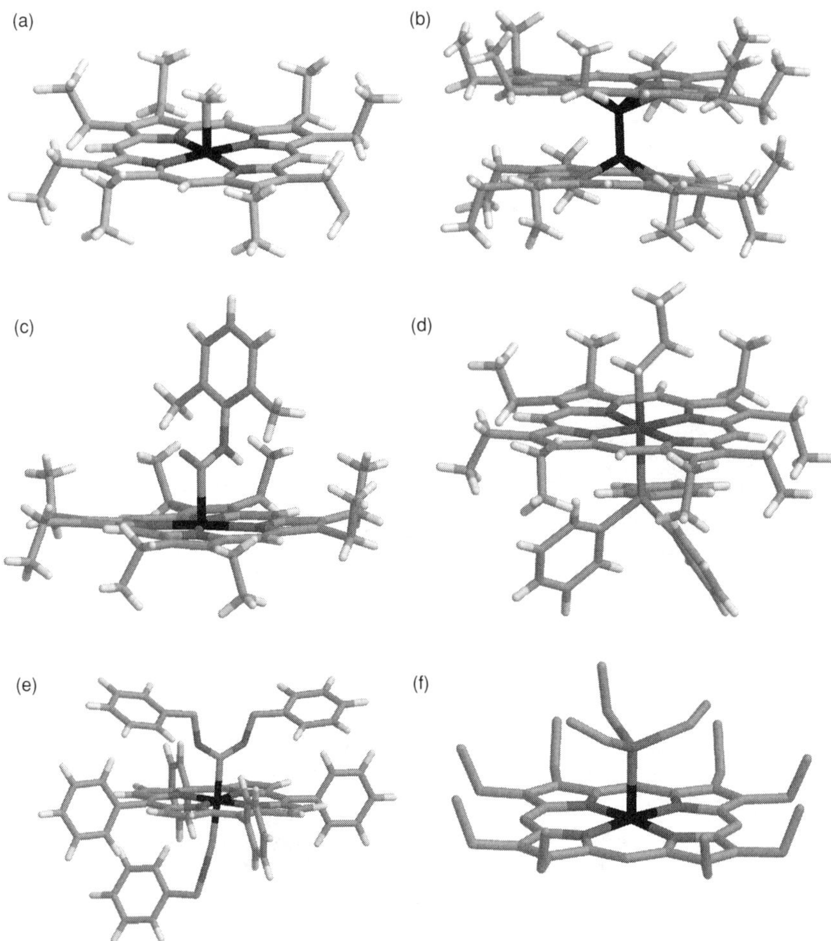

FIG. 7. Molecular structures of selected organometallic rhodium and iridium porphyrin complexes: (a) $Rh(OEP)CH_3$,[234,280] (b) Rh(OEP)In(OEP),[284] (c) $Rh(OEP)C(O)NH(2,6\text{-}C_6H_3Me_2)$,[263] (d) $Ir(OEP)(C_3H_8)(PPh_3)$,[233] (e) $[Rh(TPP)(=C(NHR)_2)(CNR)]^+$($R=CH_2$ Ph),[279] (f) $Rh(OEP)SiEt_3$.[269]

Radiolytic reduction has been investigated as a means of producing transient Rh(II) porphyrin products, and as in the above study, the observed products were strongly dependent on pH and solvent. Radiolytic reduction of Rh(TMP)Cl in alcohol formed transient Rh(TMP)· which was prevented from dimerization by the bulky TMP ligand. In alkaline 2-propanol the product is $[Rh(TMP)]^-$, in weakly acidic 2-propanol the hydride Rh(TMP)H is formed, and in strongly acidic 2-propanol the alkylated rhodium(III) porphyrins $Rh(TMP)CH_3$ and Rh(TMP) $(C(CH_3)_2OH)$ are observed. The alkyl products result from reaction of Rh(TMP)· with CH_3· and $\cdot C(CH_3)_2OH$ formed by radiolysis of the 2-propanol solvent.[242]

The electrochemistry of a series of σ-alkyl rhodium porphyrins, Rh(TPP)R, where R = C_nH_{2n+1} ($n = 1 - 6$) or $(CH_2)_nX$ (X = Cl, Br, I; $n = 3 - 6$), has been studied in detail, and in each case the site of the initial, reversible oxidation or reduction was concluded to be at the porphyrin ring rather than either the rhodium center or the alkyl ligand. In the case of reduction, the initially formed singly reduced species can then undergo Rh—C cleavage or, in some cases where R = $(CH_2)_nX$, loss of halide while retaining the Rh—C bond intact.[243] A similar study has investigated the electrochemistry of Rh(TPP)R(L) containing a neutral donor ligand.[244] Radiolytic reduction of Rh(TMP)CH_3 also leads to initial reduction at the porphyrin ring.[242]

C. *Reactions of the Rhodium(II) Porphyrin Dimer and Rhodium(III) Porphyrin Hydride*

The addition of metal hydrides to C—C or C—O multiple bonds is a fundamental step in the transition metal catalyzed reactions of many substrates. Both kinetic and thermodynamic effects are important in the success of these reactions, and the rhodium porphyrin chemistry has been important in understanding the thermochemical aspects of these processes, particularly in terms of bond energies.[245] For example, for first-row elements, M—C bond energies are typically in the range of 25–30 kcal mol^{-1}. M—H bond energies are usually 25–30 kcal mol^{-1} stronger, and as a result, addition of M—CH bonds to CO or simple hydrocarbons is thermodynamically unfavorable.

The reaction of Rh(OEP)H with CO or styrene (CH_2=CHPh) to form Rh(OEP)-CHO or Rh(OEP)CH_2CH_2Ph, respectively, and the related reaction of styrene with $[Rh(OEP)]_2$ to give (OEP)Rh—CH_2CH(Ph)—Rh(OEP), were rationalized in a landmark paper in 1985 which proposed that they proceed by radical chain reactions mediated by Rh(OEP)· (Scheme 17).[229]

$$2\text{Rh-H} \rightleftharpoons \text{Rh-Rh} + H_2$$

$$\text{Rh-Rh} \rightleftharpoons 2\text{Rh}\cdot$$

$$\text{Rh}\cdot + \text{L} \rightleftharpoons \text{Rh(L)}\cdot \quad (\text{L} = \text{CO}, \text{CH}_2\text{=CHPh})$$

$$\text{EITHER} \quad \text{Rh(L)}\cdot + \text{Rh-H} \rightleftharpoons \text{RhLH} + \text{Rh}\cdot$$

$$\text{OR} \quad \text{Rh(L)} + \text{Rh-Rh} \rightleftharpoons \text{Rh-L-Rh} + \text{Rh}\cdot \quad (\text{L} = \text{CH}_2\text{=CHPh})$$

SCHEME 17.

The key to understanding these reactions is appreciating the relative Rh—Rh, Rh—C, Rh—H, and Rh—O bond energies, and these have been developed by a series of studies from Wayland's research group.[245] Overall, the driving force for the radical chain process is the relatively weak Rh-Rh bond and the unusually strong Rh—C bond. This is nicely illustrated by looking at the formation of the formyl complex. For the overall reaction shown in Eq. (17), the enthalpy change ΔH° is related to the sums of the bond energies broken and formed (Eq. 18). Inserting the values for (C≡O) − (C=O) = 70 kcal mol^{-1} and (C—H) = 97 kcal mol^{-1} (for an aldehyde) gives the overall expression shown in Eq. (19). In other words, the enthalpy change for Eq. (17) will be negative when the M—H bond energy is less than 17 kcal mol^{-1} larger than the M—C bond energy. Given that M—H bond energies are 25–30 kcal mol^{-1} stronger than M—C bond energies, it is easy to see why formyl formation is usually thermodynamically unfavorable. Including an estimate for the entropy term (at 298 K) results in an even more stringent condition, in which ΔG° will be negative only when the M—H bond energy exceeds the M—C bond energy by no more than 9 kcal mol^{-1}.[245]

$$\mathrm{M{-}H + CO \rightleftharpoons M{-}CHO} \tag{17}$$

$$\Delta H^\circ = (\mathrm{M{-}H}) + (\mathrm{C{\equiv}O}) - (\mathrm{M{-}C}) - (\mathrm{C{=}O}) - (\mathrm{C{-}H}) \tag{18}$$

$$\Delta H^\circ = (\mathrm{M{-}H}) - (\mathrm{M{-}C}) - 17\ \mathrm{kcal\ mol^{-1}} \tag{19}$$

In the specific example involving the rhodium porphyrin hydride and formyl complexes, Rh(OEP)H and Rh(OEP)CHO, the Rh—H and Rh—C bond energies (62 and 58 kcal mol^{-1}, respectively) have been determined by equilibrium measurements. This is an unusual example where the Rh—H bond energy is in normal range but the Rh—C bond energy is unusually large. As a result the reaction to form the formyl complex is thermodynamically favorable and the difference in Rh—H and Rh—C bond energies is less than 9 kcal mol^{-1}. Addition of Rh(OEP)H to formaldehyde, ethene, and ethyne are also thermodynamically favorable. The Rh—Rh bond dissociation energy in $[\mathrm{Rh(OEP)}]_2$ (16.5 kcal mol^{-1}) has been estimated from kinetic studies based on ^{1}H NMR line broadening. The Rh—Rh bond dissociation energy is relatively weak, permitting facile bond homolysis to generate the Rh(OEP)· radical. This also has a favorable thermodynamic effect on the addition reactions of $[\mathrm{Rh(OEP)}]_2$ with CO and alkenes.[245]

Electronic effects on the reactions of $[\mathrm{Rh(Por)}]_2$ dimers and hydrides were probed by varying the porphyrin macrocycle. OEP and TPP vary considerably in their properties, with OEP being one of the strongest and TPP one of the weakest σ-donors among porphyrin derivatives. However, $[\mathrm{Rh(Por)}]_2$, Rh(Por)H, and $[\mathrm{Rh(Por)}]^-$ showed the same reactivity in a variety of reactions for both OEP and TPP, indicating that electronic effects relating to the porphyrin ligand have little influence.[246] A more dramatic electronic effect was achieved by using the octaethyltetraazaporphyrin macrocycle OETAP (in which the *meso* CH groups of OEP have been replaced by nitrogen atoms), which is both a stronger σ-donor and a stronger π-acceptor than OEP. A comparison of the reactions of $[Rh(OETAP)]_2$ and $[Rh(OEP)]_2$ showed that both reacted with ethene, CH_3I, CNMe, and $P(OMe_3)_3$, but only $[Rh(OEP)]_2$ reacted with H_2, H_2/CO, CH_3CHO, and $CH_3C_6H_5$. Equilibrium studies indicated that the Rh-Rh dissociation enthalpy of $[Rh(OETAP)]_2$ is larger than that of $[Rh(OEP)]_2$, although it is not easy to rationalize this with the different electronic properties of the two macrocycles.[247]

Both the $[Rh(OEP)]_2$ and $[Ir(OEP)]_2$ dimers reacted with the α-CH bonds in aldehydes and ketones. This is unusual because the stronger alkyl C—H bond reacted in preference to the weaker aldehydic C—H bond. For example, 2-methylpropanal reacts with $[Rh(OEP)]_2$ to give the β-formyl complex as the kinetic product, although this rearranged to the acyl complex which is the thermodynamic product (Eq. (20)). An explanation for this is that the dimers react initially with the enol tautomers of the aldehydes or ketones to give a bridged intermediate, $M—CR_2—CR(OH)—M$ (analogous to the reaction of the dimers with alkenes) which then dissociates with hydrogen migration from the OH group to give the β-carbonyl derivative $M—CR_2C(O)R$ together with M(OEP)H.[248]

$$[Rh(OEP)]_2 + Me_2CHCHO \rightleftharpoons (OEP)Rh—CMe_2CHO$$
$$\rightleftharpoons (OEP)Rh—C(O)CHMe_2 \qquad (20)$$

Diol dehydratase is an enzyme which functions together with coenzyme B_{12} to catalyze the dehydration of vicinal diols to aldehydes. A cobalt-1,3-dihydroxyalkyl intermediate has been proposed to occur in this process, but has never been directly observed in either the cobalamin or model cobalt porphyrin systems. A consequence of the strong Rh—C bond is that α-hydroxyalkyl complexes have been observed to form from the reaction of Rh(Por)H with aldehydes (Eq. 21). This reaction has been studied using glycoaldehyde as a model for diol dehydratase, producing $Rh(OEP)CH(OH)CH_2OH$ (characterized by ^1H NMR) which slowly dehydrates to form $Rh(OEP)CH_2CHO$.[249]

$$Rh(Por)H + RCHO \rightleftharpoons Rh(Por)CH(OH)R \qquad (21)$$

The radical Rh(Por)· generated from homolytic dissociation of $[Rh(Por)]_2$ reacts with alkenes $CH_2{=}CHX$ to produce an intermediate metalloorganic radical $(Por)RhCH_2CHX\cdot$, which then reacts with $[Rh(Por)]_2$ to produce the dinuclear complexes $(Por)Rh—CH_2CH(X)—Rh(Por)$ (Eqs. 22–24). If the intermediate metalloorganic radical $(Por)RhCH_2CHX\cdot$ is intercepted by the alkene rather than by the rhodium dimer, then the result could be alkene dimers, oligomers, or polymers. This possibility was investigated for acrylate substrates, $CH_2{=}CH(CO_2X)$ ($X = H$, CH_3, CH_2CH_3). With $[Rh(OEP)]_2$, the two-carbon bridged product

$$\text{Rh–Rh} \rightleftharpoons 2\ \text{Rh}\cdot \tag{22}$$

$$\text{Rh}\cdot + H_2C{=}CHX \rightleftharpoons \text{Rh–}CH_2\text{–}\dot{C}HX \tag{23}$$

$$\text{Rh–}CH_2\text{–}\dot{C}HX + \text{Rh–Rh} \rightleftharpoons \text{Rh–}CH_2\text{–}CHX\text{–Rh} \tag{24}$$

$(OEP)RhCH_2CH(CO_2X)Rh(OEP)$ is formed. Two stereoisomers were observed, resulting from inhibition of rotation about the $CH{-}CO_2X$ bond. The bulkier complex Rh(TMP)· inhibits formation of the two-carbon bridged product, and a four-carbon bridged compound, $(TMP)RhCH_2CH(CO_2X)CH(CO_2X)CH_2Rh(TMP)$, results from head-to-head dimerization of acrylate and contains two chiral centres. The Rh(II) monomers Rh(Por)· do not initiate thermal polymerization of acrylates. However, Rh(TMP)· does catalyze a photopromoted polymerization of acrylates that has living character, with NMR evidence observed for one Rh(TMP) unit attached to an oligomer of up to 15 methyl acrylate units.[250]

Alkyl radicals R· can initiate alkene polymerization to form a new radical $RCH_2CH_2\cdot$ since formation of a strong C—C bond (85 kcal mol^{-1}) more than compensates for loss of the alkene π-bond. However, even though (Por)Rh—C bonds are unusually strong, ($Rh{-}CH_3 \approx 58$, $Rh{-}CH_2R \leq 50$, $Rh{-}CHR_2 \leq 40$ kcal mol^{-1}), reaction of Rh(Por)· with ethene to form $(Por)RhCH_2CH_2\cdot$ will still be thermodynamically unfavorable. The initial product of the reaction of Rh(Por)· with the alkene, $[(Por)Rh(CH_2{=}CHX)]\cdot$, was described as a metalloradical-alkene complex, which either underwent attack by a second Rh(Por)· unit or dimerised, forming the two- or four-carbon bridged product, respectively. $(Por)RhCH_2CH_2\cdot$ does not behave as a true carbon-centerd radical, and although $(OEP)RhCH_2CHX\cdot$ can form from Rh—C bond homolysis of the two-carbon bridged product, it cannot initiate alkene polymerization. However, photochemically induced Rh—C bond homolysis of the four-carbon bridged product (formed from Rh(TMP)· with $CH_2{=}CHX$) produces $(TMP)RhCH_2CH(CO_2X)CH(CO_2X)CH_2\cdot$ which does have carbon-centered radical character, and can thus initiate the polymerization process, as observed experimentally.[250]

Using the very bulky rhodium porphyrins Rh(TTEPP)· and Rh(TTiPP)· (which contain triethylphenyl and triisopropylphenyl groups), neither of which can dimerize, direct evidence for an alkene adduct and its subsequent dimerization to the four-carbon bridged product has been obtained. Reaction of Rh(TTEPP)· with ethene in benzene gives quantitative formation of (TTEPP)Rh—$(CH_2)_4$—Rh(TTEPP). The ethene adduct Rh(TTiPP)(CH_2=CH_2)· can be observed by EPR in frozen solution ($S = 1/2$) and slowly dimerizes in solution to give the analogous four-carbon bridged product. Higher oligomers are not observed, as this would require homolysis of the relatively strong Rh—CH_2 bond.[251]

D. *C—H Activation by Rhodium Porphyrins*

Benzylic C—H bond activation in alkyl aromatics by $[Rh(OEP)_2]$ was first reported in 1985.[252] For example, $[Rh(OEP)_2]$ reacts with toluene to produce approximately equal amounts of Rh(OEP)CH_2Ph and Rh(OEP)H. Ethyl benzene and isopropyl benzene both react at the benzylic position to give kinetic products bearing the Rh(OEP) substituent at the α-position, although at longer times the alkyl products rearrange to the compounds with rhodium bonded to a less substituted carbon atom. The importance of the initial reaction at the benzylic position is further illustrated by the fact that *t*-butylbenzene fails to react with $[Rh(OEP)_2]$. The mechanism is proposed to involve Rh(OEP)· as the attacking species, and the regioselectivity is explained on the basis that benzylic C—H bonds (85 kcal mol^{-1}) are ca. 15–20 kcal mol^{-1} weaker than aryl or unactivated alkyl C—H bonds. Formation of the relatively strong Rh—C bond makes the reaction thermodynamically favorable.[252] Benzyl alcohols PhCH(OH)R also react with $[Rh(OEP)_2]$ at the benzylic C—H to give Rh(OEP)H and an intermediate α-hydroxyalkyl, (OEP)Rh—CR(OH)Ph which then eliminates RC(O)Ph and produces a further molecule of Rh(OEP)H.[248]

Unactivated alkanes did not react because of the energy cost of Rh—Rh bond homolysis in the dimer $[Rh(OEP)]_2$ (ca. 16 kcal mol^{-1}). By introducing steric requirements on the periphery of the porphyrin ligand that reduce the Rh—Rh bond strength, but do not weaken the Rh—C bond, activation of stronger C—H bonds might be feasible. Rhodium complexes of two sterically encumbered porphyrins, TXP (tetraxylylporphyrin) and TMP (tetramesitylporphyrin), were investigated with this aim. The dimer $[Rh(TXP)]_2$ has a smaller Rh—Rh bond strength (ca. 12 kcal mol^{-1}) than the OEP congener, while TMP forms a stable Rh(II) complex, Rh(TMP)· , that does not dimerize.[253,254] Both $[Rh(TXP)]_2$ and Rh(TMP)· reacted reversibly with methane at modest temperatures and pressures to give Rh(Por)CH_3 and Rh(Por)H as products. Observation of a large kinetic isotope effect together with equilibrium constant measurements were consistent with a mechanism involving a trimolecular, 4-centered, linear transition state Rh···C···H···Rh (Fig. 8). This synchronizes breaking of the C—H bond (105 kcal mol^{-1}) with formation of the new Rh—C (57 kcal mol^{-1}) and Rh—H (60 kcal mol^{-1}) bonds, an essential feature when the bond being broken is stronger than each of the new bonds being formed. Aromatic C—H bonds do not react with $[Rh(TXP)]_2$ or

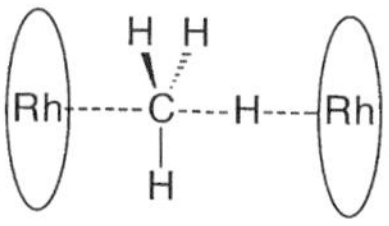

FIG. 8. Trimolecular, linear, four-centered transition state proposed for methane activation by rhodium porphyrins.[253,254]

Rh(TMP)· under these conditions, and in fact the selective activation of methane in benzene solution is a distinctive and unusual feature of this system, given that aryl C—H activation ought to be thermodynamically favored over alkyl C—H activation. The proposed linear transition state proposed in Fig. 8 is the key to this different reactivity. The corresponding trimolecular transition state for an arene would be expected to be bent, and this would be precluded by the bulky TMP ligands.[253,254] The activation of the benzylic C—H bond in toluene is believed to occur through a similar transition state, as is the reaction of Rh(TMP)· with H_2 to produce Rh(TMP)H.[255]

The proposed mechanism was further tested by the synthesis of a new, dinuclear porphyrin containing two *meso*-trimesitylporphyrin groups, each linked through the fourth porphyrin *meso* position by a $C_6H_4O(CH_2)_6OC_6H_4$ group. This new porphyrin behaves essentially as two, covalently linked TMP units, and forms a diradical containing two Rh^{II} centers, abbreviated as $Rh(PorO(CH_2)_6OPor)Rh$, that is precluded from dimerization. Linking of the two porphyrins improves the entropy term required to achieve the linear, 4-centered transition state. Activation of CH_4 proceeds with only intramolecular formation of $CH_3Rh(PorO(CH_2)_6OPor)RhH$, and with faster kinetics than either $[Rh(TXP)]_2$ or Rh(TMP)·.[256]

In classical mechanisms for C—H bond activation either C—H σ-bond donation or cyclic, 4-centered transition states are important, but these are precluded in the porphyrin systems and the mechanism proposed for activation of CH_4, toluene, and H_2 by the Rh^{II} porphyrin radicals is a new mechanistic possibility.

E. *Activation of CO and Isocyanide by Rhodium Porphyrins*

One step of the mechanism determined for the reaction of Rh(OEP)H with CO to give the formyl complex Rh(OEP)CHO involved the coordination of CO to the chain-carrying Rh(II) porphyrin Rh(OEP)·, although there had been no direct evidence observed for this species.[229] Since that time, Wayland has systematically developed the chemistry of Rh(II) porphyrins coordinated to CO, and the chemistry is now well understood.

The investigation began with the observation that two species could be observed by NMR and IR spectroscopy in solutions of $[Rh(OEP)]_2$ in toluene-d_8 exposed to CO. The first, which forms immediately, has $\nu_{CO} = 2094\ cm^{-1}$ and a broad peak at 180 ppm in the ^{13}C NMR spectrum, and was proposed to be a simple adduct, $[Rh(OEP)]_2(CO)$. The second species, which evolved more slowly, has $\nu_{CO} = 1733\ cm^{-1}$ and a triplet at 116 ppm ($^1J_{Rh-C} = 44$ Hz) in the ^{13}C NMR spectrum, and was assigned to a dimetallaketone, (OEP)Rh—C(O)—Rh(OEP). In thermodynamic terms, this reaction requires formation of sufficiently strong Rh—C bonds (ca. 50 kcal mol^{-1}) to compensate for cleavage of the Rh—Rh bond (16 kcal mol^{-1}), reduction of the C—O bond order from three to two (72 kcal mol^{-1}) and an entropy term (7.5 kcal mol^{-1}).[257] On further investigation, using lower temperatures and higher pressures of CO, a third species was identified as the double CO insertion product, (OEP)Rh—C(O)—C(O)—Rh(OEP).[258] The ^{13}C NMR chemical shift of the new product occurs at 165.5 ppm, and appears as a four-line multiplet consistent with an AA′XX′ spin system. Two ν_{CO} stretches are observed for the diketone at 1782 and 1770 cm^{-1}.[259] All four compounds (the dimer, CO adduct, and single and double CO insertion products) exist in equilibrium, and the relative proportion of products depends on the porphyrin, temperature, and pressure. The thermodynamic driving force for formation of the double insertion product is the estimated increase of ca. 5 kcal mol^{-1} in the strength of the Rh—C bond, arising from a more favorable steric arrangement. In the single CO insertion product the close proximity of the two porphyrin rings and the sp^2 hybridized ketone carbon atom results in a sterically unfavorable arrangement. In the dimetal α-diketone (the double insertion product) the increased distance between the porphyrin rings and the added degree of freedom introduced by rotation around the C—C bond allow a more favorable steric arrangement.[258] The two ν_{CO} values observed for the dimetal α-diketone, together with structure simulation, suggest that the diketone unit is not planar but is likely to have a 20–30° rotation about the C—C bond.[259]

This hypothesis was tested using a more sterically demanding porphyrin ligand. $[Rh(TXP)]_2$ reacts with CO (1 atm, 298 K) to give the dimetal α-diketone as the only observed product.[260] The reaction of Rh(TMP)· with CO permitted the observation of a paramagnetic 1:1 adduct, Rh(TMP)CO·, by EPR spectroscopy, with $S = 1/2$ and coupling of the odd electron with ^{103}Rh and ^{13}CO. The g value and ^{13}CO hyperfine coupling constant indicate that the odd electron is delocalized onto the CO ligand, and in frozen toluene solution the loss of axial symmetry suggested a bent Rh—CO unit. At low temperature in solution, the Rh(TMP)CO· species dimerizes to the dimetal α-diketone.[259,261] In the even more sterically demanding complex Rh(TTiPP)CO·, the paramagnetic CO adduct is stable and no dimerization occurs. Equilibrium studies on Ru(TMP)CO have allowed measurement of the enthalpies of dimerization for the CO adduct and dissociation for the dimetal α-diketone.[259] The two covalently linked rhodium(II) porphyrins (as used in the C—H activation studies described above), $\cdot Rh(PorO(CH_2)_6OPor)Rh\cdot$, react with CO to produce the tethered diradical $\cdot(OC)Rh(PorO(CH_2)_6OPor)Rh(CO)\cdot$ for which intramolecular coupling to form the diketone was also observed.[262]

In the TMP system the monomer Rh(TMP)CO and dimer (TMP)RhC(O)C(O)-Rh(TMP) are in equilibrium. When this mixture is treated with styrene and repressurized with CO the styrene insertion product (TMP)RhC(O)CH_2CH(Ph)C(O)-Rh(TMP) is observed by ^{1}H and ^{13}C NMR. As expected, this compound has two chemically different CO groups.[259] When CO is added to a mixture of Rh(TMP)· and Rh(TMP)H, the products consist of a mixture of the hydride, the monomeric carbonyl adduct, and the α-diketone dimer, but no formyl complex is observed. This is interesting because hydride abstraction from Rh(OEP)H by the carbonyl complex Rh(OEP)CO· is proposed as a step in formation of the formyl product in the OEP system (Scheme 17). The increased steric bulk of the TMP ligand in Rh(TMP)H must preclude this step, but Rh(TMP)CO· will react with less sterically demanding hydride sources ($HSnBu_3$ or H_2) to produce the formyl complex Rh(TMP)CHO.[259] The tethered diradical ·(OC)Rh(PorO$(CH_2)_6$OPor)Rh(CO)· reacts with water, ethanol and H_2 to form HRh(PorO$(CH_2)_6$OPor)RhCHO, H(O)CRh (PorO$(CH_2)_6$OPor)RhC(O)OEt, and H(O)CRh(PorO$(CH_2)_6$OPor)RhCHO, respectively.[262] Overall, the chemical reactivity of Rh(TMP)CO· is likened to that of the acyl radical, CH_3CO·, and represents an unusual example of a 17-electron metal carbonyl complex which exhibits carbon-centered rather than metal-centered radical reactivity.[259]

The formal relationship between CO and isocyanide ligands, CNR, sparked a comparative study of the reactions of CNR with Rh(OEP)H, $[Rh(OEP)]_2$ and Rh(TMP).[263] As with CO, the initial reaction of $[Rh(OEP)]_2$ with CNR (R = Me, Et) leads to a 2:1 adduct, $[Rh(OEP)]_2$(CNR), although the equilibrium constant for formation of the CNR adduct ($\sim 10^5$ at 298 K) is much larger than that corresponding to formation of the CO adduct ($\sim$48 at 298 K). Subsequent chemistry is different as the CNR adduct slowly transforms in solution to a mixture of Rh(OEP)R and Rh(OEP)(CN)(CNR), driven by CN—R bond cleavage. No evidence is seen for a bridging imine or diimine. When an aryl isocyanide is used, CN—R bond cleavage does not occur and reversible formation of the adduct occurs cleanly. The reaction of Rh(TMP)· with CNR (R = Me, *n*-Bu) gave the 1:1 adduct, Rh(TMP)(CNR)· although CN—R bond cleavage is also a problem. No evidence was seen for a bridging imine or diimine complexes, probably a result of the increased steric requirements of an isocyanide relative to CO.[263]

Rh(OEP)H reacts with CNR (R = Me, *n*-Bu,) to give the adduct Rh(OEP)-(H)CNR (which has no parallel in CO chemistry) which then slowly transforms to the formimidoyl insertion product, Rh(OEP)C(H)=NR. The dimer $[Rh(OEP)]_2$ reacts with CNAr (Ar = 2,6-$C_6H_3Me_2$) in aqueous benzene to give the carbamoyl product, Rh(OEP)C(O)NHAr (characterized by an X-ray crystal structure) together with the hydride, which itself reacts further with the isocyanide. This is suggested to form via a cationic carbene intermediate, formed by attack of H_2O on coordinated CNAr in concert with disproportionation to Rh(III) and Rh(I).[263]

F. *Reactions of Rh(III) and Rh(I) Porphyrins with Organic Substrates*

Aryl C—H activation by cationic Rh(III) porphyrins has also been established in the reaction of Rh(OEP)Cl with Ag^+ in benzene which gave Rh(OEP)C_6H_5. Anisole, toluene, and chlorobenzene react similarly, giving exclusively the *para*-substituted phenyl derivatives. The regiochemistry of the reaction indicates that the mechanism of this process is allied to electrophilic aromatic substitution, where the electrophile is the $[Rh(OEP)]^+$ cation. Studies on a series of substituted aryls indicate that the $[Rh(OEP)]^+$ cation has a Hammet constant similar to that of NO_2^+.[264] Similarly, unexpected arene C—H activation was also attributed to electrophilic aromatic substitution. The reaction of H_2Por (Por = OEP, TTP) with $RhCl_3{\cdot}xH_2O$ in benzonitrile gave a mixture of Rh(Por)Cl (60%) and Rh(Por)(*m*-C_6H_4CN).[265] When the highly substituted tetramesityloctaphenylporphyrin was used, the corresponding organometallic product crystallized as a coordination polymer, linked by coordination of the CN nitrogen of one Rh—C_6H_4CN group with the vacant axial site on an adjacent rhodium.[266]

$[Rh(OEP)]^+$ (formed by reaction of Rh(OEP)Cl with Ag^+) reacts with ketones at the α-CH position to give β-carbonyl products. For example, acetone reacts with $[Rh(OEP)]^+$ to give (OEP)RhCH_2C(O)CH_3. The regiochemistry indicates that the enol tautomers rather than the ketones are involved, and an initial reaction with the enol oxygen was proposed. In the case of cyclohexanone, an organorhodium product was not observed, but rather the aldol condensation product formed. The aldol condensation is catalyzed by $[Rh(OEP)]^+$, and the intermediate is believed to be the Rh—O bonded enolate rather than the Rh—C bonded β-carbonyl compound. The latter, when independently prepared, does not show any activity toward enol condensation. These reactions provide an example where the Lewis acidic Rh(III) center is used to promote enolate formation.[267] Rh(OEP)Cl reacted directly with ketones when a porphyrin containing a mild base (pyridyl or aryl alcohol) appended to the *meso* position was used. The intramolecular base assists with enolization.[268]

The anionic Rh(I) porphyrin $[Rh(OEP)]^-$ induced ring-opening reactions with 4- and 5-membered ring lactones to give organometallic products with the rhodium bonded to the alkoxide carbon rather than the carbonyl carbon.[269]

The chemistry of $[Rh(OEP)]_2$ in benzene is dominated by Rh—Rh bond homolysis to give the reactive Rh(II) radical Rh(OEP)·. This contrasts with the reactivity of $[Rh(OEP)]_2$ in pyridine, which promotes disproportionation via the formation of the thermodynamically favorable Rh(III), d^6 complex $[Rh(OEP)(py)_2]^+$ together with the Rh(I) anion, $[Rh(OEP)]^-$.[270] The hydride complex Rh(OEP)H shows NMR chemical shift changes in pyridine consistent with coordination of pyridine, forming Rh(OEP)H(py). Overall, solutions of $[Rh(OEP)]_2$ in pyridine behave as an equimolar mixture of $[Rh(OEP)(py)_2]^+$ and $[Rh(OEP)]^-$. For example, reaction of this solution with H_2 produces Rh(OEP)H, an unusual example of heterolytic activation of dihydrogen[177] where H^+ and H^- react with a metal anion and cation, respectively, to give the same product. Despite the different formulation of the species in solution, the overall reactivity of $[Rh(OEP)]_2$ in benzene and pyridine is remarkably similar, with $[Rh(OEP)(py)_2]^+$ and $[Rh(OEP)]^-$ in pyridine reacting with H_2/CO or H_2O/CO to give the formyl complex Rh(OEP)(CHO)(py). In this case activation of CO is proposed to occur through the metalloanion $[Rh(OEP)]^-$.[270]

G. *Rhodium Porphyrin Carbene Complexes and the Cyclopropanation of Alkenes Catalyzed by Rhodium Porphyrins*

In 1980 and 1982, Callot and co-workers reported that Rh(Por)I catalyzed the reaction between alkenes and ethyl diazoacetate to give *syn* cyclopropoanes as the major products (Eq. 25).[230] This was unusual as most transition metal catalysts for this reaction give the *anti* isomers as the predominant products. Kodadek and co-workers[273] followed up this early report and put considerable effort into trying to improve the *syn/anti* ratios and enantioselectivity using porphyrins with chiral substituents.

R + N_2CHCO_2Et —Rh(Por)I catalyst→ R, $CHCO_2Et$ (*syn*) + R, $CHCO_2Et$ (*anti*) (25)

Details of the mechanism of the cyclopropanation reaction were elucidated by a careful study of the reactions of Rh(TTP)I and $Rh(TTP)CH_3$ (which can also act as a catalyst) with ethyl diazoacetate in the presence and absence of an alkene, to give the overall process shown in Scheme 18.[271–273] Species which have been observed directly by spectroscopy are the alkyl diazonium and iodoalkyl complexes $[(TTP)Rh{-}CH(N_2)CO_2Et]^+$ (at −40°C) and $(TTP)Rh{-}CH(I)CO_2Et$. The latter species was determined to be the actual catalytic species participating in the cycle, as shown in Scheme 18.[273] The cyclopropanation event occurs by transfer of a carbene fragment from a rhodium carbene complex (not directly observed) to the alkene substrate.

The stereochemistry of the resulting cyclopropane product (*syn* vs *anti*) was rationalized from a kinetic study which implicated an early transition state with no detectable intermediates. Approach of the alkene substrate perpendicular to the proposed carbene intermediate occurs with the largest alkene substituent opposite the carbene ester group. This is followed by rotation of the alkene as the new C—C bonds begin to form. The steric effect of the alkene substituent determines

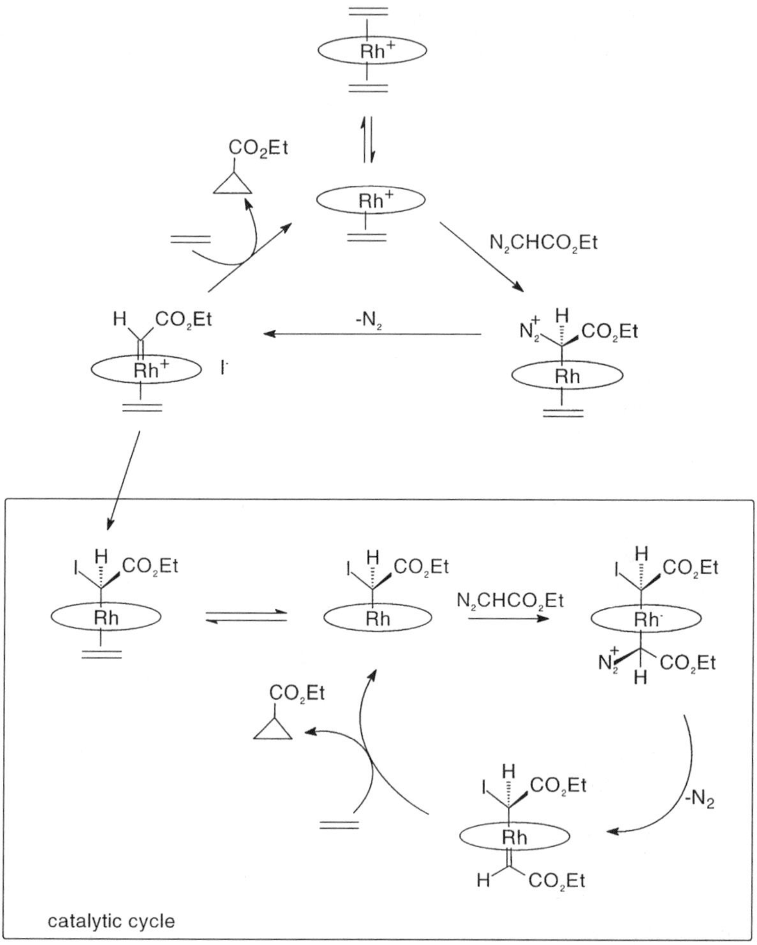

Scheme 18.

whether clockwise or anticlockwise rotation occurs, depending on whether porphyrin/substituent or ester/substituent interactions predominate, leading to either a *syn* or *anti* product.[274] Attempts to improve the *syn/anti* ratio and enantioselectivity were made by appending bulky binaphthyl or *ortho*-pyrenylnaphthyl groups to the *meso* porphyrin sites, creating a "chiral wall" or even bulkier "chiral fortress" porphyrin, respectively. Although in each case relatively good *syn/anti* ratios (in the range 2–8) were obtained for a variety of alkenes, enantiomeric excess values (ee) remained only modest, in the approximate range 10–40%, with the *anti* products generally exhibiting better ee values than the *syn* products.[275–277]

Rh(Por)I (Por = OEP, TPP, TMP) also acts as a catalyst for the insertion of carbene fragments into the O—H bonds of alcohols, again using ethyl diazoacetate as the carbene source. A rhodium porphyrin carbene intermediate was proposed in the reaction, which is more effective for primary than secondary or tertiary alcohols, and with the bulky TMP ligand providing the most selectivity.[278]

Both rhodium and osmium porphyrins are active for the cyclopropanation of alkenes. The higher activity of the rhodium porphyrin catalysts can possibly be attributed to a more reactive, cationic carbene intermediate, which so far has defied isolation. The neutral osmium carbene complexes are less active as catalysts but the mono- and bis-carbene complexes can be isolated as a result.

Although rhodium porphyrin carbene species are believed to be the key intermediates in the alkene cyclopropanation reactions, few examples of rhodium porphyrin carbenes have been fully characterized. Nucleophilic attack on coordinated isocyanide ligands to give carbene ligands is well known, but not well explored for porphyrin systems. $[Rh(Por)(CNR)_2]^+$ (Por = OEP, TPP; R = CH_2Ph, p-C_6H_4) reacts with methanol (as the nucleophile) although the carbene products are formulated as $[(Por)Rh{=}C(NHR)_2]^+$. Minor products from these reactions are the methoxycarbonyl and amide complexes (Por)Rh—C(O)OCH_3 and (Por)Rh—C(O)NHR, presumed to result from hydrolysis reactions producing the free amine, H_2NHR, which then ends up as a substituent on the carbene ligand. An X-ray crystal structure of $[(TPP)Rh{=}C(NHCH_2Ph)_2(CNCH_2Ph)]PF_6$ shows it to contain an isocyanide ligand in the position *trans* to the carbene (Fig. 7). Both the Rh-C(carbene) and Rh—C(isocyanide) bond lengths, 2.030(11) and 2.064(13) Å, respectively, are long compared to non-porphyrin rhodium carbene and isocyanide distances, and are presumed to reflect a *trans* influence.[279] In fact, both these distances are not significantly shorter than Rh—C single bonds in σ-bonded Rh(Por)R complexes, for example 1.970(4) and 2.013(6) for the Rh—C bond lengths in two different determinations of Rh(OEP)CH_3.[280] The electrochemistry of the carbene complex $[(TPP)Rh{=}C(NHCH_2Ph)_2(CNCH_2Ph)]PF_6$ has been investigated.[281]

H. *Heterobimetallic Rhodium Porphyrin Complexes*

A number of the reactions by which organometallic rhodium porphyrin complexes are prepared have parallel reactions with inorganic substrates, giving rise to "inorganometallic" rhodium porphyrin complexes. $[Rh(OEP)]_2$ reacts with a variety of silanes and stannanes, including Et_3SiH, Ph_3SiH, n-Bu_3SnH, and Ph_3SnH to give the rhodium silyl and stannyl complexes Rh(OEP)SiR_3 (R = Et, Ph) and Rh(OEP)SnR_3 (R = n-Bu, Ph) together with elimination of H_2.[269] An alternative synthesis of a rhodium silyl utilizes $[Rh(TPP)]^-$ with Me_3SiCl, giving Rh(TPP)$SiMe_3$.[282] Three complexes, Rh(OEP)$SiEt_3$,[269] Rh(OEP)$SiMe_3$,[282] and Rh(TPP)$SnCl_3$,[283] have been characterized by X-ray crystallography, and exhibit

Rh—Si or Rh—Sn bond lengths of 2.32(1), 2.035(2) and 2.450(1) Å, respectively (Fig. 7).

$[Rh(OEP)]^-$ reacts with In(OEP)Cl in THF to give the heterobimetallic diporphyrin complex (OEP)Rh—In(OEP), which has a Rh—In bond length of 2.584(2) Å, slightly shorter than the sum of the covalent radii (2.62 Å) of rhodium and indium (Fig. 7). This compound reacts with CH_3I to give $Rh(OEP)CH_3$ and In(OEP)I, and with acids HX to give Rh(OEP)H and In(OEP)X, but showed no activity toward many of the reagents that react with $[Rh(OEP)]_2$ under similar conditions (H_2/CO, styrene or acrylonitrile). Furthermore, the compound shows no tendency to coordinate donor ligands (CNBu, H_2O or $P(OEt)_3$) at the rhodium center. On the basis of its chemical reactivity the complex is considered to be a donor–acceptor complex, with rhodium retaining its Rh(I) anion character with a filled d_{z^2} orbital, and indium its In(III) character.[284] A selection of other derivatives (Por)Rh—In(Por) containing different porphyrins have been prepared, as have the thallium analogues (Por)Rh—Tl(Por) which show similar reactivity.[285–287]

VIII

THE LATE TRANSITION METALS (GROUPS 10, 11, AND 12)

A. *Overview*

Organometallic porphyrin complexes containing the late transition elements (from the nickel, copper, or zinc triads) are exceedingly few. In all of the known examples, either the porphyrin has been modified in some way or the metal is coordinated to fewer than four of the pyrrole nitrogens. For nickel, copper, and zinc the +2 oxidation state predominates, and the simple M^{II}(Por) complexes are stable and resist oxidation or modification, thus on valence grounds alone it is easy to understand why there are few organometallic examples. The exceptions, which exist for nickel, palladium, and possibly zinc, are outlined below. Little evidence has been reported for stable organometallic porphyrin complexes of the other late transision elements.

B. *Nickel*

Nickel porphyrin complexes containing a bridging carbene ligand have been known for some 25 years. Ni(TPP){μ-(Ni,N)—$CHCO_2Et$} was prepared by the reaction of the nickel(II) N-alkyl porphyrin $[Ni(TPP—NCH_2CO_2Et)]^+$ with base, and has been characterized by crystallography (selected data are given in Table V). The nickel is bonded to only three of the pyrrole nitrogens, with a long Ni$\cdots$N distance (2.610(3) Å) to the alkylated pyrrole.[288] A related vinylidene porphyrin, Ni(TPP){μ-(Ni,N)—C=CAr$_2$} (Ar = p-C_6H_4Cl), was prepared by the reaction of the N,N-vinylidene-bridged free-base porphyrin with $Ni(CO)_4$.[289] In each of these complexes, the carbene fragment has alkylated one pyrrole nitrogen, reducing the charge on the porphyrin ligand to −1, and thus the Ni^{II} center can form a bond to the carbene carbon to achieve a neutral complex. This principle has also been demonstrated for an N-alkyl porphyrin. Ni(TPP—NMe)Cl (containing the N-methyl TPP ligand) reacted with PhMgBr at −70°C to give a complex assigned by ^{1}H NMR and EPR spectroscopy to be Ni(TPP—NMe)Ph. The complex is paramagnetic ($S = 1$), in contrast to the diamagnetic bridging carbene nickel complexes. A thiaporphyrin H(SPor), in which one pyrrole NH group is replaced by a sulfur atom, thus forming a monoanionic N_3S donor ligand, behaves similarly, forming Ni(SPor)Ph from Ni(SPor)Cl and PhMgBr at −70°C, again detected by spectroscopy. Both Ni(SPor)Ph and Ni(TPP—NMe)Ph decompose upon warming above −70°C.[290] A diaoxaporphyrin (O_2Por), in which two pyrrole NH groups are replaced by oxygen atoms, is a neutral N_2O_2 donor ligand. At −70°C, Ni(O_2Por)Cl_2 reacts with one or two equivalents of PhMgBr to give Ni(O_2Por)PhCl or Ni(O_2Por) Ph_2.[291] Finally, the unusual porphyrin isomer in which one pyrrole ring is inverted, with the nitrogen atom on the periphery of the ring and a CH group pointing into the cavity forms both Ni(II) and Ni(III) complexes containing an Ni—C bond to the carbon in the inverted pyrrole ring.[292]

C. *Palladium*

Until very recently, metalloporphyrin π-allyl complexes were unknown, but this has changed with the report of one example containing palladium.[293] Like the organometallic nickel porphyrins, N-alkylation of the porphyrin is a feature of this new complex. In this case, the free-base porphyrin (TPP-NN) bears an N^{21}, N^{22}-etheno bridge, in which a PhC=CPh group bridges between two adjacent pyrrole nitrogens. The etheno-bridged porphyrin (TPP—NN) coordinates as a neutral ligand, forming a complex with a $PdCl_2$ fragment in which the palladium atom lies considerably out of the porphyrin plane and is coordinated only to the two non- alkylated pyrrole nitrogen atoms. Pd(TPP—NN)Cl_2 reacted with $AgClO_4$ followed by the allyltin reagent, CH_2=$CHCH_2SnBu_3$, to give the cationic π-allyl complex [Pd(TPP—NN)(η^3-C_3H_5)]$^+$. Two isomers were identified by ^{1}H NMR spectroscopy, presumed to differ in the orientation of the π-allyl ligand with respect to the palladium porphyrin fragment. One isomer was characterized by X-ray crystallography, and showed Pd—C distances (2.085(7) and 2.125(7) Å) similar to those observed in related π-allyl palladium bipyridine complex. Thus in this case, the porphyrin is behaving essentially as a neutral, bidentate nitrogen donor ligand.[293]

D. *Zinc*

^{1}H NMR data has been reported for the ethylzinc complex, Zn(TPP—NMe)Et, formed from the reaction of free-base N-methyl porphyrin H(TPP—NMe) with $ZnEt_2$. The ethyl proton chemical shifts are observed upfield, evidence that the ethyl group is coordinated to zinc near the center of the porphyrin. The complex is stable under N_2 in the dark, but decomposed by a radical mechanism in visible light.[294] The complex reacted with hindered phenols (HOAr) when irradiated with visible light to give ethane and the aryloxo complexes Zn(TPP—NMe)OAr. The reaction of Zn(TPP—NMe)Et, a secondary amine ($HNEt_2$) and CO_2 gave zinc carbamate complexes, for example Zn(TPP—NMe)O_2CNEt_2.[295]

IX

NON-COVALENT INTERACTIONS BETWEEN METALLOPORPHYRINS AND ORGANIC MOLECULES

An important development in the 1990s has been the growth in supramolecular chemistry, including the recognition that non-covalent interactions have an important role to play in all facets of chemistry. Chemical systems featuring these once occurred only in the realm of serendipity, but can now be achieved through careful design and synthesis. Non-covalent interactions involving metalloporphyrins, particularly where there is an interaction between an organic substrate and the metal center, form one extreme of a continuum of organometallic bonding types spanning non-covalent interactions, agostic bonding, coordination of σ-bonds (H—H or C—H), π-coordination, σ-donor bonds, and fully covalent σ-bonds.

Close approaches of arene solvents to metalloporphyrins have been observed by crystallography. For example, Mn(TPP) crystallizes as a toluene solvate with a toluene solvent molecule lying on either side of the porphyrin plane. The dihedral angle between the porphyrin plane and toluene plane is 10.7° (reduced to 6.7° for one of the pyrrole rings). The average perpendicular distance between the toluene ring and the mean 24-atom plane of the porphyrin is 3.30 Å. One aromatic C—C bond of toluene approximately eclipses one N—Mn—N axis, with Mn···C distances of 3.05 and 3.25 Å.[296] There has been considerable effort invested in the preparation of a "bare" $[Fe(Por)]^+$ cation, requiring a "least coordinating anion" as the counterion. Similar close approaches of arene solvates are seen in two examples. $[Fe(TPP)]SbF_6{\cdot}C_6H_5F$ contains a fluorobenzene solvate with a dihedral angle between the porphyrin and arene planes of 7°, and average separation between the two planes of 3.30 Å.[297] The closest Fe···C(arene) distance is 3.34 Å. A complex of $[Fe(TPP)]^+$ containing an even more weakly coordinating anion,

$[Ag(Br_6CB_{11}H_6)_2]^-$ contains a *p*-xylene molecule on each side of the porphyrin plane with dihedral angles of 13 and 5°. The closest approach of a *p*-xylene atom to the mean plane of the porphyrin is 2.89 Å, and the closest Fe$\cdots$C(*p*-xylene) distance is 2.94 Å, about 0.2 Å shorter than the sum of the covalent radii.[298] These examples blur the distinction between ligand and solvate, and in the two examples involving the $[Fe(TPP)]^+$ cations, the closely approaching arene molecules may be playing a role in compensating for the positive charge on the cations. However, whether the M(Por)/arene interaction is driven by metal/arene "coordination" or porphyin/arene π/π interactions is still an open question.

In one further example involving an iron porphyrin there is good structural evidence for heptane C—H coordination to iron.[299] The complex Fe(DAP)·heptane contains an elaborated "double A-frame" porphyrin, in which opposite pairs of phenyl rings in TPP are linked through the *ortho* positions by NHC(O)-1,4-C_6H_4—$C(CF_3)_2$-1,4-C_6H_4—C(O)NH straps, creating cavities above and below the porphyrin core. Each heptane molecule spans between two cavities on adjacent molecules, with close contacts between the iron atoms and the terminal methyl groups on the heptane molecules. The iron atom is displaced 0.26 Å from the mean N_4 plane toward the heptane, which is an indicator for five-coordination at iron. The Fe$\cdots$C(heptane) distances are 2.5 and 2.8 Å, well within the 2.5–3.0 Å range accepted for M$\cdots$C agostic interactions. The structural data were supported by density functional theory calculations which gave Fe$\cdots$C distances of 2.68–2.70 Å. This example is remarkable first because of the alkane coordination to iron(II), and second because it involves a free alkane, not one covalently tethered to the molecule as is usually observed for agostic interactions.[299]

Finally, a series of co-crystallates of fullerenes (C_{60} and C_{70}) with both free-base and metalloporphyrins show unusually short porphyrin/fullerene contacts (2.7–3.0 Å) compared with typical π-π interactions (3.0–3.5 Å).[300] The fact that these close approaches are observed for both free-base and metalloporphyrins indicates that π-π interactions rather than metal/fullerene interactions are important in these examples. For example, $H_2TPP \cdot C_{60} \cdot$3toluene crystallizes in zig-zag chains of alternating porphyrin and fullerene molecules, with an electron rich 6:6 ring juncture lying over the center of the porphyrin ring. The closest porphyrin plane–fullerene interaction is 2.72 Å. The two metalloporphyrin cocrystallates are $ZnTPP \cdot C_{70}$ and $NiTTP \cdot 2C_{70} \cdot$2toluene, which contain side-on C_{70} molecules with a carbon atom from three fused 6-membered rings lying closest to the porphyrin. The shortest Zn$\cdots$C and Ni$\cdots$C distances in these structures are 2.89 and 2.85 Å. NMR evidence indicates that the interaction persists in solution. Overall, the interactions of the planar porphyrins with the curved fullerenes are important because they demonstrate supramolecular recognition without the need for matching a concave host with a complementary convex guest. These structures help to explain the function of the TPP-appended silica stationary phases used for the chromatographic separation of fullerenes.[300]

References

(1) Smith, K. M. (Ed.) *Porphyrins and Metalloporphyrins;* Elsevier: New York, 1975.

(2) Dolphin, D. (Ed.) *The Porphyrins;* Academic: New York, 1979; Volumes 1–7.

(3) Mansuy, D.; Lange, M.; Chottard, J. C.; Guerin, P.; Morliere, P.; Brault, D.; Rougee, M. *J. Chem. Soc., Chem. Commun.* **1977,** 648.

(4) Mansuy, D.; Lange, M.; Chottard, J. C.; Bartoli, J. F.; Chevrier, B.; Weiss, R. *Angew. Chem., Int. Ed. Engl.* **1978,** *17,* 781.

(5) Brothers, P. J.; Collman, J. P. *Acc. Chem. Res.* **1986,** *19,* 209.

(6) Setsune, J. I.; Dolphin, D. *Can. J. Chem.* **1987,** *65,* 459.

(7) Guilard, R.; Lecomte, C.; Kadish, K. M. *Struct. Bond.* **1987,** *64,* 205.

(8) Guilard, R.; Kadish, K. M. *Chem. Rev.* **1988,** *88,* 1121.

(9) Kadish, K. M.; Smith, K. M.; Guilard, R. (Eds.) *The Porphyrin Handbook;* Academic: San Diego, 1999; Volume 3.

(10) Cheng, L.; Chen, L.; Chung, H. S.; Khan, M. A.; Richter-Addo, G. B. *Organometallics* **1998,** *17,* 3853.

(11) Senge, M. O. *Angew. Chem., Int. Ed. Engl.* **1996,** *35,* 1923.

(12) Fehlner, T. P. (Ed.) *Inorganometallic Chemistry;* Plenum: New York, 1992.

(13) Watanabe, Y.; Groves, J. T. In *Enzymes 3rd ed.;* Sigman D. S. Ed.; Academic: San Diego, 1992.

(14) Mansuy, D. *Pure Appl. Chem.* **1990,** *62,* 741.

(15) Meunier, B. *Chem. Rev.* **1992,** *92,* 1141.

(16) Sayer, P.; Gouterman, M.; Connell, C. R. *J. Am. Chem. Soc.* **1982,** *15,* 73.

(17) Belcher, W. J.; Boyd, P. D. W.; Brothers, P. J.; Liddell, M. J.; Rickard, C. E. F. *J. Am. Chem. Soc.* **1994,** *116,* 8416.

(18) Belcher, W. J.; Breede, M.; Brothers, P. J.; Rickard, C. E. F. *Angew. Chem., Int. Ed. Engl.* **1988,** *37,* 1112.

(19) Brothers, P. J. *Adv. Organomet. Chem.* in press

(20) Buchler, J. W. In *The Porphyrins;* Dolphin. D. Ed.; Academic: New York, 1978; Vol 1, p. 389.

(21) Scheidt, W. R.; Lee, Y. J. *Struct. Bond.* **1987,** *64,* 1.

(22) Collman, J. P.; Arnold, H. J. *Acc. Chem. Res.* **1993,** *26,* 586.

(23) Arnold, J. *J. Chem. Soc., Chem. Commun.* **1990,** 976.

(24) Arnold, J.; Dawson, D. Y.; Hoffman, C. G. *J. Am. Chem. Soc.* **1993,** *115,* 2707.

(25) Brand, H.; Arnold, J. A. *Coord. Chem. Rev.* **1995,** *140,* 137.

(26) Kim, H. J.; Whang, D.; Do, Y.; Kim, K. *Chem. Lett.* **1993,** 807.

(27) Huhmann, J. L.; Corey, J. L.; Rath, N. P. *Acta Cryst.* **1995,** *C51,* 195.

(28) Kim, K.; Lee, W. S.; Kim, H. J.; Cho, S. H.; Girolami, G. S.; Gorlin, P. A.; Suslick, K. S. *Inorg. Chem.* **1991,** *30,* 2652.

(29) Buchler, J. W.; De Cian, A.; Fischer, J.; Hammerschmitt, P.; Weiss, R. *Chem. Ber.* **1991,** *124,* 1051.

(30) Ryu, S.; Whang, D.; Kim, H. J.; Kim, K.; Yoshida, M.; Hashimoto, K.; Tatsumi, K. *Inorg. Chem.* **1997,** *36,* 4607.

(31) Sewchok, M. G.; Haushalter, R. C.; Merola, J. S. *Inorg. Chim. Acta* **1988,** *144,* 47.

(32) Arnold, J.; Hoffman, C. G. *J. Am. Chem. Soc.* **1990,** *112,* 8620.

(33) Arnold, J.; Hoffman, C. G.; Dawson, D. Y.; Hollander, F. J. *Organometallics* **1993,** *12,* 3645.

(34) Schaverien, C. J.; Orpen, A. G. *Inorg. Chem.* **1991,** *30,* 4968.

(35) Schaverien, C. J. *J. Chem. Soc., Chem. Commun.* **1991,** 458.

(36) Woo, L. K.; Hays, J. A.; Jacobsen, R. A.; Day, C. L. *Organometallics* **1991,** *10,* 2102.

(37) Woo, L. K.; Hays, A. J.; Young, V. G. J.; Day, C. L.; Caron, C.; D-Souza, F.; Kadish, K. M. *Inorg. Chem.* **1993,** *32,* 4186.

(38) Wang, X.; Gray, S. D.; Chen, J.; Woo, L. K. *Inorg. Chem.* **1998,** *37,* 5.

(39) Woo, L. K.; Hays, J. A. *Inorg. Chem.* **1993,** *32,* 2228.
(40) Hays, J. A.; Day, C. L.; Young, V. G. J.; Woo, L. K. *Inorg. Chem.* **1996,** *35,* 7601.
(41) Gray, S. D.; Thorman, J. L.; Berreau, L. M.; Woo, L. K. *Inorg. Chem.* **1997,** *36,* 278.
(42) Gray, S. D.; Thorman, J. L.; Adamian, V. A.; Kadish, K. M.; Woo, L. K. *Inorg. Chem.* **1998,** *37,* 1.
(43) Wang, X. T.; Woo, L. K. *J. Org. Chem.* **1998,** *63,* 356.
(44) Shibata, K.; Aida, T.; Inoue, S. *Chem. Lett.* **1992,** 1173.
(45) Kim, H. J.; Whang, D.; Kim, K.; Do, Y. *Inorg. Chem.* **1993,** *32,* 360.
(46) Brand, H.; Arnold, J. *J. Am. Chem. Soc.* **1992,** *114,* 2266.
(47) Huhmann, J. L.; Corey, J. Y.; Rath, N. P.; Campana, C. F. *J. Organomet. Chem.* **1996,** *513,* 17.
(48) Brand, H.; Arnold, J. *Organometallics* **1993,** *12,* 3655.
(49) Ryu, S.; Whang, D.; Kim, J.; Yeo, W.; Kim, K. *J. Chem. Soc., Dalton Trans.* **1993,** 205.
(50) Ryu, S.; Kim, J.; Yeo, H.; Kim, K. *Inorg. Chim. Acta* **1995,** *228,* 233.
(51) Arnold, J. *J. Am. Chem. Soc.* **1992,** *114,* 3996.
(52) Kim, H. J.; Jung, S.; Jeon, Y.M.; Whang, D.; Kim, K. *Chem. Commun.* **1999,** 1033.
(53) Brand, H.; Capriotti, J.; Arnold, J. *Organometallics* **1994,** *13,* 4469.
(54) Brand, H.; Arnold, J. *Angew. Chem., Int. Ed. Engl.* **1994,** *33,* 95.
(55) Kim, H. J.; Jung, S.; Jeon, Y. M.; Whang, D.; Kim, K. *Chem. Commun.* **1997,** 2201.
(56) Shibata, K.; Aida, T.; Inoue, S. *Tetrahedron Lett.* **1992,** *33,* 1077.
(57) Dawson, D. Y.; Brand, H.; Arnold, J. *J. Am. Chem. Soc.* **1994,** *116,* 9797.
(58) Toscano, P. J.; Brand, H.; DiMauro, P. T. *Organometallics* **1993,** *12,* 30.
(59) Colin, J.; Chevrier, B. *Organometallics* **1985,** *4,* 1090.
(60) De Cian, A.; Colin, J.; Schappacher, M.; Ricard, L.; Weiss, R. *J. Am. Chem. Soc.* **1981,** *103,* 1850.
(61) Collman, J. P.; Barnes, C. E.; Woo, L. K. *Proc. Natl. Acad. Sci. USA* **1983,** *80,* 7684.
(62) Collman, J. P.; Garner, J. M.; Woo, L. K. *J. Am. Chem. Soc.* **1989,** *111,* 8141.
(63) Tatsumi, K.; Hoffman, R.; Templeton, J. L. *Inorg. Chem.* **1982,** *21,* 466.
(64) Berreau, L. M.; Young, V. G., Jr.; Woo, L. K. *Inorg. Chem.* **1995,** *34,* 3485.
(65) Jones, T. K.; McPherson, C.; Chen, H. L.; Kendrick, M. J. *Inorg. Chim. Acta* **1993,** *206,* 5.
(66) Brothers, P. J.; Roper, W. R. *Chem. Rev.* **1988,** *88,* 1293.
(67) Tatsumi, K.; Hoffmann, R. *Inorg. Chem.* **1981,** *20,* 3771.
(68) Shin, K.; Yu, B. S.; Goff, H. M. *Inorg. Chem.* **1990,** *29,* 889.
(69) Balch, A. L.; Hart, R. L.; Latos-Grażyński, L.; Traylor, T. G. *J. Am. Chem. Soc.* **1990,** *112,* 7382.
(70) Arasasingham, R. D.; Balch, A. L.; Olmstead, M. M.; Phillips, S. L. *Inorg. Chim. Acta* **1997,** *263,* 161.
(71) Balch, A. L.; Latos-Grażyński, L.; Noll, B. C.; Phillips, S. L. *Inorg. Chem.* **1993,** *32,* 1124.
(72) Setsune, J. I.; Ishimaru, Y.; Sera, A. *J. Chem. Soc., Chem. Commun.* **1992,** 328.
(73) James, C. A.; Woodruff, W. *Inorg. Chim. Acta* **1995,** *229,* 9.
(74) Takeuchi, M.; Kano, K. *Organometallics* **1993,** *12,* 2059.
(75) Guilard, R.; Lagrange, G.; Tabard, A.; Lançon, D.; Kadish, K. M. *Inorg. Chem.* **1985,** *24,* 3649.
(76) Lexa, D.; Savéant, J. M.; Wang, D. L. *Organometallics* **1986,** *5,* 1428.
(77) Gueutin, C.; Lexa, D.; Savéant, J. M.; Wang, D. L. *Organometallics* **1989,** *8,* 1607.
(78) Lexa, D.; Savéant, J. M.; Su, K. B.; Wang, D. L. *J. Am. Chem. Soc.* **1988,** *110,* 7617.
(79) Lexa, D.; Savéant, J. M.; Su, K. B.; Wang, D. L. *J. Am. Chem. Soc.* **1987,** *109,* 6464.
(80) Lexa, D.; Savéant, J. M.; Schafer, H. J.; Su, K. B.; Vering, B.; Wang, D. L. *J. Am. Chem. Soc.* **1990,** *112,* 6162.
(81) DeSilva, C.; Czarnecki, K.; Ryan, M. D. *Inorg. Chim. Acta* **1994,** *226,* 195.
(82) Arasasingham, R. D.; Balch, A. L.; Cornman, C. R.; Latos-Grażyński, L. *J. Am. Chem. Soc.* **1989,** *111,* 4357.
(83) Li, Z.; Goff, H. M. *Inorg. Chem.* **1992,** *31,* 1547.

(84) Balch, A. L.; Renner, M. W. *Inorg. Chem.* **1986,** *25,* 303.
(85) Guilard, R.; Bóisselier-Cocolios, B.; Tabard, A.; Cocolios, P.; Simonet, B. *Inorg. Chem.* **1985,** *24,* 2509.
(86) Lançon, D.; Cocolios, P.; Guilard, R.; Kadish, K. M. *J. Am. Chem. Soc.* **1984,** *106,* 4472.
(87) Balch, A. L.; Cornman, C. R.; Safari, N.; Latos-Grażyński, L. *Organometallics* **1990,** *9,* 2420.
(88) Tahiri, M.; Doppelt, P.; Fischer, J.; Weiss, R. *Inorg. Chem.* **1988,** *27,* 2897.
(89) Tabard, A.; Cocolios, P.; Lagrange, G.; Gerardin, R.; Hubsch, J.; Lecomte, C.; Zarembowitch, J.; Guilard, R. *Inorg. Chem.* **1988,** *27,* 110.
(90) Kadish, K. M.; Tabard, A.; Lee, W.; Liu, Y. H.; Ratti, C.; Guilard, R. *Inorg. Chem.* **1991,** *30,* 1542.
(91) Kadish, K. M.; D'Souza, F.; Van Caemelbecke, E.; Villard, A.; Lee, J. D.; Tabard, A.; Guilard, R. *Inorg. Chem.* **1993,** *32,* 4179.
(92) Kadish, K. M.; Van Caemelbecke, E.; Gueletii, E.; Fukuzumi, S.; Miyamoto, K.; Suenobu, T.; Tabard, A.; Guilard, R. *Inorg. Chem.* **1998,** *37,* 1759.
(93) Kadish, K. M.; D'Souza, F.; Van Caemelbecke, E.; Boulas, P.; Vogel, E.; Aukauloo, A. M.; Guilard, R. *Inorg. Chem.* **1994,** *33,* 4474.
(94) Kadish, K. M.; Tabard, A.; Van Caemelbecke, E.; Aukauloo, A. M.; Richard, P.; Guilard, R. *Inorg. Chem.* **1998,** *37,* 6168.
(95) Kadish, K. M.; Van Caemelbecke, E.; D'Souza, F.; Medforth, C. J.; Smith, K. M.; Tabard, A.; Guilard, R. *Organometallics* **1993,** *12,* 2411.
(96) Kadish, K.; Van Caemelbecke, E.; D'Souza, F.; Medforth, C. J.; Smith, K. M.; Tabard, A.; Guilard, R. *Inorg. Chem.* **1995,** *34,* 2984.
(97) Vogel, E.; Will, S.; Tilling, A. S.; Neumann, L.; Lex, J.; Bill, E.; Trautwein, A. X.; Wieghardt, K. *Angew. Chem., Int. Ed. Engl.* **1994,** *33,* 731.
(98) Van Caemelbecke, E.; Will, S.; Autret, M.; Adamian, V. A.; Lex, J.; Gisselbrecht, J. P.; Gross, M.; Vogel, E.; Kadish, K. *Inorg. Chem.* **1996,** *35,* 184.
(99) Doppelt, P. *Inorg. Chem.* **1984,** *23,* 4009.
(100) Balch, A. L.; Olmstead, M. M.; Safari, N.; St Claire, T. N. *Inorg. Chem.* **1994,** *33,* 2815.
(101) Olmstead, M. M.; Cheng, R. J.; Balch, A. L. *Inorg. Chem.* **1982,** *21,* 4143.
(102) Chevrier, B.; Weiss, R.; Lauge, M.; Mansuy, D.; Chottard, J. C. *J. Am. Chem. Soc.* **1981,** *103,* 2899.
(103) Balch, A. L.; Renner, M. W. *J. Am. Chem. Soc.* **1986,** *108,* 603.
(104) Balch, A. L.; La Mar, G. N.; Latos-Grażyński, L.; Renner, M. W. *Inorg. Chem.* **1985,** *24,* 2432.
(105) Latos-Grażyński, L.; Wyslouch, A. *Inorg. Chim. Acta* **1990,** *171,* 205.
(106) (a) Kuila, D.; Kopelove, A. B.; Lavallee, D. K. *Inorg.Chem.* **1985,** *24,* 1443; (b) Lavallee, D. K. *The Chemistry and Biochemistry of N-Substituted Porphyrins;* VCH Verlagsgesellschaft: Weinheim, 1987.
(107) Song, B.; Goff, H. M. *Inorg. Chim. Acta* **1994,** *226,* 231.
(108) Riordan, C. G.; Halpern, J. *Inorg. Chim. Acta* **1996,** *243,* 19.
(109) Balch, A. L. *Inorg. Chim. Acta* **1992,** *198–200,* 297.
(110) Arasasingham, R. D.; Balch, A. L.; Latos-Grażyński, L. *J. Am. Chem. Soc.* **1987,** *109,* 5846.
(111) Arasasingham, R. D.; Cornman, C. R.; Balch, A. L. *J. Am. Chem. Soc.* **1989,** *111,* 7800.
(112) Arasasingham, R. D.; Balch, A. L.; Hart, R. L.; Latos-Grażyński, L. *J. Am. Chem. Soc.* **1990,** *112,* 7566.
(113) Cocolios, P.; Laviron, E.; Guilard, R. *J. Organomet. Chem.* **1982,** *228,* C39.
(114) Arafa, I. A.; Shin, K.; Goff, H. M. *J. Am. Chem. Soc.* **1988,** *110,* 5228.
(115) Gueutin, C.; Lexa, D.; Momenteau, M.; Savéant, J. M. *J. Am. Chem. Soc.* **1990,** *112,* 1874.
(116) Hammouche, M.; Lexa, D.; Momenteau, M.; Savéant, J. M. *J. Am. Chem. Soc.* **1991,** *113,* 8455.
(117) (a) Bhugun, I.; Lexa, D.; Savéant, J. M. *J. Am. Chem. Soc.* **1994,** *116,* 5015. (b) Bhugun, I.; Lexa, D.; Savéant, *J. M. J. Am. Chem. Soc.* **1996,** *118,* 1769.

(118) Kim, Y. O.; Goff, H. M. *J. Am. Chem. Soc.* **1988,** *110,* 8706.
(119) Song, B.; Goff, H. M. *Inorg. Chem.* **1994,** *33,* 5979.
(120) Beck, W.; Knauer, W.; Robl, C. *Angew. Chem., Int. Ed. Engl.* **1990,** *29,* 318.
(121) Mansuy, D.; Lecomte, J. P.; Chottard, J. C.; Bartoli, J. F. *Inorg. Chem.* **1981,** *20,* 3119.
(122) Goedken, V. L.; Deakin, M. R.; Bottomley, L. A. *J. Chem. Soc., Chem. Commun.* **1982,** 607.
(123) Kienast, A.; Bruhn, C.; Homborg, H. Z. *Anor. Allg. Chem.* **1997,** *623,* 967.
(124) Kienast, A.; Homborg, H. Z. *Anor. Allg. Chem.* **1998,** *624,* 107.
(125) Galich, L.; Kienast, A.; Hückstädt, H.; Homborg, H. Z. *Anor. Allg. Chem.* **1998,** *624,* 1235.
(126) Rossi, G.; Goedken, V. L.; Ercolani, C. *J. Chem. Soc., Chem. Commun.* **1988,** 46.
(127) Ercolani, C.; Gardini, M.; Goedken, V. L.; Pennesi, G.; Rossi, G.; Russo, U.; Zanonato, P. *Inorg. Chem.* **1989,** *28,* 3097.
(128) Ziegler, C. J.; Suslick, K. S. *J. Am. Chem. Soc.* **1996,** *118,* 5306.
(129) Ziegler, C. J.; Suslick, K. S. *J. Organomet. Chem.* **1997,** *528,* 83.
(130) Balch, A. L.; Chan, Y. W.; Olmstead, M. M.; Renner, M. W. *J. Org. Chem.* **1986,** *51,* 4651.
(131) Mansuy, D.; Battioni, J. P.; Lavallee, D.; Fischer, J.; Weiss, R. *Inorg. Chem.* **1988,** *27,* 1052.
(132) Balch, A. L.; Chan, Y. W.; La Mar, G. N.; Latos-Grażyński, L.; Renner, M. W. *Inorg. Chem.* **1985,** *24,* 1437.
(133) Balch, A. L.; Cheng, R. J.; La Mar, G. N.; Latos-Grażyński, L. *Inorg. Chem.* **1985,** *24,* 2651.
(134) Artaud, I.; Gregoire, N.; Battioni, J. P.; Dupre, D.; Mansuy, D. *J. Am. Chem. Soc.* **1988,** *110,* 8714.
(135) Artaud, I.; Gregoire, N.; Leduc, P.; Mansuy, D. *J. Am. Chem. Soc.* **1990,** *112,* 6899.
(136) Setsune, J. I.; Iida, T.; Kitao, T. *Tetrahedron Lett.* **1988,** *29,* 5677.
(137) Wolf, J. R.; Hamaker, C. G.; Djukic, J. P.; Kodadek, T.; Woo, L. K. *J. Am. Chem. Soc.* **1995,** *117,* 9194.
(138) Simmoneaux, G.; Hindre, F.; Le Plouzennec, M. *Inorg. Chem.* **1989,** *28,* 823.
(139) Gèze, C.; Legrand, N.; Bondon, A.; Simmoneaux, G. *Inorg. Chim. Acta* **1992,** *195,* 73.
(140) Walker, F. A.; Nasri, H.; Turowska-Tyrk, I.; Mohanrao, K.; Watson, C. T.; Shokhirev, N. V.; Debrunner, P. G.; Scheidt, W. R. *J. Am. Chem. Soc.* **1996,** *118,* 12109.
(141) Camenzind, M. J.; James, B. R.; Dolphin, D. *J. Chem. Soc., Chem. Commun.* **1986,** 1137.
(142) Collman, J. P.; Brothers, P. J.; McElwee-White, L.; Rose, E.; Wright, L. J. *J. Am. Chem. Soc.* **1985,** *107,* 4570.
(143) Collman, J. P.; Brothers, P.J.; Mc-Elwee White, L.; Rose, E. *J. Am. Chem. Soc.* **1985,** *107,* 6110.
(144) Collman, J. P.; Prodolliet, J. W.; Leidner, C. R. *J. Am. Chem. Soc.* **1986,** *108,* 2916.
(145) Collman, J. P.; McElwee-White, L.; Brothers, P. J.; Rose, E. *J. Am. Chem. Soc.* **1986,** *108,* 1332.
(146) Camenzind, M. J.; James, B. R.; Dolphin, D.; Sparapany, J. W.; Ibers, J. A. *Inorg. Chem.* **1988,** *27,* 3054.
(147) Sishta, C.; Ke, M.; James, B. R.; Dolphin, D. *J. Chem. Soc., Chem. Commun.* **1986,** 787.
(148) Ke, M.; Sishta, C.; James, B. R.; Dolphin, D.; Sparapany, J. W. *Inorg. Chem.* **1991,** *30,* 4766.
(149) Seyler, J. W.; Safford, L. K.; Leidner, C. R. *Inorg. Chem.* **1992,** *31,* 4300.
(150) Leung, W. H.; Hun, T. S. M.; Wong, K. Y.; Wong, W. T. *J. Chem. Soc., Dalton Trans.* **1994,** 2713.
(151) Ke, M.; Rettig, S. J.; James, B. R.; Dolphin, D. *J. Chem. Soc., Chem. Commun.* **1987,** 1110.
(152) Alexander, C. S.; Rettig, S. J.; James, B. R. *Organometallics* **1994,** *13,* 2542.
(153) Seyler, J. W.; Leidner, C. R. *J. Chem. Soc., Chem. Commun.* **1989,** 1794.
(154) Seyler, J. W.; Leidner, C. R. *Inorg. Chem.* **1990,** *29,* 3636.
(155) Collman, J. P.; Rose, E.; Venburg, G. D. *J. Chem. Soc., Chem. Commun.* **1994,** 11.
(156) Collman, J. P.; Ha, Y.; Wagenknecht, P. S.; Lopez, M. A.; Guilard, R. *J. Am. Chem. Soc.* **1993,** *115,* 9080.
(157) Seyler, J. W.; Fanwick, P. E.; Leidner, C. R. *Inorg. Chem.* **1990,** *29,* 2021.
(158) Seyler, J. W.; Safford, L. K.; Fanwick, P. E.; Leidner, C. R. *Inorg. Chem.* **1992,** *31,* 1545.

(159) Hodge, S. J.; Wang, L. S.; Khan, M. A.; Young, V. G. J.; Richter-Addo, G. B. *Chem. Commun.* **1996,** 2283.
(160) Seyler, J. W.; Fanwick, P. E.; Leidner, C. R. *Inorg. Chem.* **1992,** *31,* 3699.
(161) Rajapakse, N.; James, B. R.; Dolphin, D. *Can. J. Chem.* **1990,** *68,* 2274.
(162) Maruyama, H.; Fujiwara, M.; Tanaka, K. *Chem. Lett.* **1998,** 805.
(163) Collman, J. P.; Rose, E.; Venburg, G. D. *J. Chem. Soc., Chem. Commun.* **1993,** 934.
(164) Galardon, E.; Le Maux, P.; Toupet, L.; Simmoneaux, G. *Organometallics* **1998,** *17,* 565.
(165) Balch, A. L.; Chan, Y. W.; Olmstead, M. M.; Renner, M. W.; Wood, F. E. *J. Am. Chem. Soc.* **1988,** *110,* 3897.
(166) Woo, L. K.; Smith, D. A. *Organometallics* **1992,** *11,* 2344.
(167) Woo, L. K.; Smith, D. A.; Young, V. G. *Organometallics* **1991,** *10,* 3977.
(168) Djukic, J. P.; Smith, D. A.; Young, V. G.; Woo, L. K. *Organometallics* **1994,** *13,* 3020.
(169) Djukic, J. P.; Young, V. G. J.; Woo, L. K. *Organometallics* **1994,** *13,* 3995.
(170) Smith, D. A.; Reynolds, D. N.; Woo, L. K. *J. Am. Chem. Soc.* **1993,** *115,* 2511.
(171) Galardon, E.; Lemaux, P.; Simonneaux, G. *Chem. Commun.* **1997,** 927.
(172) Galardon, E.; Roue, S.; Lemaux, P.; Simonneaux, G. *Tetrahedron Lett.* **1998,** *39,* 2333.
(173) Lo, W. C.; Che, C. M.; Cheng, K. F.; Mak, T. C. W. *Chem. Commun.* **1997,** 1205.
(174) Frauenkron, M.; Berkessel, A. *Tetrahedron. Lett.* **1997,** *38,* 7175.
(175) Galardon, E.; Lemaux, P.; Simonneaux, G. *J. Chem. Soc., Perkin Trans. 1* **1997,** 2455.
(176) Collman, J. P.; Wagenknecht, P. S.; Hembre, R. T.; Lewis, N. S. *J. Am. Chem. Soc.* **1990,** *112,* 1294.
(177) Brothers, P. J. *Prog. Inorg. Chem.* **1981,** *28,* 1.
(178) Collman, J. P.; Hutchison, J. E.; Wagenknecht, P. S.; Lewis, N. S.; Lopez, M. A.; Guilard, R. *J. Am. Chem. Soc.* **1990,** *112,* 8206.
(179) Collman, J. P.; Wagenknecht, P. S.; Hutchison, J. E.; Lewis, N. S.; Lopez, M. A.; Guilard, R.; L'Her, M.; Bothner-By, A. A.; Mishra, P. K. *J. Am. Chem. Soc.* **1992,** *114,* 5654.
(180) Collman, J. P.; Wagenknecht, P. S.; Lewis, N. S. *J. Am. Chem. Soc.* **1992,** *114,* 5665.
(181) Collman, J. P.; Fish, H. T.; Wagenknecht, P. S.; Tyvoll, D. A.; Chng, L. L.; Eberspacher, T. A.; Brauman, J. I.; Bacon, J. W.; Pignolet, L. H. *Inorg. Chem.* **1996,** *35,* 6746.
(182) Geno, M. K.; Halpern, J. *J. Am. Chem. Soc.* **1987,** *109,* 1238.
(183) Fukuzumi, S.; Kitano, T. *Inorg. Chem.* **1990,** *29,* 2558.
(184) Krattinger, B.; Callot, H. J. *Bull. Soc. Chim. Fr.* **1996,** *133,* 721.
(185) Cao, Y.; Petersen, J. L.; Stolzenberg, A. M. *Inorg. Chem.* **1998,** *37,* 5173.
(186) Tse, A. K. S.; Wang, R.; Mak, T. C. W.; Chan, K. S. *Chem. Commun.* **1996,** 173.
(187) Setsune, J. I.; Iida, T.; Kitao, T. *Chem. Lett.* **1989,** 885.
(188) Masuda, H.; Taga, T.; Sugimoto, H.; Mori, M. *J. Organomet. Chem.* **1984,** *273,* 385.
(189) Kastner, M. E.; Scheidt, W. R. *J. Organomet. Chem.* **1978,** *157,* 109.
(190) Summers, J. S.; Petersen, J. L.; Stolzenberg, A. M. *J. Am. Chem. Soc.* **1994,** *116,* 7189.
(191) Al-Akhdar, W. A.; Belmore, K. A.; Kendrick, M. J. *Inorg. Chim. Acta* **1989,** *165,* 15.
(192) Cao, Y.; Petersen, J. L.; Stolzenberg, A. M. *Inorg. Chim. Acta* **1997,** *263,* 139.
(193) Halpern, J. *Polyhedron* **1988,** *7,* 1483.
(194) Chopra, M.; Hun, T. S. M.; Leung, W. H.; Yu, N. T. *Inorg. Chem.* **1995,** *34,* 5973.
(195) Woska, D. C.; Wayland, B. B. *Inorg. Chim. Acta* **1998,** *270,* 197.
(196) Fukuzumi, S.; Miyamoto, K.; Suenobu, T.; Van Caemelbecke, E.; Kadish, K. M. *J. Am. Chem. Soc.* **1998,** *120,* 2880.
(197) Woska, D. C.; Xie, Z. L. D.; Gridnev, A. A.; Ittel, S. D.; Fryd, M.; Wayland, B. B. *J. Am. Chem. Soc.* **1996,** *118,* 9102.
(198) Kadish, K. M.; Han, B. C.; Endo, A. *Inorg. Chem.* **1991,** *30,* 4502.
(199) Callot, H. J.; Cromer, R.; Louati, A.; Metz, B.; Chevrier, B. *J. Am. Chem. Soc.* **1987,** *109,* 2946.
(200) Konishi, K.; Sugino, T.; Aida, T.; Inoue, S. *J. Am. Chem. Soc.* **1991,** *113,* 6487.

(201) Will, S.; Lex, J.; Vogel, E.; Adamian, V. A.; Van Caemelbecke, E.; Kadish, K. M. *Inorg. Chem.* **1996,** *35,* 5577.
(202) Zheng, G. D.; Yan, Y.; Gao, S.; Tong, S. L.; Gao, D.; Zhen, K. J. *Electrochim. Acta* **1996,** *41,* 177.
(203) Zheng, G. D.; Stradiotto, M.; Li, L. J. *J. Electroanal. Chem.* **1998,** *453,* 79.
(204) Dobson, D. J.; Saini, S. *Anal. Chem.* **1997,** *69,* 3532.
(205) Kellett, R. M.; Spiro, T. G. *Inorg. Chem.* **1985,** *24,* 2373.
(206) Kellett, R. M.; Spiro, T. G. *Inorg. Chem.* **1985,** *24,* 2378.
(207) Setsune, J. I.; Ishimaru, Y.; Moriyama, T.; Kitao, T. *J. Chem. Soc., Chem. Commun.* **1991,** 555.
(208) Setsune, J. I.; Ishimaru, Y.; Moriyama, T.; Kitao, T. *J. Chem. Soc., Chem. Commun.* **1991,** 556.
(209) Setsune, J. I.; Watanabe, J. Y. *Chem. Lett.* **1994,** 2253.
(210) Watanabe, J. I.; Setsune, J. Y. *J. Organomet. Chem.* **1999,** *575,* 21.
(211) Gridnev, A. A.; Ittel, S. D.; Fryd, M.; Wayland, B. B. *J. Chem. Soc., Chem. Commun.* **1993,** 1010.
(212) Gridnev, A. A.; Ittel, S. D.; Fryd, M.; Wayland, B. B. *Organometallics* **1993,** *12,* 4871.
(213) Gridnev, A. A.; Ittel, S. D.; Fryd, M.; Wayland, B. B. *Organometallics* **1996,** *15,* 222.
(214) Wayland, B. B.; Poszmik, G.; Mukerjee, S. L. *J. Am. Chem. Soc.* **1994,** *116,* 7943.
(215) Wayland, B. B.; Basickes, L.; Mukerjee, S.; Wei, M. L.; Fryd, M. *Macromolecules* **1997,** *30,* 8109.
(216) Gridnev, A. A.; Ittel, S. D.; Wayland, B. B.; Fryd, M. *Organometallics* **1996,** *15,* 5116.
(217) Gridnev, A. A.; Ittel, S. D. *Macromolecules* **1996,** *29,* 5864.
(218) Setsune, J. I.; Ikeda, M.; Kishimoto, Y.; Kitao, T. *J. Am. Chem. Soc.* **1986,** *108,* 1309.
(219) Setsune, J. I.; Ikeda, M.; Kitao, T. *J. Am. Chem. Soc.* **1987,** *109,* 6515.
(220) Setsune, J. I.; Ikeda, M.; Kishimoto, Y.; Ishimaru, Y.; Fukuhara, K.; Kitao, T. *Organometallics* **1991,** *10,* 1099.
(221) Setsune, J. I.; Ikeda, M.; Ishimaru, Y.; Kitao, T. *Chem. Lett.* **1989,** 667.
(222) Setsune, J. I.; Ishimaru, Y.; Saito, Y.; Kitao, T. *Chem. Lett.* **1989,** 671.
(223) Setsune, J.; Ito, S.; Takeda, H.; Ishimaru, Y.; Kitao, T.; Sato, M.; Ohyanishiguchi, H. *Organometallics* **1997,** *16,* 597.
(224) Setsune, J.; Takeda, H.; Ito, S.; Saito, Y.; Ishimaru, Y.; Fukuhara, K.; Kitao, T.; Adachi, T. *Inorg. Chem.* **1998,** *37,* 2235.
(225) Setsune, J.; Takeda, H.; Katakami, Y. *Chem. Lett.* **1998,** 527.
(226) Setsune, J. I.; Saito, Y.; Ishimaru, Y.; Ikeda, M.; Kitao, T. *Bull. Chem. Soc. Jpn.* **1992,** *65,* 639.
(227) Watanabe, J.; Setsune, J. *Organometallics* **1997,** *16,* 3679.
(228) Wayland, B. B.; Woods, B. A. *J. Chem. Soc., Chem. Commun.* **1981,** 700.
(229) Paonessa, R. S.; Thomas, N. C.; Halpern, J. *J. Am. Chem. Soc.* **1985,** *107,* 4333.
(230) Callot, H. J.; Metz, F.; Piechoki, C. *Tetrahedron* **1982,** 2365.
(231) Del Rossi, K. J.; Wayland, B. B. *J. Chem. Soc., Chem. Commun.* **1986,** 1653.
(232) Cornillon, J. L.; Anderson, J. E.; Swistak, C.; Kadish, K. M. *J. Am. Chem. Soc.* **1986,** *108,* 7633.
(233) Kadish, K. M.; Cornillon, J. L.; Mitaine, P.; Deng, Y. J.; Korp, J. D. *Inorg. Chem.* **1989,** *28,* 2354.
(234) Takenaka, A.; Syal, S. K.; Sasada, Y.; Omura, T.; Ogoshi, H.; Yoshida, Z. I. *Acta Cryst.* **1976,** *32B,* 62.
(235) Fleischer, E. B.; Lavallee, D. *J. Am. Chem. Soc.* **1967,** *89,* 7132.
(236) Hückstädt, H.; Homborg, H. Z. *Anor. Allg. Chem.* **1998,** *624,* 980.
(237) Anderson, J. E.; Yao, C. L.; Kadish, K. M. *Inorg. Chem.* **1986,** *25,* 718.
(238) Anderson, J. E.; Yao, C. L.; Kadish, K. M. *J. Am. Chem. Soc.* **1987,** *109,* 1106.
(239) Grass, V.; Lexa, D.; Momenteau, M.; Savéant, J. M. *J. Am. Chem. Soc.* **1997,** *119,* 3536.
(240) Lexa, D.; Grass, V.; Savéant, J. M. *Organometallics* **1998,** *17,* 2673.
(241) Grass, V.; Lexa, D.; Savéant, J. M. *J. Am. Chem. Soc.* **1997,** *119,* 7526.

(242) Grodkowski, J.; Neta, P.; Abdallah, Y.; Hambright, P. *J. Phys. Chem.* **1996,** *100,*
(243) Anderson, J. E.; Liu, Y. H.; Kadish, K. M. *Inorg. Chem.* **1987,** *26,* 4174.
(244) Kadish, K. M.; Araullo, C.; Yao, C. L. *Organometallics* **1988,** *7,* 1583.
(245) Wayland, B. B. *Polyhedron* **1988,** *7,* 1545.
(246) Wayland, B. B.; Van Voorhees, S. L.; Wilker, C. *Inorg. Chem.* **1986,** *25,* 4039.
(247) Ni, Y.; Fitzgerald, J. P.; Carroll, P.; Wayland, B. B. *Inorg. Chem.* **1994,** *33,* 2029.
(248) Del Rossi, K. J.; Zhang, X. X.; Wayland, B. B. *J. Organomet. Chem.* **1995,** *504,* 47.
(249) Wayland, B. B.; Van Voorhees, S. L.; Del Rossi, K. J. *J. Am. Chem. Soc.* **1987,** *109,* 6513.
(250) Wayland, B. B.; Poszmik, G. *Organometallics* **1992,** *11,* 3534.
(251) Bunn, A. G.; Wayland, B. B. *J. Am. Chem. Soc.* **1992,** *114,* 6917.
(252) Del Rossi, K. J.; Wayland, B. B. *J. Am. Chem. Soc.* **1985,** *107,* 7941.
(253) Sherry, A. E.; Wayland, B. B. *J. Am. Chem. Soc.* **1990,** *112,* 1259.
(254) Wayland, B. B.; Ba, S.; Sherry, A. E. *J. Am. Chem. Soc.* **1991,** *113,* 5305.
(255) Wayland, B. B.; Ba, S.; Sherry, A. E. *Inorg. Chem.* **1992,** *31,* 148.
(256) Zhang, X. X.; Wayland, B. B. *J. Am. Chem. Soc.* **1994,** *116,* 7897.
(257) Wayland, B. B.; Woods, B. A.; Coffin, B. L. *Organometallics* **1986,** *5,* 1059.
(258) Coffin, V. L.; Brennen, W.; Wayland, B. B. *J. Am. Chem. Soc.* **1988,** *110,* 6063.
(259) Wayland, B. B.; Sherry, A. E.; Poszmik, G.; Bunn, A. G. *J. Am. Chem. Soc.* **1992,** *114,* 1673.
(260) Wayland, B. B.; Sherry, A. E.; Coffin, V. L. *J. Chem. Soc., Chem. Commun.* **1989,** 662.
(261) Sherry, A. E.; Wayland, B. B. *J. Am. Chem. Soc.* **1989,** *111,* 5010.
(262) Zhang, X. X.; Parks, G. F.; Wayland, B. B. *J. Am. Chem. Soc.* **1997,** *119,* 7938.
(263) Poszmik, G.; Carroll, P. J.; Wayland, B. B. *Organometallics* **1993,** *12,* 3410.
(264) Aoyama, Y.; Yoshida, T.; Sakurai, K. I.; Ogoshi, H. *Organometallics* **1986,** *5,* 168.
(265) Zhou, X.; Li, Q.; Mak, T. C. W.; Chan, K. S. *Inorg. Chim. Acta* **1998,** *270,* 551.
(266) Zhou, X.; Wang, R. J.; Xue, F.; Mak, T. C. W.; Chan, K. S. *J. Organomet. Chem.* **1999,** *580,* 22.
(267) Aoyama, Y.; Tanaka, Y.; Yoshida, T.; Toi, H.; Ogoshi, H. *J. Organomet. Chem.* **1987,** *329,* 251.
(268) Aoyama, Y.; Yamagishi, A.; Tanaka, Y.; Toi, H.; Ogoshi, H. *J. Am. Chem. Soc.* **1987,** *109,* 4735.
(269) Mizutani, T.; Uesaka, T.; Ogoshi, H. *Organometallics* **1995,** *14,* 341.
(270) Wayland, B. B.; Balkus, K. J.; Farnos, M. D. *Organometallics* **1989,** *9,* 950.
(271) Maxwell, J.; Kodadek, T. *Organometallics* **1991,** *10,* 4.
(272) Maxwell, J. L.; Brown, K. C.; Bartley, D. W.; Kodadek, T. *Science* **1992,** *256,* 1544.
(273) Bartley, D. W.; Kodadek, T. *J. Am. Chem. Soc.* **1993,** *115,* 1656.
(274) Brown, K. C.; Kodadek, T. *J. Am. Chem. Soc.* **1992,** *114,* 8336.
(275) O'Malley, S.; Kodadek, T. *Tet. Lett.* **1991,** *32,* 2445.
(276) Maxwell, J. L.; O'Malley, S.; Brown, K. C.; Kodadek, T. *Organometallics* **1992,** *11,* 645.
(277) O'Malley, S.; Kodadek, T. *Organometallics* **1992,** *11,* 2299.
(278) Hayashi, T.; Kato, T.; Kaneko, T.; Asai, T.; Ogoshi, H. *J. Organomet. Chem.* **1994,** *473,*
(279) Boschi, T.; Licoccia, S.; Paolesse, R.; Tagliatesta, P. *Organometallics* **1989,** *8,* 330.
(280) Whang, D.; Kim, K. *Acta Cryst.* **1991,** *C47,* 2547.
(281) Kadish, K. M.; Hu, R.; Boschi, T.; Tagliatesta, P. *Inorg. Chem.* **1993,** *32,* 2996.
(282) Tse, A. K. S.; Wu, B. M.; Mak, T. C. W.; Chan, K. S. *J. Organomet. Chem.* **1998,** *568,* 257.
(283) Licoccia, S.; Paolesse, R.; Boschi, T. *Acta Cryst.* **1995,** *C51,* 833.
(284) Jones, N. L.; Carroll, P. J.; Wayland, B. B. *Organometallics* **1986,** *5,* 33.
(285) Lux, D.; Daphnomili, D.; Coutsolelos, A. G. *Polyhedron* **1994,** *13,* 2367.
(286) Coutsolelos, A. G.; Lux, D. *Polyhedron* **1996,** *15,* 705.
(287) Daphnomili, D.; Scheidt, W. R.; Zajicek, J.; Coutsolelos, A. G. *Inorg. Chem.* **1998,** *37,* 3675.
(288) Chevrier, B.; Weiss, R. *J. Am. Chem. Soc.* **1976,** *98,* 2985.
(289) Chan, Y. W.; Renner, M. W.; Balch, A. L. *Organometallics* **1983,** *2,* 1888.
(290) Chmielewski, P. J.; Latos-Grażyński, L. *Inorg. Chem.* **1992,** *31,* 5231.
(291) Chmielewski, P. J.; Latos-Grażyński, L. *Inorg. Chem.* **1998,** *37,* 4179.

(292) Chmielewski, P. J.; Latos-Grażyński, L. *Inorg. Chem.* **1997,** *36,* 840.
(293) Takao, Y.; Takeda, T.; Watanabe, J. Y.; Setsune, J. I. *Organometallics* **1999,** *18,* 2936.
(294) Inoue, S.; Murayama, H.; Takeda, N.; Ohkatsu, Y. *Chem. Lett.* **1982,** *317,*
(295) (a) Murayama, H.; Inoue, S.; Ohkatsu, Y. *Chem. Lett.* **1993,** 381 (b) Inoue, S.; Nukui, M.; Kojima, F. *Chem. Lett.* **1984,** 619.
(296) Kirner, J. F.; Reed, C. A.; Scheidt, W. R. *J. Am. Chem. Soc.* **1977,** *99,* 1093.
(297) Shelley, K.; Bartczak, T.; Scheidt, W. R.; Reed, C. A. *Inorg. Chem.* **1985,** *24,* 4325.
(298) Xie, Z.; Bau, R.; Reed, C. A. *Angew. Chem., Int. Ed. Engl.* **1994,** *33,* 2433.
(299) Evans, D. R.; Drovetskaya, T.; Bau, R.; Reed, C. A.; Boyd, P. D. W. *J. Am. Chem. Soc.* **1997,** *119,* 3633.
(300) Boyd, P. D. W.; Hodgson, M. C.; Rickard, C. E. F.; Oliver, A. G.; Chaker, L.; Brothers, P. J.; Bolskar, R. D.; Tham, F. S.; Reed, C. A. *J. Am. Chem. Soc.* **1999,** *121.*

Index

D

E

F

G

R

Cumulative List of Contributors for Volumes 1–36

Cumulative Index for Volumes 37–46

ISBN 0-12-031146-1